ECOLOGY OF WEEDS
AND INVASIVE PLANTS

BICENTENNIAL
1807
⊛WILEY
2007
BICENTENNIAL
BICENTENNIAL
BICENTENNIAL

THE WILEY BICENTENNIAL—KNOWLEDGE FOR GENERATIONS

$\mathcal{E}$ach generation has its unique needs and aspirations. When Charles Wiley first opened his small printing shop in lower Manhattan in 1807, it was a generation of boundless potential searching for an identity. And we were there, helping to define a new American literary tradition. Over half a century later, in the midst of the Second Industrial Revolution, it was a generation focused on building the future. Once again, we were there, supplying the critical scientific, technical, and engineering knowledge that helped frame the world. Throughout the 20th Century, and into the new millennium, nations began to reach out beyond their own borders and a new international community was born. Wiley was there, expanding its operations around the world to enable a global exchange of ideas, opinions, and know-how.

For 200 years, Wiley has been an integral part of each generation's journey, enabling the flow of information and understanding necessary to meet their needs and fulfill their aspirations. Today, bold new technologies are changing the way we live and learn. Wiley will be there, providing you the must-have knowledge you need to imagine new worlds, new possibilities, and new opportunities.

Generations come and go, but you can always count on Wiley to provide you the knowledge you need, when and where you need it!

WILLIAM J. PESCE
PRESIDENT AND CHIEF EXECUTIVE OFFICER

PETER BOOTH WILEY
CHAIRMAN OF THE BOARD

ECOLOGY OF WEEDS AND INVASIVE PLANTS

RELATIONSHIP TO AGRICULTURE AND NATURAL RESOURCE MANAGEMENT

Third Edition

STEVEN R. RADOSEVICH
Oregon State University
Corvallis, Oregon

JODIE S. HOLT
University of California
Riverside, California

CLAUDIO M. GHERSA
University of Buenos Aires
Buenos Aires, Argentina

WILEY-INTERSCIENCE
A JOHN WILEY & SONS, INC., PUBLICATION

Published by John Wiley & Sons, Inc., Hoboken, New Jersey
Published simultaneously in Canada.

For general information on our other products and services or for technical support, please contact our Customer Care Department within the United States at (800) 762-2974, outside the United States at (317) 572-3993 or fax (317) 572-4002.

Wiley also publishes its books in a variety of electronic formats. Some content that appears in print may not be available in electronic formats. For more information about Wiley products, visit our web site at www.wiley.com.

Wiley Bicentennial Logo: Richard J. Pacifico.

Library of Congress Cataloging-in-Publication Data:

Radosevich, Steven R.
  Ecology of weeds and invasive plants: relationship to agriculture and natural resource management / Steven R. Radosevich, Jodie S. Holt, Claudio M. Ghersa.—3rd ed.
     p. cm.
    Rev. ed. of: Weed ecology / Steven Radosevich, Jodie Holt, Claudio Ghersa. 1997.
    Includes index.
    ISBN 978-0-471-76779-4 (cloth)
    1. Weeds—Ecology. 2. Weeds—Control.   I.   Holt, Jodie S.
   II.   Ghersa, Claudio.   III.   Radosevich, Steven R. Weed ecology.   IV.   Title.
   SB611.R33 2007
   632'.5—dc22

                                                                                        2007001705

Printed in the United States of America

10 9 8 7 6 5 4 3 2 1

CONTENTS

PREFACE   xv

BURDOCK by Charles Goodrich   xvii

INTRODUCTION   1

Chapter 1:   Weeds and Invasive Plants   3

Weeds, 4
  Definitions, 6
    Agrestals, 6
    Invasive Plants, 9
  Terminology, 10
  Classification Systems of Weeds and Invasive Plants, 11
    Taxonomic Classification, 11
    Classification by Life History, 13
    Classification by Habitat, 14
    Physiological Classification, 15
    Classification According to Undesirability, 16
    Ecological Classification, 16
    Classification by Evolutionary Strategy, 17
Weeds and Invasive Plants in Production Systems, 20
  Weeds on Agricultural Land, 20
    Reasons for Weed Control, 21
  Weeds in Managed Forests, 24
    Forest Regeneration, 25

Weeds in Rangelands, 26
   Original Vegetation and Early Land Use History of
      Great Basin, 28
   Introduction of Cheatgrass and Fire, 28
Invasive Plants in Less Managed Habitats and Wildlands, 30
   Local versus Regional Perspectives about Weeds, 30
   Weeds in Regional and Global Context, 31
Summary, 32

Chapter 2:   Principles   **35**

Ecological Principles, 35
   Interrelationship of Biology and Environment, 35
      Environment, 36
   Scale, 38
      Scale in Ecological Systems, 39
      Scale in Human Production Systems, 43
   Community Differentiation and Boundaries, 46
   Community Structure, 47
   Succession, 49
      Mechanisms of Succession, 50
      Succession in Production Systems, 52
      Niche Differentiation, 54
   Invasion Process, 56
      Introduction Phase, 57
      Colonization Phase, 59
      Naturalization Phase, 62
   Genetics of Weeds and Invasive Plants, 62
      Fitness and Selection, 63
      Patterns of Evolutionary Development of Weeds and
         Invasive Plants, 63
   Plant Demography and Population Dynamics, 67
Management Principles, 69
   Assessing Risk from Weeds and Invasive Plants, 69
      Management Priorities Based on Risk and Value, 71
   Market-Driven Management Considerations, 73
      Cost–Benefit Analysis, 73
      Assessing Economic Risk, 74
   Management Options in Relation to Invasion Process, 76
Social Principles, 77
   Societal Aims versus Individual Objectives, 78
      Social Conflict and Resolution, 79
      Precautionary Principle, 79
   Weed and Invasive Plant Management in Modern Society, 80
Summary, 81

Chapter 3: Invasibility of Agricultural and Natural Ecosystems 83

Plant Invasions over Large Geographical Areas, 84
  Habitat Invasibility, 86
  Community Invasibility, 87
Local Invasions, 87
  Safe Sites, 88
    Safe Site Example, 89
Factors That Influence Invasibility, 89
  Evolutionary History, 89
  Community Structure, 90
    Role of Plant Size in Species Dominance and Richness, 92
  Propagule Pressure, 93
    Relationship of Propagule Pressure to Invasion Process, 93
    Relationship of Dispersal to Propagule Pressure, 94
    Relationship of Human and Animal Transport to Propagule Pressure, 94
    Relationship of Seed Banks to Propagule Pressure, 95
  Disturbance, 95
    Disturbance and Land Use, 96
    Relationship of Disturbance and Succession, 97
    Relationship of Stress and Disturbance, 98
Invasibility and Exotic Plant Invasiveness, 99
Summary, 101

Chapter 4: Evolution of Weeds and Invasive Plants 103

Evolutionary Genetics of Weeds and Invasive Plants, 104
  Heritable Genetic Variation, 105
    Hybridization and Polyploidy, 105
    Epistatic Genetic Variance, 109
    Epigenetic Inheritance Systems, 110
Adaptation Following Introduction, 111
  Responses to Environmental Gradients, 112
    Selection in Barnyardgrass, 112
    Selection in St. Johnswort, 113
  Responses to Resident Plant Species, 113
  Release from Pests, Predation, and Herbivores, 114
Breeding Systems of Weeds and Invasive Plants, 114
  Sexual Reproduction, 115
    Self-Pollination versus Outcrossing, 115
    Founder Effects, 117
    Exceptions to Baker's Rule, 117
  Asexual Reproduction, 117
    Advantages of Asexual Reproduction in Weeds, 118

Influence of Humans on Weed and Invasive Plant Evolution, 119
  Weeds and Invasive Plants as Strategists, 119
    Competitive Ruderals, 119
    Stress-Tolerant Competitors, 121
  Adaptations of Weeds and Invasive Plants to Human Activities, 122
    Weeds, Domesticates, and Wild Plants, 122
    Crop Mimics, 122
    Shifts in Plant Species Composition, 126
Summary, 126

Chapter 5:   Weed Demography and Population Dynamics   **129**

Principles of Plant Demography, 129
  Natality, Mortality, Immigration, and Emigration, 130
  Life Tables, 131
  Modular Growth, 133
  Models of Plant Population Dynamics, 134
    Models Based on Difference Equations, 134
    Transition Matrices, 138
  Metapopulations, 139
    Risk of Extinction, 140
    Metapopulation Dynamics Applied to Invasive Species, 141
Dynamics of Weed and Invasive Plant Seed, 142
  Seed Dispersal through Space, 142
    Estimates of Dispersal Distance, 144
    Agents of Spatial Seed Dispersal, 146
  Seed Banks, 149
    Entry of Seed into Soil, 150
    Longevity of Seed in Soil, 152
    Density and Composition of Seed Banks, 157
    Fate of Seed in Soil, 162
    Weed Occurrence in Relation to Seed Banks, 165
  Dormancy: Dispersal through Time, 166
  Descriptions of Seed Dormancy, 166
    Physiological Dormancy, 167
    Physical Dormancy, 168
    Combinations of Physiological and Physical Dormancy, 170
    Seed with Underdeveloped Embryos, 171
  Using Seed Dormancy to Manage Weed Populations, 171
Recruitment: Germination and Establishment, 171
  Seed Germination, 171
    Light Requirement for Germination, 172
  Risk of Mortality, 176
    Seedlings, 176
    Vegetative Propagules of Perennial Plants, 177

Epidemics of Weeds and Invasive Plants, 178
Predictive Models of Weed Reproduction, Dispersal, and Survival, 179
    Example: Predictions of Changes in Weed Abundance in
      Agricultural Fields, 180
Summary, 181

Chapter 6: Plant–Plant Associations   183

Neighbors, 184
Interference, 184
  Effect and Response, 185
  Is it Competition?, 187
Modifiers of Interference, 188
  Space, 189
    Density, 191
    Species Proportion, 199
    Spatial Arrangement, 200
Methods to Study Interference (Competition), 201
  Additive Designs, 201
  Substitutive Designs, 202
    Replacement Series, 203
    Nelder Designs, 205
    Diallel Designs, 206
  Systematic Designs, 207
    Addition Series and Additive Series Designs, 207
  Neighborhood Designs, 209
  Approaches Used to Study Plant Interference (Competition) in
    Natural and Managed Ecosystems, 212
    Descriptive Studies, 213
    Retrospective Studies, 214
    Case Studies, 214
    Gradient Studies, 216
Intensity and Importance of Competition, 216
  Intensity of Competition, 217
    Competition Intensity Indices, 218
    Relative Yield, 218
    Relative Yield Total, 219
    Intra- versus Interspecific Competition, 221
  Importance of Competition, 222
  Competition in Mixed Cropping Systems, 223
    Weed Suppression in Mixed Planting Systems, 223
Competition Thresholds, 224
  Thresholds in Agriculture, 224
    Damage (Density/Biomass) Thresholds, 225
    Critical-Period Thresholds, 227

Thresholds in Natural Ecosystems, 228
Mechanisms of Competition, 230
  Theories, 230
    Theories of Grime and Tilman, 230
    Role of Plant Traits, 231
    Plant Growth Rates and Components of Growth, 233
Other Types of Interference than Competition, 237
Negative Interference in Addition to Competition, 237
  Allelopathy, 237
    Responses of Plants to Allelochemicals, 241
    Methods to Study Allelopathy, 242
  Microbially Produced Phytotoxins, 243
  Parasitism, Predation, and Herbivory, 243
    Parasitism, 244
    Predation, 246
    Herbivory, 248
Positive Interference, 250
  Facilitation, 250
    Commensalism, 251
    Protocooperation, 252
    Mutualism, 253
Summary, 255

**Chapter 7:   Weed and Invasive Plant Management Approaches,
Methods, and Tools**   **259**

Prevention, Eradication, and Control, 259
Weed Management in Agroecosystems, 260
  Economics and Biology of Weed Control: Whether to
    Control Weeds, 260
    Weed Response to Control, 261
    Opportunity to Improve Productivity: Crop Response
      to Weeds, 261
    Profitability: Value of Weed Control, 262
  Influence of Weed Control on Agricultural Crops and Weed
    Associations, 265
    Reduction in Weed Density, 265
    Alteration in Species Composition, 265
  Influence of Weed Control on Other Organisms, 267
Management of Invasive Plants in Natural Ecosystems, 269
  Approaches to Prioritize Management, 269
  Documenting Invasions, 271
    Terms Used by Land Managers, 271
  Incorporating Risk Assessment into Invasive Plant Management, 272

Individual Species Approach, 272
Plant Community or Habitat Approach, 273
Risks Associated with Action and Inaction, 275
Framework to Combine Research and Management of Invasive Plants, 277
Methods and Tools to Control Weeds and Invasive Plants, 279
Physical Methods of Weed Control, 279
Hand Pulling and Hoeing, 279
Fire, 280
Flame, 281
Tillage (Cultivation)/Disturbance, 281
Mowing and Shredding, 286
Chaining and Dredging, 289
Flooding, 289
Mulching and Solarization, 289
Cultural Methods of Weed Control, 290
Weed Prevention, 290
Crop Rotation, 291
Competition, 292
Smother Crops, 293
Living Mulches and Cover Crops, 294
Harvesting, 294
Biological Control: Using Natural Enemies to Suppress Weeds, 295
Procedures for Developing Biological Control, 296
Grazing, 301
Mycoherbicides, 301
Allelopathy, 301
Chemical Control, 302
Herbicides, 303
Summary, 305

Chapter 8: Herbicides   **307**

Herbicides as Commercial Products, 307
Laws for Herbicide Registration and Use in United States, 308
Information on Herbicide Label, 309
Voluntary and Legislative Restrictions on Herbicide Use, 309
Properties of Herbicides that Affect Human, Animal, and
Environmental Safety, 311
Toxicity, 312
Biological Magnification, 314
Persistence, 315
Voluntary Selection Criteria for Herbicide Use, 315
Chemical Properties of Herbicides that Affect Use, 318
Chemical Structure, 318
Water Solubility and Polarity, 319

Volatility, 321
Formulations, 321
  Carriers and Adjuvants for Herbicide Applications, 322
Herbicide Classification, 322
  Classification Based on Chemical Structure, 322
  Classification Based on Use, 323
    Soil-Applied Herbicides, 323
    Foliage-Applied Herbicides, 324
    Soil Residual Herbicides, 324
    Soil Fumigants, 326
    Aquatic Herbicides, 326
  Classification Based on Biological Effect in Plants, 326
Herbicide Symptoms and Selectivity, 327
  Symptoms, 327
    Abnormal Tissues and Twisted Plants, 327
    Disruption of Cell Division, 327
    Chlorosis, Necrosis, and Albinism, 327
    Altered Geotropic and Phototropic Responses, 328
    Reduced Leaf Waxes, 328
  Selectivity, 328
    Plant Factors of Herbicide Selectivity, 329
    Chemical Factors of Herbicide Selectivity, 332
    Environmental Factors of Herbicide Selectivity, 332
Herbicide Application, 333
  Proper Rate (Dose), 334
  Proper Distribution, 334
  Application Equipment, 334
Fate of Herbicides in Environment, 335
  Herbicide Displacement in Environment, 336
    Herbicide Movement in Air, 337
    Herbicides in Soil, 338
    Herbicide Movement with Water, 341
  Herbicide Decomposition in Environment, 342
    Photochemical Decomposition, 343
    Chemical Decomposition, 344
    Microbial Decomposition, 344
  Reduction of Herbicides in Agriculture and Natural Resource
    Production Systems, 345
Summary, 345

**Chapter 9: Systems Approaches for Weed and Invasive
Plant Management**   **349**

Cycles of Land Use, Expansion, and Intensification for Production, 350
  Evolution of Modern Integrated Pest Management, 351

Evolution of Weed Science, 352
Approaches for Pest and Weed Management, 353
  Integrated Weed Management, 354
    Levels of Integrated Weed Management, 354
    Ecological Principles to Design Weed Management Systems, 355
    Future Directions in Integrated Weed Management, 357
Novel Ecosystems, 366
Novel Weed/Invasive Plant Management Systems, 368
  Agriculture, 369
  Managed Forests and Forest Plantations, 369
    When Limited Herbicide Use Is Acceptable, 371
  Rangeland, 372
Value Systems in Agricultural and Natural Ecosystem Management, 374
  Role of Human Institutions in Weed Management, 375
    The 2,4,5-T Controversy, 375
    Atrazine and Water Quality, 376
    Herbicide-Resistant Crops, 377
  Consequences of Human Values on Weed and Invasive Plant
    Management, 378
  Simplification, Deterioration, and Loss of Biological Regulation
    in Agriculture, 379
  Weeds and Invasive Plants as Symptoms of Ecosystem Dysfunction, 380
    Weed Occurrence on Deteriorating Soil Base, 380
    Other Examples of Ecosystem Deterioration, 380
  Socioeconomic Influences on Weed and Invasive
    Plant Management, 381
  Future Challenges for Scientists, Farmers, and Land Managers, 381
Summary, 382

References   385

Index   439

PREFACE

This book, now in its third edition, began almost 25 years ago when *Weed Ecology: Implications for Vegetation Management* was published in 1984. That text concentrated on the need for farmers, foresters, rangeland managers, and the researchers who advised them to understand better the biology of weeds and the role people play in creating and maintaining weeds in agriculture and other production systems. We were assisted in that first effort by the writings of many early scientists, such as J. L. Harper, H. G. Baker, and E. J. Salisbury, who studied the biology of weeds as a class of vegetation. We continue to be grateful for their pioneering work and theoretical perspectives that they provided.

Our focus on the biology of weeds continued though the second edition, which was published in 1997. We described the many empirical findings that had emerged since our first edition about the biology of weeds and discussed these findings within an ecological framework to explain how weed invasions occur, how weed communities continue to exist, and even how agroecosystems and other natural ecosystems work. We also added three chapters about the technology of weed control which had developed over the previous four decades and had become part of the general knowledge about weeds in farms, forests, or rangelands. Our emphasis, however, continued to be the ecological underpinnings of the discipline of weed science. We believed then and continue to believe that better management results from the understanding of how plants interact with each other and their environment and management to create and maintain weed populations.

We find with this latest edition, *Ecology of Weeds and Invasive Plants: Relationship to Agriculture and Natural Resource Management*, that weeds are now at the forefront of many ecologists' minds. Their recent interest in weed and invasive plant ecology has generated new understanding about the concepts of invasibility and in the disciplines of genetics and plant population dynamics. In addition, considerable research has incorporated the principles of integrated pest management (IPM) and ecological thresholds into weed and invasive plant management. While many new and enlightening papers have been written about weeds, invasive plants, and their management over the last decade, we found as we updated our text that in some instances little had changed or that seminal papers on a subject had already been published, often decades earlier. Thus, we cite both new and vintage papers in our third edition.

Ecologists have a history of working predominantly in natural ecosystems and only recently have incorporated disturbance and human impacts into their research on a large scale. Weed scientists, on the other hand, have traditionally worked in agricultural and managed ecosystems and focused on the applied disciplines, with less emphasis on basic science. With the recognition of the impacts of invasive plants and their weedy attributes, the two disciplines, ecology and weed science, have begun to converge on the study of weeds and invasive plants. Thus, we hope that ecologists will examine carefully and apply the approaches and tools of weed science while weed scientists continue to embrace the principles of ecology. In this way, we believe both disciplines can move forward together toward better understanding and land management.

We suggest humbly, while also reminding ourselves, that there is never epiphany in the unprepared mind.

STEVEN R. RADOSEVICH
JODIE S. HOLT
CLAUDIO M. GHERSA

Corvallis, Oregon
Riverside, California
Buenos Aires, Argentina
October 2006

Charles Goodrich

Burdock

Few seeds as tenacious as burdock,
clutching the dog's fur
tight as ticks. The leaves aren't as plush as mullein,
but will pass for Kleenex in a pinch.

We haven't tried digging it up,
roasting the roots in an open pit, then
grinding the mess together with berries and fat
for pemmican.

but I own a sharp spade.
I'm not afraid to eat
bitter, woody plants,
or creatures that wiggle and squeal.

When I pull the burrs out of her fur,
I toss them to the dog
and she eats them.

Good dog.

INTRODUCTION

In one of his early texts on weed control, A. S. Crafts begins by saying, "in the beginning there were no weeds." What Dr. Crafts meant was that even though plants have existed for a long time, weeds did not exist before humankind. Now, with the ever-increasing movement of people across the globe and the occurrence of worldwide trade, weeds are no longer locally restricted to agricultural and managed lands and the problem of exotic invasive plants has become widespread. Still, however, weeds and invasive plants exist because of our human ability to judge and select among the various species of the plant kingdom. This anthropomorphic perspective of weeds and invasive plants provides little insight into their evolution, biological characteristics, or interactions that occur so markedly in managed and natural ecosystems. In this text, our focus is on these biological features of weeds and invasive plants, especially as they exist in agriculture, forests, rangelands, and natural ecosystems. By considering weeds foremost as plants and by relying heavily on the concepts of plant ecology, we hope to provide a better understanding about this vegetation and therefore better management of the ecosystems so often invaded by them.

Ecology of Weeds and Invasive Plants. By Steven R. Radosevich, Jodie S. Holt, and Claudio M. Ghersa
Copyright © 2007 John Wiley & Sons, Inc.

1

WEEDS AND INVASIVE PLANTS

Weeds exist as a category of vegetation because of the human ability to select desirable traits from among various members of the plant kingdom. Just as some plants are valued for their uses or beauty, others are reviled for their apparent lack of these characteristics. Weeds are recognized worldwide as an important type of undesirable, economic pest, especially in agriculture. However, the value of any plant is unquestionably determined by the perceptions of its viewers. These perceptions also influence the human activities directed at this category of vegetation.

Harlan, in the middle of the last century, described how vegetation evolved under the impacts of humans. He suggested that vegetation, in relation to the degree of human involvement with it, exists as three categories: wild plants, weeds, and crops. Crops were domesticated from wild plants while weeds evolved from wild plants as an unintentional consequence of growing crops. Some crops also were once weeds and some have again escaped from domestication. In Harlan's concept neither weeds nor crops can permanently displace wild plants from wild habitats over time (DeWet and Harland 1975).

Invasive plants, unlike agricultural weeds, are those that can successfully establish and spread to new habitats after their introduction, seemingly without further assistance from humans. These plants can spread into new areas already occupied by a native flora and displace those species. Such invasions from the intentional or unintentional transport of plants to new regions now seriously threaten the biodiversity, structure, and function of many of the world's ecosystems. Invasive

Ecology of Weeds and Invasive Plants. By Steven R. Radosevich, Jodie S. Holt, and Claudio M. Ghersa

plants are thus weeds in the broadest sense because they evoke human dislike and often some form of management to eradicate or contain them in their new environments. Not all weeds are invasive, however. In this text, the term *weed* will be used in the broad sense and to describe undesirable plants in agricultural systems, while *invasive plant* will be used for those weeds that can spread beyond their point of introduction, often in natural ecosystems.

WEEDS

A *"plant growing out of place,"* that is, plants growing where they are not wanted, at least by some people, is a common, accepted explanation for what weeds are. This notion of undesirability imparts so much human value to the idea of weediness that it is usually necessary to recognize who is making the determination as well as the characteristics of the plants themselves. For example, certain plants growing in a cereal field or pasture or along a fence row may be unwanted by a farmer or rancher, but they also may be wildflowers or a valuable wildlife cover to other people. Vine maple, *Acer circinatum*, is a valued source of deer browse in the spring and a spectacular source of coloration in the Cascade Mountains of Oregon and Washington in the United States, during autumn, but it also is known to hamper forest regeneration. It can be argued that many weeds in agricultural fields, forest plantations, and rangelands are not "out of place" at all but are simply not wanted there by some people.

In Table 1.1 we list many of the "human" characteristics that have been used to describe weeds. Most of these characteristics are based on some judgment of

TABLE 1.1 Definitions and Descriptions of Weeds

Definition	Description
Growing in an undesirable location	A plant growing where it is not desired (Weed Science Society of America 1956)
Competitive and aggressive behavior	A plant that grows so luxuriantly or plentifully that it chokes all other plants that possess more valuable properties (Brenchley 1920)
Persistence and resistance to control	The predominance and pertinacity of weeds (Gray 1879)
Useless, unwanted, undesirable	A plant not wanted and therefore to be destroyed (Bailey and Bailey 1941); a plant whose virtues have not yet been discovered (Emerson 1878)
Appearing without being sown or cultivated	Any plant other than the crop sown (Brenchley 1920); a plant that grows spontaneously in a habitat greatly modified by human action (Harper 1944)
Unsightly	A very unsightly plant of wild growth, often found in land that has been cultivated (Thomas 1956)

Source: Adapted from King (1966).

worth, success, or other human attribute, like aggressiveness, harmfulness, or being unsightly or ugly. Since this anthropomorphic view of weeds is so prevalent (Table 1.1), it may be that weeds are little more than plants that have aroused a level of human dislike at some particular place or time. Unfortunately, the anthropomorphic view of weeds provides little insight into why and where they exist, their interactions and associations with crops, native plants, and other organisms, or even how to manage them effectively. Weeds are found worldwide and have proven to be successful organisms in the environments that they inhabit. Therefore, it is important to explore whether weeds posses common traits that distinguish them from other plants or whether they are only set apart by local notions of usefulness.

A list of biological characteristics that describe weeds was proposed in the 1970s and continues to be used today (Table 1.2) (Baker 1974), but it seems unlikely that any plant species could possess all of those "ideal" weedy traits. However, Herbert Baker, botanist and originator of the list, suggests that a species might possess various combinations of the characteristics in Table 1.2, resulting in a range of weediness from minor to major weeds (Baker 1974). In the latter case, Baker believes that evolutionary processes would compound specific adaptations into highly successful (weedy) individuals, which constitutes an "all-purpose genotype." It must be stressed, however, that ecological success in the form of weediness cannot be measured solely from the perspective of noxiousness. The number of individuals, the range of habitats occupied, and the ability to continue the species through time must be considered foremost when evaluating success of a species. The obvious limitation of the list in Table 1.2 is that almost every plant species has some "weedy" characteristics, but, of course, not all plants are weeds.

TABLE 1.2   Ideal Characteristics of Weeds

Germination requirements fulfilled in many environments
Discontinuous germination (internally controlled) and great longevity of seed
Rapid growth through vegetative phase to flowering
Continuous seed production for as long as growing conditions permit
Self-compatibility but not complete autogamy or apomixis
Cross-pollination, when it occurs, by unspecialized visitors or wind
Very high seed output in favorable environmental circumstances
Production of some seed in a wide range of environmental conditions; tolerance and plasticity
Adaptations for short-distance dispersal and long-distance dispersal
If perennial, vigorous vegetative reproduction or regeneration from fragments
If perennial, brittleness, so as not to be drawn from the ground easily
Ability to complete interspecifically by special means (rosettes, choking growth, allelochemicals)

Source: Baker (1974). *Annu. Rev. Ecol. Syst.* 5:1–24. Copyright 1974 by Annual Reviews, Inc., Palo Alto, CA.

Definitions

As we have just observed (Tables 1.1 and 1.2), weeds can be described in either anthropomorphic or biological terms. Weeds emerge from such descriptions as organisms that may possess a particular suite of biological characteristics but also have the distinction of negative human selection. Thus, a definition of a weed as *any plant that is objectionable or interferes with the activities or welfare of man* (Weed Science Society of America 1956) seems to describe sufficiently this category of vegetation. A sample of definitions of weeds published over the past century was presented by Randall (1997), who also argued that the most important criterion was problem-causing plants that interfere with land use.

Other authors, for example Zimmerman (1976), Aldrich (1984), and Rejmánek (2000), define weeds in more specific terms than the simple definition given above. Zimmerman believes that the term "weed" should be used to describe plants that (1) colonize disturbed habitats, (2) are not members of the original plant community, (3) are locally abundant, and (4) are economically of little value (or are costly to control). Aldrich defines weeds as plants that originated under a natural environment and, in response to (human) imposed or natural conditions, are interfering associates of crops and human activities. Each of these definitions implies that weeds have some common biological traits but also a level of relative undesirability as determined by particular people. Whether or not a plant is a weed depends on the context in which someone finds it and on the perspectives and objectives of those involved in dealing with it. Rejmánek, on the other hand, believes that weeds, colonizers, and naturalized species (including invasive plants) reflect three overlapping concepts. In his view (Figure 1.1), weeds are plants growing where they are not desired (anthropomorphic definition), colonizers occur early in succession (ecological definition), and invasive plants are plants that become locally established and spread to areas where they are not native (biogeographical definition).

The most important criterion for weediness is interference at some place or time with the values and activities of people—farmers, foresters, land managers, and many other segments of human society. However, the abundance of weeds is often of more concern than the mere presence of them. For instance, farmers and land managers are usually less concerned about the occurrence of a few isolated plants in a field, even noxious ones, than the occupation of land by vast numbers of weeds. Therefore, the relative abundance of plants, their location, and the potential use of the land they occupy should also be considered in weed definitions. When abundance is applied as a criterion for weediness, it implies a condition of the land as well as a class of vegetation (Table 1.2) and a form of human discrimination (Table 1.1). Weed abundance also may be an indicator or symptom of land mismanagement or neglect.

Agrestals. Agrestals are weeds of tilled, arable land. They require the nearly continual disturbance of agriculture to occupy the land. Holzner et al. (1982) indicate that every cropping system, for example, cereals, root crops, and orchards, also

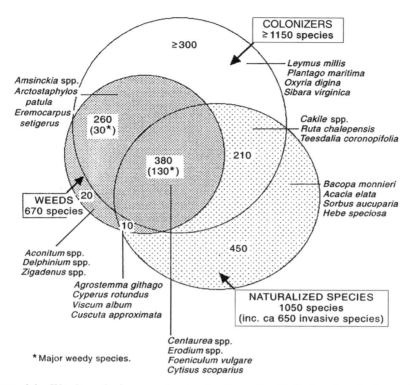

Figure 1.1 Weeds, colonizers, and naturalized species (including invaders) are three overlapping but not identical concepts reflecting three different viewpoints: anthropomorphic (weeds), ecological (colonizers), and biogeographical (naturalized species). Invaders are a subset of naturalized species, namely those nonnative species that are spreading. Estimated species numbers and examples of species representing seven resulting categories of the California vascular flora are given. (From Rejmánek 2000, *Aust. Ecol.* 25:497–506. Copyright 2000, Blackwell Publishing Ltd., reproduced with permission.)

has its special complement of weeds, which may be either native plants or exotics that have been naturalized into the local flora. A list of the 76 worst agricultural weeds in the world was developed by Holm and his associates (1977) and has become the standard by which agrestals are compared. The top 18 weeds on this list are given in Table 1.3. An additional 104 of the weeds that cause the greatest impacts on agriculture was reviewed by Holm et al. in 1997. As a group these 180 agricultural weeds are estimated to cause over 90% of the loss of crop productivity worldwide (Holm et al. 1997).

Holzner and his associates (1982) suggest that agrestals have evolved as either specialists or colonizers during the course of agricultural history. Specialized weeds (*specialists*) have evolved a narrow adaptation to a single crop or sometimes crop cultivar and its particular growing conditions. Perhaps the most extreme example of how human activities influence weed species distribution and

TABLE 1.3   Scientific and Common Names of Certain
Annual Weed Species Considered the World's 18 Worst

Species	Common Name
Amaranthus hybridus	Smooth pigweed
Amaranthus spinosus	Spiny amaranthus
Avena fatua	Wild oat
Chenopodium album	Common lambsquarters
Convolvulus arvensis	Field bindweed
Cynodon dactylon	Bermudagrass
Cyperus esculentus	Yellow nutsedge
Cyperus rotundus	Purple nutsedge
Digitaria sanguinalis	Large crabgrass
Echinochloa colonum	Junglerice
Echinochloa crus-galli	Barnyardgrass
Eichhornia crassipes	Waterhyacinth
Eleusine indica	Goosegrass
Imperata cylindrica	Cogongrass
Paspalum conjugatum	Sour paspalum
Portulaca oleracea	Common purslane
Rottboellia exaltata	Itchgrass
Sorghum halepense	Johnsongrass

Source: Adapted from Holm et al. (1977, 1997).

composition are *crop mimics*. These are weeds that have evolved life cycles or morphological features so similar to a crop that the two species cannot be distinguished or separated easily. Chapter 4 considers the influence of humans on the evolution of weed species, including crop mimicry, in much more depth. Since agrestals that are specialists have evolved along with the cultural practices of a particular crop, any change in practices usually disfavors the weed. *Colonizers*, on the other hand, are plants with characteristics that allow them to rapidly occupy and dominate disturbed areas. These species follow the general characteristics listed in Table 1.2 and Figure 1.1.

Weeds are major constraints to crop production, yet as primary producers, they also can be important components in an agroecosystem. It is in this context that weeds are sometimes perceived as an ecological "good" (Gerowitt et al. 2003). Awareness of the importance of weeds on arable land for their role in other trophic levels is growing as natural landscapes become rare or disappear due to the expansion of human-occupied landscapes. The weed flora in many parts of the world has changed over the past century, with some species declining in abundance while others have increased (Haas and Streibig 1982, Marshall et al. 2003, de la Fuente et al. 2006). These changes in the weed flora reflect improved agricultural efficiency, the use of different crops in arable rotations, and the use of more broad-spectrum herbicide combinations (Marshall et al. 2003, de la Fuente et al. 2003). Many weed species of arable land support a high diversity of insects, so the reduction in abundance of weed host plants can affect associated insects and, therefore, the abundance of other taxa. For example, in the United Kingdom a number of insect groups and farmland-associated birds (notably the grey partridge, *Perdix perdix*) have undergone marked population decline, which is associated with changes in agricultural practices over the past 30 years (Marshall et al. 2003). Thus, it seems that weeds may have a general role in supporting biodiversity within agroecosystems.

Invasive Plants. Invasive plants, unlike agricultural weeds, are generally defined as those that can successfully establish, become naturalized, and spread to new natural habitats apparently without further assistance from humans (Randall 1997). They are also generally nonnative or exotic in the new habitat and are often relatively new introductions to an ecoregion (Mashhadi and Radosevich 2003). Invasive plants respond readily to human-induced changes in the environment such as disturbance but also may initiate environmental change through their dominance on the landscape (Pyke and Knick 2003, Hobbs et al. 2006). In addition, the spatial and temporal extent of their impact may be expressed at scales ranging from local to global. Some ecological impacts believed to be caused by invasive plants are as follows (Parker et al. 1999, Alien Plant Working Group 2002):

- Reduction of biodiversity
- Loss or encroachment upon endangered and threatened species and their habitats
- Loss of habitat for native insects, birds, and other wildlife
- Loss of food sources for wildlife
- Changes to natural ecological processes such as plant community succession
- Alterations to the frequency and intensity of natural fires
- Disruptions of native plant–animal associations such as pollination, seed dispersal, and host–plant relationships

It is widely believed that the most effective way to limit plant invasions is to prevent the introduction of exotic species, which may be difficult because of the ongoing expansion in global travel and trade, changes in environments at all scales (local to global), and increasing development of land for human use (Kolar and Lodge 2001).

Although the traits of an "ideal weed" (Baker 1974) have also been ascribed to invasive plants, few empirical studies have tested this concept (Kolar and Lodge 2001). The biological characteristics of invasive plants appear in many cases to be dependent upon the habitat in which they occur (Sakai et al. 2001). Thus, general descriptions of invasive plants remain inconclusive. Some useful generalizations have been made, however, from reviews of empirical evidence or broad-scale analyses of floras or databases. For example, Reichard and Hamilton (1997), using a regression tree analysis of biological and environmental traits of invasive plants, suggest that species known to be invasive elsewhere should be limited in

TABLE 1.4   Biological Characteristics Responsible for Invasiveness

- Fitness homeostasis or the ability of an individual or population to maintain relatively constant fitness over a range of environments. This is equivalent to Baker's (1974, 1995) "general-purpose genotype."
- Small genome size—usually associated with short minimum generation time, short juvenile period, small seed size, high leaf area ratio, and high relative growth rate.
- Dispersed easily by humans and animals.
- Ability to vegetatively propagate. This is an especially important characteristic in aquatic environments (Auld et al. 1983, Henderson 1991) and at high latitude (Pyšek 1997).
- Alien plants belonging to exotic genera are more invasive than are alien species with native congeners. This may be partly because of an absence or limited number of resident natural enemies for that species (Darwin 1859, Rejmánek 1999).
- Plant species without dependence on specific mutualisms (root symbiosis, pollinators, seed dispersers, etc.) (Baker 1974, Richardson et al. 2000).
- Tall plants tend to invade mesic plant communities.
- Persistent seed banks—seeds with different inherent dormancies that provide a random appearance through time and guarantee their survival and persistence.

Source: Adapted from Rejmánek and Pitcairn (2002).

introduction to a new area with a similar environment, where they might also be invasive. Reichard and Hamilton further suggest that a species related to one that is already "invading" a site may share invasive traits through a common ancestor. From a retrospective review of literature, Rejmánek (2000) lists several biological characteristics related to invasiveness, including constant fitness, small genome size, effective dispersal and vegetative propagation, and absence of strong interactions with other taxa (e.g., natural enemies, pollinators, seed dispersers) (Table 1.4). Sutherland (2004) reviewed databases for nearly 20,000 plant species in the United States and concluded that invasive exotic species were more likely to be perennial, monoecious, self-incompatible, and trees than noninvasive exotic species. A broad-scale analysis of the flora of the Czech Republic over 500 years showed that life-form and competitiveness were related to invasiveness (Pyšek et al. 1995). Similarly, an analysis of global datasets revealed some common traits of invasive plants, including nitrogen fixation and clonal growth (Daehler 1998). Other traits that have been shown to be related to invasiveness are described in later chapters.

Terminology

Massive amounts of money, time, and energy are expended on weeds and invasive plants because of their economic and ecological costs and impacts on agricultural and natural systems. Because of the magnitude of these effects, it is important that scientists and land managers consider carefully the metaphors they use to describe these two categories of vegetation. Larson (2005) points out that metaphors allow people to understand abstract or perplexing subjects in term of something they already know about, a common referent. Thus, weeds and especially invasive plants are often described in militaristic terms, which probably date to Elton's (1958) classic *The Ecology of Invasions by Animals and Plants*. Davis (2005) points out that such terms as *alien, exotic, invader,* and *invasion* commonly used by invasion ecologists contrast markedly to the less evocative terms such as *colonizer, founding population, introduced plant, nonnative, spread,* or *migration,* which could be used to describe weeds and "invasive" plants. It should be noted that a similarly militaristic terminology has been used for decades in the pest management field.

From a management point of view, there is little doubt that the "invasion" terminology and metaphors have been useful in pointing out the significance of weeds to land managers and policymakers. From a strictly scientific point of view, however, it is difficult to argue against returning to the more value-neutral terminology used by Baker and Stebbins (1965) in their early classic, *The Genetics of Colonizing Species* (Davis 2005). Since this text is designed to fulfill a dual role for both scientists and land mangers and because the notion of "weed" is itself value laden, we have chosen to use the language of both scientists and managers that is in conventional use to discuss this important class of vegetation.

Classification Systems of Weeds and Invasive Plants

Botanical classification is the systematic grouping of plants using criteria that distinguish among types of vegetation. These criteria may be biologically meaningful, based on phylogenetic or evolutionary evidence, or artificial and based on structural or other visible or functional attributes. Some common methods used to classify weeds are by taxonomic relationships, life history, habitat, physiology, and degree of undesirability. Weeds and invasive plants can also be classified by ecological behavior related to invasion and evolutionary strategies related to carbon allocation.

Taxonomic Classification. Systematics is the scientific study of biological organisms and their evolutionary relationships. Ideally, organisms are classified systematically according to their presumed genetic relationships, although often this information is unknown. The basis of modern classification is taxonomy, the identification, naming, and grouping of plants according to their traits in common. The accepted taxonomic system used today classifies organisms into a hierarchy of categories: kingdom, phylum (also called division in some botany texts), class, order, family, genus, and species. Recent evidence has shown that an additional category, the domain, occurs above the level of the kingdom; the three recognized domains are Bacteria, Archaea, and Eukarya. All land plants are placed in the domain Eukarya and the kingdom Plantae. Most weeds occur in the phylum Anthophyta (angiosperms, flowering plants), although notable exceptions occur (e.g., some ferns, which are seedless, and conifers, seed plants that have no flowers, are considered weeds). Angiosperms are further divided into the classes Dicotyledones (dicots) and Monocotyledones (monocots).

The next level of classification is the order. Although systematists do not agree on the exact number of orders, the commonly accepted Cronquist system recognizes 64 orders of dicots and 19 orders of monocots (Cronquist 1988). The orders are divided further into families, which, like classes and orders, are comprised of plants whose morphological similarities are greater than their differences. Approximately 383 angiosperm families are currently recognized (318 dicot and 65 monocot). The level of genus includes plants that have common characteristics and that are presumed to be genetically related. The narrowest category of classification is the species, which consists of plants that can interbreed freely (the biological species concept). For practical purposes, however, most species are grouped largely on the basis of anatomical and morphological characteristics (the morphological species concept).

At this point in taxonomic classification, the plant group is given a name, called a scientific name or *Latin binomial*, which consists of both the genus and species names of the plant. For example, Table 1.3 is a list of common agricultural weed species and their Latin binomials. This method of classification is the basis for the organization of all taxonomic texts and many books used to identify weeds.

There are approximately 250,000 species of flowering plants in the world (depending upon which authority is used). However, less than 250 of these, about 0.1%, are troublesome enough to be called major agricultural weeds throughout the world (Holm et al. 1977). It is far more difficult to estimate the number of invasive plant species in nonagricultural habitats worldwide. In the United States, by one estimate, introduced invasive plants comprise from 8 to 47% of the total flora of most states (Rejmánek and Randall 1994). Of the 250 recognized major agricultural weeds, nearly 70% occur in only 12 plant families and over 40% are found in only two families, Poaceae (grass family) and Asteraceae (aster or composite family). Although these observations are fruitful areas of speculation for plant evolutionary biologists, it should be noted that about 75% of world food production is provided by only a dozen crops: barley, maize, millet, oats, rice, sorghum, sugarcane, wheat, cassava, soybean, sweet potato, and white potato. Eight of these crops (the first eight in the list above) are also members of the grass family. The distribution of both the world's worst agricultural weeds and its major crops is quite taxonomically restricted, again pointing to the extreme discrimination and selection that humans apply to vegetation.

It is sometimes necessary to distinguish only broadly among weed species, for example when broad-scale methods of weed control are used. In such situations, distinction among grasses and sedges (monocot) and broadleaf (dicot) plants may be sufficient, and a much abbreviated system of classification is satisfactory. Such a system was once in common use by weed control specialists; a typical description of weeds by this method is shown below (Ross and Lembi 1985, 1999):

Dicots. Plants whose seedlings produce two cotyledons or seed leaves. Usually typified by netted leaf venation and flowering parts in fours, fives, or multiples thereof. Examples include mustards (*Brassica* spp.), nightshades

(*Solanum* spp.), and morningglory (*Convolvulus* spp.). Commonly called broadleaved plants.

Monocots. Plants whose seedlings bear only one cotyledon. Typified by parallel leaf venation and flower parts in threes or multiples of three. Most weeds are found in only two groups, grasses and sedges, although other groups exist.

Grasses. Leaves usually have a ligule or at times an auricle. The leaf sheaths are split around the stem with the stem being round or flattened in cross section with hollow internodes.

Sedges. Leaves lack ligules and auricles and the leaf sheaths are continuous around the stem. In many species the stem is triangular in cross section with solid internodes.

Classification by Life History. Another method used to classify weeds is by the life cycle of the plant. The length of life, season of growth, and time and method of reproduction are used to classify weeds in this way.

Annuals. An annual plant completes its life cycle from seed to seed in one year or less (Figure 1.2). Annuals are often divided into two groups, winter and summer, according to the plant's time of germination, maturation, and death:

Winter Annuals. These plants usually germinate in the fall or winter, grow throughout the spring, and set seed and die by early summer.

Summer Annuals. These plants germinate in the spring, grow throughout the summer, set seed by autumn, and die before winter.

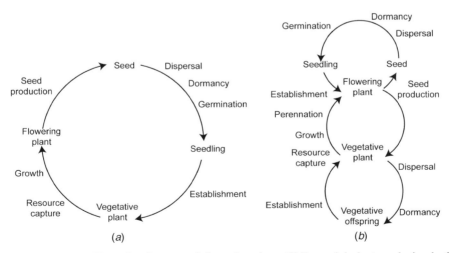

Figure 1.2 (*a*) Life cycle of an annual flowering plant. (*b*) Perennial plant producing both seed and vegetative progeny. (Adapted from Grime 1979, *Plant Strategies and Vegetation Processes.* Copyright 1979 by John Wiley & Sons, Chichester.)

In mild climates, however, it is usual for some winter annuals to germinate in late summer or autumn and for some summer annuals to live throughout the winter. Annual plants are the largest single category of weeds.

Biennials. These plants live longer than one but less than two years. During the first growth phase, biennials develop vegetatively from a seedling into a rosette. Because of this growth habit, biennials sometimes can be confused with winter annuals. After a cold period, vegetative growth resumes, and floral initiation, seed production, and death occur. Biennials are often large plants when mature and have thick fleshy roots. Relatively few weed species are biennials, but some annual plants may behave as biennials under certain conditions and some biennials may behave as short-lived perennials in mild climates.

Perennials. Perennial plants live for longer than two years and may reproduce several times before dying (Figure 1.2). These plants are characterized by renewed vegetative growth year after year from the same root system:

  Simple Herbaceous Perennials. Simple herbaceous perennials reproduce almost exclusively from seed and normally do not reproduce vegetatively. However, if the root system of these plants is injured or cut, each piece usually regenerates into another plant. Dandelion (*Taraxacum officinale*), plantain (*Plantago lanceolata*), and sulfur cinquefoil (*Potentilla recta*) are examples of simple herbaceous perennials.

  Creeping Herbaceous Perennials. Creeping herbaceous perennials survive over the winter and produce new vegetative structures (ramets) from asexual reproductive organs such as rhizomes, tubers, stolons, bulbs, corms, and roots. These plants also reproduce sexually from seed (genets). Most aquatic weeds, except algae, are creeping perennial plants.

  Woody Plants. This is a special category of perennial weed. Plants in this group are characterized by stems that have secondary growth, producing wood and bark, which results in an incremental increase in diameter each year. Some tree, some shrub, and many vine species are considered to be woody weeds.

Classification by Habitat. Weeds can be classified according to where they grow. Most weeds are terrestrial, that is, found on land, but some are restricted to the aquatic environment. Some weeds only infest a particular crop or cropping system, complex of plant communities, or growing condition. Therefore, it is common to find lists and descriptions of weeds that are usually found in particular environments, such as arable land, pastures and rangeland, forests, rights-of-way, or wildlands. These classifications can also be land uses and are described in a following section of this chapter:

  Aquatic Weeds. Aquatic weeds are plants that are modified structurally to live in water. They have been categorized further based on their location in the

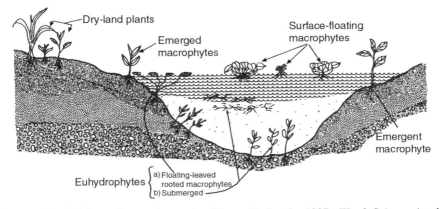

Figure 1.3 Habitats of aquatic weeds. (From Akobundu 1987, *Weed Science in the Tropics: Principles and Practices.* Copyright 1987 by John Wiley & Sons, New York. Reproduced with permission.)

aqueous environment. These categories are depicted in Figure 1.3 as *floating*, *emergent*, and *submerged*. Algae are also considered to be aquatic weeds.

Floating Weeds. These plants rest upon the water surface. Their roots hang freely into the water or sometimes attach to the bottom of shallow ponds or streams.

Emergent Weeds. These typical plants of natural marshlands are often found along the shorelines of ponds and canals. They stand erect and are always rooted into very moist soil.

Submerged Weeds. Although a few floating stems or leaves may exist on the water surface, these plants grow completely under water.

Some weeds and invasive plants occur mainly in riparian habitats, along rivers, streams, or other watercourses. These terrestrial plants, such as Japanese knotweed (*Polygonum cuspidatum*), Himalaya blackberry (*Rubus armenicus*), reed canarygrass (*Phalaris arundinacea*), and saltcedar (*Tamarix* spp.), require the frequent disturbance or high water table associated with rivers, streams, lakes, or ponds. These plants can alter the hydrology of an area and also reduce human access to areas where they occur.

Physiological Classification. Plants differ in their responses to temperature, light, day length, and other factors of the environment. These differences in plant physiology and biochemistry have also been used as a basis for weed classification.

Photosynthetic Pathway. Most plants, called C_3 plants, use the Calvin–Benson cycle exclusively as a method of fixing carbon dioxide, water, and light energy into sugars. This terminology is used because the first stable product of photosynthesis in such plants (phosphoglyceric acid) has three carbon atoms. In some plants, called C_4 plants, the first stable photosynthetic products are four-carbon atom sugars, such as oxaloacetate, malate, and aspartate. This physiological distinction may not seem significant as a means of categorizing weeds. However, these differences in photosynthetic pathway result in substantial biochemical, anatomical, and morphological variation among species. Because of these differences, C_4 weeds are often more efficient at photosynthesis and can be more competitive than C_3 weeds and crops, especially in hot, dry climates. Of the 18 worst weeds in the world noted by Holm et al. (1977), 14 have the C_4 pathway of carbon fixation.

Day Length. Classification by day length is based on a photoperiodic response of flower initiation in plants. Three distinct classes of day length response are known: short day, long day, and day neutral. Although these responses are named for the length of the light period, it is now known that plants detect and respond to the length of the dark period (e.g., short-day plants are actually long-night plants). Weeds that have a short-day response to day length, such as lambsquarters *(Chenopodium album)* and cocklebur *(Xanthium* spp.), are stimulated to flower when days are short and maintain vegetative growth when days are long. Long-day weeds, like henbane *(Hysocyamus niger)* and dogfennel *(Eupatorium capillifolium)*, maintain vegetative growth when days are short but are induced to flower under long-day conditions. Other weeds (e.g., nightshades) remain vegetative or flower irrespective of the photoperiodic condition.

Classification According to Undesirability. The term *noxious* weed is a legal term that refers to any plant species capable of becoming detrimental, destructive, or difficult to control. Legally, a noxious weed is any plant designated by a federal, state, or county government as injurious to public health, agriculture, recreation, wildlife, or property (Sheley et al. 1999). Many states, provinces, and countries maintain at least one official list of such weeds so that their introduction can be prevented or restricted. Noxious weeds usually create a particularly undesirable condition in crops, forest plantations, grazed rangeland, or pastures. For example, the presence of noxious weed seed in seed crops can prevent the sale and distribution of that crop across national and international boundaries. Poisonous weeds, which can be landscape ornamentals or occur in pastures and rangeland, represent a special kind of undesirability, since they can be a direct threat to human or animal health.

Ecological Classification. Weeds, and in particular invasive plants, are often classified using ecological categories related to population behavior. As shown in Figure 1.1, the flora of California includes many weeds, which may also be colonizers (taxa appearing early in vegetation succession) or naturalized species (exotic species that form sustainable populations without direct human assistance). By this classification scheme, invasive plants are a subset of naturalized species that are spreading. Not all naturalized taxa are invasive, however, nor are all colonizers considered to be weeds.

Groves (1986) and Cousens and Mortimer (1995) divide the process of invasion by an exotic species into the phases of *introduction, colonization*, and *naturalization*. These three phases of invasion are defined as follows:

Introduction. As a result of dispersal, propagules arrive at a site beyond their previous geographical range and establish populations of adult plants.

Colonization. The plants in the founding population reproduce and increase in number to form a colony that is self-perpetuating.

Naturalization. The species establishes new self-perpetuating populations, undergoes widespread dispersal, and becomes incorporated into the resident flora.

Richardson et al. (2000), however, argue that colonization as used by Cousens and Mortimer is a component of naturalization, and the term *invasion* should be distinguished from naturalization and used to describe widespread dispersal and incorporation of an exotic species into the resident flora. Such differences of opinion on terminology pertaining to invasion will likely diminish as further knowledge is gained about the ecological processes involved. The steps of the invasion process are discussed later in Chapters 2 and 3.

Classification by Evolutionary Strategy. Weed species can be organized according to evolutionary strategies that are based on genetically determined patterns of carbon resource allocation. One prevalent theory holds that two fundamental external factors limit the amount of plant material (vegetation) that can accumulate within an area. These factors are *stress* and *disturbance* (Grime 1979). When the extremes of these factors are considered (Table 1.5 and Figure 1.4), the following possible strategies of evolutionary development emerge (see Chapter 2 and Figure 2.10 for a more thorough explanation of this classification approach):

Stress Tolerators. These are plants that survive in unproductive environments by reducing their biomass allocation for vegetative growth and reproduction and increasing their allocation to maintenance and defense. They exhibit characteristics that ensure the endurance of relatively mature individuals in

TABLE 1.5 Plant Evolutionary Strategies Resulting from Disturbance and Stress

Intensity of Disturbance	By Intensity of Stress	
	High	Low
High	Plant mortality	Ruderals
Low	Stress tolerators	Competitors

Source: Grime (1979). *Plant Strategies and Vegetation Processes.* Copyright 1979 with permission of John Wiley and Sons, Inc.

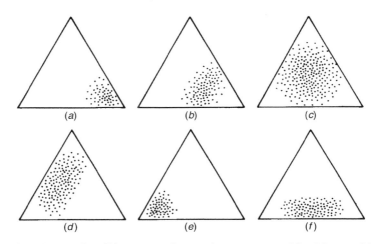

Figure 1.4 Diagram describing range of strategies encompassed by (*a*) annual herbs, (*b*) biennial herbs, (*c*) perennial herbs and ferns, (*d*) trees and shrubs, (*e*) lichens, and (*f*) bryophytes. For the distribution of strategies within a triangle, see Figure 2.10. (From Grime 1977, *American Naturalist* 111:1169–1194. Copyright 1977 by the University of Chicago.)

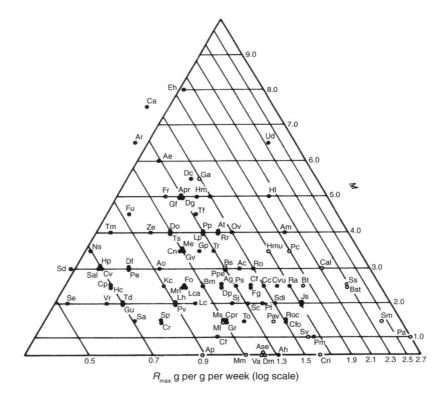

R_{max} g per g per week (log scale)

Figure 1.5

harsh, limited environments. The environmental limitation may be caused by physical factors, such as reoccurring drought or flood, or biotic factors, such as use of resources by neighboring plants or herbivory. Species with these characteristics are prevalent in continually unproductive environments or during the late stages of succession in fertile environments.

Figure 1.5   Triangular ordination of herbaceous species. (○), annuals; (●), perennials (including biennials). The morphology index (M) was calculated from the formula $M = (a + b + c)/2$, where a is the estimated maximum height of leaf canopy (1, <12 cm; 2, 12–25 cm; 3, 25–37 cm; 4, 37–50 cm; 5, 50–62 cm; 6, 62–75 cm; 7, 75–87 cm; 8, 87–100 cm; 9, 100–112 cm; 10, >112 cm); b is the lateral spread (0, small therophytes; 1, robust therophytes; 2, perennials with compact unbranched rhizome or forming small (<10 cm diameter) tussock; 3, perennials with rhizomatous system or tussock attaining diameter 10–25 cm; 4, perennials attaining diameter 26–100 cm; 5, perennials attaining diameter >100 cm); c is the estimated maximum accumulation of persistent litter (0, none; 1, thin discontinuous cover; 2, thin continuous cover; 3, up to 1 cm depth; 4, up to 5 cm depth; 5, >5 cm depth (Grime 1974). Key to species: Ac, *Agrostis canina* ssp. *canina*; Ae, *Arrhenatherum elatius*; Ag, *Alopecurus geniculatus*; Ah, *Arabis hirsuta*; Am, *Achillea millefolium*; Ao, *Anthoxanthum odoratum*; Ap, *Aira praecox*; Apr, *Alopecurus pratensis*; Ar, *Agropyron repens*; As, *Agrostis stolonifera*; Ase, *Arenaria serpyllifolia*; At, *Agrostis tenuis*; Bm, *Briza media*; Bs, *Brachypodium sylvaticum*; Bst, *Bromus sterilis*; Bt, *Bidens tripartita*; Ca, *Chamaenerion angustifolium*; Cal, *Chenopodium album*; Cc, *Cynosurus cristatus*; Cf, *Carex flacca*; Cfl, *Cardamine flexuosa*; Cfo, *Cerastium fontanum*; Cn, *Centaurea nigra*; Cp, *Carex panicea*; Cpr, *Cardamine pratensis*; Cr, *Campanula rotundifolia*; Cri, *Catapodium rigidum*; Cv, *Clinopodium vulgare*; Cvu, *Cirsium vulgare*; Dc, *Deschampsia cespitosa*; Df, *Deschampsia flexuosa*; Dg, *Dactylis glomerata*; Dm, *Draba muralis*; Do, *Dryas octopetala*; Dp, *Digitalis purpurea*; Eh, *Epilobium hirsutum*; Fg, *Festuca gigantea*; Fo, *Festuca ovina*; Fr, *Festuca rubra*; Fu, *Filipendula ulmaria*; Ga, *Galium aparine*; Gf, *Glyceria fluitans*; Gp, *Galium palustre*; Gr, *Geranium robertianum*; Gu, *Geum urbanum*; Gv, *Galium verum*; Hc, *Helianthemum chamaecistus*; Hl, *Holcus lanatus*; Hm, *Holcus mollis*; Hmu, *Hordeum murinum*; Hp, *Helictotrichon pratense*; Js, *Juncus squarrosus*; Kc, *Koeleria cristata*; Lc, *Lotus corniculatus*; Lca, *Luzula campestris*; Lh, *Leontodon hispidus*; Lp, *Lolium perenne*; Me, *Milium effusum*; Ml, *Medicago lupulina*; Mm, *Matricaria matricarioides*; Mn, *Melica nutans*; Ms, *Myosotis sylvatica*; Ns, *Nardus stricta*; Ov, *Origanum vulgare*; Pa, *Poa annua*; Pav, *Polygonum aviculare*; Pc, *Polygonum convolvulus*; Pe, *Potentilla erecta*; Pl, *Plantago lanceolata*; Pm, *Plantago major*; Pp, *Poa pratensis*; Ppe, *Polygonum persicaria*; Ps, *Poterium sanguisorba*; Pt, *Poa trivialis*; Pv, *Prunella vulgaris*; Ra, *Rumex acetosa*; Rac, *Rumex acetosella*; Ro, *Rumex obtusifolius*; Rr, *Ranunculus repens*; Sa, *Sedum acre*; Sal, *Sesleria albicans*; Sc, *Scabiosa columbaria*; Sd, *Sieglingia decumbens*; Sdi, *Silene dioica*; Sj, *Senecio jacobaea*; Sm, *Stellaria media*; Sp, *Succisa pratensis*; Ss, *Senecio squalidus*; Sv, *Senecio vulgaris*; Td, *Thymus druceri*; Tf, *Tussilago farfara*; Tm, *Trifolium medium*; To, *Taraxacum officinalis*; Tr, *Trifolium repens*; Ts, *Teucrium scorodonia*; Ud, *Urtica dioica*; Va, *Veronica arvensis*; Vr, *Viola riviniana*; Ze, *Bromus erectus*. Estimates of R_{max} are based on measurements during the period 2–5 weeks after germination in a standardized productive controlled environment conducted on seedlings from seeds collected from a single population in Northern England. (In Grime 1974, from Grime 1979, *Plant Strategies and Vegetation Processes*. Copyright 1979 with permission of John Wiley & Sons Inc.)

Competitors. These are plants that have evolved characteristics that maximize the capture of environmental resources in productive but relatively undisturbed conditions. These plants have extensive vegetative growth and are abundant during the early and intermediate stages of succession.

Ruderals. Ruderals are plants that are found in highly disturbed but potentially productive environments. These plants are usually herbs, characteristically having a short life span, rapid growth, and high seed production. They occupy the earliest stages of succession.

Grime (1979) suggests that most herbaceous weed species fall into one of two combined strategies, *competitive ruderals* or *stress-tolerant competitors*. Plants possessing the competitive ruderal strategy have rapid early growth rates and competition between individual plants occurs before flowering. Such plants occupy fertile sites and periodic disturbance (e.g., annual tillage) favors their abundance and distribution. Many annual, biennial, and herbaceous perennial weed species found on arable land fit the criteria for the competitive ruderal tactic (Figure 1.4 and Chapter 2).

Stress-tolerant competitors are primarily trees or shrubs, although some perennial herbs also fall into this category (Figure 1.5). Common characteristics of these weeds are rapid dry-matter production, large stem extension, and high leaf area production.

WEEDS AND INVASIVE PLANTS IN PRODUCTION SYSTEMS

There are many books that describe and identify weeds. Some weed species have even achieved worldwide prominence (Table 1.3) (Holm et al. 1977, 1997). Most weeds are important, however, from a more local perspective. The local distribution of weeds is influenced by biotic and abiotic environmental factors that determine habitat types and human activities. Abiotic factors that affect weed occurrence are soil type, soil pH, soil moisture, light quantity and quality, precipitation pattern, and variation in air, soil, and water temperatures. Disturbed areas (either by natural or human causes) also are higher in susceptibility to invasion than habitats that exist for long periods of time in late succession. Biological factors, such as the incidence of insects and diseases on either weeds or associated crops, grazing activities of animals, and plant competition, also can influence the distribution of weeds. It is for all of these reasons that human land uses, such as farming, forestry, range management, and recreation, are major causes of local and regional patterns of weed distribution. Plant species react in different ways when their habitats are disturbed by humans; some species flourish because of the disturbance, whereas others migrate or die and are replaced.

Weeds on Agricultural Land

Many textbooks about modern weed control are quick to point out that weeds have been with us since settled agriculture began, perhaps 10,000 years ago. Weeds must have been known to early farmers because hoes and other "grubbing" implements, artifacts of those ancient times, have been found at archeological sites. In addition, many references account for the detrimental effects of weeds on crop yields, from the early writings of Theophrastus and the Bible to more recent books. These writings have shaped our ideas and definitions of weeds as we saw earlier in this chapter (Table 1.1). Even today, weeds are considered to be just an incidental part of food production in most parts of the world, where farmers are simply people with hoes. The use of modern mechanical and chemical tools to control weeds is actually little more than a century old, even though weeds have been associated with humans since agriculture began. The many reasons to control weeds are described below, while methods and tools for weed management are discussed in detail in Chapters 7 and 8.

Holzner and Immonen (1982) and Marshall et al. (2003) indicate that human action is the most important factor determining the occurrence and distribution of agricultural weed species. They note that many agrestals that accompanied crops for centuries in Europe have now become locally extinct, retreating to their climatic optimum where most survive outside cultivated fields. Haas and Streibig (1982) also note that other weeds have increased in both prominence and abundance as agricultural practices change. Holzner and Immenon suggest several causes for such changes in weed species composition:

- Improved seed cleaning, which results in the local eradication of "specialists" that are unable to grow outside arable land and depend on being sown with the crop
- Abandonment of crops, which leads to loss of specialized weeds
- "Leveling" of environmental conditions, which results in a uniform weed flora
- Increased reliance on crop monocultures, which tends to simplify the weed flora
- Combine harvesting, which allows some weed species to shed seed in the field and distributes the seed of others
- Reduced-tillage and "no-tillage" operations, which promote perennial species
- Reduced competitive ability of short-stature crops and crops treated with chemical growth regulators
- Extensive use of herbicides, which causes sensitive species to become locally extinct or to evolve resistance to the chemical

Reasons for Weed Control. A goal of agriculture for the last half century or more has been to develop efficient methods of weed control in crops, forest plantations, rangelands, and noncrop situations. The search for cost-effective ways to control weeds has often focused on tillage and herbicides as a means to reduce labor requirements and production costs or increase yields. Below are some reasons to control weeds in cropland.

Improve Crop Production. The threat of weeds to crop productivity accounts for most of the human effort devoted to weed control. It is estimated that 10–15% of the total market value of farm products in the United States is lost because of weeds. This loss amounts to about $8 billion to $10 billion per year. Direct losses to forests and rangeland are more difficult to estimate than agricultural losses. Walstad and Kuch (1987) believe that nearly 30% reduction in wood productivity could result because of weed occupation during the early stages of forest plantation formation. The U.S. Forest Service estimates that about 3.5 million acres of National Forest System lands are infested with invasive plants (U.S. Forest Service 2001).

Enhance Product Quality. Weeds have a detrimental effect on crop quality as well as quantity, especially crops that must meet size, color, nutrient content, or contamination-free standards. For example, yields of alfalfa hay in California are often highest during the first cutting when annual weeds are present. However, hay quality is also low when weeds are present in the crop. For example, protein content can fall from over 20% to below 10% when the hay contains large amounts of weeds. Such decreases in grade or quality often mean lowered revenue for growers, since a premium price is usually paid for commodities of high quality.

In some cropping systems, the crop seed and weed seed are so similar in weight and shape that separation at harvest is difficult. Examples are alfalfa and dodder (*Cuscuta* spp.) seed, soybean seed and nightshade fruits, and pea seed that are mixed with the immature flowers of Canada thistle (*Cirsium arvensis*). If the weed material is not removed from these crops by screening, lower price for the commodity will result. For seed crops, the presence of a few noxious weed seed, even less than 1%, usually makes the commodity unmarketable.

Reduce Costs of Production. Weed control is a major reason for many cultural practices associated with crop production. For example, weeds are killed during plowing and cultivation (tillage) to prepare seedbeds for planting. A report by the U.S. National Research Council (2000) indicates that 92–97% of the acreage planted to corn, cotton, soybean, and citrus are treated with herbicides each year. In addition 87% of all citrus acreage and 75% of potato and vegetable crops acreage in the United States are chemically treated for weed control. According to the U.S. Environmental Protection Agency, 60% of the total pesticide sales in the United States in 1999 was for herbicides. There is no doubt that weed control is a costly endeavor in the production of most crops.

Weeds also interfere with harvesting operations, often making harvest more expensive and less efficient. For example, weeds sometimes get wrapped around rollers or cylinders of mechanical harvesters, causing equipment breakdowns and longer harvest times. Up to 50% loss in efficiency and 20% loss of yield can result from weed presence at harvest time.

Reduce Other Pests. Some weed species act as alternate hosts or harbor insects, pathogens, nematodes, or rodents that are crop pests. Numerous specific examples exist of various pest organisms that benefit from the presence of weeds. For example, aphids and cabbage root maggots live on wild mustard, later attacking cabbage and other cole crops. Nightshades are hosts of the Colorado potato beetle. Disease organisms, such as maize dwarf mosaic and maize chlorotic dwarf virus, use Johnsongrass (*Sorghum halepense*) rhizomes to overwinter. Black stem rust uses barberry (*Berberis thunbergii*), quackgrass (*Agropyron repens*), and wild oat (*Avena fatua*) as hosts prior to infesting cereal crops. Rodent damage to orchards can be prevented by weeding around trees before winter.

It also is possible for weeds to aid in the prevalence or spread of certain beneficial organisms that are used to control other pests. In such cases, the weeds act as an alternate source of food or cover for the beneficial organisms, allowing them to survive when the preferred host is not available.

Improve Animal Health. Some weeds are poisonous to animals. However, plants toxic to one species of animal may be harmless to others. For example, larkspur (*Delphinium* spp.) will kill cattle if eaten in sufficient quantity, but sheep and horses are relatively unaffected by this rangeland weed. In contrast, fiddleneck (*Amsinckia* spp.) is highly toxic to horses, while other livestock are relatively tolerant of it. It is estimated that up to 10% of range-grazing livestock may become afflicted by poisonous plants at some time during each growing season.

In addition to direct poisoning, animals may experience other discomforts from association with certain weed species. Some plants [e.g., St. Johnswort (*Hypericum perforatum*), buckwheat (*Eriogonum longifolium*), and spring parsley (*Alchemilla arvensis*)] contain chemicals that make animals abnormally sensitive to the sun, a phenomenon called photosensitization. Other plants contain teratogenic materials that result in fetal malformations. For example, malformed lambs can result if false hellebore (*Veratrum californicum*) is ingested by sheep around the fourteenth day of gestation. Bracken fern (*Pteridium aquilinum*) causes a disease of cattle called "red water" because of the blood-colored urine that is its symptom. This weed causes cancer of the bladder if eaten in sufficient quantities.

Enhance Human Activities. Weeds affect a number of human activities that are difficult to assess in monetary terms. The presence of weeds can reduce real estate values because of the unkempt and unsightly appearance of the property. Dense moisture-holding weed growth aids the deterioration of wooden and metal structures and machinery, further reducing property value. In fire-prone ecosystems, weeds can provide fuel to carry fire, further endangering structures and property. Access and enjoyment of recreation areas are also reduced by weed presence. Other weed impacts and nonmonetary reasons to control weeds are noted in the section on wildlands later in this chapter.

Reduce Effects on Transportation. Some rivers and lakes in the tropics and subtropics are clogged by aquatic weeds, making travel on them nearly impossible. Ross and Lembi (1985) provide an interesting example of how weeds influence transportation costs. They indicate that in 1969 and 1970, 487,000 tons of wild oat seed were inadvertently transported from Canada to the United States along with 16 million tons of grain. The transportation costs for the wild oat were estimated at $2 million, which did not include the $2 million cost for cleaning the grain to remove contamination.

Weeds are kept free from highway intersections to prevent accidents. Airports and railways also keep signs and lights free of weeds so that maximum visibility can be maintained. Power line rights-of-way are kept free of tall growing vegetation to prevent power outages if trees contact power lines during storms and to increase access to downed power lines.

Reduce Risks to Human Health. Toxicants or irritants produced by weeds can cause serious health problems for some people. These discomforts or illnesses include hay fever, dermatitis, and direct poisoning. Hay fever afflicts millions of people each year. It is caused by an adverse effect of proteins associated with the pollen of certain plants on the respiratory system of susceptible people. Ragweed (*Ambrosia* spp.) is best known for causing hay fever. However, pollen from many other broadleaved plants, grasses, trees, and shrubs causes similar allergic reactions. Each year, many people are troubled by poison ivy (*Rhus radicans*), poison oak (*R. diversiloba*), and poison sumac (*R. vernix*). These plants produce and store a toxic substance called urushiol that causes intense itching and rash upon contact with the skin. Many plants contain toxic substances that when ingested cause sickness or death to humans. Toxic substances in weeds include alkaloids, glycosides, oxalates, resins and resinoids, volatile oils, acrid juices, phytotoxins (toxalbumens), and minerals. There are few poisons, including synthetic substances and minerals, that approach the strength and violence of illnesses caused by some plant-produced toxins.

Weeds in Managed Forests

There are many natural conditions such as climate, soil type and fertility, topography, and events like hurricanes and wildfire that shape forested landscapes. Following "catastrophic" disturbances, it is common for forests to undergo a sequence of vegetation changes that result in a forest nearly identical to the one previously destroyed. This process of natural forest reestablishment through successive changes in vegetation composition is called secondary succession (Chapter 2). Following a radical disturbance, like a fire or clearcut, a new patch in the physical environment is once again available for colonization by plants. In such situations, "pioneer" tree (e.g., poplar, birch, alder, and some conifers) or shrub species (e.g., ceanothus or manzanita) (Figure 1.6) are quick to colonize the disturbed areas and can dominate them for years to decades. This rapid recolonization by usually native pioneer species, although a normal stage in succession, can delay the revegetation of disturbed sites with more economically desirable trees (Balandier et al. 2006).

The major disturbance to forests of any region is the harvesting of wood by humans. It was estimated in 1989 that each year the world loses 37 million acres of forest in this manner (Perlin 1989) and current estimates remain unchanged [Food and Agriculture Organization (FAO) 2001]. In temperate conifer forests

Figure 1.6   A young Douglas-fir plantation following logging and artificial regeneration. (Photograph by S. R. Radosevich, Oregon State University.)

logging, especially without any follow up reforestation activities, led to the gradual replacement of conifers by less desirable herbaceous, shrub or hardwood species. Sutton (1985) pointed out that in Canada large-scale weed problems have occurred due to exploitation forestry, which strives to maximize profits and minimize costs. Weed problems were exacerbated by poor choice of forested stands to harvest, season and method of harvesting, intensity of utilization, and lack of attention to regeneration (Sutton 1985). Walstad and his associates (1987) similarly indicated that hardwoods occupy 32% of the prime timberland in western Oregon that was once dominated by conifers.

Forest Regeneration. Most forests regenerate naturally following disturbance given enough time. However, logging activities and land clearing are the principal disturbance factors that both set up and modify the natural patterns and time frames of succession so that native and exotic weed species are favored and even dominate many forest types (Balandier et al. 2006, Wagner et al. 2006). The ability of a site to regenerate, as well as the composition of species following such disturbances, is most dependent on the type, frequency, and severity of the tree removal operation (Kimmins 1997). In the coastal Douglas-fir forests of the

U.S. Pacific Northwest, the impact of both native and exotic plants is currently restricted to the earliest stages of forest succession that follow logging and fires. Ruderal exotic forbs, such as Canada thistle or woodland groundsel (*Senecio jacobaea*), and some exotic shrubs, such as Scotch broom (*Cytisus scoparius*), displace native early seral vegetation in some locations and reduce tree regeneration in others. Though exotic plants are typically eliminated from the plant community after a few years to a decade of forest stand development, exotic shade-tolerant species are capable of persisting and/or invading forest understories if relatively open stand conditions are maintained through clearcutting or severe silvicultural thinning. In particular, false-brome (*Brachypodium sylvaticum*) poses a serious threat to forest understory communities in that region (Zouhar et al. 2007).

Several techniques, collectively known as *artificial regeneration*, have been used successfully to replant many logged-over areas in many countries. This method usually involves collecting seed of preferred tree species, germinating and growing the seedling trees in nurseries, outplanting them to field sites, and following this by intensive chemical weed control. Wagner et al. (2006), surveying 60 studies, found that the most intensive vegetation management treatments always improved crop tree growth, although results varied by location, tree species, and length of time from experiment initiation. Despite these successes in projected crop tree biomass yield, important questions still remain about the ecological (Balandier et al. 2006), social, and economic desirability of converting vast acreages of naturally regenerated forests into tree farms.

Weeds in Rangelands

The destruction and replacement of vegetation by humans are now common occurrences over most of the world, with a loss in primary productivity and floristic diversity often being the result. The invasion of exotic plants is both a cause and a consequence of such environmental manipulation. However, it is rare that invaders cause the replacement of most or all of the plant and animal species in a disturbed ecosystem (Billings 1990). A possible exception to this generalization is rangeland weeds. In this system of production, species replacement following disturbance has been so complete that only a sketchy picture of predisturbance conditions remain. We offer the sagebrush (*Artemisia tridentata*)–cheatgrass (*Bromus tectorum*) steppe as an example (Figure 1.7).

The chance introduction of cheatgrass before the turn of the last century to the Great Basin of North America altered the entire native shrub ecosystem of that region. D'Antonio and Vitousek (1992) after Billings (1990) indicate that its introduction provides a classical case of biological impoverishment where the concomitant environmental change allows successful replacement of indigenous vegetation. In this case, native perennial bunchgrasses and shrubs, particularly sagebrush, were first grazed by large herbivores, then invaded by cheatgrass, and subsequently subjected to range fires (Figure 1.8).

Figure 1.7 Cheatgrass (*B. tectorum*) in former sagebrush–bunchgrass range. (Photograph by S. R. Radosevich, Oregon State University.)

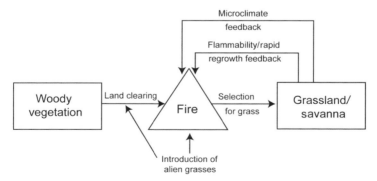

Figure 1.8 Conceptual diagram of land clearing and grass–fire cycle (modified from Fosberg et al. 1990) to illustrate influence of alien grass invasion. In some cases grass invasion itself is sufficient to initiate grass–fire positive feedbacks; more often, it interacts with human-caused land use change. (From D'Antonio and Vitousek 1992, *Annu. Rev. Ecol. Syst.* 23:63–87. Copyright 1992. Annual Reviews Inc.)

Original Vegetation and Early Land Use History of Great Basin. Billings
(1990) and others (Klemmedson and Smith 1964, Mack 1984) indicate that the
western Great Basin was not part of the bison range of the North American Great
Plains because the rhizomatous C_4 grasses on which the bison thrived cannot
grow on the summer-dry steppes of this region. Rather, perennial C_3 bunchgrasses
of the genera *Poa*, *Festuca*, *Agropyron*, and *Stipa* dominated the grass stratum
of this sagebrush ecological formation. Apparently, the native bunchgrasses of
the region also did not carry fire well because range fires in the sagebrush–
bunchgrass steppe, in contrast to the Great Plains, were rare.

The native ungulate herbivores were antelope, deer, desert bighorn sheep, and
elk which, because of their smaller size and numbers than bison, created a relatively light impact on the sagebrush–grass community. During the 1840s and
1850s, the first overland wagon trains to Oregon and California introduced domestic livestock to the region. Thus, the first grazing impacts in the area appeared
along the Oregon and California Trails. For example, Beckwith, (1854) (in
Billings 1990) pp. 305–306 an early explorer, noted the following in June 1854:
"Fine droves of cattle, which had been wintered near Great Salt Lake, passed
today on their way to California, and one or two large flocks of sheep are but a
few miles behind them. The more experienced stock-drovers to California send
their cattle back from the river to feed on the nutritious grass of the hills."

Watson (1871) made one of the first good botanical descriptions of the area,
listing 59 species of Poaceae. Cheatgrass was not among the species listed,
suggesting that it had not yet arrived to the intermountain region of North America.
In the summer of 1902, Kennedy (1903) made the first survey of range conditions
in northern Nevada and 50 years later Robertson (1954) retraced Kennedy's route.
Billings (1990) compared the writings of both men and noted the following
differences in range conditions that occurred over that 50-year time period:

- Desirable livestock browse shrubs decreased.
- Bluebunch wheatgrass (*Agropyron spicatum*), a prime forage bunchgrass,
  decreased from "abundant" to "generally absent" or "less than 5% density."
- Annuals, notably cheatgrass, not present in 1902 had increased to an
  "extreme degree."
- Burn scars were "absent or unimportant" in 1902. In 1952 much of the route
  was bordered or crossed by "burned-off range" and covered by cheatgrass or
  little rabbitbrush (*Chrysothamnus viscidiflorus*).
- Big sagebrush replaced "bluegrass meadows" at lower elevations.
- "Stream channels had eroded deeper and wider."

All of the conditions in the above list indicate heavy grazing, cheatgrass invasion,
and occurrence of repeated fires.

Introduction of Cheatgrass and Fire. According to Mack (1981, 1986), the first
collections of cheatgrass in the Great Basin were from Spenses' Ridge, British

Columbia, in 1889; Ritzville, Washington, in 1893; and Provo, Utah, in 1894. Each location is in a wheat-growing area, which suggests that cheatgrass seed may have arrived as a contaminant of crop seed. From these beginnings, the species spread throughout eastern Washington and Oregon, southern Idaho, northern Nevada, and Utah. By the 1930s, it was abundant throughout the entire sagebrush steppe. Billings believes that the rapid spread of the grass across the region was aided by railroad stock cars and grazing animals that were subsequently driven onto the rangelands. In addition, the climate of the Great Basin was ideal for the new weed, which, being a winter annual, requires moist soils during the cold season and cold winter weather while vegetative in order to flower the following spring.

Billings (1990) and D'Antonio and Vitousek (1992) indicate that once cheatgrass became established, the region was set for wildfires (Figure 1.8). Cheatgrass usually sets seed, dies, and dries up by June in most areas of the region. Thus, a supply of fuel that was nonexistent in the original open sagebrush–bunchgrass ecosystem became available. Without fire, cheatgrass simply invades the overgrazed sagebrush range, where it forms an ephemeral annual stratum in that community. However, once this plant community experiences either lightning or human-caused fire, the sagebrush is killed. Since this shrub cannot sprout following fire, the native shrublands of the Great Basin have been replaced by vast expanses of annual grassland. Upland areas of the Great Basin, notably the pinyon pine–juniper biome, are now increasingly threatened by a similar process of vegetative change.

Management of Cheatgrass. The assemblage of features that allowed cheatgrass to invade the Great Basin have also largely prevented its eradication. For example a plant capable of germinating over an eight-month period, as cheatgrass can, is nearly impossible to control completely in the seedling stage. Mowing or grazing in early spring makes little difference since developing seed of the species are readily viable and capable of germinating the following autumn. Even when fire removes all vegetative plants, new ones emerge from seed reserves in the soil and, of course, further reentry is always possible. Furthermore, as we will see in Chapter 4, the plant has little problem adapting to the wide variety of environmental conditions of both rangeland and cultivated fields. While cheatgrass is clearly a permanent member of the Great Basin vegetation, it may now be possible to restore local areas of cheatgrass infestation to a more pristine and desirable state (Briske et al. 2003, Sheley and Krueger-Mangold 2003).

On the other hand, and from the standpoint of volume of herbage produced and extent of area covered, cheatgrass is unquestionably the most important forage plant in the Great Basin now (Klemmedson and Smith 1964). It provides the bulk of early spring grazing for all classes of livestock on millions of acres in the arid West. While it is not easy to comprehend the economic importance of this ecological change, the extent and permanence of it are readily comprehensible.

INVASIVE PLANTS IN LESS MANAGED HABITATS AND WILDLANDS

Certain forests, deserts, prairies, beaches, marshes, estuaries, and riparian areas have been protected from disturbance or designated as wilderness throughout the world. Wilderness and similarly managed natural areas, such as national parks and monuments, provide many benefits to society. These benefits include the preservation of biodiversity, unique natural features, and watersheds as well as opportunities for recreation and personal fulfillment. Although land management agencies place a high priority on protection of natural ecosystems and wilderness areas, some of these benefits are threatened by increasing levels of human activity within and outside areas designated for protection. The introduction of exotic species into such areas is of particular concern due to the potential for irreversible impacts on the natural ecosystems that such areas represent (Aldo Leopold Wilderness Research Institute 2003, D'Antonio et al. 2004).

Three research areas were identified by the Aldo Leopold Wilderness Research Institute to address the question of exotic plant invasion into wilderness:

- Understanding the introduction, spread, and distribution of exotic species within wilderness
- Understanding the effects of exotic species on wilderness values
- Identifying and evaluating management options and their consequences

Parks and her associates (2005a,b) examined the patterns of invasive plant diversity in mountainous ecoregions of the northwestern United States. Their analysis found that altered riparian systems and disturbed forests were especially vulnerable to exotic plant invasion. Conversely, alpine areas, forests, and grasslands designated as wilderness were still relatively unaffected by invasive plants, with introductions often being restricted to campsites, roads, or trails. The predominance of wilderness throughout much of the western United States is believed to contribute to the lower incidence of invasive plants in mountainous ecoregions of that area compared to other regions. Human settlement and intense land use at low elevations were identified as factors that enhance invasive plant introductions (Parks et al. 2005a,b, Mack et al. 2000).

Local versus Regional Perspectives about Weeds

Most of the previous discussion has focused on weeds and invasive plants at the individual plant, species, or field level of scale. However, weeds may extend much farther than individual fields and the benefits and costs of weed control may extend much further than to individual farmers, foresters, or land managers. For example, consumers of agricultural products or users of natural resource areas may benefit from lower priced food, more abundant lumber, or greater access to recreational areas as a result of weed control. These same people also may have legitimate concerns about the presence of chemical residues in food or water, public safety, soil erosion, or other impacts that weed control techniques have on them or their environment. Others may be concerned about the overall vitality of an industry or profession as new technologies are introduced and others are regulated. All of these issues extend beyond the aims of individuals to the needs, wants, and expectations of a society. These issues are explored further in Chapters 2 and 9.

Weeds in Regional and Global Context

There are many examples of the widespread regional or even global distribution of weeds. One of the earliest examples is that of Hitchcock and Clothier (1898), which describes the distribution of native and introduced weeds in Kansas as that land was being developed for agriculture. A similar study was accomplished by Mason (1932), who described the occurrence of wild oat throughout several provinces of central Canada (Figure 1.9). These studies are augmented by more recent descriptions of widespread infestations of weed species, for example, leafy spurge (*Euphorbia esula*), purple loosestrife (*Lythrum salicaria*), downy brome (also known as cheatgrass) (*B. tectorum*), Paterson's curse (*Echium platagineum*) (Auld and Tisdel 1988), and lantana (*Lantana camara*) (Figure 1.10) (Cronk and Fuller 1995). The ability to disperse widely is a common characteristic of many weed and invasive plant species, which has been exacerbated in recent decades by increasingly global movement of humans and goods. Any harmful organism that is spreading or has the capacity to spread poses a threat to uninfested areas without regard for ownership boundaries. Thus, a spreading species represents a problem to more people than just those whose land it currently occupies. Such situations make a strong case for legislation (weed laws), quarantine districts, or other governmental interventions to reduce or slow the spread of weeds and

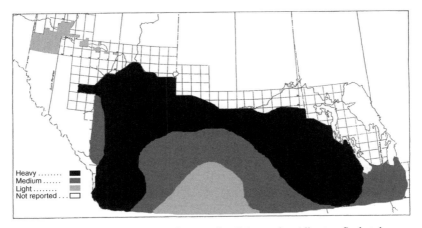

Figure 1.9 Distribution and prevalence of wild oat in Alberta, Saskatchewan, and Manitoba in 1931. (Modified from Mason 1932.)

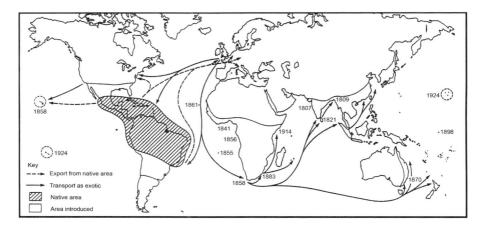

Figure 1.10 Some invaders, such as the shrub *L. camara*, have been introduced repeatedly in new ranges, the result of global human colonization and commerce. As the array of estimated years indicates, lantana was introduced throughout the nineteenth and twentieth centuries in many subtropical and tropical areas. In each new range it has become highly destructive, both in agricultural and natural communities. (Cronk and Fuller 1995, from Mack et al. 2000, *Ecol. Appl.* 10:179–200. Copyright 2000. Ecological Society of America.)

invasive plants. Furthermore, governmental objectives for weed suppression may be less constrained by cash flow than those of individual farmers, ranchers, or forest land owners (Auld and Tisdel 1988).

SUMMARY

Weeds are a category of vegetation that exists because of the human ability to select among plant species. In most cases, the value of a weed is determined by the perception of its viewer. Weeds have been described and defined in both anthropomorphic and biological terms. They also may describe a condition of the land or environment and they affect almost everyone at some time or place. Some of the negative aspects of weeds are lowered crop yields, animal discomfort and death, poor product quality, increased costs of production and harvest, higher incidence of other pests, and reduced human health and activities. Invasive plants, unlike many agricultural weeds, can successfully occupy and spread to new "natural" habitats apparently without further assistance from humans.

Weeds and invasive plants have been classified in numerous ways. Some methods used to classify weeds are by taxonomic relationships, life history (annuals, biennials, perennials, etc.), habitat, physiological differences, degree of undesirability, ecological behavior, and evolutionary tactic. Weeds are distributed widely throughout the world, inhabiting most agricultural, and managed forest and rangeland systems. However, weeds account for less than 0.1% of the flowering plants of the world. Many environmental, biological, and human factors influence distribution of weed species, although humans are the main factor for the continued evolution of weeds and spread of invasive plants into new regions of the world.

2

PRINCIPLES

A fundamental goal of plant ecology is to explore the underlying order of vegetation, to understand the processes of nature. Some ecologists are most concerned with the overall relationship of vegetation to environment while others study the biology of certain plant species in relation to local conditions. Many ecologists view plants as only a component of nature and are most interested in the interactions, interdependencies, and cycles that occur among its parts. Applied plant ecologists seek to use fundamental information to address vegetation management problems. Although the approaches differ among these scientists, the manner in which plants adapt to and exist in their respective environments is of common interest. It is through ecological principles and concepts that farmers, foresters, and other land managers begin to understand the nature of weediness and plant invasions. Once this basic foundation is established, it is possible to explore the relationships and interactions that exist among environment, weeds and crops in agriculture, and invasive plants in forest, rangelands, and other wildland systems. In the process, less costly, more efficient, or environmentally sound vegetation management and improved efficiency or profitability can result.

ECOLOGICAL PRINCIPLES

Interrelationship of Biology and Environment

Weeds and invasive plants, by definition (Tables 1.1 and 1.4), are aggressive and can persist in the ecosystems they inhabit. Such plants have particular life history

Ecology of Weeds and Invasive Plants. By Steven R. Radosevich, Jodie S. Holt, and Claudio M. Ghersa

traits that enable rapid colonization; for example, escape from native predators, small seed size, short juvenile growth period, persistent seed bank, and young reproductive age are associated with aggressive traits of weeds and displacement of native plants (Rejmánek and Richardson 1996, Sakai et al. 2001). However, these characteristics are not generalizable to plant invasions worldwide, nor do they account for weed–ecosystem interactions. For instance, significantly more invasive plants occupy grassland and semiarid areas in contrast to forests and marshes (Parks et al. 2005a). Hence, vulnerability to invasion varies by plant community. Disturbed habitats and agroecosystems have higher susceptibility to invasions than do systems that spend long periods of time in late successional phases. One explanation for semiarid and grassland vulnerability is that these areas have spatially open niches that are devoid of vascular plants for some or most of the year (Figure 1.8) (Billings 1990). On the other hand, plants with life-forms dissimilar to the native vegetation also have invaded some ecosystems. The conversion to annual grasslands from tussock grasslands in California and the invasion of prickly pear (*Opuntia stricta*) into Australia (where no members of Cactaceae existed previously) are examples where biological characteristics played an important role in the invasion process. The intrinsic biology of the species and the extrinsic nature of the area being occupied or invaded (environment) are equally important in determining the success and expansion of weeds.

Environment. Central to plant ecology is a recognition that plants exist in and therefore respond to a wide array of environments. The *environment* is the summation of all living (biotic) and nonliving (abiotic) factors that can affect the development, growth, or distribution of plants. It is now recognized, however, that biotic and abiotic factors are closely interrelated and often hard to separate. For example, soil microorganisms often determine the nature of soils and plant canopies have profound effects on quality and quantity of radiation. Environment is often divided into two components, macroenvironment and microenvironment. The *macroenvironment* is the broad-scale regional environment that includes many aspects of soil and climate, such as overall light intensity (radiation or irradiance), rainfall, humidity, wind, and temperature. The *microenvironment* occurs on a smaller scale and is that aspect of the macroenvironment that is influenced by the presence of objects (rocks, trees, etc.), chemicals (organic matter, nutrients), and topography. Although both macroenvironment and microenvironment can be measured and therefore expressed in similar terms, it is the microenvironment to which individual plants respond to form the mosaic of vegetation over a local or regional landscape.

The factors most responsible for growth and persistence of weeds and invasive plants are soil, climate, and land use. It is well known that the distribution of agricultural weeds (Aldrich 1984, Zimdahl 1993) and occurrence of plant invasions fall within the ranges of certain soil types (Huenneke et al. 1990). Climate also defines biotic and abiotic thresholds for floristic growth in ecosystems. Climate can change habitat suitability over short time scales through such events as drought, frost, or flooding (Nielson and Muller 1980). Invasive plants may adapt to a variety of habitats but are usually found in areas with similar climates to their native range first, then adapt to other climates later (Panetta and Mitchell 1991, Reichard and Hamilton 1997); however, notable exceptions also exist (Mack 1995). Topography and elevation influence climate and thus the species that can grow at a given location. Land use is a third general driver of environmental suitability for weeds and invasive plants. Changes in land use are thought to be the single most important factor in species extinction and to have strong influence on invasible sites (Cousens and Mortimer 1995, Elton 1958, Hodkinson and Thompson 1997). Humans have modified most of the world's ecosystems to some degree, and this has direct effects on environmental suitability for weed growth and plant invasions. Although some ecosystems are altered by the presence of weed species through their effects on fire frequency, nitrogen depletion or addition, or allelochemicals, other weeds/invasive plants are adapted for change in land use whereas often native species are not.

Resources and Conditions. Factors in the environment that influence plant growth are usually divided into two categories, resources and conditions. Environmental *resources* are consumable and include radiation, CO_2, water, nutrients, and oxygen. In contrast to resources, environmental *conditions*, such as temperature, soil pH, and soil bulk density (compaction), are not directly consumed but nonetheless affect plant growth. The general relationships of plant responses to both types of environmental factors are shown in Figure 2.1. Plant responses to environmental resources increase through a resource-limited phase to a saturation level, at which point another factor generally becomes limiting. For example, nitrogen response curves generally have the shape shown in Figure 2.1a. At saturation, further addition of the resource does not increase the response significantly. Higher than optimum levels of resource may cause the response to decline. Conditions usually limit plant response by either absence or abundance until a threshold level is reached. Plant responses to environmental conditions generally have a bell shape and three cardinal points can be identified (Figure 2.1b). The minimum and maximum points occur where the process ceases, while the optimum range is where the highest rate can be maintained if no other factors are limiting. A temperature optimum for seed germination is an example of an environmental factor that markedly influences plant response but is not consumed.

Space. Some ecologists have chosen to consider the impact of resources on plant growth as a single conceptual unit, called *space*. Thus, space refers to the composite of all resources necessary for plant growth as well as their interactions. The concept of space as an integrative resource allows scientists to study the effects of proximity among individual plants without concern for the actual source of the interaction. In this manner, the effect of individuals upon each other that share the same environment can be measured, since each must act as the biological indicator of "space" utilization by the other. Whether to consider resource availability as a composite factor, such as space, or to consider resources independently depends on the information desired. For example, there are situations in which

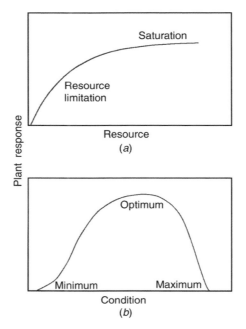

Figure 2.1 Idealized curves showing typical responses to varying levels of (*a*) environmental resource and (*b*) condition. In (*a*), the resource levels at which limitation and saturation occur are shown. In (*b*), the three cardinal points are shown: minimum, optimum and maximum. (From Holt 1991. *Weed Sci.* 39:521–528. Copyright 1991. Weed Science, Society of America. Reprinted by permission of Alliance Communications Group, a division of Allen Press, Inc.)

the identification and supplementation of a single limiting resource or condition has corrected a growth deficiency that may have been aggravated by the presence of neighboring plants. Thus, at times it is advantageous to consider space and its implied resources as the object of study, whereas at other times the influence of separate resources should be considered.

Scale

Ecological patterns and processes occur at many different *scales*, or spatial and temporal dimensions. Scale is often referred to in terms of a hierarchy or set of levels of organization (Gurevitch et al. 2002). Hierarchy theory (O'Neill et al. 1986) as an ecological concept was developed within the context of general systems theory (GST), a holistic scientific theory and philosophy of the hierarchical order of nature which is thought to consist of open systems with increasing *complexity* (number of interconnections). Biological systems have two subsets within GST: (1) living systems, such as individual plants or animals, and (2) ecological systems. In ecological systems, combinations of organisms and environments are arranged, or arrange themselves, in a nested hierarchy of sub-systems (levels) according to differences in process rates. At high levels complexity is high and process rates (e.g., life span, turnover rates, rates of activity) are slow, while at low levels process rates are fast and complexity is generally lower. Thus, temporal and spatial scales are determined not only by their dimensions but also by the differences in complexity and process rates among levels.

Scale in Ecological Systems. A common ecological hierarchy such as that depicted in Figure 2.2 includes the following levels: *biome, ecosystem, community, guild, species, population,* and *organism,* although the hierarchy could be extended downward to include tissues, cells, and enzymes. The components within any particular level are always linked and perturbations are characterized by feedbacks both within and among levels. Generally, even slight changes in a higher level result in substantial impacts at lower levels, but impacts over large areas or long times are necessary for lower levels to influence higher ones. Within the concept of scale, levels of organization are thought to have *emergent properties*, which occur due to properties, processes, and interactions unique to that

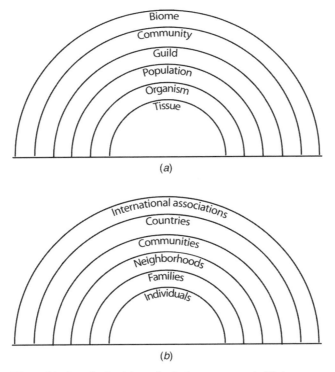

Figure 2.2 Hierarchical scale in (*a*) ecological systems and (*b*) human social systems arranged according to size and complexity. (From Ghersa et al. 1994c. *Bioscience* 44:85–94. Copyright 1994. American Institute of Biological Sciences.)

level (Gurevitch et al. 2002, O'Neill et al. 1986). Most plant ecologists are concerned with the levels from ecosystems to organisms because at these levels interactions can be recognized easily. The everyday activities of agriculture, forestry, and resource management also operate within these levels of organization.

A *population* is a group of organisms within a *species* that co-occur in time and space. A species is usually composed of several to many populations. The dynamics of plant populations will be discussed in more depth later in this chapter and in Chapter 5. The term *community* refers to all the populations that occupy a particular site at a particular time and is a concept used primarily to examine or explain the biological interrelationships of organisms. Although "all" populations includes soil organisms, insects, birds, plants, and so on, the focus of plant ecologists is usually on the principal higher plants, for example, a corn field and its associated weeds or a stand of ponderosa pine and its associated plants. However, as noted by Booth et al. (2003), interactions among taxa can be significant determinants of pattern in plant communities and should not be overlooked.

The basic structural attributes of a plant community, as described by Barbour et al. (1999), are as follows:

- Relatively consistent floristic composition
- Relatively uniform physiognomy (structure, height, cover, etc.)
- Characteristic distribution in a particular type of environment or habitat

Communities are also dynamic and are characterized by the processes that create their structure, including succession and assembly. Some of the specific traits that characterize plant communities are listed in Table 2.1. Although the idea of a community as a discrete ecological unit is controversial, the concept is useful because it allows ecologists to define and spatially delineate the vegetation with which they are concerned. Thus, communities are typically defined by the characteristics of the dominant plants in an assemblage, which are then related to larger areas of land classification (environment or habitat type). Often a community can be characterized by a *keystone species*, one whose impact on the community is disproportionately large relative to its abundance (Power et al. 1996, Perry 1997).

Communities typically appear as discrete units in the landscape (stands) that are delineated by common environmental constraints or boundaries (e.g., climatic or physical limits) or by biotic factors (e.g., competition or herbivory). One early view of communities, the holistic or organismic hypothesis proposed by Clements (1916, 1936), described communities as more than the sum of their species and function like supraorganisms. In this view species assemblages evolve over time into predictable associations with strong interactions. In contrast, Gleason (1926) proposed the continuum or individualistic hypothesis, in which the links among species are thought to be relatively loose and the structural and functional characteristics of plant communities are considered idiosyncratic and variable. In this latter view, communities are formed through stochastic events and are a random collection of species with similar environmental tolerances. Thus, factors such as

TABLE 2.1 Some Characteristics of Plant Communities

Physiognomy	Nutrient cycling
Architecture	Nutrient demand
Life-forms	Storage capacity
Cover, leaf area index (LAI)	Rate of nutrient return to the soil
Phenology	Nutrient retention efficiency of nutrient cycles
Species composition	Change or development over time
Characteristic species	Succession
Accidental and ubiquitous species	Stability
Relative importance	Response to climatic change
(cover, density, etc.)	Evolution
Species patterns	Productivity
Spatial arrangement	Biomass
Niche breadth and overlap	Annual net productivity
Species diversity	Allocation of net production
Richness	Creation of, and control over, a microenvironment
Evenness	
Diversity (within stands	
and between stands)	

Source: Table 8.1 p. 187 from *Terrestrial Plant Ecology*, 3rd ed., by M. G. Barbour et al. Copyright 1999 by Addison Wesley Longman, Inc. Reprinted by Permission of Pearson Education, Inc.

resource availability, soil characteristics, species evolutionary history, species relative abundance, and colonization history determine how particular assemblages occur over space and time. Both schools of thought are supported by empirical evidence, although Gleason's hypothesis has gained more acceptance over the last half century (Booth et al. 2003, Walker 1999). The present view of communities is a synthesis of the two hypotheses above and the newer concepts of scale and complexity such that no single model is likely to explain all patterns of vegetation (Barbour et al. 1999).

A complete group of communities and their environments, when considered together, forms an *ecosystem*. Ecosystems are defined by their functional outputs, which change over time from early to mature stages. Odum (1971) summarized the temporal changes in ecosystem outputs as follows:

- Increases in biomass (physical structure)
- Increasing number of feedback loops (recycling of energy and matter)
- Increases in respiration, although respiration relative to biomass decreases
- Dominance of larger organisms
- Increases in then stabilization of entropy
- Increases in information (gene transfer, biochemicals, etc.)

Based on these characteristics, Jørgensen (2006) has proposed a hypothesis for ecosystem development, *the ecological law of thermodynamics*, whereby all components of the ecosystem organize to generate the highest flux of useful energy and store it in a system. The spatial dimensions of ecosystems are also scaleless because their limits are defined by the interactions among the various biotic and abiotic components that compose them, rather than by location.

The different organisms within ecosystems are classified by their function or energy flow, that is, who eats whom. The four trophic levels that function in an ecosystem to form the "food chain" are as follows:

Producers—Green photosynthetic plants (autotrophs) that provide the basic food resource to all other organisms (heterotrophs).

Primary Consumers—Herbivores that feed directly on green plants.

Secondary Consumers—Predators that feed on the primary consumers, for example, birds that feed on grasshoppers. Many secondary consumers also act occasionally as herbivores.

Decomposers—Bacteria and fungi that contribute to decay and are important for nutrient cycling, breakdown of chemical residues, and soil formation.

Food or trophic chains can be constructed to describe the relationships among organisms in a particular community or ecosystem. For example, the chain in a simple agroecosystem might be

Grass/clover		cattle		humans
(producer)	$\longrightarrow$	(primary consumer)	$\longrightarrow$	(secondary consumer)

These chains become complex as more communities and populations are introduced into the ecosystem to create a trophic web, as shown below for a hypothetical alfalfa–weed plant community:

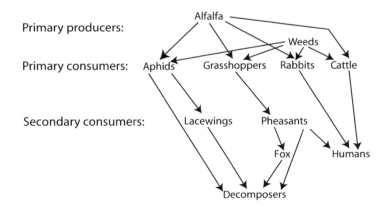

Each arrow in such schemes represents transfers of food, or energy, from one organism to another. These transfers are important because they define the efficiency of the system in the flow of mass (carbon), energy, and nutrients.

Scale in Human Production Systems. The perceptions of scale determine relevant questions for weed scientists and land managers. For example, weed control practices cause other effects besides simple changes in a local flora. Management practices that seemingly affect only crop–weed or native–exotic plant composition might eventually influence an entire agricultural, forested, or rangeland landscape of a region. Cultivations or herbicide applications certainly reduce weed competition and change vegetative composition, but they also may cause losses in soil fertility through erosion or nutrient leaching that are larger scale issues. In addition, weeding stimulates feedback within a plant community that increases the probability of invasion by new weeds. This feedback may explain why relative and absolute abundance of the weed flora in the United States increased steadily from 1900 to 1980 (Forcella and Harvey 1983), despite enormous local efforts to control weeds.

Just as ecological systems can be arranged through GST into a hierarchical structure (Figure 2.2*a*), social or human systems also may be arranged similarly according to function and scale (Figure 2.2*b*). A common hierarchy of human social systems is *individual, family, neighborhood, community, country*, and *global* or *international association.* As in ecological systems, levels in human systems can be determined by actual differences in process rates (e.g., adoption of new technology, cultural assimilation, education) that define functional boundaries of scale as opposed to arbitrary ones.

Human actions and impacts during the early evolution of agriculture probably operated at a spatial and temporal scale similar to that of the natural ecosystem (Figure 2.2). That is, individual humans and human populations manipulated individual plants and plant populations, and the reaction times and feedbacks of information were probably similar for both social and ecological processes. As agriculture has evolved, however, the capacity for humans to use energy (Figure 2.3) and acquire and transmit information across greater distances and to

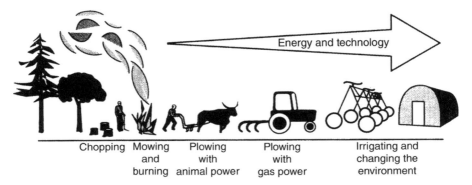

Figure 2.3 Evolution of human activities and the agroecosystem. Width of the arrow indicates relative amount of energy and technology. (From Ghersa et al. 1994c. *Bioscience* 44:85–94. Copyright 1994. American Institute of Biological Sciences.)

future generations has led to a divergence of scale. In other words, human actions now have an impact over larger space, greater time frames, and longer response times than is possible for ecological systems to accommodate easily. Thus, the functional levels of human systems and agroecological production systems have diverged as modern agriculture has evolved.

A better understanding of the agroecosystem should assist in the recoupling of human systems to ecosystems they impact. However, the healthy coevolution of human systems and agroecosystems will require minimizing human impacts. For example, in modern industrial agriculture, just replacing herbicides with cultural approaches, breeding weed-suppressive or allelopathic intercrops, or using biological control with insects or pathogens may not be enough to reduce large-scale impacts. Rather, it will also require the maximum use of information about human values and biotic interactions when designing new or different weed management strategies. Levins (1986) uses three generalized models of pest management to demonstrate this point (Table 2.2). He calls these models the *industrial*, *IPM* (*integrated pest management*), and *ecological agricultural* approaches to pest management. Of the three models, only the third, which is still theoretical, requires an understanding of fundamental processes in ecological and human systems (Figure 2.2 and Table 2.2) to manage weeds, invasive plants, as well as other pests. More recently, Cardina and colleagues (1999) described five levels of increasing integration in weed IPM that advance from the individual plant or field to the regional or global scale. In this model, as the level of integration increases, so does the spatial and temporal scale and the degree of complexity of methods, approaches, and concepts needed. Similar to Levins, these authors point out that current weed science research today falls at level III (the farm–landscape scale), while levels IV and V (landscape-to-global scales) are still theoretical. Further discussion of these topics in weed management is found in Chapter 9.

Weeds as Components of Complex Production Systems. As seen above, the components of complex systems maintain links that allow the flow of matter, energy, and information within and between systems both above and below (Figure 2.2). These rates of flow among components determine the behavior of the entire system and the regulation of any component by positive- and negative-feedback responses. A *positive-feedback* response occurs when a process increases in relation to its previous rate. In contrast, *negative feedback* results in a reduced process rate in relation to its previous rate. Such regulation in ecological systems follows cybernetic principles (Checkland 1981); that is, when negative-feedback responses are absent, the population experiences exponential growth. All populations have intrinsic potential for exponential growth when environmental regulation is lacking (Begon and Mortimer 1986, Cousens and Mortimer 1995). This growth causes an invasion process often called an *infestation*, *infection*, or *epidemic*, depending on the organism and the perspective of the observer. From the cybernetic viewpoint, a weed infestation is a plant population lacking negative-feedback control to compensate for the positive response of reproduction and growth. Food webs (herbivory), nutrient cycling, and competition are all forms of

TABLE 2.2   Approaches to Pest Management

	Industrial	Present IPM	Ecological Agriculture
Goal	Eliminate or reduce pest species	Maximize profits	Multiple economic, ecological, and social goals
Target	Single pest	Several pests around a crop and the predators	Fauna and flora of a cultivated area
Single for intervention	Calendar date or presence of pest	Economic threshold	Multiple criteria
Principal method	Pesticide	Prevention by plant breeding and crop timing, careful monitoring and multiple interventions	System design to minimize outbreaks and mixed strategies
Diversity	Low	Low to medium	High
Spatial scale	Single farm	Single farm or small region defined by pest	Agrogeographic regions
Time scale	Immediate	Single season	Long-term steady-state or oscillatory dynamics
Boundary conditions	Everything as is: crops, cropping system, land tenure, microeconomics, decision rules, social organization	Major crops, land tenure, and decision rules	Societal goals
Research goal	Improved pesticides	More kinds of interventions	Minimize need for intervention

Source: Levins (1986), in Kogan (1986). *Ecological Theory of Integrated Pest Management Practice.* Copyright 1986 reprinted with permission of John Wiley & Sons, NY.

negative regulation. Negative regulation is lost when weed seed are dispersed from year to year by machinery, for example (Ghersa and Roush 1993), or when the soil nutrient cycle is bypassed by fertilization. In these situations, a plant population outbreak occurs because of positive feedback or the absence of negative feedback. If the plant population has human value, we may call it a high-yielding crop; if it is a weed, we call it a problem.

The lack of negative regulation in a human production system also may occur if a new element is incorporated into it. When a new plant population immigrates into an ecosystem, it often lacks links to other regulatory components. For example, animal or insect grazers may not immediately recognize the new, exotic plant as food. This view of invasive plants using cybernetic theory emphasizes understanding and correcting what is causing the outbreak of an exotic plant population, rather than concentrating on eliminating the invader. While these ideas are most appropriate for ecological systems they are applicable to socioeconomic systems, as well. For example, a farmer or land manager is usually willing to spend more money for weed control in relation to how densely invaded his or her fields or land is.

Community Differentiation and Boundaries

As noted above, questions still remain about whether plants occur as discrete associations (communities) where members have similar distribution limits within a particular habitat or whether species distributions occur in an overlapping fashion along continuous environmental gradients (ecosystems). Interestingly, changes in ecosystems are studied by examining the characteristics of plant community structure and function over time and space. In such cases, finite land area units are identified that are acceptably homogeneous in plant composition. But in other cases, this spatial approach cannot be applied easily because the structure and composition of the community changes along an environmental gradient. In either case, however, it is clear that groups or stands of vegetation exist whenever there are discontinuities in the environment, such as changes in soil type, topography, or disturbance (Figure 2.4). Thus, plant communities can be viewed as a mosaic of ever-changing patches of vegetation that occupy a larger landscape. Uniform landscapes with little variation in soil characteristics or disturbance exhibit more extended stands of vegetation, whereas mountainous areas with topographic and soil variation present smaller stands. The size, shape, abundance, and connection of these stands play an important role in how plant communities function and respond to environmental change (Forman 1995). The translation from

Figure 2.4 Secondary succession after logging activity. Various stages of community development are evident depending on the time (years) after canopy removal. (Photograph by S. R. Radosevich, Oregon State University.)

physical structure to either plant communities or ecosystems is difficult when environments are so homogeneous that differences in species assemblage are only detectable over large distances (Barbour et al. 1999).

Most plant communities exhibit both vertical and horizontal differentiation; that is, different species occur at various heights above the ground and also are distributed differently along the ground surface. The vertical distribution of species is usually determined by a gradient of sunlight, with the upper canopy being in full sun and lower canopies occurring in diminished radiation. Horizontal distribution is more complex. Whittaker (1975) identified four ways in which species in a community (also individuals in a population) can be distributed horizontally (Figure 2.5). In natural communities, species often appear to be scattered at random (Figure 2.5a) and, indeed, regular spacing of plants (Figure 2.5c) is usually rare. This observation is not true in the agricultural plant community or forest plantation since at least one species, the crop, is usually planted in rows. Departures from randomness also are known to occur in natural plant communities. In these cases, the species are concentrated into patches (Figures 2.5b and d). Patchy distribution may result from the dispersal pattern of parent plants, gradients in microenvironment, disturbance, or species interactions, that is, positive or negative associations of one species with another.

It is generally held by ecologists that each species within a plant community has a unique pattern of distribution. These patterns may be correlated with those of other species, but they are not identical to them. Thus, a plant community is a composite of numerous species distributions each superimposed upon the other and sometimes confounded by disturbance such that a myriad of subtle interactions exist and can be seen across a landscape.

Community Structure

Community attributes can be grouped into those that describe the appearance and functioning of the community as well as those that describe the numbers and abundance of the species present (Table 2.1). Communities can be described by

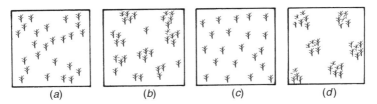

Figure 2.5 Four ways in which individuals of a population can be distributed in horizontal space in a community: (a) Random dispersion (note its apparent irregularity); (b) clumped or contagious distribution; (c) regular or negatively contagious distribution; (d) combinations of strong clumping of individuals into colonies and regular distribution of the colonies. (Modified from R. H. Whittaker 1975, *Communities and Ecosystems.* Copyright 1975 by Robert Whittaker, Macmillan Publishing Co., N.Y.)

their spatial and temporal appearance, which together comprise physiognomy. The spatial components of physiognomy include such features as architecture, life-form of dominant species, canopy cover, and LAI. Temporal structure includes phenology, or timing of life-cycle events, as well as life history traits such as life span of the dominants. The collection of species in a community can be described by their functional attributes, as well, including their physiological or biochemical behavior. Thus, communities can also be described by such traits as the metabolic pathway and leaf characteristics (e.g., evergreen or deciduous) of the dominant species. Many of the physical and functional attributes of plant communities are somewhat independent of their floristic composition. For example, communities in different regions of the world with similar macroclimates can have similar physiognomies but very different species composition, demonstrating the process of convergent evolution. Communities are commonly described and compared using these structural attributes, which can also serve as indicators of community health or change.

Communities can also be described by their floristic composition, or the list of species they contain and their abundance. In order to study and compare communities, species composition is typically quantified as *diversity*, generally considered to be the number of species in a given area. More specifically, diversity is comprised of species *richness*, or the number of species in a particular area, and *evenness*, or the distribution of individuals among those species (equitability of abundance). Species diversity, therefore, is a combination of species richness weighted by species evenness (Barbour et al. 1999). Numerous approaches have been used to measure and describe diversity of plant communities, from simple species per-unit-area calculations to more complex mathematical diversity indices, such as Simpson's index and Shannon–Wiener's index, which consider both richness and evenness. These and other indices are described in Gurevitch et al. (2002) and other basic ecology texts.

The term diversity can be used to describe not only species composition, as above, but also genetic or other levels of taxonomic diversity. Biological diversity, or *biodiversity*, of living organisms has become a widely recognized descriptor of the status of communities and ecosystems worldwide. Ecologists view diversity as an important and positive attribute for its role in community stability, as discussed below. In addition, many environmental and economic benefits of biodiversity have been described, including soil formation, organic waste disposal, biological nitrogen fixation, biological pest control, plant pollination, and providing a source of crop and livestock genes as well as pharmaceuticals (Pimentel et al. 1997). These ecological services apply to both natural ecosystems and agroecosystems, where even weeds and other pest organisms are considered a component of biodiversity (Altieri 1999). However, in natural ecosystems, exotic species are generally not considered a desirable component of biodiversity, since their presence is often correlated with a reduction in native species diversity (Randall 1996). Similar to the concept of a weed, therefore, biodiversity can be a subjective concept that varies with who is defining it. Regardless of the point of view, changes in biodiversity are usually seen as a significant and irreversible impact of human activity in natural and managed ecosystems (Chapin et al.

1998). Assessing the consequence of human impacts on biodiversity is an area of active research and profound debate. Further discussion of the impacts of weeds and invasive plants on biodiversity is found in Chapter 3.

Succession

The species composition of a plant community changes unidirectionally over time in a continuous process called *succession*. Traditionally, plant communities were thought to develop over time into a specific stable stage with constant species composition, the *climax* community. The *seres* or stages of community change over time include the pioneer (initial) and intermediate *seral* stages, which lead to the climax stage. Odum (1971) and Whittaker (1975) codified the features of the classical successional model, which assumed that climax plant communities exist in a state of "dynamic equilibrium" where change was small and random. Such communities were thought to be controlled primarily by competition (Connell and Slatyer 1977), and species coexistence was dependent on niche differentiation and resource partitioning (Booth et al. 2003). While *primary succession* occurs on newly formed substrate that was not previously vegetated, *secondary succession* occurs following disturbance on land previously vegetated. In the classical view, that of the "balance of nature," a plant community would return to its original climax stage following secondary succession.

Current models of succession dispel the concept of change leading toward a single climax community. While the terminology of succession has not changed, communities are now thought to exist in a state of nonequilibrium, termed the "flux of nature" (DeAngelis and Waterhouse 1987, Pickett et al. 1992). It is now recognized that while some communities may be at equilibrium at some scales, change in communities is continuous and the classical view of a single climax community is unrealistic. Thus, no single climax community exists on a site, and different types of communities can develop following disturbance. Although a predictable floristic composition does not result, some general patterns occur during succession. Plant cover, biomass, and species diversity tend to increase over time, although they may not be maximal in the climax community. Certain life history traits have been associated with early seral stages, such as rapid growth, production of many small seed, and short life span (Booth et al. 2003, Bazzaz 1979, Huston and Smith 1987). These traits are common in *r-selected* or *ruderal* (R) species, discussed later in this chapter, while traits of *K-selected* or *stress-tolerant* (S) species (slow growth, production of few large seed, and long life span) have been associated with later seral stages (Grime 1977, Pianka 1970).

Much research has focused on the concept of *stability* in plant communities, which can be defined simply as the lack of change (Barbour et al. 1999). More specifically, stability of a community is determined by how it resists change and how it recovers in response to disturbance and stress (Pimm 1991, Leps and Hadincova 1992). The components of stability are persistence, resistance, and resilience (Table 2.3). *Persistence* is the ability to remain relatively unchanged over time, which might occur when a community exists in a protected

TABLE 2.3   Terms Associated with Community Stability

Term	Definition	Source
Persistence	The ability of a community to remain relatively unchanged over time	Barbour et al. (1999)
Resistance	The ability of a community to remain unchanged during a period of stress	Barbour et al. (1999)
Resilience	The ability of a community to return to its original state following stress or disturbance	Barbour et al. (1999)
Elasticity	The speed at which the system returns to its former state following a perturbation	Putman (1994)

Source: Walker (1999) in Walker (1999) *Ecosystems of Disturbed Ground*. Copyright 1999 with permission of Elsevier.

environment. *Resistance* is the ability of a community to remain unchanged during a period of stress or disturbance, characteristic of climax communities. The ability of a community to return to its original state following stress or disturbance is *resilience*, which often characterizes early seral stages. *Elasticity* is a related term defined as the speed at which the system returns to its former state following a perturbation (Barbour et al. 1999, Putman 1994). Although the combination of the three components determines its stability, a community does not necessarily possess all of these components. For example, a plant community can persist in a buffered environment but not be resistant or resilient to disturbance. If a disturbance threshold is passed in this case, an alternative community with a new species composition could result (Orians 1975, Kimmins 1997, Perry 1997, Booth et al. 2003). Community resilience and elasticity are frequently related to the complexity of a community, such as species- or trophic-level interactions (Elton 1958), although complex communities are not necessarily the most stable.

The size, shape, abundance, and connections among stands of a particular community (Figure 2.4) are important factors determining plant community responses to environmental change and disturbance. For example, a plant community that is fragmented into small, independent stands that are separated by relatively large distances (hundreds of meters) is less stable than when the same total area is clustered into larger, connected stands. Fragmentation affects all stability components of the plant community (Table 2.3). In landscapes dominated by production systems, stand dimension, form, and connectivity are molded by human activities and are probably major drivers that influence functional and structural attributes of plant communities (Forman and Gordon 1981, Ghersa 2006).

Mechanisms of Succession. By definition species replace one another during succession, although much debate has occurred regarding the mechanisms that drive this change. In 1977 Connell and Slatyer synthesized current hypotheses and proposed several alternative mechanisms through which species may replace each other during succession (Figure 2.6). After a major perturbation of the environment, "opportunistic" species with broad dispersal powers, rapid growth, and short

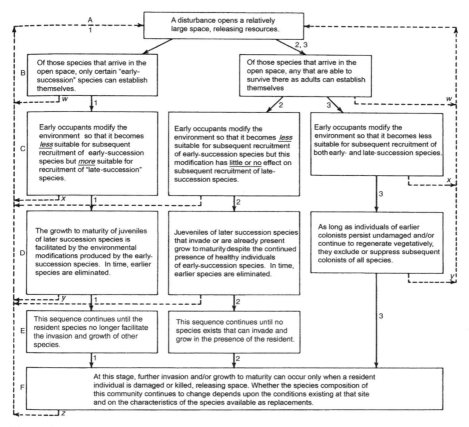

Figure 2.6   Three models of mechanisms producing sequence of species succession. The dashed lines represent interruptions of the process, in decreasing frequency, in the order *w*, *x*, *y*, and *z*. The numbers 1, 2, and 3 refer to facilitation, tolerance, and inhibitory models, respectively, which are discussed in the text. (From Connell and Slatyer 1977, *American Naturalist* 111:1119–1144. Copyright by the University of Chicago.)

life spans usually arrive first and occupy the empty space. These species usually do not invade or grow in the presence of living adults of their own or other species. According to Connell and Slatyer (1977), the species that replace these earliest occupants may be determined by one of three mechanisms (Figure 2.6).

In the first situation, the *facilitation* model, the entry and growth of the later species require the earlier species to "prepare the ground" for them. Only after a suitable change occurs in the microenvironment can later species colonize the area. This is the traditional model of Clements and his followers but may pertain mainly to certain primary successions. The second, *tolerance*, model suggests that a predictable successional sequence occurs because of the existence of species that have evolved different strategies for exploiting environmental resources. Later species are those that are able to tolerate lower levels of resources than earlier ones. Thus, later species are able to invade and grow to maturity in the presence of those that preceded them. In the third, *inhibition*, model all species, even the earliest, resist invasion by competitors. The first occupants preempt the space and continue to exclude or inhibit later species until the former die or are damaged, thus releasing environmental resources. Only then can later colonizers become established and eventually reach maturity. In the majority of natural communities and certainly agricultural ones, succession is often interrupted by disturbance, which starts the process over again. The pattern of these changes depends on whether individuals are more likely to be replaced by members of their own or another species.

While the Connell and Slatyer (1977) models are still considered valid explanations for the mechanisms driving succession, others have extended and refined these theories in the context of nonequilibrium models of communities. For example, Walker and Chapin (1987) proposed that the processes underlying successional change can vary with successional stage, the type of succession (primary or secondary), and resource level. Pickett et al. (1987) proposed a hierarchical framework for examining succession in which three main factors—site characteristics, species availability, and species traits—determine the nature of succession. Each of these factors is in turn affected by particular processes as well as modifiers of those processes. For example, disturbance creates available sites for succession and is influenced by the extent and magnitude of the disturbance, while species availability for colonization depends on the pool of propagules and their dispersal. More recently, the concept of community *assembly* has been extended to the study of succession (Booth et al. 2003). This approach considers natural communities to follow *rules* of organization such that species assemblages differ from a random selection in the species pool. Community assembly is influenced by habitat type and species behavior and interactions (Drake et al. 1999). Although the concept of community assembly is still largely theoretical, it has potential to explain plant invasion in managed ecosystems, where regular disturbance provokes repeated episodes of secondary succession (Booth and Swanton 2002).

Succession in Production Systems. Production systems are characterized by the establishment and management of a modified and simplified plant community, often comprising exotic species. This changes the ecosystem by altering the composition and activities of associated herbivores, predators, symbionts, and decomposers (Swift and Anderson 1993). The composition, diversity, structure, and dynamics of production systems differ in many respects from those of the ecosystem that dominated the landscape before the onset of agriculture or other human activities. Little information exists on how shifts in plant flora affect function in these systems, and it is uncertain whether succession can actually occur in such highly disturbed environments. Most information in this respect is skewed by the perception that weeds, pests, and diseases are invaders (Williamson 1996) and that they do not follow the generally accepted stages of succession.

The cultivated field represents a special example of secondary succession because it is continually being disturbed. Succession on abandoned cropland is called *old-field* succession. Once agricultural operations cease, the systematic replacement of early and intermediate seral stages occurs through time until a climax community similar but not identical to the original pristine one appears. If disturbance occurs repeatedly, as in the agricultural system by tillage and in forests by frequent logging or fire, succession may become cyclic such that earlier stages are favored. For example, after logging or severe forest fire, first herbaceous pioneers then intermediate shrub communities may dominate a site for decades or more, especially if it is frequently disturbed (Figure 1.6). If disturbance is frequent and thus severe enough, succession may become arrested (Figure 2.6) with a new plant community becoming established. Similarly, in agriculture, frequent disturbance and dominance by herbaceous annual species usually delay the establishment of perennial species that normally would succeed annuals. In these situations, herbaceous perennials appear to have lower competitive ability than annuals in the seedling stage and therefore have difficulty becoming established as seed. Once established, however, perennial species tend to replace annuals in the community because of their greater resilience (stability) associated with vegetative reproduction.

In human production systems, land is logged, grazed, burned, or cultivated in a cyclic way. This means that land is exposed to regular disturbances and has periods with low soil cover but high resource availability. Because during these periods nutrient absorption is low and mineralization of organic matter is high, a large proportion of the mineralized nitrogen from organic matter is lost by leaching or denitrification (Tivy 1990, Swift and Anderson 1993, Smith et al. 1997). High resource availability and a simplified biotic system make human production systems, such as crop–weed communities, forest plantations, and grazed rangelands, susceptible to plant invasion. Each invasion of such plant communities creates a new scenario of instability, which Williamson (1991, 1996) calls "press" perturbation—a situation in which the structural and functional properties of the community are modified by new species rather than by the extinction of ones already present. Hobbs et al. (2006) also indicate that it is possible for entirely new ecosystems that contain new combinations of species to arise because of human action, environmental change, and the impacts of deliberate or accidental introductions of species from other regions. These novel ecosystems (Hobbs et al. 2006) result when species occur in combinations and relative abundances that have not occurred previously within a biome.

Under the disturbed regime typical of most human production systems, it is unnecessary for plant species to have different life history characteristics in order to replace earlier residents. Direct control of weeds is common on most farms and forest plantations, which provides an environment of continued disturbance. Thus, no difference in competitive ability needs to occur among weed species, since the earliest stage of succession is constantly being recycled. Replacement of weed species over time may occur at random or due to subtle year-to-year changes in meteorological conditions or management practices. However, the entire weed–crop community can respond to such management manipulations. These responses are usually short term owing to the transitory nature of most cropping systems, but under some conditions long-term responses are possible. Once they have occurred, neither short- or long-term responses are easily reversed, and they can have significant impacts on continued weed and crop management.

Niche Differentiation. *Niche* is a term used to describe a species' place within a community, including its place in space, time, and function of that community. The concept of a niche denotes specialization. As Whittaker (1975) points out in his analogy of a niche to human society, an individual may gain from professional specialization to acquire the resources (income) needed to live. Two or more individuals may gain by following different specialties since they are not in direct competition, and society at large may gain if the specialization of one individual satisfies the needs of another. Thus, considerable evolutionary advantage must underlie the specializations of the species within any plant community. Through differential specialization, species avoid at least some degree of direct competition. A niche is defined in terms of the biotic and abiotic requirements of a species. In reality, the potential range of conditions under which a species can grow, its *fundamental niche*, is broader than its *realized niche*, where it is actually found due to the influences of competition and predation.

In order to understand the importance of niche separation in natural and managed plant communities, we must consider the logistic equation of Lotka (1925) and Volterra (1926). As described by Whittaker (1975), if environmental resources are not limiting, a population may increase exponentially, that is,

$$\frac{dN}{dt} = rN \qquad (2.1)$$

in which the rate of growth in numbers of individuals per unit time (dN/dt) equals the number of individuals (N) in the population at a given time multiplied by r, the *intrinsic rate of increase* for that population in the absence of crowding or effects of competition on growth. If environmental resources are limiting, the growth rate of the population is continually lessened by competition as the number of individuals approaches the maximum number the environment can support. This maximum number is the *carrying capacity* of the environment, K. The logistic curve (Figure 2.7) generated from the following equation is a convenient first approximation for growth rate of a population to a ceiling level set by a limiting environment:

$$\frac{dN}{dt} = \frac{rN(K - N)}{K} \qquad (2.2)$$

where $(K - N)/K$ specifies that population growth will be reduced as population number N approaches carrying capacity K and will be zero when $N = K$; the population is then stabilized at carrying capacity.

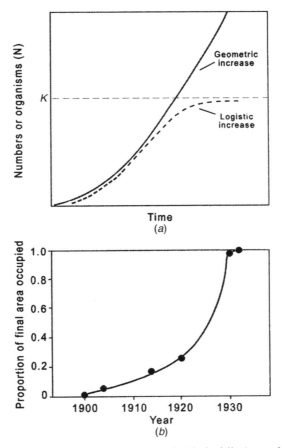

Figure 2.7 (*a*) Exponential (solid line) and logistic (dashed line) population growth over time. *K* is the carrying capacity of the environment for a population showing density-dependent logistic growth (dashed line). (*b*) Increase in area occupied *by Bromus tectorum* in western North America. (Modified from Radosevich et al. 1997.)

The logistic equation now may be applied to two competing populations:

$$\frac{dN_1}{dt} = r_1 N_1 \frac{K_1 - N_1 - \alpha N_2}{K_1} \tag{2.3}$$

$$\frac{dN_2}{dt} = r_2 N_2 \frac{K_2 - N_2 - \beta N_1}{K_2} \tag{2.4}$$

where N_1 and N_2 are the populations of species 1 and 2 at a given time, r_1 and r_2 are their intrinsic rates of population increase, K_1 and K_2 are the environmental resource limits (carrying capacities) for each species in the absence of the other, and α and β are competition coefficients that express, through αN_2 and βN_1, the effects of the population level of one species on the population change of the other species. The equations imply that, for most values of α and β, one species increases while the other competitor declines until at equilibrium the latter is extinct. This idea that two species cannot coexist permanently in the same niche is known as *Gause's competitive exclusion principle.*

Species that divide a shared resource among themselves are collectively called a *guild.* According to Gause's principle, if two species in a guild are direct competitors, one species should approach extinction (Gause 1934). This suggests that the competitive relationships that nearly always develop between weeds and crops or exotic and native plants might be regulated to some extent by natural (competitive exclusion) processes. Where seed or propagules are repeatedly introduced, such as in agricultural fields, the competitive exclusion principle would be manifest as extreme dominance or suppression of one species by another rather than local extinction. If, however, the species differ in their requirements or specializations, then it is possible for them to coexist. Because of niche separation, many natural systems are typified by a high degree of species diversity, coexistence, and uniform total productivity. In rangeland and forest systems, species diversity and uniform productivity are acceptable when coexistence of several species is the ultimate goal. In addition, because of the range of vegetation types often present in rangelands and forests, some spatial and temporal specializations are evident among particular weed and crop plants that would allow coexistence without significant reductions in productivity of desirable species.

In contrast, when productivity of a single species is of concern, most of the environmental potential (resources) is directed toward the crop and weed suppression rather than coexistence is the desired goal. Although some agricultural crops are superior competitors to weeds, it is not enough simply to allow them to compete with the hope of eventual weed suppression or even extinction since some loss in crop yield would inevitably occur over the time frame of a typical production season. Furthermore, the niche differences between weeds and agricultural crops usually are not great enough to allow maximum crop productivity to occur without some human intervention for weed control.

Invasion Process

The area occupied by an invasive plant will tend to expand exponentially across a region much like the general process depicted in Figure 2.7*a* and the actual expansion of cheatgrass (*Bromus tectorum*) across western North America shown in Figure 2.7*b*. Groves (1986) and Cousens and Mortimer (1995) divide the process of invasion (range expansion) by plants into three phases: introduction, colonization, and naturalization. These three phases of the invasion process were defined previously (Chapter 1) and are shown in Figure 2.8 in relation to the general exponential and logistic curves for plant population growth (Figure 2.7). Although the phases of plant invasion can be defined differently (e.g., invasion is often separated from naturalization; Richardson et al. 2000), all definitions or

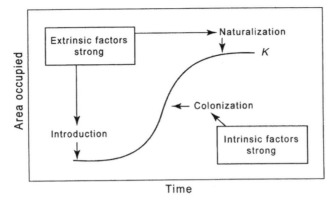

Figure 2.8   Growth curve depicting phases of expanding populations. (From Radosevich et al. 2003. *Weed Sci.* 51:254–259. Copyright 2003. Weed Science Society of America. Reprinted by permission of Alliance Communications Group, a division of Allen Press, Inc.)

models recognize that invasions occur at different geographic and biological scales and can be organized according to the stages of population development, as shown in Table 2.4. Sauer (1988) and Forman (1995) also indicate the processes of population development are scale dependent and can range from individual plants to metapopulations, as indicated in Table 2.4.

Introduction Phase. Theoretical population growth curves can be generated for species given assumptions about their environment, initial population size, intrinsic growth rate, and time (Figures 2.7 and 2.8). The earliest phase of such population curves must result from introduction of the species. However, small populations are often undetected during the introduction phase, and plant invasions are most likely to fail at this point due to randomness or stochasticity (Mack 1995) or to a lack of a minimum critical patch size (Latore et al. 1998). Seedlings from newly introduced seed must compete with the established flora that is well adapted to the site and

TABLE 2.4   Ecological Processes, Patterns, and Scales at Different Phases of Plant Invasion

Phase of Invasion	Ecological Process	Ecological Pattern	Scale
Introduction	Dispersal, immigration, survival	Species recruitment	Individual
Colonization	Birth, death, immigration, emigration	Patch expansion	Population
Naturalization	Birth, death, immigration, emigration	Range expansion	Metapopulation

Source: Radosevich et al. (2003). *Weed Sci.* 51:254–259. Copyright 2003. Weed Science Society of America. Reprinted by permission of Alliance Corporation Group, a division of Allen Press, Inc.

occupies space, so chance of establishment is low. Thus few introductions proceed to the colonization phase. Most factors, aside from suitability to the ecosystem, that influence the rate of introduction are extrinsic to the disseminating seed. Prediction of the introduction phase is strongly related to these environmental factors.

Role of Disturbance. Disturbance is believed to be a major factor favoring plant introductions. Grime (1979) defines disturbance simply as the removal or damage of plant biomass. Pickett and White (1985) provide a more detailed definition of disturbance as "any relatively discrete event in time that disrupts ecosystem, community or population structure and changes resources, substrate availability, or the physical environment". Plant introduction may result from direct destruction of vegetation or indirectly from changes in resource levels (Davis et al. 2000, Booth et al. 2003) or other conditions that subsequently affect the development of plants. Disturbance may be caused by large-scale fire, floods, and storms or smaller scale events like soil turnover or vegetation removal by animals or humans (Hobbs 1991).

Disturbance does not always lead to the introduction of exotic species, but it may provide a temporary location or "safe site" (Harper 1977) for a potential invasive species to establish a founding population. Some form of disturbance usually accompanies the success of many invasive species. Humphries et al. (1991) lists a number of major "environmental" weeds in Australia along with their association to at least one type of disturbance. Multiple disturbances usually increase the chance of successful invasion. For example, Hobbs and Atkins (1988) conducted an experiment in two woodland and three shrubland sites in Australia. They planted wild oat (*Avena fatua*) seed into four types of disturbance treatments at each site: undisturbed, disturbed, fertilized, and disturbed and fertilized. They observed that disturbance in general and multiple disturbances in particular increased the density and biomass of the introduced species (Hobbs 1991).

Rule of Tens. The inflow of exotic plants into new regions is a continuous process and many believe it is happening at an ever-faster rate. However, research suggests that only a few exotic species that make it to a new region actually become established there, and then only a small percentage of the established species become invasive. Williamson (1996) estimated that only 10% of all introduced species into the British flora actually became established and then only 10% of those were invasive enough to be considered pests. He presented this finding as the "tens" rule for plant invasions. Kowarik (1995) similarly reported that of 3150 woody species introduced into Brandenburg and Berlin, Germany, 10% spread beyond the initial site of introduction, 2% became established, and half of those naturalized. Kowarik also noted this ratio (10:2:1) for the Central European flora of vascular plants (12,000 species). Weeda (1987), examining a number of "neophytes" and exotic species in the Netherlands, estimated that approximately 1% penetrated the natural vegetation, supporting the later estimates of Williamson and Fitter (1996) and Kowarik (1995). Similarly, Kornas (1990) found that of the 799 plant species introduced around Montpelier, France, 692 species failed to become established (a success rate of around 13%).

Colonization Phase. Colonization of invasive plants is best characterized by an often prolonged lag time after introduction followed by exponential population growth (Figures 2.7 and 2.8). During colonization, the species typically becomes noticeable and control efforts to halt its spread begin.

Lag Phase of Population Growth. Although some invasive plants may experience rapid population growth after introduction (usually insects, like the African bee), most invasive plants have a long lag phase between initial introduction and subsequent rapid population growth (Figures 2.7 and 2.8). According to Crooks and Soule (1999), three categories of population lag are recognized:

1. Inherent lag times caused by the nature of population growth and range expansion from an initially small population.
2. Prolonged lag times caused by environmental factors unsuitable for the organism. When a disturbance or other environmental event occurs to change the ecological conditions, such as soil nutrient enrichment, climate change, altered dispersal vectors, or intraspecific interactions, the population can expand.
3. Lag times caused by genetic factors related to fitness. Some exotic species have a so-called general-purpose genotype (Baker 1965, 1974), which enables them to grow over a wide range of environments. If an introduced species lacks such characteristics, it will be unable to expand until a genetic change occurs through such events as recombination, hybridization, introgression, or to a lesser degree mutation, and the species adapts to the new environment.

The likelihood of overcoming genetic lag or fitness deficit is proportional to the population size and the rate of genetic adaptation (Crooks and Soule 1999) (Chapter 4). A recent example is bur chervil (*Anthriscus caucalis*) in Idaho (T. Prather 2006, personal communication). This exotic herbaceous species was first found only associated with hackberry shrubs. Its distribution did not move from under hackberry for almost two decades. Now bur chervil is becoming widely distributed in many of the plant communities found in the Palouse Prairie, one of the most endangered habitats in the United States.

Past performance of an exotic species is a poor predictor of its lag time in a new habitat; however, some plant characteristics like length of time to reproduce have a profound effect on lag time. Kowarik (1995) in a historical reconstruction of the invasion dynamics of 184 exotic woody species near Bradenburg, Germany, found that only 6% of the species spread within 50 years after their first introduction to the area. A total of 25% lagged up to 100 years, 51% lagged for 200 years, 14% lagged for up to 300 years, and 4% did not expand for more than 300 years.

Exponential Phase of Population Growth. The speed at which colonization proceeds is closely related to the intrinsic rate of increase for the species. Hence, predictions of colonization rates and selection of management options during this phase of the invasion process should focus on the intrinsic biology of the species with less emphasis on other factors.

For example, Maxwell et al. (1988) used a demographic approach to examine population growth and management options for leafy spurge (*Euphorbia esula*). Following Watson (1985), they divided the life history of leafy spurge into five stages: seed, buds, seedling, vegetative shoots, and flowering shoots (Figure 2.9, top). By identifying these stages, the process of population development was determined. It was found that three important transitions, basal buds to vegetative shoots (G2), number of basal buds that flowering shoots produced (V5), and number of basal buds that vegetative shoots produced (V4), were sensitive to their own density. When these three density-dependent functions were simultaneously included in the model, exponential growth followed by decline and eventual stabilization of the simulated population was predicted (Figure 2.9, middle). They then subjected the simulated population to several management tactics: a single application of picloram (Figure 2.9, middle) and several levels of a foliage-feeding herbivore (Figure 2.9, bottom) on leafy spurge. The accuracy of the population dynamics model was striking in predicting both population growth and outcomes of actual management treatments.

Sources and Satellites. Colonization occurs when plants in a founding population reproduce and increase sufficiently to become self-perpetuating. Cousens and Mortimer (1995) describe the radial expansion of such self-perpetuating patches as an advancing front that spreads outward at a constant rate in all directions. The rate of increase in area then conforms to the equation

$$\frac{dA}{dt} = 2\pi r^2 t \tag{2.5}$$

where A equals the area occupied, r is the distance advanced outward each year, and t is time in years or generations. Thus, assuming successful recruitment continues within the founding population and the rate of expansion is constant, the relative increase in area of a patch would decline over time (Cousens and Mortimer 1995). Although this model of range expansion seems intuitively reasonable, it has received only limited examination in field studies (Moody and Mack 1988, Auld and Coote 1990). It is possible for some individuals from a founding *source* population to disperse widely and to produce new *satellite* populations of the species. In this case, dispersing individuals recruited from an already established source population should behave much the same and have the same requirements as a new introduction.

A relevant research and management question emerges from this information, which is whether to contain "source" or "satellite" populations of an invading species after successful introduction has occurred. Cousens and Mortimer (1995) and Moody and Mack (1988) suggest that the preferred containment strategy would be to remove satellite populations as they occur over space and through time, since these populations expand more rapidly and potentially cover greater

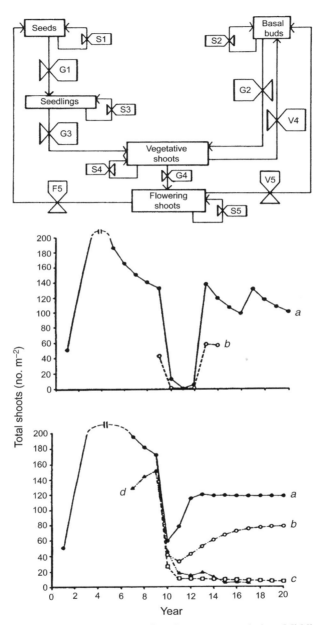

Figure 2.9 Top: Diagrammatic model of leafy spurge population. Middle: Leafy spurge population simulation with density-dependent functions and single application of (*a*) picloram simulated at year 10 and (*b*) observed effects of picloram application. Bottom: Leafy spurge population simulation with density-dependent functions simulating introduction of foliage-feeding herbivore at year 10 that removed (*a*) 40%, (*b*) 50%, and (*c*) 60% of the stems. Also shown are (*d*) observed effects of sheep feeding on leafy spurge. (In Maxwell et al. 1988 modified from Radosevich et al. 1997. Copyright 1997 with permission of John Wiley & Sons Inc.)

area than the front of a source population. Ghersa et al. (2007) determined that satellite populations of Johnsongrass (*Sorghum halepense*) that were uniformly distributed over a previously vacant area occupied that area more quickly than the advancing front of an adjacent source population. Nonetheless, this strategy of expansion has received relatively little experimental attention by field scientists. Land managers continue to control source rather than satellite populations of invading species (Moody and Mack 1988, Cronk and Fuller 1995, Zamora and Thill 1999), perhaps because such populations are much more obvious than the smaller satellite populations or control tactics seem more effective.

Naturalization Phase. A species becomes naturalized in its new environment when it successfully establishes new self-perpetuating populations, is dispersed widely throughout a region, and is incorporated into the resident flora. At some carrying capacity K, the population approaches a quasi-threshold density where its population growth may remain near 1, that is, stabilize and not expand very quickly (naturalization) (Figures 2.7 and 2.8). The K density occurs when niche occupancy and available resources limit the rate of spread. This phase is controlled most by environmental factors so predictions of risk for populations approaching K should also be focused on parameters extrinsic to the biology of the species. Most agencies remove weeds from their target list (e.g., noxious weed lists) in this phase of invasion since they are too difficult or expensive to control or eradicate.

Genetics of Weeds and Invasive Plants

According to Darwin's theory of evolution, whenever heritable variation exits within populations and is associated with differential survival and reproduction of the variants, individuals in a population will change over generations because of natural selection acting on that variation. Natural selection is the differential reproduction of genotypes, although it is often described in terms of the factors in the environment that favor one genotype over another. Genetic variation is the basic structure through which natural and human selection act, providing weeds, invasive plants, or any other population of organisms with traits that can enhance survival in a new or novel environment. Both genotypic (genetic architecture of individual plants) and phenotypic (the expression of the genetic architecture, i.e., chemical, structural, or behavioral characteristics of individuals) variation are necessary for natural selection to operate. Thus, invasion success could be facilitated by the presence of a genetic substrate within a source or founder population upon which natural selection could act. Adaptive selection operates whenever heritable variation exists among the genotypes in a population and selection results in changes in the phenotypic expression of individuals that make them better suited to the environmental selector. A consequence of selection, either by natural or human causes, is an unavoidable reduction in the amount of genetic variation among the subsequent individuals in the resulting population. The evolutionary genetics of weeds and invasive plants is yet to be explored completely. However, the genetic architecture of such species has been studied and researchers are now able to determine some of the genetic attributes that are important for weed selection and invasion success. These attributes of weeds and invasive plants include additive genetic variance, epistasis, hybridization and polyploidy, genetic trade-offs, and the action of small numbers of genes and genomic rearrangements (Mayr 1982, Lee 2002). These topics are explored further in Chapter 4.

Fitness and Selection. Because survival and reproduction are demographic processes (Chapter 5), natural selection is also a demographic process. *Fitness* is a single value of relative evolutionary success that combines both survival and reproduction and refers to the relative success with which a particular genotype transmits its genes to the next generation (Silvertown and Charlesworth 2001). Fitness is not a fixed value; rather, it is determined within a particular environment or suite of ecological conditions and is relative to the success of other phenotypes that also exist in the same population. Fitness is an important factor determining the ecological success of many, if not most, weed and invasive plants.

Significant evolutionary change in a population occurs when three criteria are met: (a) there is phenotypic variation among individuals, (b) some of this variation is heritable, and (c) individuals possessing this variation differ in fitness (Silvertown and Charlesworth 2001). Under these conditions, natural or artificial selection act upon the range of genotypes present. This potential for evolutionary change is easily demonstrated in the supermarket or plant nursery. The wide array of cole crops (crops in the mustard family, Brassicaceae, including cabbage, broccoli, cauliflower, brussel sprouts, and kohlrabi) is a result of artificial, human-directed selection of one ancestral species, wild mustard *Brassica campestris*. Similarly, all of the strains and varieties of roses originated from a common wild plant. Natural selection can produce results just as dramatic as artificial selection; it is just not as rapid. Natural selection occurs when one phenotype leaves more descendants than others because of its superior ability to survive or produce offspring in a particular environment.

Patterns of Evolutionary Development of Weeds and Invasive Plants. All organisms are capable of budgeting energy and resources in order to successfully complete their life cycle. This process is called *resource allocation*. Allocation is closely linked to survival, and the pattern of resource allocation in a species, called its *strategy* or life history pattern, is generally viewed as an inherited set of adaptations that minimize extinction. The resources available to a species are limited and divided among three primary functions—growth, maintenance (survival), and reproduction. According to the *principle of allocation*, these limited resources are allocated so as to maximize fitness over the lifetime of the organism (Barbour et al. 1999). In plants, the amount of photosynthetic energy (actual resources) allocated to root, shoot, leaf, and reproductive organs and the amount of time (implied resources) spent in dormancy, growth, maintenance, and reproduction are important attributes that govern species success. Figure 1.2 illustrates those major activities performed by plants that require resource allocation. Several points of view are possible concerning the patterns of resource allocation that exist among species; however, these theories all recognize the importance of resource allocation for species survival and plant community development.

r and K Selection. The most widely held concept dealing with patterns of evolutionary development is that of *r* and *K* selection. This idea was derived from the logistic equation of population growth (Lotka 1925, Volterra 1926). As shown in Figure 2.7, population growth in an ideal (limitless) environment increases exponentially according to the intrinsic rate of population growth, *r*, whereas in real (limited) environments growth declines as the population approaches *K*, or carrying capacity.

The concept of *r* and *K* selection, first proposed by MacArthur (1962) and later Pianka (1970, 2000), is that organisms lie on a continuum between two extremes of resource allocation, which represent two strategies for survival. In the extreme cases, species may be *r* or *K* selected. Table 2.5 lists various traits associated with each strategy. Extreme *K*-selected species tend to be long lived, have a prolonged

TABLE 2.5   Traits of *r* and *K* Selection

Trait	*r* Selection	*K* Selection
Climate	Variable and/or unpredictable; uncertain	Fairly constant and/or predictable; more certain
Mortality	Often catastrophic; density independent	Density dependent
Survivorship	Mortality at early age	Continuous mortality through life span or more as age increases
Population size	Variable in time; not in equilibrium; usually well below carrying capacity of the habitat; recolonization each year	Fairly constant in time; in equilibrium; at or near carrying capacity of the habitat; no recolonization necessary
Intraspecific and interspecific competition	Variable; often lax	Usually keen
Life span	Short, usually less than one year	Long, usually more than one year
Selection favors	Rapid development; early reproduction; small body size; single reproduction period in life span	Slower development; greater competitive ability; delayed reproduction; larger body size; repeated reproduction periods in life span
Overall result	Productivity	Efficiency

Source: Pianka (1970) *American Naturalist* 104:592–597. Copyright 1970 by the University of Chicago.

vegetative stage, allocate a small portion of biomass to reproduction, and occupy late stages of succession. The population size is near carrying capacity and is regulated by biotic factors. Extreme *r* selection leads to a short-lived plant that occurs in open habitats and early stages of succession. A large portion of biomass is allocated to reproduction and the population is regulated by physical factors. It should be noted that few plant species, if any, are entirely *r* or *K* selected. Most species represent a compromise between the two strategies. Weeds associated with agricultural lands and highly disturbed sites in forests and rangelands fit most closely the characteristics of *r* selection noted in Table 2.5.

C, R, and S Selection. Another concept concerning plant resource allocation and evolutionary pattern was proposed by Grime (1979), although this view is an extension of the more widely acknowledged *r* and *K* continuum. Grime proposes that there are two basic external factors that limit the amount of plant material in an environment: stress and disturbance. He defines *stress* as external factors that limit production, such as reduced or limiting radiation, water availability, nutrients, or suboptimal temperature. *Disturbance* is the partial or total disruption of plant biomass, for example by mowing, tillage, grazing, logging, or fire. As with the *r* and *K* continuum, the spectrum of these two factors can vary widely, but if only the extremes of high and low stress and disturbance are considered, four possible combinations occur (Table 1.5). Of these four combination, only three possible evolutionary strategies are apparent: *ruderals (R)*, *stress tolerators (S)*, and *competitors (C)*. The fourth possible combination, high stress and high disturbance (Table 1.5), creates an environment unsuitable for plant survival. Plants that fall into each of these strategies can be classified according to their common adaptations (Table 2.6).

Grime arranges the three evolutionary strategies into a triangular model (Figure 2.10) to describe the various equilibria between stress (I_s), disturbance (I_d), and competition (I_c). In this model C, R, and S represent the three extremes of specialization. Since few species have all the characteristics listed in Table 2.6, Grime "maps" the species according to certain traits using triangular ordination. Although the indices for stress, disturbance, and competition are difficult to establish quantitatively, this procedure provides a tool to categorize plants according to life history and successional stage (Figures 1.4 and 1.5).

In terms of evolutionary strategy, many weeds possess characteristics common to both competitors and ruderals (Table 2.6). From Figures 2.10 and 1.5, it appears that many herbaceous annuals, biennials, and certain herbaceous perennials follow a pattern of *competitive ruderals.* Trees and shrubs most closely follow the pattern of *stress-tolerant competitors* (see Chapter 1). Although Grime describes many other patterns of vegetation in relation to both life-form and evolutionary strategy, it seems that these two classes warrant further investigation to characterize the nature of weediness. This aspect of weed evolution also will be explored in Chapter 4.

TABLE 2.6   Some Characteristics of Competitive, Stress-Tolerant, and Ruderal Plants

	Competitive	Stress Tolerant	Ruderal
	Morphology		
1. Life forms	Herbs, shrubs, and trees	Lichens, herbs, shrubs, and trees	Herbs
2. Morphology of shoot	High dense canopy of leaves; extensive lateral spread above and below ground	Extremely wide range of growth forms	Small stature, limited lateral spread
3. Leaf forms	Robust, often mesomorphic	Often small or leathery or needlelike	Various, often mesomorphic
	Life History		
4. Longevity of established phase	Long or relatively short	Long to very long	Very short
5. Longevity of leaves and roots	Relatively short	Long	Short
6. Leaf phenology	Well-defined peaks of leaf production coinciding with period(s) of maximum potential productivity	Evergreens, with various patterns of leaf production	Short phase of leaf production in period of high potential productivity
7. Phenology of flowering	Flowers produced after (or more rarely, before) periods of maximum potential productivity	No general relationship between time of flowering and season	Flowers produced early in life history
8. Frequency of flowering	Established plants usually flower each year	Intermittent flowering over a long life history	High frequency of flowering
9. Proportion of annual production devoted to seeds	Small	Small	Large
10. Perennation	Dormant buds and seeds	Stress-tolerant leaves and roots	Dormant seeds
11. Regenerative strategies[a]	V, S, W, B_s	V, B_r	S, W, B_s

(*Continued*)

TABLE 2.6 *Continued*

	Competitive	Stress Tolerant	Ruderal
	Physiology		
12. Maximum potential relative growth rate	Rapid	Slow	Rapid
13. Response to stress	Rapid morphogenetic responses (root–shoot ratio, leaf area, root surface area) maximizing vegetative growth	Morphogenetic responses slow and small in magnitude	Rapid curtailment of vegetative growth, diversion of resources into flowering
14. Photosynthesis and uptake of mineral nutrients	Strongly seasonal, coinciding with long continuous period of vegetative growth	Opportunistic, often uncoupled from vegetative growth	Opportunistic coinciding with vegetative growth
15. Acclimatization of photosynthesis, mineral nutrition and tissue hardiness to seasonal change in temperature, light, and moisture supply	Weakly developed	Strongly developed	Weakly developed
16. Storage of photosynthate and mineral nutrients	Most are rapidly incorporated into vegetative structure with some storage for growth the following season	Storage systems in leaves, stems, and/or roots	Confined to seeds

Source: Grime (1979). *Plant Strategies and Vegetation Processes.* Copyright 1979 with permission of John Wiley & Sons Inc.
[a]Key to regenerative strategies: V, Vegetative expansion; S, seasonal regeneration in vegetation gaps; W, numerous small wind-dispersed seeds or spores; B_s, persistent seed bank; B_r, persistent seedling bank.

Plant Demography and Population Dynamics

Stages in the life history of plants provide the opportunity to assess how changes in population size or structure occur over time. These basic changes in plant life history are shown in Figure 2.11. The population of seed in the soil is generally referred to as a *seed bank*, or reservoir. Some seed in this population germinate, transition to the next stage, and become *seedlings* while others remain dormant in the reservoir or die. Seedlings that germinate at nearly the same time are a

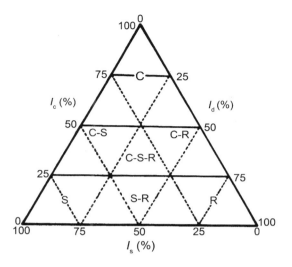

Figure 2.10 Model describing various equilibria between competition, stress, and disturbance in vegetation and location of primary and secondary strategies: I_c, relative importance of competition; I_s, relative importance of stress; I_d, relative importance of disturbance. (From Grime 1977, *American Naturalist* 111:1169–1194. Copyright by the University of Chicago.)

cohort, although agriculturists often refer to the phenomena as "flushes" of germination. *Recruitment* is the transition from juvenile stages, seed and seedlings, to adult form where independent existence and reproduction are possible. Seed are a primary method of recruitment, but vegetative reproduction also occurs in many plants. These vegetative offshoots, called *ramets* or *clones*, may remain attached to the "mother" plant or be separated from it. *Genets*, in contrast, are genetically distinct individuals that arise from seed.

Plant demography is the statistical study of population changes and their causes throughout the life cycle (e.g., Figures 1.2 and 2.11). There are four basic demographic processes that determine how a population of plants changes over time. These are birth (B), death (D), immigration (I), and emigration (E). Population ecologists describe how these factors change the size of a plant population (N) between one time interval (t) and another ($t + 1$) with the following difference equation:

$$N_{t+1} = N_t + B - D + I - E \tag{2.6}$$

All experiments or analyses about the population dynamics of plants, weeds, crops, or natural systems ultimately come back to the above simple equation, which is explored more fully in Chapter 5.

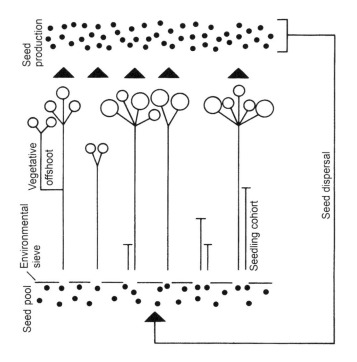

Figure 2.11 Idealized plant life history. (Adapted from Harper and White 1971. In Harper 1977, *The Population Biology of Plants*, 1977. Copyright with permission from Elsevier.)

MANAGEMENT PRINCIPLES

Farmers and land managers routinely use a variety of strategies to manage weeds and invasive plants. The most common strategies used are prevention, early detection and eradication, and containment or control. A major focus of weed scientists is to work with practitioners to control individual weed or invasive plants, usually in limited geographical areas. Various tools and tactics are used to reduce the prevalence and impact of weeds and invasive plants once they are detected. These strategies, tactics, and tools are discussed in more depth in Chapters 7 and 8.

Assessing Risk from Weeds and Invasive Plants

Anticipating which plants may become agricultural weeds or invasive plants in particular landscapes and what potential harm they might cause (*risk assessment*) is an important aspect of weed and invasive plant management. Most decisions about weed management in agriculture are based on assessing the likelihood that weeds already present will have a negative impact on the crop. In agroecosystems, weed management typically emphasizes control based on the following three elements: weed responsiveness to tools, opportunity to improve crop yields, and profitability. In contrast, in natural systems a major focus of management is assessing the risk that new or narrowly distributed plant species might become invasive in order to prevent their spread. Thus, management of weeds in natural systems focuses primarily on detection and eradication of potentially invasive plants that are not yet widespread. Byers et al. (2002) identify four levels of risk assessment associated with the biological stages of exotic species invasion: arrival (risk associated with entry pathways), establishment (risk of forming viable, reproductive populations), spread (risk of expanding the range or extent), and impact (risk of having a measurable effect on existing species or communities). In both agricultural and natural systems, the economics of prevention, detection, eradication, or control of weeds and invasive plants is a necessary and important consideration (Auld et al. 1987, Leung et al. 2002, Pitafi and Roumasset 2005). In natural systems, the costs of ecosystem restoration also impact decisions to manage invasive plants (Hobbs and Humphries 1995).

Methods to detect, survey, and monitor the presence and abundance of weeds and invasive plants are considered in Chapter 7, as are the tools for their management. The assessment of impact or risk of potential impact from weeds and invasive plants, however, has proven more difficult than simply finding, counting, and controlling them. Farmers and land managers also need tools to better quantify the present and future impacts of weeds and invasive plants in order to justify the effort and expense of management, particularly in areas where no direct economic benefit can be shown (e.g., improved profit from land use) (Hobbs and Humphries 1995, Radosevich et al. 2003).

Modeling the impacts of weeds in various production systems and real or potential impacts of weed invasions has been attempted in numerous studies over the last decade with varying success. Weed–crop models based on threshold levels of weeds that reflect yield or other forms of economic loss have been valuable additions to available weed management tools (Radosevich et al. 1997, Bastiaans et al. 2000). These models are often used to predict crop yield improvement or other economic gains from various levels of weed management. Such models are typically simulations of weed and crop dynamics in an array of crop, forest plantation, or rangeland systems; in each case the robustness of the model depends on the quality of the biological and economic data used. Examples of such models include the well-developed INTERCOM model (Kropff and van Laar 1993), HERB (Wilkerson et al. 1991), SELOMA (Stigliani and Resina 1993), APSIM (Keating and Carberry 2003, Knowe et al. 2005), and others (Deen et al. 2003). Another approach uses economic principles to develop decision models to help farmers and land managers control weeds with fewer inputs (Bennett et al. 2003, Finnoff et al. 2005). However, such models have not been widely accepted for use in real situations, perhaps because of the concern that even very low weed densities can cause crop loss (Wilkerson et al. 2002). The challenge in using the modeling approach for weeds in agro- and natural ecosystems remains one of making the models more practical and acceptable to practitioners.

Although models of range expansion from source populations have been developed for invasive plants, as noted above, this approach has not been incorporated into management decisions to any extent and models to predict invasive plant impact or risk have proven more difficult. A commonly used approach to predict potential risk is screening exotic species for their propensity to invade elsewhere and, by inference, their impact in other invaded areas (Mack 1996, Pyšek et al. 2004, Rejmánek 2000). Another approach to predicting invasiveness uses biological characteristics related to invasiveness (Goodwin et al. 1999, Reichard and Hamilton 1997, Rejmánek 2000), but this approach also does not address impact directly. Quantitative predictive models have often used analytical diffusion equations to examine plant spread over a generalized area (Skellam 1951, Holmes et al. 1994). Climate matching or homoclime analysis assesses the risk of species from other regions invading a new region with a similar climate (Panetta and Mitchell 1991), which has been extended to incorporate other environmental components in ecological niche models (Peterson and Vieglais 2001, Thuiller et al. 2005). Only recently, simulation models of invasion that incorporate management options and outcomes have been developed (Goslee et al. 2006, Kriticos et al. 2003). As for weeds in crops, however, these models are not yet in widespread use.

The future of invasion modeling lies in the interplay between biology, potential habitats, and human impacts on land. However, much of the information necessary for developing reliable predictive models is scattered throughout the scientific literature or is only now becoming available for invasive plants and their likely habitats. A need for local as well as global databases of invasive plants is clear (Wade 1997, Ewel et al. 1999), but the infrastructure for such a database is currently not present either locally or globally.

Management Priorities Based on Risk and Value. Hobbs and Humphries (1995) suggest a framework for setting priorities for management of invasive plants based on land value and the degree of site disturbance (risk of invasion). Hiebert (1997), following Stubbendieck et al. (1992), proposes a similar approach for considering the possible impact of an exotic species versus its feasibility for control. The planning approach proposed by Hobbs and Humphries (Figure 2.12) depicts four distinct categories of management based on the characteristics of the region. These are as follows:

- *Sites of high value that are relatively undisturbed*; that is, the risk of invasion is low. Such sites could be free of weeds and invasive plants and, according to Hobbs and Humphries, the management objective should be to keep them that way. Wilderness areas are good examples of such locations where monitoring and localized containment around trailheads should keep such areas free of invading species. Hobbs and Humphries suggest that such locations should be treated as "fortresses," and management resources should be directed at minimizing human-induced disturbance and the dispersal and establishment of invasive plants.
- *Locations of high value that are subject to greater levels of disturbance (risk) and, hence, are more susceptible to invasion.* Hobbs and Humphries

believe the management objective in this case should be to manage such sites by reducing or removing any disturbance factors, controlling current populations, and preventing further introductions. Such areas also could be likely locations for restoration.

• *Sites of low value that are subject to low levels of disturbance.* These sites should require little or no management input but should require constant monitoring so that local colonies of invasive plants do not spread.

• *Sites of low value that are subject to high levels of disturbance (risk).* Grasslands of California are good examples of such locations where invasive exotic plants have spread so extensively that they have naturalized, replacing the native bunchgrass plant community. Although such locations may be subject to rapid change and extensive plant invasion, they should be regarded as low priority for management because attempts to restore the native vegetation to its pristine condition are unlikely to succeed.

The four categories depicted in Figure 2.12 establish clear management priorities that consider potential impact or risk and the real or potential value of land. However, in many locations the determinations of management priorities are more difficult. Unfortunately, the prevailing trend is one of transition from the

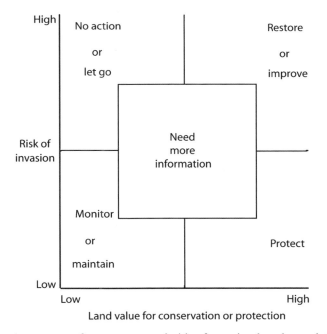

Figure 2.12 Assessment of management priorities for region based on relative value of different sites for conservation and/or production and their relative degree of risk of invasion. (Modified from Hobbs and Humphries 1995, and Mashhadi and Radosevich 2003.)

bottom right (protect) to the top left (let go) of Figure 2.12 as environmental degradation continues and plant communities come into higher risk of invasion by exotic plants.

Market-Driven Management Considerations

Weed control can have both widespread economic benefits and costs. However, private profits gained from weed control may not always or fully reflect community gains or losses (net benefits) from those practices. Auld et al. (1987) identify two main reasons why the private profits gained by agriculturists from weed control may not fully reflect social net benefits from such procedures:

- Even within a fully efficient system of markets, some of the gains from weed control are likely to be distributed to consumers or purchasers of commodities where these commodities register an increase in yield or quality following weed treatment. Market competition may cause the price of such commodities to fall, and this will benefit the consumer or purchaser of them as well as those who supply tools to accomplish the weed control activity. Thus, to assess the overall benefits of weed control in this context, the industrywide gains to consumers or purchasers as well as to suppliers (farmers and their suppliers) must be considered.
- Some costs or benefits of economic activities involving weed control may not be taken into account in the market system. For example, herbicide drift from one property may damage crops or other attributes on another property but the herbicide user may pay no compensation. Similarly, widespread herbicide use or tillage for weed control may result in reduced water quality, litigation, environmental assessments, and remediation. In all such cases, the actual costs to society are much greater than those borne by the user.

These types of market factors and *externality* costs, that is, those that occur beyond the domain of the individual, need to be taken into account when considering the value of weed control.

Cost–Benefit Analysis. The simplest way to make assessments of net benefit is to express all values in common economic terms, that is, a *cost–benefit* ratio. Numerous economic analyses have been performed by state and federal agencies to examine the benefits of weed containment and damages and production losses for various noxious weeds and invasive plants (Pimentel et al. 1999). Unfortunately, such analyses are often not straightforward, as the following example of serrated tussock (*Nassella trichotoma*) provided by Auld et al. (1987) indicates.

Serrated tussock is a naturalized weed of grasslands and pastures in southeastern Australia, and in the early 1980s various options to control the species and protect serrated tussock–free areas from invasion and reinvasion were explored. Vere et al. (1980), using discounted cash flow analysis, evaluated these procedures and estimated the overall change in social benefits resulting from serrated tussock control to be approximately $60 million per year. This amounted to a national social benefit–cost ratio of 11:1, a very favorable ratio from an economic viewpoint. However, Edwards and Freebairn (1982) pointed out that almost all of Australia's wool was exported in 1980. Therefore, it would be foreign consumers who benefited most from serrated tussock control through lower wool prices, not Australians. This work demonstrates the difficulties associated with economic assumptions that arise when only benefit–cost analysis is used for assessment of impacts of weeds and costs of control.

Assessing Economic Risk. Many of the environmental and biological hazards associated with weeds and the benefits of weed control (Chapter 1) are a matter of human perception, biologically difficult to measure, or are economically intangible. This makes assessment of risks and benefits from weeds or the activities to control them difficult (Figure 2.13). Although no weed, invasive plant, or form of weed control is exempt from such assessment difficulties, herbicides are particularly vulnerable because the externalities that can result from the use of this tool are particularly obvious (Chapter 8).

Impacts of Externality Costs. Auld et al. (1987) provide an approach for how such externalities may be taken into account and perhaps corrected (Figure 2.13). They suggest that farmers or land managers, guided only by their perceptions of individual gains (profits), could treat weeds in ways that that either impose a net

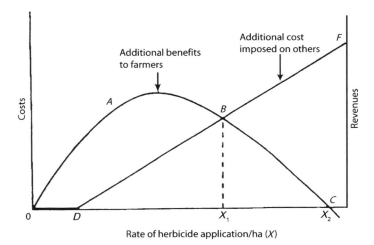

Figure 2.13   From a social point of view, account should be taken of losses imposed on others by tool use. In this case, farmers would maximize their benefits by applying X_2 of the herbicide per hectare, whereas losses imposed on others (*DBF*) would suggest that an application rate of X_1 is socially optimal. (Reprinted from Auld et al. 1987. *Weed Control Economics.* Copyright 1987 with permission of Elsevier.)

loss to society or act to society's advantage. For example, curve 0ABC in Figure 2.13 represents the profits realized by farmers from increasing the amount of herbicide used, by either increased frequency of application or higher rates. They maximize profits when X_2 amount of herbicide is applied.

However, suppose that others in society are damaged by such high rates or frequencies of chemical use (e.g., through reduction in water quality or increased risks to biodiversity or human health) so that restitution or restoration must be made or the probability of morbidity is increased. In that case, the additional externality costs are represented by curve DBF and the optimal herbicide rate or frequency of application is represented by X_1. Above a total application rate of X_1 the extra gains to farmers (land managers) are insufficient to compensate for the extra externality costs or damages imposed on them.

This discussion, however, ignores the question of how risk and uncertainty resulting from weeds and any form of weed management are assessed and valued. It also does not address whether the risks from weeds or tool use are voluntary or involuntary. If risks are voluntary, they can be taken into account without undue difficulty (Auld et al. 1987, Strauss et al. 2000, Martinez-Ghersa et al. 2003). If, however, individuals are subject to greater risk than they perceive as acceptable or if they are subjected to risk against their will, difficulties of assessment arise because those people are made worse off and are not fully compensated.

Bioeconomic Risk Analysis. Some weeds and invasive plants threaten people because they contribute to environmental damage (Kareiva 1996) and losses in biodiversity (Elton 1958, Randall 1997). Finnoff et al. (2005) suggest that the economic theory of *endogenous risk*, coupled with the population ecology of invasive species, provides a way to determine the economic cost effectiveness of prevention and control strategies. Endogenous risk is the idea that people, land managers in this case, influence the risk they face through their behaviors (Ehrlich and Becker 1972, Shogren and Crocker 1991). Thus, land managers and farmers choose the level of risk they want to avoid through their efforts and investments while accounting for the trade-offs involved in those decisions (Finnoff et al. 2005).

Finnoff et al. (2005) and Pitafi and Roumasset (2005) developed models to examine the resource economics of invasive species, including plants. These models identify optimal allocations of investment resources (e.g., labor and capital) to prevention *versus* control and acceptable invasion risk (Leung et al. 2002). Figure 2.14 provides an analysis for risk-neutral (RN), mildly risk adverse (RA1), moderately risk adverse (RN2), and highly risk adverse (RN3) managers when considering four monetary discount rates. These results indicate that that more risk adverse managers select less risky alternatives, which means activities with less prevention and more control (Finnoff et al. 2005). Control is intuitively more attractive to these managers because it removes existing invaders from the system, whereas prevention only eliminates the chance of invasion but does not eliminate invaders. Finnoff et al. indicate that since prevention and control are substitutes, under this management scenario prevention is used rarely and

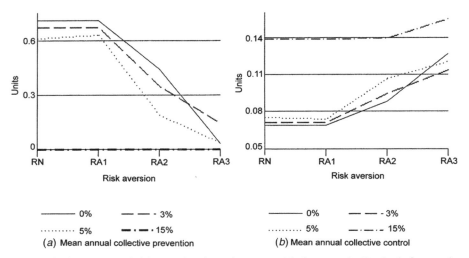

Figure 2.14 Impact of risk aversion in endogenous risk framework. For both figures the horizontal axes are increasing levels of risk aversion. Units of collective prevention on the left are the average number of prevention events that take place on an annual basis, whereas units of collective control on the right are the average number of control events (e.g., molluscicide applications) on an annual basis. (From Finnoff et al. 2005. *Rev. Agric. Econ.* 27:475–482. Copyright 2005, Blackwell Publishing Ltd., reproduced with permission.)

implementation of control is delayed, which increases the probability of invasions occurring and an overall decline in ecosystem welfare.

Management Options in Relation to Invasion Process

Farmers, land managers, and management agencies tend to direct resources more toward control of already major weed species and much less to the prevention, early detection, or containment of new exotic plants. Most major weed control programs get underway only after a particular species is an obvious problem and most management and biological control research is directed at these recognized problem species (Hobbs and Humphries 1995, Oregon Department of Agriculture 2000). Understanding the causes for the appearance of an invasive species in a habitat will help land managers undertake preventive actions. Changing prevailing land management regimes, such as overgrazing, nonsustainable logging practices, or overreliance on particular cropping practices, also may be necessary to prevent the occurrence of an invasive species episode. However, socioeconomic factors of the area often prevent such changes from occurring readily. Therefore only small or incremental changes in causal factors for weed and invasive plant occurrence may be possible. Figure 2.15 (Chippendale 1991) demonstrates the course of a plant invasion from early introduction to its development into a major weed problem. At each stage of the invasion, human activities may act to encourage

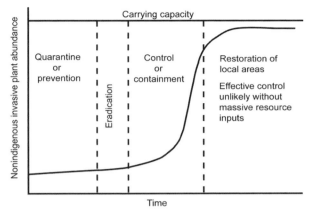

Figure 2.15 Phases of weed invasion and priorities for action at each phase. Ease of treatment of an invasion problem declines from left to right. (After Chippendale 1991, from Hobbs and Humphries 1995. *Conserv. Biol.* 9:761–770. Copyright 1995, Blackwell Publishing Ltd., reproduced with permission.)

spread, and changes in human behavior are usually required to reduce the impact of current plant invasions or minimize future problems.

SOCIAL PRINCIPLES

A new environmental awareness has been rising since the 1960s in the United States, and the establishment of a new environmental ethic is seen as at least a partial solution to the environmental problems brought upon ourselves (List 2000). The assumption behind this perceived need for a new ethic is that philosophical beliefs and attitudes make a difference in behavior; that is, the way we think about the earth and its many ecosystems influences our behavior toward them.

Value systems are the way people think about activities (what we do and how we do it) and technologies (what we do it with). They are actually the underlying principles through which we judge our own and other people's actions. Although invasive plant management and weed control are only a part, arguably a small part, of land management decisions, it is important to examine where they are placed within this larger societal context. Castle (1990) indicates that there are four fundamental value systems that influence decisions about the management of agricultural and natural resources:

- *Material Well-being.* Any activity or technology should be of benefit or utility to society. Implicit in this belief is that society will be better off with than without the new tool, tactic, approach, or activity.
- *Sanctity of Nature.* An activity should not proceed if risk or damage to the environment is likely to result. Nonintervention is a key ingredient of this

value system. While some human interventions into nature are accepted as necessary, they should be minimally disruptive to avoid adverse consequences.

• *Individual Rights*. Individual liberty and property entitlement are the primary concerns in this value system. The marketplace is often posed as the best way for society to accept or reject activities, through the products produced and used. Government interventions are often seen as a problem, not a solution.

• *Justice as Fairness*. All people have equal use of the earth's resources and benefits from them should be distributed equitably. The issue is not bounded by time or geography. Thus, impacts of activities and technologies on other parts of the world or on future generations are of concern.

Each of these value systems raises different questions about our efforts to manage land, raise crops, or manipulate vegetation (Radosevich and Ghersa 1992). However, the value system most accepted by Occidental cultures seems to be the first one in the above list, material well-being or utilitarianism (Ferre 1988). Explicit in that value system is that people will be better off with than without a particular tool or activity. Nevertheless, people usually do not view such benefits identically, so an approach has been devised to assess the relative benefits and costs in such situations. This procedure is called *cost–benefit* or, more recently, *risk–benefit* analysis, which was discussed earlier. Another relevant example of risk–benefit analysis occurs during herbicide registration and cancellation proceedings. In these cases, the potential benefits of a chemical product for weed control are weighed against its potential risks to human health and harm to the environment.

Societal Aims versus Individual Objectives

Auld et al. (1987) indicate that differences in societal and individual goals arise when some benefits or costs of an activity occur outside the domain of the individual decision maker. These externalities result in consequences to others that would not normally be taken into account when making private decisions. In relation to weeds and invasive plants, an external benefit arises if a farmer or land manager controls a readily dispersed plant, thereby reducing the incidence of such weeds on neighboring farms or areas. On the other hand, if one land owner fails to control that weed species, then other people may be subjected to an external risk of infestation and suffer an additional cost for weed control, loss of crop yield, or decline in land value. In a similar analogy, a weed control tactic may be employed by a farmer that has no immediate or obvious impact on the land other than to control weeds. If, however, many farmers use that tactic, the impact on water quality or public safety may be substantial, with the additional costs being borne primarily by consumers or other users of the land resource. A primary reason why externalities are important is that weeds often occur over vast areas of farm, forest, or rangeland or have the propensity to spread over large areas. The same can be said for techniques to control weeds.

Social Conflict and Resolution. Conflicts about weeds and weed control arise when the positive effects of a species or control measure are felt by one group of people but the negative effects are felt by another group. Such conflicts sometimes occur between people who make their living in rural areas (farmers or natural resource managers), the people or professions that provide tools or production information to them, and other people who do not live or make their living from the agriculture/natural resource sector. These divergent populations are sometimes separated by distance into rural and urban populations, with the latter being generically described as "the public." Until the last several decades, the public left farmers, foresters, and land managers alone to make their own decisions about how to grow crops, produce wood, or manage grazing lands. These decisions included weed control and weed control procedures.

The situation now is quite different, with virtually every sector of agriculture, forest, and range management and even management of natural lands being influenced by public and now more than ever by political views. Weed control and vegetation management are no exception to this generalization. In fact, weed control tactics, especially herbicide spraying in forests, marked some of the earliest conflicts over land management policies in the United States and Canada. One such episode in the history of weed science, the public debate about the controversial herbicide 2,4,5-T, is discussed in Chapter 9. It is clear that public education or public relation campaigns to turn around sentiment about any land management practice, including weed control, have been singularly unsuccessful (Breton and Tremblay 1990, Mater 1992, Perrin et al. 1993). Similarly, education activities by the U.S. Forest Service in numerous settings have had almost no impact on public opinion about herbicide use. If anything, Perrin et al. (1993) note, these educational attempts destroyed whatever credibility the agency had and firmed up public opposition to current practices. There is now general movement away from the use of education or promotion to solve social conflicts and movement toward approaches of conflict resolution. These approaches involve identifying the beliefs, values, interests, concerns, and desired benefits of various segments of the general population. It then involves showing how what one has to offer can be of benefit from their perspective, rather than attempting to "sell" a belief or preconceived message. There is much written on this subject of natural resource conflict resolution (Wilson and Morren 1990, Checkland 1981) and the serious student of this subject is directed to that literature.

Precautionary Principle. Typically scientists and land managers defer to government, politicians, or at times economists to dictate how technologies should be handled once they are developed (Verhoog 1996). Ethics and science are generally regarded as separate entities throughout the academic, agricultural, and natural resource communities unless scientists choose to take part in a technological debate (Martinez-Ghersa et al. 2003). However, the fact is that technology is so embedded in modern society that without dialogue about it those who enjoy the benefits or suffer harms from technology have no choice in the matter or no understanding about its impacts. At times, people would rather abandon technology altogether than accept the uncertainty that often goes with it. Also, of course, with understanding does not necessarily come acceptance (Strauss et al. 2000).

The *precautionary principle* is one way to guide discussions about technology and management. Ferre (1988) indicates that human activities regarding the implementation of new technology can be summarized as follows: *Do not destroy good, try to create good, and be fair.* It seems plausible that even more basic than creating new good is the obligation to be careful that one's actions do not leave the world or others worse off than before. Adopting this principle also necessitates discussion about the creation and implementation of policies about technological advances or management opportunities. It forces discussion before harm is done and in this way protects good science from becoming bad technology (Martinez-Ghersa et al. 2003).

Weed and Invasive Plant Management in Modern Society

Current discussions about change in science, technology, and society itself suggest that a new era in which old, albeit successful, concepts based on assumptions of reductionism and linearity are yielding to a paradigm that emphasizes holism, circuitry, and connections. The origins of this shift in values were recognized as early as the late-nineteenth century (Steiner 1983) and were developed in science by the mid-twentieth century (Pauly 1987). Is it possible to take a different look at weeds, invasive plants, and their management from this new perspective? Weeds pose a dual dilemma for agriculturists and land managers because they directly reduce crop yields, land values, or biodiversity. Weed control is also often difficult and expensive and sometimes creates other undesirable side effects. Human social and ecological systems are often represented as nested hierarchies of function and scale (Figure 2.2). Designing weed management strategies now requires working through the dilemma of how weeds, ecosystems, and human institutions have evolved, and continue to evolve, together.

Perhaps the study of multilevel interactions as suggested by Levins (1986) will assist in the recoupling of the biological and social components of human production systems (Figure 2.2 and Table 2.2). What is certain, as far as weeds and invasive plants are concerned, is that simply replacing herbicides with cultural approaches to control weeds (e.g., using crop rotation to avoid specialized weeds; planting weed-suppressive crops, intercrops, or native plants; or using biological controls) will not be enough. Neither will the use of herbicide-resistant crops or excessive rates and frequencies of chemicals and tillage be enough to achieve coevolution of weed management with healthy ecosystems. The recoupling of biological and social systems requires recognition within our social institutions (businesses, governments, scientific and land management agencies, and educational organizations) that such coevolution exists and is desirable. It also requires institutional change to minimize overconsumption, optimize labor and energy inputs for production, and maximize the use of biotic interactions and social values when designing new weed management strategies. This is the seemingly formidable, yet challenging, task facing invasive plant ecologists, land managers, and weed scientists.

SUMMARY

Plant ecology is the study of relationships between plants and their environment and is the scientific field in which information about weeds, invasive plants, and their associated plants is generated. Through the application of plant ecological principles, weed scientists, agriculturists, and land managers can begin to understand the nature of weediness and develop appropriate and sustainable management approaches and methods. The study of weed and invasive plant ecology requires understanding of general principles pertaining to the environment, scale and hierarchical structure in both ecological and human systems, community and niche differentiation, succession, the invasion process, genetics and evolution, and plant demography. Methods to select and prioritize management approaches for weeds and invasive plants have been proposed based on land value and risk of invasion. However, weeds and invasive plants and the methods to manage them often extend beyond individual fields and land managers to broader aspects of society. Invasive plants sometimes spread over vast areas and affect many people, and weed control also can have widespread benefits and costs. Externalities are consequences of weeds or weed control that extend beyond the area of infestation or domain of the individual person making a management decision. Since such consequences can be felt or perceived differently, conflicts can arise among people affected differentially by weed control decisions and tactics. It is usually best to resolve such conflict, rather than try to teach or educate others who have an alternate point of view. Value systems are the way people think about activities and technologies. There are four fundamental value systems which affect natural resource management: material well-being, sanctity of nature, individual rights, and justice as fairness. Each of these value systems raises different questions about human efforts to raise crops and manage land, control weeds, and contain invasive plants. The most common way to assess differences among value systems is to use a risk–benefit analysis; another is the precautionary principle.

3

INVASIBILITY OF AGRICULTURAL AND NATURAL ECOSYSTEMS

An increasingly global economy, worldwide transport of biological commodities, and opportunities for transworld travel have all promoted the introduction and subsequent colonization of exotic plants in many parts of the world. Rejmánek (2000) indicates that over 21% of the 22,000 vascular plants found in North America are nonnative or exotic. Atkinson and Cameron (1993) also report that at least 50% of the existing vascular plants in New Zealand are exotic, while 40% of the total flora of the British Isles is introduced from other areas of the world (Ellis 1994). In Australia, 1500–2000 plant species have been introduced since European settlement, of which over 200 species are now noxious weeds (Humphries et al. 1991, Parsons and Cuthbertson 1992). If this magnitude of plant introductions continues at its current pace, the earth's flora could eventually homogenize to only a few highly successful species (Luken and Thieret 1997, Ewel et al. 1999, McNeely 1999).

Invasive plants, after successful introduction, can apparently spread into new areas already fully occupied by native vegetation and displace native species (Randall 1996). Plant invasions may occur across broad landscapes and also can be locally abundant. Knowing the susceptibility of different habitats and plant communities to invasion provides insight into how weeds and invasive plants spread. It also can help in designing programs to control weeds in agriculture, manage invasive plants, and protect and restore native habitats.

Plant invasion or the invasion process (Chapter 2) is generally divided into a biological component, or the capacity of a plant to spread beyond the site of

Ecology of Weeds and Invasive Plants. By Steven R. Radosevich, Jodie S. Holt, and Claudio M. Ghersa
Copyright © 2007 John Wiley & Sons, Inc.

introduction and become established in new sites (*invasiveness*), and an environmental component, which is the susceptibility of a habitat to the colonization and establishment of individuals from species not currently part of the local community (*invasibility*) (Davis et al. 2005). However, these two elements of invasion interact strongly (see Figures 2.7 and 2.8). Environmental differences among habitats and communities contributing to invasibility are often easier to identify than the biological traits associated with invasiveness (Table 1.2) (Reichard 1997, Lonsdale 1999), although certain habitats, such as those of mature forests and dense grassland, tend to have relatively few exotic plant species (Richardson et al. 1994, Harrison 1999, Perelman et al. 2003, Parks et al. 2005a). Evolutionary history, community structure, propagule pressure, disturbance, and stress are all factors that account for differences in invasibility (Alpert et al. 2000, MacDougall and Turkington 2005).

PLANT INVASIONS OVER LARGE GEOGRAPHICAL AREAS

The geographical spread of cheatgrass (*Bromus tectorum*) from its introduction into western North America late in the nineteenth century to about 1930 is shown in Figure 3.1. From this modest beginning it has spread throughout all of the Great Basin (Chapter 1) and is now one of the most successful weeds in the world. Mack (1981) describes the pattern of geographic and population increase

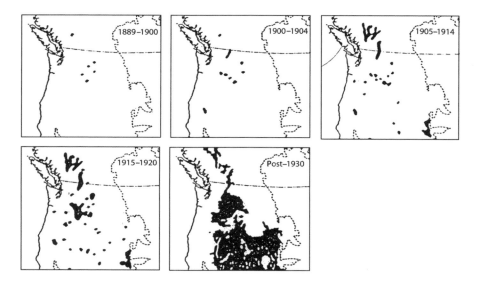

Figure 3.1 Spread of *B. tectorum* in western North America. Dotted line is Canadian–U.S. border. (From Hengeveld 1989, after Mack 1981, in *Agro-Ecosystems* 7:145–165. Invasion of *Bromas tectorum* into Western North America: An ecological perspective. Copyright 1981 with permission from Elsevier.)

of cheatgrass over this time period (Figure 2.7), which follows the generalized curves for the invasion process of any invasive species (Figure 2.8 and Equations 2.1 and 2.2). The geographic expansion of an invasive plant occurs in two ways, as *patches* that grow as a front and as individual plants (satellites, see Chapter 2) that create new patches.

The opportunity for biological invasion begins with dispersal, and many weeds of both arable and natural ecosystems possess well-adapted appendages to assist in long-distance movement of their seed (Figure 5.5). Because such appendages enhance the ability to move, they markedly increase the likelihood of seed and seedling survival by removing the individual from sources of parental-associated mortality (Figure 3.2). The greatest concentration of seed falls below or only a short distance from the parent plant (on-site production) and decreases with increasing distance away from it. This absence of seed and seedlings under parents is a function of initial seed dispersal and seedling mortality. As shown in Figure 3.2, the product of the two factors, dispersal and mortality, is a seedling recruitment surface that indicates the optimum distance between neighboring plants of the same species (Cook 1980). Thus, most seed, even those with special adaptive features for long-distance dispersal, tend to migrate as an advancing front. The result of this interaction is a patch that creeps across a given area. This recruitment effect is also noticeable with seed adapted for wind dispersal. Furthermore, plants from seed that are widely dispersed tend to survive well and colonize as isolated individuals (Figure 3.2), and after high densities are reached, they begin to spread as fronts.

Cousens and Mortimer (1995) describe the dynamics of range expansion of a plant species using geometry (Equation 2.5). They indicate that the area of range expansion from a patch should occur as rings and decrease over time to about

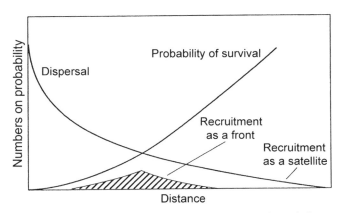

Figure 3.2 Recruitment of new genotypes as function of number of dispersed seed and probability of juvenile survival. [From Cook 1980, in O. T. Solbrig (Ed.), *Demography and Evolution in Plant Populations.* Copyright 1980. Blackwell Publications, Oxford, reproduced with permission.]

one-half the distance of the previous year. With the occurrence of satellite populations that also colonize as patches, the ripples of invasive plants across a landscape would appear as if they were pebbles cast into a pond. Since satellite populations tend to grow faster than well-established patches (*sources*), it stands to reason that the area of increase of an invasive plant will expand more slowly if all satellites are eradicated during the first year of expansion than if only the main patch is controlled (Auld et al. 1978, Moody and Mack 1988, Cousens and Mortimer 1995).

Habitat Invasibility

A species is considered invasive if it expands its geographical range beyond the areas previously occupied, that is, when individual plants immigrate into a new habitat or community and establish new populations there (Chapter 5). While expanding their geographical range, invasive plants can alter biodiversity by changing the composition of species assemblages and can also alter the habitat into which they are spreading. Thus, invasibility can be a characteristic of habitats as well as of particular plant communities. Habitats and plant communities, however, are not the same. In fact, similar habitat types may be occupied by quite different plant communities.

Although the term *habitat* is critically important in conservation biology, the term has been defined and used inconsistently in the literature over the past 20 years (Hall et al. 1997). A habitat is most commonly defined as the environment, biotic and abiotic, where a particular organism (species) exists and can survive and reproduce. Habitats include the sum of resources needed by a species and can be more than a single vegetation type or environment (Franklin et al. 2002). Thus, habitats are characterized as follows:

- Community type (a particular assemblage of plant species)
- Physiognomic characteristics of the dominant species or life-form (e.g., grassland, woodland, forest)
- Environment type (e.g., wetland, marsh, dune)
- Environmental attributes (conditions where the species is found, e.g., shade, acid soils, temperature range, or soil depth).
- Disturbance type and level (natural, grazed, tilled, etc.)

Similar to (but broader than) a niche, a habitat is defined for a particular species. Thus, a habitat can be invaded by species with similar requirements or when habitat quality is degraded for some reason. For example, disturbance, pollution, or fragmentation may cause the environment to change sufficiently to allow occupation by another species. Invasive species can also affect habitat quality and thereby alter habitat invasibility. For example, the habitat of a particular native bird might become unsuitable because of structural changes caused by an invasive plant, which could further reduce available habitat for other native organisms.

Community Invasibility

When habitat is characterized by its species assemblages, it is equivalent to a community. The ability of a species to invade a particular habitat depends on the structural and functional characteristics (Table 2.1) of at least some of the plant communities that compose it (Mack et al. 2000). As plant communities change over time through succession (Chapter 2), some environmental conditions could occur that match those of the original habitat of a newly introduced species. The necessity for matching environmental conditions could continue to be met as weeds and invasive plants adapt to their new environment over time. This evolution also would allow the new species to spread into many seres or across an entire geographical area.

Colonization of natural ecosystems by weeds often differs from that of agroecosystems. In natural ecosystems, for example, areas suitable for occupancy can be few and separated widely. Agricultural land, on the other hand, is often exposed to routine, even annual, disturbances that result in periods of high resource availability and low plant cover (Auld and Coote 1990, Ghersa and Roush 1993). These times of substantial disturbance tend to reduce environmental heterogeneity and are particularly well suited for establishment of agricultural weeds.

LOCAL INVASIONS

Environment is made up of all the biotic and abiotic factors that can affect the development and distribution of plants (Chapter 2). For new or migrating species, environment is often thought of as a sieve through which some species pass and survive but most fail. This notion of environmental porosity is based on diffusion theory and has its origin in the Fisher–Skellam equations that date prior to the 1950s (Fisher 1937, Skellam 1951, Holmes et al. 1994). These equations have many derivations; one such equation is

$$\frac{z}{t} = (4rD)^{1/2} - b(t^{-1}\log t) \tag{3.1}$$

where z is the distance traveled, t is time, r is the intrinsic rate of population increase, D is diffusability (a diffusion coefficient that reflects the ability of the site to absorb seed or propagules), and b is a constant. The Fisher–Skellam equations, as noted above, recognize "r" (Equation 2.1), which can be considered the innate ability of a species to expand into an open environment. Thus, this equation indicates that spread of a plant is influenced by the biological traits of the species as well as environmental factors. Diffusability (D) is dependent on the density (N) of all plants in the area of concern (Equation 2.2). Thus, population growth declines as the density of a population approaches the carrying capacity (K) of a site and will be zero when $N = K$. This explains why patch growth of invasive plants declines as they become denser and older.

The concept of diffusability also explains why disturbance has such dramatic effects on both introduction and naturalization stages of the invasion process (Figure 2.8). In ecological terms, disturbance reduces the impact of high plant density by creating vacancy. By imposing change, disturbance also increases the probability of finding a *safe site*, a set of suitable environmental conditions (Harper 1977), for some species. In contrast, environmental vacancies or safe sites may be closed to the occupancy of a new species by the high density of other invading plants. This concept of propagule pressure is discussed later in this chapter.

Safe Sites

Regeneration from seed usually occurs in an intensely hostile environment for most plants (Fenner 1985, 1994). In this respect, many studies demonstrate that seed germination is highly responsive to fine scale differences in the physical environment, especially at the soil surface (Silvertown and Charlesworth 2001). Harper and his associates (Harper 1977) derived the concept of safe sites based on their observation that most seed in the soil seed bank (Chapter 5) do not germinate, and of those that do, few survive. Harper (1977) describes a safe site as a zone that provides the following:

- Stimuli for seed dormancy breaking
- Conditions for germination to proceed
- Availability of resources for seedling growth
- Absence of hazards

Factors included in the definition of a safe site are the placement of seed in relation to the microtopography of the soil surface and the availability of water and other resources and conditions necessary for germination. Also contributing to successful germination and establishment are various adaptations for seed to be buried in the soil and seedlings to acquire resources. Thus, seed placement and germination in an appropriate safe site should enhance the survival of the resulting seedlings (Figure 3.2). Unfortunately, it is difficult to determine the specific criteria for a safe site until the seedling is actually present there. However, a variety of techniques have been employed successfully to demonstrate the occurrence and variability of safe-site requirements among species (Harper 1977).

Humans create safe sites to enhance seed germination and seedling survival of crop species. Agricultural weeds, however, must either be selected for adaptations to the safe sites of crops or, like wild plants, evolve mechanisms to avoid mortality. Some of these mortality-avoiding features are considered later when seed dispersal, seedling recruitment (Chapter 5), and plant competition (Chapter 6) are discussed. Harper (1977) describes studies demonstrating the occurrence of safe sites and specific seed characteristics that increase survival. These studies show that the ability of a seed to locate an environmental safe site varies on a species-to-species basis, but in general, it is a chance event. The success of any species on a site seems to be predominantly a function of the success of its ancestors in leaving well-adapted progeny.

Safe Site Example. Sea rocket (*Cakile maritima*) is a weed species that can disperse widely yet also remain in an already secure safe site. The plant was first introduced into San Francisco Bay in the United States, probably around Marin County, from Europe in 1936 and has dispersed effectively both north and south along the Pacific Coast since then. Sea rocket inhabits the foredune of beaches and has two-segmented fruits. The distal segments dehisce readily and can float long distances in seawater; the proximal segments tend to shed their seed while attached to the maternal plant (Davy et al. 2006). Winter storms easily dislodge the distal segments and their seed, which float off with the ocean currents and eventually wash to shore. The seed dehisced from the attached, proximal segments are often covered over by sand in the storms. Nearly all of these seed germinate within a meter of the parent, while the others germinate and survive only if they happen to find, by chance, a suitable place.

FACTORS THAT INFLUENCE INVASIBILITY

The impact of exotic weeds and invasive plants on productivity and other ecological processes has become recognized over the last several decades (Vitousek et al. 1996, Radosevich et al. 1997, Mack et al. 2000). It is also recognized that few plant communities are impenetrable to invasion by exotic plants (Gordon 1998, Sakai et al. 2001, Parks et al. 2005a). A plant may be invasive because it shares traits with resident species or because it possesses traits that allow it to occupy vacant niches in a habitat or community. In the following sections, we discuss factors that affect the susceptibility of plant habitats and communities to occupation by exotic species. A list of such factors is given in Table 3.1.

Evolutionary History

Alpert et al. (2000) indicate that past intensities of competition and human disturbance affect the invasibility of habitats. They suggest that plant communities in which competition has been intense over evolutionary time are low in invasibility because the native species were selected for high competitiveness and can thus outcompete newly introduced, potentially invasive plants. The low competitive ability of native plants on islands (Loope and Mueller-Dombois 1989) explains the high invasibility of those habitats. Parks et al. (2005a) note that past and current patterns of human land use often account for the presence or absence of invasive plants. Their analysis, which was conducted in mountainous ecoregions of the Pacific Northwest in the United States indicates that altered riparian systems and disturbed forests have relatively high levels of invasive plants, while alpine and designated wilderness areas are still relatively unaffected by invasion.

TABLE 3.1   Factors that Might Decrease Invasibility of Habitats by Nonnative Plant Species

Factor	Evidence
Evolutionary history	
Long history of human disturbance	Invasion from Old World to New World
Long history of intense competition	High invasibility of islands
Community structure	
High species diversity	Mostly negative
Strong indirect species interactions	Theoretical
Weak competition between plants	Effects of disturbance
Absence of mutualists	Effects of mycorrhizae, nitrogen-fixing
Presence of herbivores	bacteria, seed dispersers; effectiveness of biological control
Propagule pressure	
Weak dispersal agents	High invasibility of stream sides
Absence of fragmentation	High invasibility of fragments and edges
Disturbance	
Maintenance of typical regime	Manipulation of fire, grazing, and gaps
Stress	
Low nutrient availability	Increased invasibility after resource addition
Low water availability	Low invasibility of resource-poor areas
Low light availability	Competition experiments
Extreme conditions	Little

Source: Alpert et al. (2000). *Perspect. Plant Ecol. Evol. Syst.* 3:52–66. Copyright 2000 with permission from Elsevier.

Similarly, Alpert et al. (2000) indicate that as human disturbance increases worldwide, habitats with a long history of human land use should have lower invasibility since the resident native and exotic species of those habitats are selected to perform well under human-disturbed conditions. This observation might explain why there are more successful plant introductions into the New World from the Old World than in the opposite direction (Lonsdale 1999). Selection by human use might also explain the apparent long-term success of many agrestals in Europe (Holzner and Numata 1982) and the United States (Forcella and Harvey 1983).

Community Structure

Features of plant community structure that influence invasibility include species richness and the interactions among species that result in such factors as dominance and evenness. These characteristics comprise biodiversity, as described in Chapter 2, and are known to vary with scale (Whittaker 1975, Magurran 1988, Forman 1995). In spite of much research on the topic, the interactions between invasive plants and biodiversity and therefore the role of community structure in invasibility are under active debate. Hypotheses that have been proposed about the relationship between plant biodiversity and invasion include the following:

1. Species-poor communities are more susceptible to plant invasion than species-rich ones.
2. Species-rich communities facilitate plant invasion because of high resource availability.
3. No consistent relationship between biodiversity and exotic plant invasion exists because it is dependent upon the scale at which observations are made.

Ecologists have long assumed that diverse landscapes are resistant to invasion by exotic plants since complex communities are thought to be more efficient than simple ones in partitioning environmental niches (Chapter 2), using available environmental resources (Trenbath 1974), or accommodating more intense competition (Elton 1958, Levine and D'Antonio 1999). However, species-rich plant communities can experience both high and low levels of invasion by exotic plants. Experiments sometimes report negative relationships between biodiversity and invasibility (Tilman 1997, Knops et al. 1999, Naeem et al. 2000, Prieur-Richard et al. 2000), whereas positive relationships are found in other studies (Stohlgren et al. 1998, 1999, Wiser et al. 1998, Lonsdale 1999, Kalkhan and Stohlgren 2000). Davis et al. (2000) and Wardle (2001) propose that the underlying mechanism for invasibility is net resource availability in a plant community. They suggest that any increase in net resources due to disturbance or direct fertilization facilitates invasion, independent of plant species diversity. These ideas are consistent with those of Huston (1994) and Stohlgren et al. (1999), who predict that highest native plant diversity and exotic plant invasion should occur on productive sites, or "hot spots" of native diversity (Stohlgren et al. 1999), where moderate disturbance frees resources for invasive plants.

Levine (2000) suggests that the inconsistency in experimental results concerning the relationship of invasibility to biodiversity is a consequence of scale. He examined the occurrence of native and exotic plants on islands along a 7-km stream. A positive correlation was found at the community scale that he attributed to the abundance of plant propagules entering that riparian community. The overwhelming influence of propagule pressure relative to the influence of diversity or disturbance was also demonstrated by Von Holle and Simberloff (2005). However, when Levine experimentally added propagules to microenvironments he created on one of the islands, at the small spatial scale (0.035 m^2) of the microsite additions high species richness was correlated with greater resistance to invasion by exotic species. These results are consistent with those of Watkins and Wilson (1994) and Wilson et al. (1995), who found that invasive plant expansion in grasslands operated only at small neighborhood scales, since the plants in that system competed directly and there was little space available for substitution among ecological equivalents. Scott et al. (2006) add to this body of evidence for small geographical scale; however, the opposite pattern was found at intermediate to large spatial scales (Scott et al. 2006).

The findings of Levine and D'Antonio, Wilson et al., and Scott et al. agree with those of many others (Lonsdale 1999, Stohlgren et al. 1999, 2003, 2005, Levine 2000, Stadler et al. 2000, Sax 2001, 2002, Davis et al. 2005, Gilbert and Lechowicz 2005), who indicate that exotic and native plants share at least some similar life history traits and therefore may respond similarly to environmental resources and conditions. Thus, the factors that regulate biodiversity are likely the same as those that regulate invasion (Levine and D'Antonio 1999). At landscape and regional scales, therefore, biodiversity and invasion will depend on native and exotic plant traits as well as on other ecological factors that may covary with biodiversity (e.g., propagule pressure, resources, and disturbance). However, the degree to which exotic and native plants are similar remains an open question because, unlike native species, some exotic invaders have major impacts even though high numbers of native plants can reduce the amount of open niches available for occupation (Elton 1958, Levine and D'Antonio 1999). In cases of significant impact by invasive plants, plant traits could overwhelm the role of native biodiversity in determining invasion resistance (Ortega and Pearson 2005).

Role of Plant Size in Species Dominance and Richness. During succession both species richness and the body size of dominant species increase (Holling 1986). In addition, the evenness of the plant community changes over time, varying from strong dominance in early succession when environmental resources are abundant to weaker dominance later in succession when resources are more evenly distributed among community components (Chapter 2). Dominant species are also relatively small early in succession and tend to be larger during later stages. Although most studies examine how total species richness or resource availability affects the relationship between biodiversity and invasibility (Lonsdale 1999, Gilbert and Lechowicz 2005), the size of dominant plants has scarcely been studied in this context. According to Ritchie and Olff (1999), small and large organisms explore the environment differently. By using simple scaling rules, these authors examined the relationship between size and the ability of organisms to explore patches (microenvironments) that differ in both area and resource concentration. They found that large organisms occupy high proportions of habitat volume and use large amounts of resources but are only able to efficiently use large patches with low resource concentration. Organisms with small bodies occupy smaller volume and use less resource that is concentrated into small patches of habitat. Thus, size of dominant plants in a community could be an important variable controlling community richness (Ritchie and Olff 1999).

These results suggest that plant assemblages where size of dominants is large should be susceptible to invasion by smaller plants that can more efficiently use higher rather than lower resource concentration patches in the habitat (Gilbert and Lechowicz 2005). This observation also suggests that size of dominants could be a surrogate for species richness that may, in turn, allow opening or closing of environmental "windows" (niches) for invasion. The body size hypothesis seems complementary to Harper's safe sites, since both explain how barriers to invasion may be altered throughout succession. These ideas are also compatible with Davis et al. (2000) and Wardle (2001), who believe that net resource availability is the underlying mechanism for invasibility of plant communities.

Propagule Pressure

As noted above, propagule pressure can be a major factor influencing invasibility of communities that can overcome other ecological factors, such as biodiversity. The influence of propagule pressure on invasibility has been evaluated theoretically using models (Casagrandi and Gatto 1999, Neubert and Caswell 2000, Bouquet et al. 2002) and empirically (Tilman 1997, Wiser et al. 1998, Harrison 1999, Soons and Heil 2002, Honnay et al. 2002). As will be seen in Chapter 5, reproductive structures of plants are morphologically, physiologically, and genetically different from one another, which affects how plants disperse seed and survive. *Propagule pressure* describes the probability that seed, fruit, and vegetative reproductive structures will disperse, establish, and survive in a suitable place (Figure 3.2) and in sufficient numbers to maintain a species (Williamson 1996). The number of successful propagules is, thus, an important factor affecting the range expansion of any species.

Because propagule pressure involves both plant dispersal and survival on a site, it is influenced by landscape factors such as habitat fragmentation, human land use, disturbance, and overall species richness. It can also be affected by other features like topography, wind speed, wind direction, or any factor that influences mass flow and source–sink relationships of plant populations. An example of high propagule pressure is a riparian corridor where floodwaters collect and carry many seed from habitats located throughout a watershed (Johansson et al. 1996). (See also the discussion on source and satellite populations in Chapter 2.)

Relationship of Propagule Pressure to Invasion Process. As defined in Chapter 1 and further described in Chapter 2, the process of invasion includes three phases—introduction, colonization, and naturalization (Cousens and Mortimer 1995)—although the specific terminology may vary (e.g., introduction, naturalization, and invasion; Richardson et al. 2000). The introduction phase is when propagules arrive at a new site. Differences in propagule pressure during introduction are caused by the rate at which propagules arrive. Arrival, in turn, is related to the frequency, quantity, and kind of propagules during each introduction event. Since seed, fruit, and vegetative structures are morphologically, physiologically, and genetically different from one another, they contribute to successful introduction in specific ways, depending on the environmental conditions present in the particular plant assemblage being entered.

Mechanisms for seed to "sense" openings in plant canopies (Chapter 5) contribute significantly to the colonization phase of the invasion process, assuring successful seedling establishment in environments with high competition. For example, invasibility of dense grasslands is often low, especially by seed of species without dormancy, that is, poor ability to perceive canopy openings (Harper 1977, Tomback and Linhart 1990, Martinez-Ghersa et al. 2000a, Martinez-Ghersa and Ghersa 2006). However, invasibility in the same grasslands is high by species that reproduce by vegetative propagules or produce large amounts of seed that germinate under an array of environmental conditions (Warwick and Black 1983). Johnsongrass (*Sorghum halepense*) is such a species. This weed can overcome a wide range of environmental barriers during seed germination and maintain successful genotypes through clonal reproduction. Thus, it invades grasslands and crop fields by dispersal and establishment of particularly suitable propagules during the introduction and colonization phases of the invasion process (Ghersa and Roush 1993, Martinez Ghersa and Ghersa 2006).

Relationship of Dispersal to Propagule Pressure. The temporal and spatial distribution of weed propagules, and thus propagule pressure, is determined by morphological characteristics of seed and a species' agents of dispersal (Chapter 5). Propagules may be randomly dispersed over an area, transported into particular plant communities, or accumulate in particular topographic locations (van der Pijl 1982). The pattern of seed dispersal determines the potential area for plant recruitment and also influences biological processes such as predation, competition, and mating (Figure 3.2). Dispersal reduces the chance of seed being eaten or attacked by pathogens and for competition between parent and offspring or among siblings. In addition, dispersal decreases the likelihood of inbreeding depression (Nathan and Muller-Landau 2000, Soons and Heil 2002). Hence, seed dispersal contributes to colonization of sites by affecting gene flow (Chapter 4) within and among plant populations (Caswell et al. 2003).

Relationship of Human and Animal Transport to Propagule Pressure. Propagule distribution of weeds at small geographical scales is rarely random; rather it usually appears as clumps (Figure 2.5) (Harper 1977). Humans are often a major vector of weed seed spread. For example, Vibrans (1999) analyzed the seed of 50 important weed species in a maize-growing area of Mexico for morphological adaptations for long-distance dispersal. He found that most species had no visible adaptations for wind dispersal but were transported in mud. He concluded that the most likely dispersers of the weeds were humans and that people transported relatively large amounts of seed to other favorable habitats. Nonrandom distribution of propagules also may occur at much larger scales. For example, propagule flow among continents is concentrated along commercial routes (Figure 1.10), which increases propagule pressure of exotic plants near markets, ports, and transportation corridors. Kartesz and Farstad (1999) note that many exotic plants were first introduced into the Pacific Northwest between 1850 and 1920 through seaports such as Portland, Oregon, and Seattle, Washington.

Dispersal is often directed toward particular locations or features, especially if seed are dispersed by animals. Recently, Purves and Dushoff (2005) demonstrated that the probability of plant invasion increased when propagule pressure was directed to target habitats. For example, they noted the importance of directed seed dispersal of Brazilian waterhyacinth, *Eichhornia paniculata*, an aquatic plant restricted to ephemeral pools, by waterfowl.

Relationship of Seed Banks to Propagule Pressure. Seed banks (Chapter 5) increase propagule pressure (Williamson 1996) and are important in the colonization and naturalization phases of the invasion process (Figures 2.7 and 2.8). A large fraction of seed in the seed bank is typically dormant or in a state of arrested development (Chapter 5). Dormancy disperses seed through time because it improves the chances for germination and survival if suitable environmental conditions are likely to exist in subsequent years (Mitchell et al. 1998, Stöcklin and Fischer 1999). Seed banks, through seed dormancy, assure adequate conditions for establishment, which help to reduce environmental risks to seedlings (Harper 1977, Fenner 1985, Simpson 1990, Benech Arnold et al. 2000). Furthermore, if a plant fails to produce seed in one year, the presence of a seed bank also assures that the species will persist at that site over time.

Baskin and Baskin (1998) define a persistent seed bank as seed that live in the soil until at least the second germination season (Chapter 5). Seed persistence is strongly related to seed size and shape. For example, Bekker et al. (1998) demonstrate that seed longevity is estimated best by integrating seed size, shape, and depth of distribution in the soil. Similarly, Thompson et al. (1998) observe that gradients of habitat disturbance are accompanied by predictable changes in seed persistence and usually parallel shifts in seed size. Large seed accumulate in shallow soil layers and have a greater death risk than smaller seed or seed that accumulate in deeper layers of the soil. These authors also link persistence with physiological and morphological traits inherent in seed. Agricultural land or other disturbed habitats where soil layers are mixed are an exception to this general rule because in such production systems seed are distributed throughout the entire plow depth regardless of size, physiology, or morphology (van Esso et al. 1986).

Disturbance

Disturbance is the total or partial destruction of vegetative cover. It is an episodic, discontinuous change in a plant community that is caused by exogenous factors that kill or remove plant biomass. For that reason disturbance often stops succession or modifies the diversity and complexity of plant communities (Glenn-Lewin and van der Maarel 1992, D'Antonio et al. 1999). Recovery of a plant community from disturbance is called secondary succession (Figures 1.6 and 2.6). It is during this recovery period when vegetative cover is low that invasibility of plant communities is increased (Hobbs and Huenneke 1992, D'Antonio et al. 1999, Smith and Knapp 1999, Chaneton et al. 2002, Ghersa 2006).

Disturbance does not always enhance invasibility, however. Plant invasions into some plant communities occur without detectable disturbance (Tilman 1997, Wiser et al. 1998, D'Antonio et al. 1999), while some disturbances even decrease invasibility. For example, in some grassland ecosystems fire can decrease invasion while grazing can increase it (Figure 1.8), but in other production systems the opposite has been observed (Ghersa and León 1999, Smith and Knapp 1999, Mazía et al. 2001). Grassland invasion by pines may be either promoted or curtailed by fire (Richardson and Bond 1991). The invasion by exotic species into the pampas grasslands of Argentina is promoted by grazing or tilling but reduced by floods (Chaneton et al. 2002, Ghersa 2006). The inconsistencies between disturbance and invasibility noted above are best understood when examined with the additional perspective of evolutionary history. For example, D'Antonio et al. (1999; Figure 1.8) and Alpert et al. (2003; Figure 3.3) indicate that disturbance probably increases invasibility if it departs from the natural disturbance regime of an ecosystem.

Disturbance and Land Use. Parks et al. (2005a) observed that land cover change, particularly as it relates to grazing, logging, and fire/fuel management, is the underpinning for the successful establishment of exotic plants in the mountains of the Pacific Northwest region. They found three major land use (disturbance) categories in their analysis of invasibility. The first category is *anthropic systems*, which have a high degree of human use and invasibility. These systems include farms, forests near towns, roads, homesites, and managed parks and campsites. The second category of land use, *human-impacted natural systems*, encompasses areas that have experienced intense or prolonged grazing or logging, areas with an altered fire regime, or old fields from past farming activity. Roads and forests that are extensively logged create significant pathways for plant introduction in this land use category. In such areas, exotic species can be locally

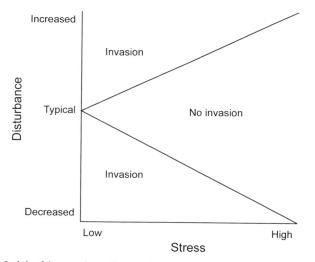

Figure 3.3 Model of interactive effects of stress and disturbance on habitat invasibility. (From Alpert et al. 2000. *Perspect. Plant Ecol. Evol. Syst.* 3:52–66. Copyright 2000 with permission from Elsevier.)

abundant depending on the degree of disturbance, rate of succession to native shrubs and trees, and amount of connecting roads. *Wilderness areas* and some national parks comprise the third category of land at risk of exotic plant invasion. The intact vegetation common in such areas often limits the intrusion of exotic plants to only a few meters from trails or campsites. In this analysis, riparian areas were especially vulnerable to invasion by exotic plants, probably because of their high natural and human-caused disturbance level (e.g., floods and agriculture) and propagule pressure due to transport in water. In contrast, wilderness habitats, particularly in alpine zones, were most resistant to invasion, although high visitor number to these areas could increase disturbance and accidental introductions of invasive plants to them (Lonsdale 1999).

Relationship of Disturbance and Succession. If ecosystems evolve under a recurrent disturbance regime, secondary succession occurs and the resulting plant assemblages increase in biomass and complexity over time (Chapter 2). In such ecosystems the destruction of native plant biomass is an internal (endogenous) part of the system, but additions of other plants or animals also influence the production, growth form, and complexity of the new communities and should also be considered when examining the impact of disturbance on succession. This concept is illustrated by changes in plant community structure in the Argentine rolling pampas that have occurred from grazing and agriculture over the last four centuries. Similar effects on succession have occurred from human impacts in the *Artemisia*-dominated grasslands of the Pacific Northwest.

Rolling Pampas of Argentina. The original landscape of the pampas was flat and covered with grass, and trees were completely absent (Tecchi 1983) except along a few major rivers. The grassland communities evolved in the absence of large herbivores. Cattle and horses were introduced by Europeans in the early 1800s and formed extraordinarily large herds (Soriano et al. 1992). Before the late-nineteenth century, plant invasions primarily of thistle species, probably due to cattle and horse grazing, became conspicuous and affected the function and structure of the grassland communities (Soriano et al. 1992).

In the first decade of the twentieth century, when still less than one-third of the vast grassland was converted to agriculture, the flora of the rolling pampas was comprised of about 1000 species of vascular plants. Grass fields at that time consisted of about 220 species, of which 60% were native. In agricultural fields, species richness was reduced to 53 species, with 32 from the original grassland flora and 21 new species of cropland weeds. Burning and plowing excluded many native species and generated the conditions for invasion by annual weeds that were excluded from the grassland but were adapted to soil disturbance by plowing. Surveys carried out in agricultural fields later in the century indicated that native plants had increased to 35 species by 1960 and 54 species by 1990 (Ghersa and León 1999). Exotic plant richness remained invariant over this time period at about 45 species, while the grassland cover of the rolling pampas was exponentially reduced since most of the area was plowed into croplands

(Hall et al. 1992). Agricultural land use also intensified in the region with the past mixed annual cropping–grazing production system being replaced by double cropping of wheat and soybean (Michelena et al. 1989, Maddonni et al. 1999).

Despite these changes in land use, the plant assemblages in the annually cropped fields continued to gain in native species richness, probably due to adaptive genetic changes in some of the species, but little variation in exotic species was observed (Ghersa and León 1999). The introduction of zero-tillage cropping systems again changed the structure of plant communities in the now cropped fields, allowing for invasion of woody species and reduction of the native species previously described there (Ghersa et al. 2002, de la Fuente et al. 2006). The absence of tillage and invasion of woody plants represent still another disturbance to the vegetation of the rolling pampas and in the annual cropping system that has been evolving there. This latest disturbance is different but analogous to that created when the native grasslands of the rolling pampas were first plowed and turned toward annual cropping 150 years ago.

Artemisia-Dominated Grasslands of North America. The bunchgrass ecosystems of the Pacific Northwest have also undergone significant change during the past 150 years (Bunting et al. 2002). These shrublands/grasslands were typically dominated by sagebrush (*Artemisia tridentata*) and perennial bunchgrasses. Cultivation, grazing, altered fire regimes, and the introduction of exotic plants have resulted in their transition to grasslands dominated by exotic grasses and forbs (Johnson and Swanson 2005). The mix of exotic plants includes cheatgrass, medusahead (*Taeniatherum caput-medusae*), *Ventenata dubia*, knapweeds (*Centaurea* spp.), toadflax (*Linaria* spp.), and sulfur cinquefoil (*Potentilla recta*). Endress et al. (2007) believe that the distribution and dominance of the exotic plants in the region can, at least in part, be attributed to past human-caused disturbances, such as the presence of abandoned agricultural old fields.

Relationship of Stress and Disturbance. Stress is defined as a condition in which a plant's physiological functioning is reduced below its maximal level. Environmental factors that create plant stress function much as a disturbance and thereby can affect invasibility of habitats or communities. Three types of environmental stress are hypothesized to affect invasibility: low resource availability, conditions that limit resource acquisition such as extreme temperature, and presence of toxins. Alpert et al. (2003) propose that stress due to reduced levels of nutrients, water, and light reduces invasibility because (1) exotic plants cannot tolerate the maximum levels of stress to which native species are adapted or (2) in stressful conditions the competitive balance between invasive and native plants is altered. However, if the native flora is stress tolerant, low stress may favor invasive species if they are better able than natives to take advantage of high resource availability (Grime 1979, Dukes and Mooney 1999).

Alpert et al. (2000, 2003) propose that invasion in a plant community should occur when stress is low, when disturbance departs positively or negatively from the natural level of disturbance (the historical level under which an ecosystem was self-assembled), or when relatively low stress is combined with large departures from the natural level of disturbance (Figure 3.3).

INVASIBILITY AND EXOTIC PLANT INVASIVENESS

The ability of exotic plants to competitively suppress native plants is often cited to explain the local dominance of exotic plants (Chapter 6). However, recent studies by Milton (2003), Seabloom et al. (2003a,b), and Corbin and D'Antonio (2004) suggest that exotic plant dominance is caused by the interactions among exotic plant dispersal, disturbance, and land use (changes in plant cover). Kimmins (1997) indicates that it is possible for some native plant ecosystems to become so altered that it is impossible for them to return easily to a relatively unaltered state or composition of plants (Figure 3.4). Thus, transformation of native plant communities may be the consequence of fundamental environmental changes that limit native flora, and exotic plants may simply be "passengers" of these changes in environment rather than driving the process (MacDougall and Turkington 2005).

Hobbs et al. (2006) believe that novel (exotic-dominated) ecosystems arise from human impacts that result in the following:

- Local extinction of most original plant, animal, and microbial populations and the introduction of new species not previously present in the biogeographical region.
- Urban, cultivated, or degraded landscapes that create dispersal barriers for native species recolonization.

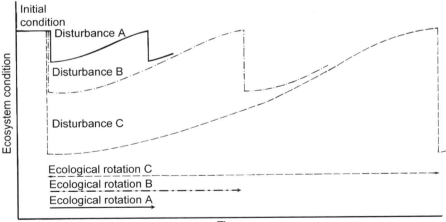

Figure 3.4 Graphical representation of concept of ecological rotation—period required for an ecosystem to recover to its original, or some new condition. (Modified from Kimmins 1997. *Forest Ecology: A Foundation for Sustainable Management.* 2nd Ed. Prentice Hall, Upper Saddle River, New Jersey.)

- Major changes in the abiotic environment or a decrease in the original species pool of propagules. These impacts can be from either direct (removal of soil, harvest, pollution) or indirect (erosion for lack of vegetative cover or overgrazing) causes.

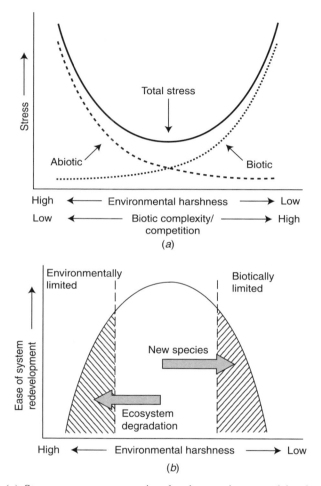

Figure 3.5 (*a*) Stress on an ecosystem is related to environmental harshness and biotic complexity: In harsh environments the constraints to establishment and/or growth are primarily abiotic, while in more benign environments the constraints are mainly biotic, arising from the preexisting mix of species present. Total stress is greatest at either end of the gradient. The inverse image of this graph in (*b*) portrays the ease with which an ecosystem will redevelop following disturbance or human modification. Ecosystem degradation leads to more abiotic stress, while the addition of new species leads to more biotic stress, and ecosystem redevelopment is less likely in both cases. [From Hobbs et al. 2006, after Ewel et al. 1999. In Hobbs et al. 2006, *Global Ecol. Biogeogr.* 15:1–7. Copyright 2006. Blackwell Publishing Ltd., reproduced with permission. See also Bertness and Callaway (1994) and Menge and Sutherland (1987).]

These authors suggest that these types of novel ecosystems (Figure 3.5) lay somewhere in the middle of a gradient between "wild" ecosystems and intensively managed systems.

Interactions among the factors of site invasibility and plant invasiveness are not well understood, although empirical studies have addressed each factor separately. From a management perspective, two important questions regarding exotic-plant-dominated ecosystems need to be addressed: (1) Under what conditions do such ecosystems form? (2) What mechanisms (e.g., dispersal of exotic propagules, competition, disturbance, lack of native propagules) maintain dominance by exotic plants? The development of persistent exotic-species-dominated plant communities is problematic for land managers because most tools for weed control do not address the underlying ecological mechanisms whereby novel communities are created. Thus, weedy and invaded sites can remain susceptible to invasions by the same or new exotic species, even after weed control treatments have yielded short-term reductions in invasive plant abundance.

SUMMARY

Plant invasion has two components, biological (invasiveness) and environmental (invasibility). These two components interact, often making it difficult to separate the influence of one from the other. Plant invasions can occur across large geographical areas or be more restricted to local sites. In either case, the process of plant invasion begins with propagule dispersal, and the resulting population enters an exponential phase of growth. Expanding populations move as fronts from an existing patch and as individual plants (satellites) that begin new patches. These patterns of movement are determined by the interaction of seed dispersal and seedling mortality in relation to distance from a parent plant. Habitat invasibility usually encompasses very large areas of susceptibility to invasion while community invasibility is smaller in scale and is associated with succession and environmental change. The concept of diffusability relates to local invasions and can be thought of as an environmental sieve that describes the openness of an area to the addition of other plants. The concept of safe sites is relevant to invasion since they represent places (niches) where new species can germinate, survive, and reproduce. Disturbance generally increases diffusability and therefore the invasibility of an area. Factors that influence invasibility are evolutionary history of habitats and plant communities, plant community structure or biodiversity, propagule pressure from either native or introduced plants, and disturbance, land use history, and stress. It is important to understand how exotic-dominated plant communities form and how they are maintained to accomplish long-term management of these ecosystems.

4

EVOLUTION OF WEEDS AND INVASIVE PLANTS

From an evolutionary perspective, agricultural weeds and invasive plants in natural ecosystems probably arose from stochastic events that affect the ability of small populations to survive and quickly adapt to new habitats (Gray et al. 1986, Lee 2002). These plants acquired traits to overcome a range of biotic and abiotic constraints or to adapt to changing selection pressure in their native habitats (Silvertown and Charlesworth 2001). Broad environmental tolerance and phenotypic plasticity are often used to explain the success of weeds and invasive plants. In contrast, some weeds, such as crop mimics or other species that respond directly to cultural practices, do not acclimate to all of the environmental conditions that are possible across a region. Instead, such species experience strong selection pressure and express heritable traits in environmental tolerance only after they are introduced into or invade new habitats. Even plants that share the same population source may either adapt widely to the environments in a region or become specialized and adapt to only a narrow range of environments (Dekker 2003). *Autogamy* (self-fertilization) and the combination of sexual and clonal reproduction is a particularly important trait for weeds and invasive plants (Table 1.2). This particular hypothesis has become so prevalent that it is sometimes called *Baker's rule*. According to Baker (1974):

> A notable feature of most weeds, especially annuals, is their ability to set seed without the need for pollinator visits, either by autogamy (self-fertilization) or agamospermy. Even when outcrossing does take place, wind or generalized flower

Ecology of Weeds and Invasive Plants. By Steven R. Radosevich, Jodie S. Holt, and Claudio M. Ghersa
Copyright © 2007 John Wiley & Sons, Inc.

visitors are adequate. The advantages of autogamy or agamospermy for a weed include providing for starting a seed-reproducing colony from a single immigrant or regeneration of a population after weed-clearing operations have removed all but a single plant. In addition, they allow rapid build-up of the population by individuals virtually as well adapted as the founder. Where the weed is a perennial, self-compatibility is less certain to be found (and some such weeds are even dioecious), but an extra emphasis upon vegetative reproduction here achieves the same end, i.e., the rapid multiplication of individuals with appropriate genotypes.

However, there are many exceptions to this rule, since weeds and invasive plants display a wide range of possible evolutionary pathways. For example, many successful invasive plants, whether annual herbaceous species such as Italian ryegrass (*Lolium multiflorum*) or perennial woody species such as pines or legume shrubs and trees, are outcrossers without specialized means for vegetative propagation. It is unlikely, therefore, that studies of the shared attributes of successful weeds or invasive plants will provide as many insights into the evolution of weediness as focused experiments on the genetic changes that plants undergo to colonize a new area. It is important to understand the genetic consequences of introduction and colonization because during biological invasions significant genetic change can occur in species that are no longer limited by their native environment (Gray et al. 1986).

EVOLUTIONARY GENETICS OF WEEDS AND INVASIVE PLANTS

The evolutionary genetics of most weeds and invasive plants is still under intensive study. However, the examination of plant populations using reciprocal transplant experiments, common gardens, and long-term in situ manipulations provide convincing cases where natural selection operates on these species and results in genetic differentiation. The heritable genetic variation that is necessary for selection to act is gained through several different mechanisms (below) and is frequently maintained by *gene flow* among complexes of related species. Gene flow is the result of processes such as pollen or seed movement and alters allele frequency through (1) immigration of external genes into a population, (2) breeding systems and *genetic drift* (random changes in allele frequencies) that limit the flow of genes within a population, and (3) the presence of a seed bank and dormancy that slow the loss of genes from a population (Levin and Kerster 1974).

Trait differentiation is documented for most important features of plant structure and function in nearly every plant taxon, which includes herbaceous annuals and perennials, woody perennials, aquatics, narrow endemics, as well as widely distributed plant species. While there are exceptions (Lee 2002), most studies demonstrate that physiological and morphological traits of weeds and invasive plants arise from selection imposed by biotic (e.g., competitors or predators) or abiotic factors (e.g., temperature, photoperiod, herbicides, tillage, or harvesting) that enable a plant species to produce more seed or disperse better into new habitats.

Heritable Genetic Variation

Sufficient heritable genetic variation is essential for selection and evolutionary adaptation to occur in response to environmental change. Since most plant invasions are initiated by stochastic events (Chapter 3), genetic variation is expected to be low during introduction and subsequent colonization. For this reason a successful plant invasion is determined most by how a species accumulates heritable or *additive genetic variance* (*AGV*) (summation of effects of all genes influencing a trait), rather than individual plant reproductive output or population density. It is during such "founder" events, when only a small number of seed or plants arrive at a new area that presumably carry a fraction of the total AGV, that multiple introductions are necessary for total AGV to contribute to plant invasion. This situation was probably the case for many agricultural weeds introduced to the British Isles before the beginning of the twentieth century, when seed were recurrently introduced by wool imports from different parts of the world (Crawley 1989, Williamson 1996).

Genetic and phenotypic variation contribute to AGV, and both forms of variation are highest when the geographical area occupied by a plant species is large. For example, plant species that invaded the New World and became major weeds range geographically throughout Asia, Africa, and Europe and are known to be variable in many plant characteristics such as seed and leaf traits, phenology, physiological and biochemical activity, parasite resistance, competitive ability, breeding system, and life history (Sauer 1988, Dekker 2003). High AGV increases the probability that individual plants of a potentially invasive species carry traits that confer fitness in environments where the population has never existed and whose appearance in new habitats, therefore, seems unpredictable. Populations distributed over large areas, extending across regions or continents, have a better chance of colonizing new areas (or persisting after catastrophic environmental change) because of their greater AGV than species that are restricted to small areas.

Weeds and invasive plants acquire AGV in several ways, such as hybridization and polyploidy, epistasis, and epigenetic inheritance. These acquisitions of AGV involve the action of small numbers of genes (Mayr 1982) and are discussed below.

Hybridization and Polyploidy. Weed and invasive plant populations can originate through intraspecific or interspecific crossing, or *hybridization*, among wild and cultivated plant species (Zossimovich 1939, Ashton and Abbott 1992) or from changes in ploidy (de Wet and Harlan 1975, Hanfling and Kollmann 2002). Both hybridization and *polyploidy* (a condition where individual genotypes possess an excess of entire sets of chromosomes) improve competitive ability and sometimes plasticity of individual plants. Polyploids generally grow faster, occupy larger areas than diploid plants (Baker 1995), and are frequently plant species with invasive or weedlike characteristics (Table 4.1).

Hybridization of invasive plant populations with native or other nonnative populations alleviates loss of AGV during founder events and generates novel genotypes in the new environment. *Intraspecific hybridization* is gene flow among

TABLE 4.1   Comparison of Gramineae (Poaceae) and Compositae (Asteraceae) Plant Families with Weedy Species of These Families in California

	Weedy Gramineae of California (%)	Gramineae in General (%)	Weedy Compositae of California (%)	Compositae in General (%)
Diploids	33	34	65	67
Polyploids	67	66	35	33
Annuals	64	24	57	35
Perennials	36	76	43	65
Annual diploid	40	59	67	81
Annual polyploid	60	41	33	19
Perennial diploid	21	34	63	56
Perennial polyploid	79	66	37	44

Source: Heiser and Whitaker (1948). *Am. J. Bot.* 35:179–186. Copyright 1948. American Journal of Botany.

individuals of different populations of the same species and has been reported for both open- and self-pollinated weed species. For example, in the normally self-pollinated foxtail species group (*Setaria* spp.), natural outcrossing among plants in the group is an important source of new variants. Specifically, in *Setaria viridis*, spontaneous outcrossing rates among plants are between 0 and 7.6% of the auto-gamous rates (Darmency et al. 1987, Prasada Rao et al. 1987). Species sometimes acquire adaptive traits by hybridization with related species (*interspecific hybridiz-ation*) or *introgression* (repeated backcrossing of an interspecific hybrid with one of its parents). Milne and Abbott (2000) found that introgression occurred in *Rhododendron ponticum*, an introduced invasive plant growing extensively in the British Isles. They found that genes from *Rhododendron catawbiense*, a species from Spain, improved cold tolerance and allowed *R. ponticum* expansion into Britain's coldest region in eastern Scotland. This is also an example of adaptation to an environmental gradient (Abbott et al. 2000; see below).

Intraspecific Hybridization. Ellstrand and Schierenbeck (2000) discuss how hybridization within a species leads to invasiveness. They develop a theoretically optimal level of relatedness, similar to that of Waser (1993), which yields geno-types most likely to become invasive (Figure 4.1). According to this model, hybridization among closely related populations should not yield different results from mating within a population. However, distantly related populations probably evolve cross-incompatibility or otherwise unfit progeny (*outbreeding depression*) (Figure 4.1). Only a fraction of interpopulation combinations yield progeny with superior fitness as compared to their parents (*hybrid vigor* or *heterosis*). They suggest that if hybridization among populations of the same taxa is important in the evolution of invasiveness, then invasiveness would be most likely after mul-tiple introductions of a species since multiple introductions would provide geno-types from disparate population sources.

Ellstrand and Schierenbeck (2000) also indicate that (1) species intentionally introduced should have an invasive advantage and (2) invasiveness should occur

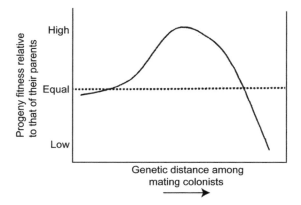

Figure 4.1   As genetic distance between mating colonists increases, so too should heterosis in their progeny—up to a point. Then progeny fitness declines as outbreeding depression becomes important. (From Ellstrand and Schierenbeck 2000. *Proc. Nat. Acad. Sci.* USA 97:7043–7050. Copyright 2000 National Academy of Sciences USA.)

after a lag time, during which hybridization and selection create and increase invasive genotypes. Both phenomena occur frequently (Chapter 2). Finally, Ellstrand and Schierenbeck challenge the commonly held view that weeds and invasive plants are genetically depauperate because of the genetic bottleneck imposed during founder events (i.e., acquisition of AGV). They point out that invasive species often originate from multiple foci, each with an independent origin that can spread, coalesce, and create opportunities for hybridization among the independent lineages. They further suggest that if the evolution of invasive traits follows hybridization between well-differentiated populations, the resulting populations should be more genetically diverse than their progenitors and leave relatively high levels of within-population polymorphism as a "signature."

These authors provide two examples of invasive plants that support their arguments. Paterson's curse (*Echium plantagineum*) was intentionally and unintentionally introduced several times in Australia (Burdon and Brown 1986, Piggin and Sheppard 1995). Invasive populations of this species are more diverse than populations genetically analyzed in its native European range. A second example is cheatgrass, *Bromus tectorum*, an exotic invasive plant introduced in North America (Chapter 1). Similar to Paterson's curse, North American populations of cheatgrass originated through multiple introductions and are known to have increased within-population genetic variation as compared to populations from its source range in Europe and northern Africa (Novak et al. 1991, Novak and Mack 1993, Novak et al. 1993).

Interspecific Hybridization. Interspecific hybrids resulting from introgression among different species of weeds or invasive plants are rare but have important consequences when they occur. Fertilization between species usually yields sterile progeny. When fertile progeny are produced, the new genetic variants are often unfit for the competitive conditions encountered in natural ecosystems. However, disturbance opens an array of ecological conditions that usually are better suited for hybrids than for their parents (Stebbins 1959, Young and Evans 1976). The importance of interspecific hybridization was studied by Ellstrand et al. (1996), who determined the frequency and distribution of natural hybridization between species in five biosystematic floras. They found that interspecific hybridization was nonrandomly distributed among taxa, concentrated in certain families and genera, and often found at a frequency that was not in proportion to the size of the family or genus (Table 4.2). Most common introgressed species were

TABLE 4.2   Six Families and Four Genera with Most Hybrids in Five Biosystematic Flora

Flora	Families (Rank)[a]	Hybrids	Genera	Family	Hybrids
British Isles	Scrophulariaceae (6)	88	*Euphrasia*	Scrophulariaceae	71
	Salicaceae (20)	55	*Salix*	Salicaceae	55
	Rosaceae (3)	53	*Epilobium*	Onagraceae	43
	Onagraceae (25)	46	*Rosa*	Rosaceae	36
	Poaceae (2)	45			
	Asteraceae (1)	41			
Scandinavia	Cyperaceae (4)	30	*Carex*	Cyperaceae	25
	Poaceae (2)	25	*Salix*	Salicaceae	15
	Asteraceae (1)	18	*Viola*	Violaceae	7
	Salicaceae (17)	15	*Calamagrostis*	Poaceae	5
	Rosaceae (3)	13			
	Dryopteridaceae (31)	9			
Great Plains[b]	Asteraceae (1)	29	*Amaranthus*	Amaranthaceae	12
	Poaceae (2)	20	*Aster*	Asteraceae	10
	Rosaceae (7)	15	*Rosa*	Rosaceae	9
	Fabaceae (3)	14	*Verbena*	Verbenaceae	8
	Amaranthaceae (31)	13			
	Verbenaceae (34)	8			
Intermountain[b]	Asteraceae (1)	43	*Penstemon*	Scrophulariaceae	10
	Scrophulariaceae (3)	19	*Carex*	Cyperaceae	9
	Poaceae (1)	19	*Castilleja*	Scrophulariaceae	7
	Cyperaceae (4)	11	*Oryzopsis*	Poaceae	7
	Boraginaceae (5)	7	*Stipa*	Poaceae	7
	Orchidaceae (15)	6			
Hawaii	Gesneriaceae (9)	67	*Cyrtandra*	Gesneriaceae	67
	Asteraceae (1)	49	*Dubautia*	Asteraceae	24
	Campanulaceae (4)	12	*Bidens*	Asteraceae	10
	Rubiaceae (7)	9	*Clermontia*	Campanulaceae	8
	Euphorbiaceae (12)	4			
	Lamiaceae (5)	4			

Source: Ellstrand et al. (1996 *Proc. Natl. Acad. Sci. USA* 93:5090–5093. Copyright 1996 National Academy of Sciences, USA.)

[a]Numbers in parentheses represent the rank in terms of species number.
[b]Floras of North America.

outcrossing perennials, species with reproductive modes such as *agamospermy* (asexual formation of embryos and seed without fertilization), species that spread vegetatively, and permanent polyploids.

The acquisition of herbicide resistance in agricultural weeds from genetically engineered crops, such as oilseed rape (*Brassica napus*) and its wild relative bird's rape (*B. rapa*), provides new examples of adaptation through hybridization (Andow and Zwahlen 2006). These cases also demonstrate unequivocally the importance of introgression among species for gaining AGV. Although it was accepted that oilseed rape could produce viable hybrids with bird's rape, early research emphasized the barriers to gene flow and the low likelihood of hybrid survival (Downey et al. 1980, Miller 1991). Contrary to these expectations, however, Jørgensen and Andersen (1994) found that crop genes were transmitted readily from oilseed rape to bird's rape, and later, herbicide resistance genes, including transgenes, were found in bird's rape (Mikkelsen et al. 1996, Hall et al. 2000). This gene flow enabled the weed to withstand herbicide applications and continue invading heavily sprayed fields.

Polyploidy. Certain phylogenetic groups of plants are biologically predisposed to form and maintain hybrids. Recently formed *allopolyploid* hybrids (a hybrid between different species in which chromosomes of both parental species are retained) typify many widespread and successful weed species. For example, all of the 18 species recognized by Holm et al. (1977) as the world's worst agricultural weeds (Table 1.3) are polyploids. Eleven of the 20 plant species described by Crawley (1987) as the most successful British exotic species are also polyploids.

Polyploid hybrids in plants tend to be more fit than diploid hybrids, possibly because of increased heterozygosity and reduced inbreeding depression. In addition, much genetic variation could arise from multiple origins of polyploidy within allopolyploid plants. In sterile and asexual allopolyploids, additional benefits arise from fixed heterosis where favorable genotypes cannot recombine, thus maintaining hybrid vigor in a population. For sterile allopolyploids, trade-offs between the benefits of fixed heterosis and the costs of lowered AGV are poorly understood. Levels of AGV vary considerably among sterile allopoly-ploids. For example, AGV is low in smooth cordgrass, *Spartina anglica*, but unexpectedly high in a triploid hybrid formed from the cross among individuals of diploid and tetraploid populations of the dandelion (*Taraxacum officinale*). The high variation in the triploid population was caused by recombination between chromosomal homologues (van Baarlen et al. 2000).

Epistatic Genetic Variance. *Epistatic* genetic variance, or *epistasis*, appears in plants when two or more gene loci have joint effects on a phenotype and the result is greater than the sum of each separate locus. Epistasis is an alternative solution for low AGV in invading plant populations. Following plant introduction or recolonization after a catastrophic event when the number of individual plants is small, a trade-off exists between the loss of AGV and a gain in variation by epistatic genetic variance. A two-locus epistatic interaction is either *synergistic* or

TABLE 4.3   Types of Epistasis Produced by Two-Locus Interaction

Genotype	Trait Values	AB	Ab	aB	Ab	Type of Epistasis
No epistasis (additive across loci)	$AB = Ab + aB - ab$	2	1	1	0	No epistasis, additive inheritance
Synergistic epistasis	$AB > Ab + aB - ab$	3	1	1	0	Synergistic epistasis
Antagonistic epistasis	$AB < Ab + aB - ab$	1	1	1	0	Antagonistic epistasis

antagonistic. For example, in Table 4.3 and Figure 4.2, plants with genotypes *AB*, *Ab*, *aB*, and *ab* at two loci are represented where relatively higher values suggest greater expression of a trait. In synergistic epistasis each mutation has a disproportionately large effect on the plant's fitness. In antagonistic epistasis, the addition of new mutations decreases fitness more than expected.

Allard (1996) indicates that assembly of favorable epistatic combinations of genetic loci is the most important factor for plant adaptation. For example, seed dormancy, an adaptive trait promoting plant survival and an important characteristic for invasion success (Chapter 5), is a genetically complex trait controlled by *polygenes* (multiple genes that affect the same trait) (Johnson 1935, Anderson et al. 1993), although its effects can be modified by genetic background and environment. Changes in temperature or light requirements for germination are acquired through epistasis, which affects the environmental conditions for germination and the ability to secure quality microsites (e.g., safe site; Chapter 3) (Martinez Ghersa et al. 2000a, Martinez-Ghersa and Ghersa 2006). Epistatic dormancy traits are postulated for rice, wheat, and wild oat on the basis of Mendelian genetics (Seshu and Sorrells 1986, Jana et al. 1988, Bhatt et al. 1993, Fennimore et al. 1999). Epistasis also has been demonstrated by comparing relationships between observed phenotypic and allelic variation (Alonso-Blanco et al. 2003, Gu et al. 2004).

Epigenetic Inheritance Systems. *Epigenetics* is a relatively new field of genetics where molecular pathways that regulate how genes are packaged in chromosomes

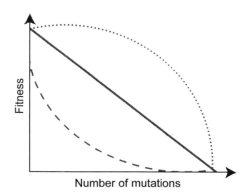

Figure 4.2   Relationships between numbers of mutations and fitness. *Synergistic* epistasis, dotted line; *antagonistic* epistasis, dashed line.

and expressed in cells are studied. It is the study of reversible, heritable change in gene function that occurs without changes in nuclear deoxyribonucleic acid (DNA) or the processes involved in development of an organism. The object of study is to determine how gene regulatory information, not expressed in nuclear DNA sequences, is transmitted from one generation of cells or organisms to the next. Thus, *epigenetic inheritance* is the transmission of information from a cell or multicellular organism to its descendants without the information being encoded in the nucleotide sequence of a nuclear gene. Maternally inherited traits are a form of epigenetic transmission of information from one generation to the next. Epigenetic inheritance also includes phenotypic polymorphism as well as how environmental factors affecting a parent influence the way genes are expressed and development unfolds in the offspring.

Epigenetic mechanisms often appear anomalous to Mendelian models of genetics. However, Grant-Down and Dickinson (2005) believe that these mechanisms explain how the entire genome operates and evolves and how introduced species escape the internal and external constraints on their genetic systems that should curtail success during founder effects. Although carrying or generating an "epimutational load" may have advantages and disadvantages in terms of species success, epigenetic change seems to explain why hybrids between native and nonnative plant species are unusually adaptable and become highly invasive (Grant-Down and Dickinson 2005).

Recently, significant epigenetic change was found in Italian ryegrass, a successful invasive plant in temperate-climate grasslands that has become naturalized in many regions of the world. Vila Aiub and Ghersa (2005) induced herbicide resistance in several lines of cloned ryegrass using recurrent sublethal applications of diclofop-methyl herbicide. The resistance was maintained through mitosis but reversed during meiosis. In addition, two of the isolines exposed to herbicide (herbicide resistant and herbicide susceptible) produced epigenetic variation in seed dormancy and germination relative to untreated controls (Mendoza et al. 2005). Epigenetic variation was also induced in Italian ryegrass by infection with fungal endophytes (*Neotyphodium* spp.). The endophyte is only transmitted maternally through seed and also may alter the plant's phenotype, thus modulating gene expression (Grant-Down and Dickinson 2005) and increasing resistance to diclofop-methyl (Vila Aiub and Ghersa 2001, Vila Aiub et al. 2003, Vila Aiub and Ghersa 2005).

ADAPTATION FOLLOWING INTRODUCTION

Changes in genetic architecture that are expressed as particular traits under new environmental conditions are as important as the generation of heritable genetic variation during plant invasion. The new traits appear as morphological or physiological changes in phenotypes that lead to adaptation, affect demographic characteristics, or connect contrasting trade-offs among the traits themselves. For example, traits conferring stem elongation to avoid competitors negatively impact plant defense against herbivory (Izaguirre et al. 2006). Seed size, which increases the ability of seedlings to grow in conditions of low resource availability, is nega-tively correlated with dispersal capacity and positively correlated with the risk of being eaten by granivores (Martinez-Ghersa and Ghersa 2006).

With such perspective, it is assumed that, following introduction, selection initially acts on the dispersal ability or physiological tolerance of introduced plants to stresses imposed by the new habitat. In recently introduced plants, there-fore, adaptation is believed to proceed in response to the following selective pressures in a new environment:

- Environmental gradients, such as temperature, photoperiod, or moisture (Chapters 2 and 3)
- Resident species acting as competitors or facilitators (Chapter 6)
- Pests, predation, and herbivores (Chapters 5 and 6)

Responses to Environmental Gradients

Several case studies provide strong inferential evidence that selection along environmental gradients produces the genetic differentiation needed among popu-lations for plant invasions to progress.

Change in flowering time in response to latitude is reported for species of goldenrod (*Solidago altissima* and *S. gigantea*) that were introduced into Europe from North America approximately 250 years ago. These introductions now exhibit a cline in flowering time that resembles the cline in their native range. Common garden experiments reveal the genetic basis for this phenological correspondence, which led Weber and Schmid (1998) to speculate that the cline of introduced popu-lations resulted from selection on existing variants and new mutations. Similarly, shepherd's purse (*Capsella bursa-pastoris*) is composed of a range of genetically determined ecotypes with striking differences in flowering time (Linde et al. 2001). Allozyme data indicate that multiple preadapted ecotypes of shepherd's purse were introduced into California from Europe and that selection resulted in early-flowering ecotypes in the desert and late-flowering ecotypes in coastal and snowy-forest regions (Neuffer and Hurka 1999). Slender wild oat (*Avena barbata*) is an introduced annual plant in California that exhibits significant genetic differentiation in allozyme pattern and morphology between cool, mesic northern California and hot, xeric southern parts of the state. When examined across a single hillside, genetic differentiation of slender wild oat was demonstrated among locations that were only 5–50 m apart. Genotype and allele frequencies characteristic of mesic regions were common in the mesic sections of the hillside bottom, while genotype and allele frequencies characteristic of drier southern California were found along the xeric hilltop (Hamrick and Holden 1979).

Selection in Barnyardgrass. Selection by environmental gradients was reported for barnyardgrass (*Echinochloa crus-galli*) that recently invaded the colder climate of Quebec, Canada, from more southern regions of North America (Roy et al. 2000). Barnyardgrass has the C_4 system of carbon fixation during photosyn-thesis that confines it, like many other C_4 plant species, to warm geographical regions. Hakam and Simon (2000) found that the Quebec population of barnyardgrass evolved an enhanced catalytic efficiency of the glutathione reductase enzyme, which serves an antioxidant function. This species also responds to the selection pressures of agriculture, resulting in crop mimics in rice. For example, rice-selected variants in California differ in flowering time, seed dispersal, and requirements for seed germination from other agricultural biotypes of barnyardgrass (see crop mimics, below). Furthermore, when barnyardgrass is under strong selection from intensive hand weeding, plants change morphologically into forms that resemble rice. Plants can also evolve herbicide resistance when selection is imposed by chemical weed control (Barrett and Seaman 1980).

Selection in St. Johnswort. Maron et al. (2004) compared the size, fecundity, and leaf area of St. Johnswort (*Hypericum perforatum*) from native European and introduced western and central North American populations in common gardens in Washington, California, Spain, and Sweden. They also determined genetic relationships among the plants by examining variation in amplified fragment length polymorphism (AFLP) markers. There is substantial genetic variation among introduced populations and evidence for multiple introductions of St. Johnswort into North America. Across common gardens, introduced plants were neither universally larger nor more fecund than natives. However, both introduced and native populations in common gardens exhibited significant latitudinal clines in size and fecundity. Clines among introduced populations broadly converged with those of native populations. Introduced and native plants originating from northern latitudes generally outperformed those originating from southern latitudes when grown in northern latitude gardens of Washington and Sweden. Conversely, plants from southern latitudes performed best in southern gardens in Spain and California. Clinal patterns in leaf area, however, did not change among gardens. The authors observed that the introduced plants did not always occur at similar latitudes as their most closely related native progenitor; thus, they believe that preadaptation (climate matching) was not the only explanation for clinal patterns among introduced populations. Instead, they suggest that introduced plants are evolving adaptations to broad-scale environmental conditions in their introduced range.

Responses to Resident Plant Species

Interspecific interactions in newly colonized habitats pose many challenges for invasive plants (Chapter 6). Some species evolve plastic growth responses to accommodate such unpredictable conditions. For example, velvetleaf (*Abutilon theophrasti*) was introduced to the United States from southeast Asia before 1700, but it became an aggressive weed in cultivated fields of the midwestern United States only during the last century. Velvetleaf evolves different life history strategies depending on the nature of competition in the fields it occupies. According to Weinig (2000), velvetleaf populations evolve plastic growth in response to light quality when grown with soybean, which enables plants to outgrow the crop. In contrast, growth plasticity is of little advantage when competing with corn, which led to phenotypes that avoid corn competition by delaying growth relative to that of the crop.

The evolution of phenotypic plasticity in competitive environments often involves concurrent responses of several characters to environmental variation. When the pattern of genetic variation differs among traits, genetic expression depends on the environment (Donohue et al. 2000, Grant-Down and Dickinson 2005). For example, genetic variation in plastic responses of plants to neighborhood density involves variation in response to the ratio of red to far-red wavelengths (R : FR) of radiation and decreased light availability in dense canopies (Smith 1982, Ballaré et al. 1990). Gene flow among related species (introgression) can also be responsible for evolutionary adaptations that enable weeds to adapt to agricultural environments. When chemical control is used, for example, genetic exchange among weeds and genetically engineered crops has resulted in herbicide resistance in several weed species (Spencer and Snow 2001).

Release from Pests, Predation, and Herbivores

The selective regime acting on weeds and invasive plants may be increased for adapted genotypes or relaxed because of the absence of coevolved natural enemies or plant competitors in the newly colonized area (Blossey and Nötzold 1995). For example, in an enemy-free space such as some agricultural fields that are maintained by insecticides, herbicides, or tillage, plant resources previously used for herbivore or competition defenses by a weed can be reallocated to growth and reproduction. The change in plant-level resource allocation also impacts the weed population by increasing reproduction and, thus, the amount of the area occupied as more propagules are dispersed. The absence of pests and pathogens, as well as the increase in plant density and invaded area, can also impact phenotypic variation through epistatic or epigenetic mechanisms, adding genetic variation for traits that impart competitiveness and dispersal ability (Sauer 1988).

BREEDING SYSTEMS OF WEEDS AND INVASIVE PLANTS

Plants are remarkably variable in their breeding systems, which is well represented among weeds and invasive plants (Prinzing et al. 2002, Dekker 2003). Breeding systems include agamospermy, obligate self-pollination, mixed mating systems that provide a combination of selfing (*autogamy*) and outcrossing (*allogamy*), obligate outcrossing, and complete dioecy. Sometimes genetic traits vary facultatively, that is, the breeding system changes depending on resource availability or environmental conditions, and therefore show epigenetic variation (Grant-Down and Dickinson 2005). Genetic information contained in genome segments can be transmitted biparentally through sexual reproduction, while other information is passed along strictly maternally or paternally. Chromosome numbers in weeds and invasive plants range from a few to several hundred, and restrictions for genetic recombination in plants are mild. Thus, everything from a very conservative copying of parental genotypes to complete genetic reshuffling every generation is possible in weeds and invasive plants (Linhart and Grant 1996).

Two fundamental types of breeding systems are predominant for weeds and invasive plants: (1) sexual reproduction that increases genetic variation, involving autogamy, allogamy, and hybridization, and (2) asexual reproduction that restricts variation (apomixis), including vegetative propagation and agamospermy.

Sexual Reproduction

Flowering plants, depending on their evolutionary histories, have a wide range of mechanisms to assure sexual reproduction regardless of their life cycle (Figure 1.2 and Chapter 5). Pollination and the ability to produce offspring from a single individual are crucial for short-lived annual plants living in hazardous environments. Probably for this reason, self-pollination is a widespread phenomenon among annual plants. Stebbins (1970) indicates that the transition from predominantly outcrossing to predominantly selfing is more common in angiosperms than any other evolutionary change.

Self-Pollination versus Outcrossing. Breeding systems are an integral part of plant life history and evolve from a complex of traits that are subject to selection and are highly variable within taxa. They are often considered to be simply another suite of characters subject to selection (Brown 1990) or change through epigenetic inheritance. A clear relationship exists between disturbance and the transition from outcrossing to selfing (Symonides 1988). According to Cruden (1977), the evolutionary shift from open pollination (*xenogamy*) to obligate autogamy and *cleistogamy* (selfing within a flower that does not open) is associated with a significant decrease in the mean pollen–ovule ratio, which is low in disturbed ecosystems and high at late successional stages. Most authors conclude that the evolution of self-pollination resulted from strong selection pressure to ensure seed production under the conditions that make outcrossing difficult (Solbrig and Rollins 1977, Gouyon et al. 1983). Competition for pollinators, for example, may have led to self-pollination in morning glory (*Ipomoea hederacea*) and oneflower stitchwort (*Arenaria uniflora*) (Stucky 1984, Wyatt 1984).

The scale of gene flow and population divergence in weeds and invasive plants is strongly affected by their breeding systems. For example, small plants with inconspicuous flowers often have more limited gene flow and tend to be associated with self-pollination, in contrast to species that have large, showy flowers or are self-incompatible. Also, there are certain plant families that are characterized by predominantly outbreeding systems (e.g., many taxa in the Asteraceae, Pinaceae, Primulaceae, and Rubiaceae). Plants in these families frequently exhibit less differentiation among populations than taxa with high levels of self-pollination. Genetic variation within populations of "selfers" is generally low but is relatively high among outcrossed populations (Hamrick et al. 1992). In addition, species with pollen or seed capable of long-distance dispersal often evolve population features that produce landscape-scale homogeneity (Linhart and Grant 1996).

Richards (2000) experimentally tested the importance of inbreeding in white campion (*Silene alba*), a colonizing weed. Reduced fitness of isolated populations resulted from sibling crosses and outcrosses within patches, but elevated fitness occurred with between-patch crosses. In addition, patches of full siblings were more likely to gain pollen from distant sources than were patches composed of unrelated plants. Thus, pollen-mediated gene flow was genetically important in "rescuing" inbred populations and also responded to the genetic structure of local populations.

The problems of cross-pollination and inbreeding depression in flowering plants are sometimes solved by genome duplication (*autopolyploidy*). Gene duplication creates a buffering effect, reducing inbreeding depression (Figure 4.3), which supports the observation that polyploids are more successful weeds than diploids (Crawley 1987, Baker 1995). There is strong association between polyploidy and apomixis, as well as self-fertilization in annual or monocarpic plants. In general, however, as compared to the total flora of a region, the most successful weeds are mainly annuals that have a relatively lower proportion of polyploidy than do herbaceous perennials (Stebbins 1970).

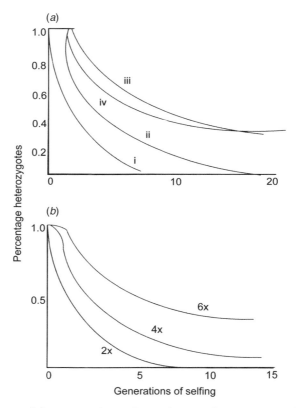

Figure 4.3 Loss of heterozygosity under various mating systems and ploidy levels. (*a*) Loss of heterozygosity in selfing diploids with (i) self-fertilization, (ii) sib mating, (iii) double first-crossing mating, and (iv) circular half-sib mating. The ordinate is the heterozygosity relative to the starting population; the abscissa is the time in generations. (*b*) Loss of heterozygosity in selfing diploids, tetraploids, and hexaploids. Note by comparing (*a*) and (*b*) that selfing in tetraploids is equivalent to sib-mating in diploids. (From A. J. Gray et al. 1986. Do invading species have definable genetic characteristics? *Philos. Trans. R. Soc. Lond.* 314:655–674. Copyright 1986, The Royal Society.)

Founder Effects. The breeding system carried by a founder population should play at least some role in the success of an invasive species. The founder population, which contains only individual traits of introduced plants, determines the total genetic diversity present in the species in the new environment. It is likely that a founder population of a self-fertilizing plant species, even with multiple introductions, carries less total genetic variation than an outcrossing species, a probability that has been confirmed by studies of self-pollinating cocklebur (*Xanthium strumarium*) (Moran and Marshall 1978) and the outbreeder Paterson's curse (Brown and Burdon 1983). Both plant species are introductions to Australia.

Exceptions to Baker's Rule. Weediness has been linked to uniparental reproduction (autogamy or agamospermy) with occasional genetic recombination (Baker's rule). However, the ecological consequences of this system of reproduction in annuals should not be summarized into a simple pattern but be cautiously interpreted within the ecological context of each species. There are examples of reductions in seed production and quality for self-fertilized relative to cross-fertilized plants (Loyd 1965) as well as examples where individuals of self-fertilized populations have higher reproductive effort than those of cross-fertilized populations (Wyatt 1984). Evolutionists are now exploring whether inbreeding in wild populations of outcrossing species depresses fitness (Keller and Waller 2002). The absence of pollinators or unrelated individuals can be especially critical for outbreeding species because of inbreeding depression. Yellow starthistle (*Centaurea solstitialis*), for example, is an aggressive outcrossing species whose expansion rate is determined by the availability of honeybees, the only pollinator that visits its flowers (Barthell et al. 2001).

Asexual Reproduction

Many agricultural weeds and invasive plants have the ability to reproduce asexually through vegetative means. This characteristic extends to almost all taxonomic classes of plants, especially herbaceous perennials. Plants are metameric organisms, producing single, repeating reproductive units of genetically homogeneous clonal growth (*ramets*). *Genets* are individuals that are genetically distinct (Harper 1977). Vegetative reproduction, which does not involve flowers, occurs through stolons and runners, rhizomes, tubers, bulbs, corms, roots, stems, or fragments of these organs (Abrahamson 1980). There are some taxa that reproduce mostly as clones and thus are genetically uniform. However, even in those species some sexual reproduction may occur, which provides important genetic variation (Hughes and Richards 1988).

As discussed in Chapter 2, plant invasion is a multistep process. A minimum number of founders must be introduced and survive both chronic and stochastic features of the environment in the new range (Mack et al. 2000). In addition, natural selection likely operates by eliminating unfit phenotypes and shaping the genetic composition and structure of each founder's descendants (Brown and Marshall 1981). Under this scenario, species with little or no genetic variation would be least likely to persist unless by chance the founders or their descendants produce phenotypes that are preadapted to the new range.

Exotic plants with asexual reproduction (vegetative propagation or agamospermy) are expected to possess extremely low genetic variation (Williams 1975, Janzen 1977) and have been characterized as evolutionary "dead ends" (Ellstrand and Roose 1987, Bayer 1990). Similar to sexual inbreeders, the likelihood of their persistence in a new range would seem further diminished when only a few individuals from a local area immigrate, because the population's risk from stochastic forces is increased (Mack 1995) while the species' total genetic variation is minimized (Barrett and Husband 1990). High levels of local differentiation in the native range would further minimize any representation of the species' genetic variation among such immigrants (Brown and Marshall 1981, Novak and Mack 1993).

The above characterization may not, however, be fully justified. First, investigations with enzyme electrophoresis reveal that plant species with predominantly asexual reproduction vary widely in magnitude and distribution of diversity within and among populations, similar to the wide variation among sexually reproducing taxa (Ellstrand and Roose 1987, Hamrick and Godt 1990). Thus, lack of genetic variation may not pose a limit to invasion by clonal plants. Some adaptive features among clonally reproducing species enhance the likelihood of establishment, including the potential for population establishment from a few individuals (Penfound and Earle 1948), high reproductive output (Dean et al. 1986), maintenance of genetic variation from generation to generation, fixation of heterozygous genotypes, and the maintenance of adaptive genotypes through gametic-phase disequilibrium (co-occurrence rather than independence of alleles) (Barrett and Richardson 1986). In addition, genetic variation among clonal immigrants could increase with multiple introductions, particularly if they are drawn from different parts of the species' native range (Novak and Mack 1993). Such variation could enhance the likelihood of persistence if it were followed by seed dispersal or gene flow among the founders' descendants (Novak et al. 1993).

Clonal species could also respond to founder effects, genetic drift, and selection in a manner similar to the response among sexual species (Parker and Hamrick 1992). These features, along with possession of at least some genetic variation, may explain the paradox whereby most invasive perennial plants possess clonal reproduction (Barrett and Shore 1989). Several authors indicate that the facultative ability of many weed species to reproduce through asexual and sexual means provides enormous advantages over either mode alone (Ellstrand and Roose 1987, Gill et al. 1995).

Advantages of Asexual Reproduction in Weeds

Species with the capacity to reproduce vegetatively have proven to be more fit or to persist in particular cropping systems because clonal reproduction:

- Allows successful genotypes to be replicated repeatedly, perhaps without limit

- Produces individual plants that can spread and acquire resources scattered over large areas (Grant et al. 1992, Mitton and Grant 1996), with the concomitant result that small-scale spatial heterogeneity can be either exploited or minimized
- Enhances the viability of the entire genotype with demonstrable advantages to the entire clone through transfer of water and nutrients from a microsite where they are abundant to an area of clonal growth where they are limited (Stuefer et al. 1994, Wijesinghe and Handel 1994)
- Produces physiologically independent individuals through fragmentation, which then can successfully exploit their own particular microenvironment
- Spans environments temporally, occasionally covering periods of time estimated to be up to 17 millennia (Vasek 1980) or possibly even a million years long (Grant 1993)

INFLUENCE OF HUMANS ON WEED AND INVASIVE PLANT EVOLUTION

Weeds and invasive plants often share particular evolutionary life history strategies, which can result from adaptive responses to the selection pressure imposed by humans. These attributes of weeds and invasive plants are considered in the following sections.

Weeds and Invasive Plants as Strategists

As already discussed in Chapters 1 and 2, stress and disturbance are environmental factors that can shape the patterns of resource allocation in plants and thus selection of life history strategies. Many weeds and invasive plants possess characteristics common to both competitors and ruderals (Table 2.6) and exist predominately in two combined strategies, competitive ruderals and stress-tolerant competitors (Figure 1.5) (Grime 1979). It is also tempting to consider the pattern that Grime (1979) calls C–S–R strategists as appropriate for weeds (Figure 2.7). However, he notes that environments that would select this strategy are usually subject to pronounced seasonal and temporal variation in disturbance, stress, and competition. For example, in temperate zones in certain unfertilized pastures that experience constant grazing pressure, the vegetation is usually composed of species mixtures that develop characteristics intermediate to competitors (C), ruderals (R), and stress tolerators (S). Weeds are highly specialized and successful organisms of productive environments. Thus, it seems unlikely that many weed species would adapt in such a broad manner as the C–S–R category suggests.

Competitive Ruderals. Plant species with the adaptations of competitive ruderals are found on productive sites where dominance by true competitors is diminished by some level of disturbance. Only occasional disturbance is expected in this case, however, since very frequent or severe disturbance would favor strictly ruderal vegetation. Conditions favorable for competitive ruderals might result when damage to vegetation occurs once or twice annually or during the life cycle but does not affect or eliminate all the individuals from the plant community. Examples of this type of habitat include fertile meadows and grasslands subject to seasonal damage (e.g., grazing), floodplains, eroded areas, and margins of lakes and ditches. Arable land also is included in this habitat type.

Plants with the competitive ruderal evolutionary strategy are expected to have rapid early growth rates, and the onset of competition between individuals should occur before the initiation of flowering. Annual dicot herbs that Grime places in this category, such as common ragweed (*Ambrosia artemisiifolia*), Pennsylvania smartweed (*Polygonum pennsylvanicum*), and velvetleaf, are characterized by a relatively long vegetative phase. Grasses, such as Italian ryegrass and others, are also capable of rapid and large dry-matter production. Optimization of resource capture and seed production are also important criteria for competitive ruderal species. Many crops, such as barley, rye, corn, and sunflower, are annuals with rapid early growth rates and the capacity to produce a high leaf area index. Grime believes these species also should be classified as competitive ruderals.

The 18 annual weed species considered by Holm et al. (1977, 1997) to be of greatest worldwide importance (Table 1.3) are all found primarily on productive arable land. Holm et al. (1977, 1997) also describe times from germination to maturity, vegetative growth (shoot length and biomass), and reproductive output for these and other troublesome weed species. Although the data are of diffuse origin, the 18 weed species listed in Table 1.3 generally are characterized by high plasticity in vegetative growth, rapid early growth rates, and a prolonged vegetative phase prior to and during reproduction. Holm et al. repeatedly emphasize the ability of these plants to compete against crop species and to allocate a large proportion of resources to seed production. Many of the annual species listed by Holm et al. as the world's worst weeds appear to fit the category of competitive ruderals.

Although many weed species classified as competitive ruderals are annuals, others are not. Notable exceptions are herbaceous perennial species like Canada thistle (*Cirsium arvense*), quackgrass (*Agropyron repens*), coltsfoot (*Tussilago farfara*), and Johnsongrass (*Sorghum halepense*), which are common weeds of agricultural fields and rangelands. Such species tend to be strongly rhizomatous or stoloniferous with a high capacity for vegetative growth. Many also maintain high seed production. They are often noted for their competitive behavior but can be displaced as seedlings by more competitive annual species, especially under a frequently tilled or disturbed regime. However, tillage can promote proliferation of growth from vegetative fragments once herbaceous perennials become established. Thus, the establishment and spread of these species is also enhanced by occasional disturbance.

Most herbaceous weeds common to arable land have adapted to the combined strategy of competitive ruderals. As ruderals, these species require the soil disturbance associated with agriculture for establishment and growth. Since it usually is not possible to maintain a completely disturbed environment and still grow a crop, plants of this strategy evolved competitive characteristics as well.

Perhaps arable weeds were selected initially as ruderals in their native habitats and, with the advent of agriculture, a relatively recent event on the evolutionary scale of time, were also selected for characteristics that allowed success in more competitive or less disturbed environments.

Stress-Tolerant Competitors. In order to understand the role of stress-tolerant competitors as weeds, the process of succession, particularly in productive environments, must be revisited. As seen in Chapter 2, this process involves initial colonization followed by progressive microenvironmental modification and replacement of the vegetation over time. The role of stress-tolerant competitors, primarily shrubs and trees (Figure 1.4), is particularly evident within the time frames common to forest succession. In this case, shrubs often dominate an early to intermediate seral stage for a prolonged period of time.

The shrubs and trees that appear early in forest succession on productive habitats are similar in many ways to competitive-ruderal herbs. The common characteristics include rapid dry-matter production in comparison to other shrubs and trees, rapid stem extension and leaf production through most of the growing season, and rapid phenotypic responses of leaf or shoot morphology to shade. Grime (1979) indicates that such features are particularly conspicuous among deciduous trees (*Ailanthus*, *Betula*, *Populus*) and shrubs that occur in the early phases of natural reforestation in disturbed woodlots of eastern North America. Species that assume a similar role in the forested areas of northwestern North America occur in the genera *Arctostaphylos*, *Ceanothus*, *Rubus*, *Alnus*, and *Acer*. These species are often deterrents to forest regeneration efforts in western United States.

Most of the woody species that initially colonize an area following disturbance by fire or logging do so from long-term seed reserves in the soil. Initial seedling growth is usually slow, but once the plant is established, extensive and rapid vegetative growth results. In addition, many species, including greenleaf manzanita (*Arctostaphylos patula*), bigleaf maple (*Acer macrophyllum*), and tanbark oak (*Lithocarpus densiflora*) in western North America, sprout readily from root crowns or stumps if another disturbance occurs to remove top growth. In that event, total canopy coverage can approach the predisturbance levels in only a few years. Maximum photosynthate production and usually vegetative growth are coincidental with periods of low-moisture stress. These species are notably shade intolerant, however, and the ultimate dominance of more shade tolerant trees is assured.

It appears that many shrub and tree species that are considered weeds on disturbed forest lands follow the combined strategy of stress-tolerant competitors. These species usually dominate the vegetation of early and intermediate seral communities following a major disturbance to the forest. They are long-lived, often existing for decades, and tend to allocate a significant amount of their resources to stems and branches for canopy (foliage) support. However, as competitors these species also possess rapid early rates of vegetative growth, especially if sprouting occurs following top removal. Because of the competitive ability of these plants, succession may proceed beyond the intermediate stages only as more stress tolerant species gradually outlive them.

Adaptations of Weeds and Invasive Plants to Human Activities

Weeds, Domesticates, and Wild Plants. De Wet and Harlan (1975) describe three classes of vegetation based on the degree of association with human-caused disturbance (also see Radosevich and Holt 1984). De Wet and Harlan believe that weeds have evolved in response to human-caused disturbances in three principal ways: (1) from wild colonizers through adaptation to and selection for continuous habitat disturbances, (2) as derivatives of hybridization between wild and cultivated races of domesticated species (crops), and (3) from abandoned domesticates by selection toward a less intimate association with humans. A review of the genetic relationships and gene flow among crops and their wild relatives, with numerous examples of interactions among plants in various crop–weed–wild plant complexes, can be found in Ellstrand (2003).

De Wet and Harlan (1975) propose that most weeds have evolved directly from wild species that invaded human-disturbed habitats. As evidence, many weed species, for example, dandelion, henbit (*Lamium amplexicaule*), and crabgrass (*Digitaria sanguinalis*), have been distributed far beyond their native range and also have wild races in their native Old World habitats. In addition, most domesticated plant species also have weed and wild races, supporting the second avenue of weed evolution (Ellstrand 2003). De Wet and Harlan (1975) indicate that hybrids between wild and cultivated forms rarely invade successfully the natural habitat of the species but are common in disturbed habitats associated with cultivation. Ellstrand (2003) reviews numerous examples of hybridization between a crop and its wild relative followed by the evolution of inceased weediness in the progeny, such as in weed beets (*Beta* sp.) in Europe (Boudry et al. 1993), weedy rye (*Secale cereale*) in California (Sun and Corke 1992, Burger et al. 2006), and wild radish (*Raphanus* sp.) in California (Panetsos and Baker 1968, Hegde et al. 2006). Domesticated races also can revert to a weedy growth habit when they are no longer cultivated. However, eventual replacement by other races or species less dependent upon cultivation than crops is likely. In cases of reversion to a weedy form from a domesticate, seed dispersal mechanisms of the weed race are often similar to those found in wild types. Cultivated forms of cereals, for example, are characterized by spikelets that persist on the inflorescence at maturity, whereas wild-type cereals disperse seed by means of an abscission layer that forms between the rachis and the spikelet. Weed races of cereals (e.g., wild oat) are similar to the wild type regarding the method of seed dissemination.

Crop Mimics. The selective forces created by agricultural practices often result in the evolution of agricultural races or *agroecotypes* of weeds (Barrett 1983). Some of these agroecotypes are intimately associated with a specific crop. The more closely a weed ecotype resembles the crop, generally the more difficult it is to control without crop damage. Barrett (1988) indicates that although mimicry has been studied extensively for over a century in animals and insects, there are few studies of this type of coevolutionary phenomenon in weeds. Rather, most accounts of mimicry in weeds that occur in the literature exist as reports that document the form of resemblance between a crop and a weed. Weins (1978) suggests that the following terms are central to the concept of mimicry:

- *Model*—the animate or inanimate object or function that is being imitated
- *Mimic*—the imitating organism
- *Operator*—the organism that is unable to discriminate effectively between the model and the mimic

In the case of weed mimicry, the operators or selective agents are usually mechanical devices or chemicals designed and operated by humans.

There are many examples of weeds and crops that grow in conjunction with one another and resemble each other in form or function. For example, barnyardgrass occurs wherever rice is grown and is a weed of major importance in that crop. Wild oat (*Avena fatua*) is considered to be nearly cosmopolitan wherever cereal crops occur. In Europe, large-seeded false flax (*Camelina sativa*) evolved several ecotypes that are closely associated with flax production. In such cases, these crops and associated weeds may have evolved together under similar cultural practices or environments so that growth and reproduction of the weed match the life cycle of the crop.

Mimics in Flax. One of the finest examples of crop mimicry involves weeds associated with flax. Baker (1974) suggests that is because flax is one of the oldest plants grown as a crop in Eurasia. In open-grown situations the weed, *C. sativa* var. *sativa* is a generalized annual plant with a wide-branching growth habit. When found with flax, however, this weed takes on a taller, less branched form that resembles a flax plant. In some areas where flax cultivation is very intensive, *C. sativa* var. *sativa* is replaced by another ecotype, *C. sativa* var. *linicola*, which is even more specialized. In the variety *linicola*, the life cycles of the weed and flax are so closely aligned that the weed is always harvested with the crop. As in flax, the fruit resists shattering at harvest and the seed so closely resemble flax seed that they cannot be separated readily by winnowing. Thus seed of both plants are often sown together. In his review of this subject, Baker (1974) indicates that *C. sativa* var. *linicola* undoubtedly evolved from var. *sativa* and that crop mimicry in this case seems to be fixed genetically. Perhaps crop mimics once possessed a more general habit of growth or development that became more croplike as selection for specialization increased.

Mimics in Rice. Another striking example of crop mimicry by a weed species is barnyardgrass in rice. Barrett (1988) indicates that although mimetic forms of barnyardgrass certainly originated under "primitive" agricultural systems, they are also present and presumably evolved under modern mechanized rice culture as well. In California, *E. crus-galli* var. *oryzicola* has replaced *E. crus-galli* var. *crus-galli* as the major weed in rice fields. According to Barrett and Seaman (1980) two distinct races of *E. crus-galli* var. *oryzicola* were introduced into

TABLE 4.4 Barnyardgrasses in California Rice Fields

Taxon[a]	Ploidy	Rice Weed Status	Distribution in California	Microhabitat in Rice Agroecosystem
E. crus-galli (L.) Beauv. var. *crus-galli* Barrett 1201, TRT	6x	Worldwide	Widespread	Rice field edges, levees, and shallow-water areas in fields
E. crus-galli (L.) Beauv. var. *oryzicola* (Vasing) Ohwi: (a) early flowering form (*E. oryzoides* [Ard.] Fritsch) Barrett 1202, TRT	6x	Europe, Asia, Australia, Argentina	Generally restricted to central valley	Rice fields
E. crus-galli (L.) Beauv. var. *oryzicola* (Vasing) Ohwi; (b) late flowering form (*E. phyllopogon* [Stapf.] Koss) Barrett 1203, TRT	4x	Europe, southeast Asia, India, Russia, China, Japan	Restricted to rice-growing regions of central valley	Rice fields
E. muricata (P. Beauv.) Fern. (*E. pungens*) Barrett 1204, TRT	4x	U.S. and Australia	Scattered localities in central and northern California	Rice field edges, levees, and drainage ditches
E. colona (L.) Link Barrett 1294, TRT	6x	Worldwide (mostly in upland rice)	Scattered localities mostly in southern California	Rice field edges and levees

Source: Modified from Barrett (1983), after Barrett and Seaman (1980).
[a]Frequently used synonyms are given in parentheses.

California as rice seed contaminates in 1912–1915 when rice culture was just beginning in the state. These two races now behave as distinct biological species and differ in chromosome number, morphology, flowering, and distribution (Table 4.4). Despite the occurrence of populations containing millions of individuals of *E. crus-galli* var. *oryzicola* in California rice fields each year, as well as their continued spread to uninfested rice-producing areas, few populations are found outside the rice agroecosystem (Figure 4.4). Barrett indicates that this restricted habitat preference is typical of crop mimicry and that the behavior and spread of *E. crus-galli* var. *oryzicola* contrasts significantly with those of

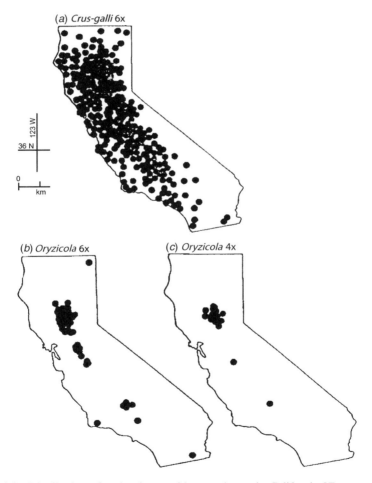

Figure 4.4 Distribution of major forms of barnyardgrass in California [*E. crus-galli* var. *crus-galli*, *E. crus-galli* var. *oryzicola* (4×, 6× races)]. (Modified from Barrett 1983. *Econ. Bot.* 37:255–282.)

E. crus-galli var. *crus-galli*. Although these two varieties are both pernicious weeds of worldwide importance, they exhibit very different adaptive strategies. *Echinochloa crus-galli* var. *crus-galli* exhibits many of the traits of Baker's general-purpose genotype (Table 1.2), whereas the rice mimic var. *oryzicola* is a specialized biotic ecotype with limited ecological amplitude (Barrett 1988). The analysis of the contrasting life histories of these two barnyardgrass varieties would be helpful in understanding the selective forces operating in production systems that cause shifts in the spectrum of weed species over time.

Shifts in Plant Species Composition. Plant communities often are perceived as being stable over short time scales (Williamson 1991) with the composition of species appearing relatively constant from season to season and year to year. Evolutionary change is inevitable in all plant communities, however, and shifts in species composition over time from human disturbance should always be expected. This topics is addressed in Chapter 7.

SUMMARY

Weedy and invasive plant species are characterized by a set of genetic, physiological, and morphological characteristics and life history traits, some of which are especially important for their success. These characteristics result from evolutionary processes and are explained by the basic concepts of adaptation and selection. The study of evolutionary genetics reveals the characteristics of successful plant invasions for several reasons:

- Plant invasions are rapid evolutionary events resulting in populations that are genetically dynamic over space and time.
- Genetic characteristics of weed and invasive plant populations impact the capacity of these species to expand their geographical range.
- Natural selection and genetic drift alter the genetic structure of invading plant populations in ways that modify their tolerance or behavior in the new environment.
- Invading populations can induce evolutionary changes in native species of plants and animals.

Although the evolutionary genetics of many weeds and invasive plants are still poorly understood, there are studies that describe the different ways that plant structure and function evolve as a consequence of environmental change. Examples of such evolution are known for seed characters, leaf traits, phenology, physiological and biochemical activities, heavy metal tolerance, herbicide resistance, parasite resistance, competitive ability, organel characters, breeding systems, and life history. Among the forces that have shaped these patterns of differentiation are environmental resources (radiation, soil moisture, fertility), environmental conditions (temperature), biotic factors (pollination vectors, parasitism, gene flow, population dynamics), and human-related activities such as toxic soils, mowing, and grazing.

Adaptive change induced by selection plays a central role in plant speciation and in molding traits of weeds and invasive plants. Adaptability of a species may be more important than its tolerance or plasticity to environmental change. There are many examples of differentiation among plant populations that occur across short spatial distances and over relatively short time periods. The importance of evolutionary adaptation is well known for agricultural weeds and invasive plants, making a strong case for natural and human-induced selection.

Genetic differentiation is produced through natural selection in response to both biotic and abiotic heterogeneity. Genetic change occurs over relatively short periods of time, and frequently a relatively small number of genes are involved in adaptive differentiation. Additive genetic variation can be gained through different mechanisms but is frequently maintained by gene flow among complexes of related species, some of which may be weeds or invasive plants. It is apparent that when plant populations go through evolutionary bottlenecks, such as those occurring in founder populations or those generated by strong selective forces (e.g., weed control activities to eradicate exotic species or reduce the area of source populations), genetic responses occur to increase adaptability of individuals.

Weeds and invasive plants do not form a homogeneous group but rather display a wide range of evolutionary pathways. In the wider context of invading species, searching for the shared attributes of successful invasive plants is unlikely to provide the information necessary to predict potential invasive plants or to explain why invasions occur. Research in this respect should be focused on understanding the genetic changes that invasive plants undergo after they are introduced and begin to colonize a new area. It is important to study and describe the genetic consequences of plant invasions, especially exploring the notion that biological invasions enable species to break their genetic molds to escape internal and external constraints on their genetic systems. The key to managing invasive biological systems, including agroecosystems, relies on being able to understand how, when, and at what rate species adapt and evolve to new or changing environments.

5

WEED DEMOGRAPHY AND POPULATION DYNAMICS

Population ecology is the branch of ecology that deals with the impact of the environment on a population, a group of individuals of the same species occurring in the same geographical area. In the case of agricultural weeds, populations are often selected by the tools or tactics designed to suppress them (Harper 1956). Populations are dynamic and changing; demography is the study of the numerical changes in a population through time. Demographic analysis can be used to evaluate reasons for changes in population size and structure over time. It can also provide insight about the range expansion (invasiveness) of recently introduced species across a region or landscape. The processes and patterns of plant invasion are scale dependent and vary from the level of the individual to metapopulations (Table 2.4), which can also be a subject of demography. In addition, the consequences of various management practices on weeds and invasive plants can be determined through demographic analysis.

PRINCIPLES OF PLANT DEMOGRAPHY

The life cycle is the fundamental descriptive unit of an organism. Specifically, an individual plant is the result of a series of processes that started at fertilization and continued through embryonic growth, seed germination, seedling establishment, development into adulthood, and finally senescence and death (Figure 2.11). Populations of individuals have an age structure, which is the relative numbers of

Ecology of Weeds and Invasive Plants. By Steven R. Radosevich, Jodie S. Holt, and Claudio M. Ghersa
Copyright © 2007 John Wiley & Sons, Inc.

individuals of different ages or in different stages of the life cycle, as well as size structure and genetic structure (Silvertown and Charlesworth 2001). As individual plants age, the various stages occur at a measurable rate in the population. These *vital rates*, which form the basis of plant demography, describe how the plants in a population pass through their life cycle (Figure 1.2). The response of the vital rates of a population to the environment determines population dynamics in ecological time and evolution of life histories over evolutionary time (Solbrig 1980).

By focusing on the vital rates of the population, demography addresses both the structure of plant populations and the dynamics of populations, or how they change over time. The potential for exponential growth of groups of individuals is fundamental to population dynamics (Figure 2.7). As described in Chapter 2, populations do not continue unrestricted growth indefinitely (Figures 2.7 and 2.8), because eventually resource limitation and biotic interactions curtail population growth until equilibrium between birth rate and death rate is attained.

Natality, Mortality, Immigration, and Emigration

Plants differ substantially among species in life-form and timing of developmental stages, but certain basic population processes are common to all (Chapter 2). A large number of plants inhabiting a field, for example, do not remain static over time. If the number of plants per unit area (*density*) increases, there has been an influx of individuals from somewhere else, new plants have been created (born), or both events have occurred. These are two of the most basic processes that affect population size: *immigration* and *birth*. Alternatively, if the number of plants in the hypothetical field declines, some of them must have died and either not been replaced or their seed left the area. These two processes that reduce plant numbers are *death* and *emigration*. Seed dispersal is an example of how plants emigrate. All four of these basic processes can occur simultaneously in a population (see below). If a population declines, then death and emigration together outweighed birth and immigration, and the opposite is true if the population increases:

Immigration (movement in)

↑

Natality ← Plant density → Mortality

(birth)   ↓   (death)

Emigration (migration out)

In Chapter 2, birth, death, immigration, and emigration were combined in a simple algebraic equation to describe the change in numbers of a population between two points in time.

$$N_{t+1} = N_t + B - D + I - E \tag{2.6}$$

where N is number of individuals, B is births, D is deaths, I is immigration, and E is emigration. When the population is so large that absolute numbers cannot be used, then the equation is constructed in terms of density, so N_t becomes, for example, the number of plants per square meter at time t. Understanding demography requires measurement of the four basic processes in Equation 2.6 and accounting for their values. Unfortunately, this task is rarely simple or straightforward because every plant passes through a series of stages in its life cycle (seed, seedling, adult plant, and so on) and each stage must be identified individually. Furthermore, environmental factors influence the rate at which a plant passes through these stages such that age and growth stage are not necessarily tightly linked. Therefore, Equation 2.6 represents a general, ideal model upon which more realistic descriptions are built.

The complexities of weed and invasive plant populations are best described using diagrammatic life-table models. This demographic approach to study weed population dynamics was first introduced by Sagar and Mortimer (1976) and has been used widely for the past 30 years. Reviews are found in Cussans (1987), Cousens and Mortimer (1995), Begon et al. 1996, Caswell (2001), Gotelli (2001), and Silvertown and Charlesworth (2001).

Life Tables

In order to conduct a demographic analysis of a population, the life cycle of the species is divided into fractions or components. For example, a plant's life cycle could be divided into an active fraction (growing plant) and a passive one (dormant seed and vegetative plant parts such as stolons, bulbs, and rhizomes). The life cycle might also be divided into the sporophyte (vascular plants as we see them) and the gametophyte (a phase that is reduced in vascular plants with seed and includes reproductive structures in the flower). The number of fractions included in a study and the level of detail necessary largely depend on the purpose of the project. The aims of some projects may only require simple and general descriptions in which a few components are considered with similar level of detail. In cases where deeper understanding is required, more fractions would be included and those determined to be critical would be studied in great detail. An approach followed by many plant demographers is to start with a simple general model to find critical components to be studied later in greater detail.

A simple diagrammatic life table of an idealized higher plant is shown in Figure 5.1. In this diagram, each stage in the life cycle is represented by a *node* (in rectangular boxes) and transitions between them are represented by rates or probabilities (in triangles or diamonds) along lines between nodes. The number of individuals (N) at the start of each developmental stage (seed, seedlings, adults) is given inside the rectangular boxes. The N_{t+1} adults alive at time $t+1$ (i.e., the next generation) come from two sources: (1) survivors of the N_t adults alive at time t and (2) those coming from birth, which in Figure 5.1 is a multistage process involving seed production, germination, and the survival and growth of seedlings. In the first source, the probability of survival (the proportion of them

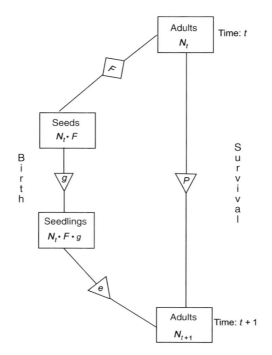

Figure 5.1 Diagrammatic life table for idealized vascular plant; F = number of seed per plant; g = chance of seed germinating $(0 \leq g \leq 1)$; e = chance of seedling establishing itself as an adult $(0 \leq e \leq 1)$; p = chance of adult surviving $(0 \leq p \leq 1)$. (From Begon and Mortimer 1986, *Population Ecology: A Unified Study of Animals and Plants*. Copyright 1986 Blackwell Publishing Ltd, reproduced with permission.)

that survive) is placed inside the triangle and noted by p in Figure 5.1. For instance, if $N_t = 100$ plants and p, the survival rate, is 0.9, then there are 100×0.9, or 90, survivors contributing to N_{t+1} at time $t + 1$ (10 individuals have died; the mortality rate, $1 - p$, between t and $t + 1$ is 0.1).

For the other source of plants in Figure 5.1, birth, the average number of seed produced per adult (the average *fecundity* of the plant population) is noted by F and placed in a diamond. The total number of seed produced is, therefore, $N_t \times F$. The proportion of these seed that germinate is denoted by g. Multiplying $N_t \times F$ by g gives the number of seed that germinated successfully. The final step of the process is the establishment of seedlings as independently photosynthesizing adults. The probability of surviving this stage of plant development is denoted by e in Figure 5.1 and the total number of births is, therefore, $N_t \times F \times g \times e$. The number of the population at time $t + 1$ is the sum of this calculation and $N_t \times p$.

It is now possible to substitute the terms of the life table (Figure 5.1) into Equation 2.6, giving a basic equation for population growth of this hypothetical species:

survival

$$N_{t+1} = \overbrace{N_t - \underbrace{N_t(1-p)}_{\text{death}}}^{} + \underbrace{N_t \times F \times g \times e}_{\text{birth}} \tag{5.1}$$

In this example, death was calculated as the product of N_t and the mortality rate $1 - p$, because survival and mortality are opposite processes whose sum is 1. However, immigration and emigration were ignored; thus this description of how a plant population may change over time is incomplete.

Modular Growth

A major distinction between species of the plant and animal kingdoms is how they grow and develop, which provides pattern to the organization and differentiation of tissues and organs. Most animals are unitary organisms that grow in a diffuse manner, while plants are modular organisms (Harper 1977, 1981, Gotelli 2001). In both animals and plants, development from the zygote to the adult involves an irreversible process of growth and differentiation, but in animals this process leads to very predictable organ development and growth form. In contrast, growth and differentiation in plants are initiated in *meristems* at the apices of shoots and roots and at nodes in the axils of leaves (Esau 1965). Cell division occurs in these meristems, which results in root and shoot elongation and the creation of more meristems. Thus, growth from meristems leads to a repetitive but unpredictable modular structure in the plant body. Botanically, a module is an axis with an apical meristem at its distal end.

Four demographic consequences arise from the modular construction of vascular plants. First, the addition of modules generates a colony of repeating units arranged in a branched structural form. The exact architecture of the plant depends on whether modules vary in form, their rate of production, and their position relative to one another. The way the modules are structured influences the size and shape of the plant and, therefore, interactions among individuals, which has demographic implications (Horn 1971, Gotelli 2001). Second, modules are relatively autonomous; therefore, herbivory or other physical damage may harm the plant but rarely kill it. The presence of relatively autonomous meristems also allows reiteration of many parts of the individual. In most plants removal of vegetative branches often leads to production of new ones, but in unitary organisms, although tissue regeneration does occur, removal of a whole organ can cause death. The third consequence of modularity is the opportunity for natural cloning of plants (Harper 1984), which is only possible when the meristems at the nodes retain the ability to produce new shoots and roots. Fragmentation of an individual into independent clones may arise through physical agents, such as tillage, trampling, and grazing by herbivores, or it may be determined genetically. Cloning is an important characteristic for the persistence and dispersal of many perennial weeds (Holzner and Numata 1982, Radosevich et al. 1997) and invasive plants. Finally, modular growth can make the age of a plant difficult to assess and in fact relatively meaningless, since some plants can remain alive for many years through vegetative reproduction, although the original parent tissue may be long gone. Similarly, plant size may be a poor indicator of a plant's age.

The fundamental equation of population biology (Equation 2.6) applies not only to *genets* (the plant as a whole or, more specifically, a plant arising from a seed) but also to *ramets*, which are modular, clonal units of the plant with the potential for independent growth. Demographic approaches for modular organisms use the same techniques as for populations of unitary organisms (Harper and Bell 1979, Jackson et al. 1985, Ebert 1999).

Models of Plant Population Dynamics

Models of how a population behaves are needed to understand how the fundamental demographic processes of birth, death, immigration, and emigration influence the stability or change in population size. We have already seen the simplest form of a demographic model (Equation 2.6), which Silvertown and Charlesworth (2001) suggest may approach being an algebraic truism. However, much more complex equations for population change arise from attempts to account for the way in which birth, death, and migration rates alter population density and age structure along with the effects of competitors, predators, pathogens, and mutualists (Silvertown and Charlesworth 2001, Caswell 2001). Although detailed mathematical treatment is not possible in this text, some of the basic modeling approaches are described below.

Models Based on Difference Equations. Two life-cycle models are needed to examine the population dynamics of vascular plants: (1) a model for species in which vegetative reproduction does not occur (Figure 5.2), such as most annual and biennial plants, and (2) a model for species in which vegetative reproduction does occur (Figure 5.3), as in perennial plants.

When Vegetative Reproduction Does Not Occur. Two routes are shown in Figure 5.2 by which a population of individuals of size A_1 in generation G_t may move to the succeeding generation G_{t+1}. The population A_2 in generation G_{t+1} may come from sexual reproduction or from survivors of the previous generation. Route I (Figure 5.2) represents sexual reproduction and plant establishment from seed, which is divided into six intermediate phases. In Figure 5.2 B is the total number of viable seed produced by the population A_1; C is the total number of viable seed falling onto the soil surface, to which are added any seed arriving by invasion (G); and D_1 is the total number of viable seed that are present in the surface seed bank. This seed bank may lose individuals to the buried seed bank (D_2), which includes the "carryover" seed from previous generations. (There also may be inputs to the surface seed bank from new introductions of invading seed

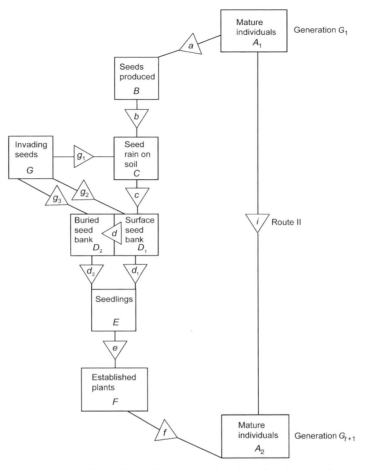

Figure 5.2 Generalized life table for higher vascular plant that does not have ramet production. Symbols are described in the text. (From Sagar and Mortimer 1976, *Ann. Appl. Biol.* 1:1–47. Copyright 1976 Blackwell Publishing Ltd, reproduced with permission.)

or to the buried seed bank from sown crop seed contaminated with weed seed.) The symbol E is the number of seedlings germinated from either seed bank and F is the number of plants established.

The model in Figure 5.2 also recognizes seven interphases (a through g), which represent the probability or number of individuals proceeding to the next phase. Thus, interphase a represents sexual reproduction by seed, interphase b represents loss that occurs between seed production and arrival of seed on the soil surface, interphase c is the fate of seed on the ground, interphase d is the probability of a seed in the seed bank germinating and giving rise to a seedling, interphase e represents the seedling fate after emergence, and interphase f is the fate after seedling establishment. Interphase g is the level of seed invasion from plants that occur

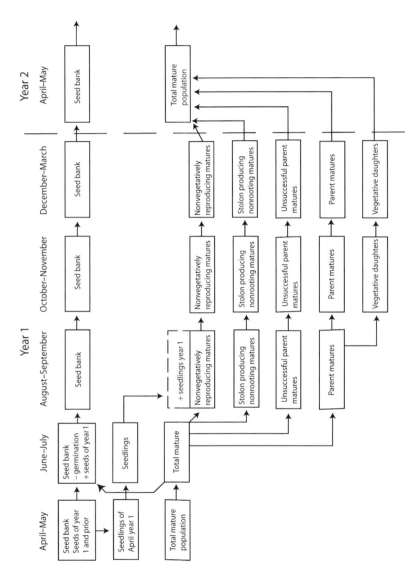

Figure 5.3 Transitions occurring throughout year in buttercup population (*Ranunculus repens*) as envisioned by Sarukhan and Gadgil (1974). [From Sarukhan and Gadgil (1974), *J. Ecol.* 62:921–936. Copyright 1974 Blackwell Publishing Ltd, reproduced with permission.]

off-site. The invasion interphase g is further subdivided to distinguish contributions to the seed rain, g_1, to the surface seed bank, g_2, and to the buried seed bank, g_3.

Route II (Figure 5.2) is found in all species except *ephemerals*, those that live for only a short time, and annuals. This route indicates an interphase probability for the fraction of the population A_1 that survives to generation G_{t+1}. For a biennial species the interphase (i) may theoretically carry a value of 0.5, since half the plants in the population A_1 would flower and die and half would remain vegetative and survive as population A_2. However, Figure 5.2 requires modification for biennial species because of the overlap of generations and is inappropriate for species that have mixed populations of genets and ramets.

When Vegetative Reproduction Occurs. Sarukhan and Gadgil (1974) used transitions depicted in Figure 5.3 to describe the population dynamics of creeping buttercup, *Ranunculus repens*, in Great Britain. This species reproduces sexually by seed and asexually by vegetative reproduction, though recruitment by these means occurs at different times during the year. Seed germinate in late spring and early summer, while new clonal progeny become established in late summer as separate plants from shoots borne at nodes along creeping stolons. In essence, this complex flow diagram (Figure 5.3) is an age–stage classification in which the fluxes from one stage to another are also defined chronologically. This approach makes an additional distinction from those in Figures 5.1 and 5.2 in that asexually produced progeny (ramets) are classified separately from sexually produced seedlings (genets), at least during the first year of life. This demographic approach has been used to describe the dynamics of many other perennial weeds and invasive plants.

Age-Specific Models. Mortality and fecundity are usually age specific. In order to solve complex problems due to varying plant ages and stages of development in real populations, it is necessary to increase the complexity of the general model (e.g., Figure 5.1) into a diagrammatic life table such as that shown in Figure 5.4 (Begon and Mortimer 1986, Begon et al. 1996). Here the population is divided into four age groups: a_0, a_1, a_2, and a_3; a_0 represents the youngest adults and a_3 the oldest, B represents age-specific fecundity, and p represents age-specific survivorship. In a single time step, t_1 to t_2, individuals from group a_0, a_1, and a_2 pass to the next respective age group; each age group contributes new individuals to a_0 (through birth); and the individuals in a_3 die. This model clearly assumes that the population consists of discrete age groups and has discrete age-specific survivorship and birth statistics, in contrast to the reality of a continuously aging population.

It is possible to now write a series of algebraic equations to express the changes that might occur in Figure 5.4:

$$t_2 a_0 = t_1 a_0 \times B_0 + t_1 a_1 \times B_1 + t_1 a_2 \times B_2 + t_1 a_3 \times B_3$$
$$t_2 a_1 = t_1 a_0 \times p_0$$
$$t_2 a_2 = t_1 a_1 \times p_1$$
$$t_2 a_3 = t_1 a_2 \times p_2$$

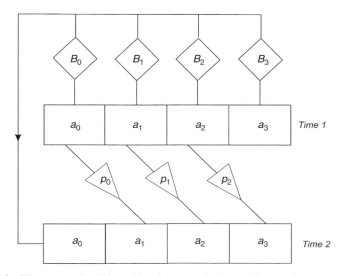

Figure 5.4 Diagrammatic life table for population with overlapping generations: a = number of different age groups; B = age-specific fecundities; p = age-specific survivorships. (From Begon and Mortimer 1986, *Population Ecology: A Unified Study of Animals and Plants.* Copyright 1986 Blackwell Publishing Ltd, reproduced with permission.)

where the numbers in the age group are subscripted t_1 or t_2 to identify the time period to which they refer. There are four equations in this model because there are four age groups, and the equations specifically state how the numbers in the age groups are determined over the time step t_1 to t_2. An example of how to use this type of life table is given in Appendix 1 in Radosevich et al. (1997).

Transition Matrices. Another way to describe the behavior of populations with overlapping generations that have individuals that fall into different age or size classes, that is, have different rates of reproduction and death depending upon age or size, is with *matrix models*. These models are generally simpler to use and more realistic than those using difference equations.

The matrix model was introduced to population biology by P. H. Leslie in 1943 to describe changes in population size due to reproduction and mortality and is often called the *Leslie matrix* (Leslie 1945). In general form, for n age groups, it is written as

$$
\begin{bmatrix}
B_0 & B_1 & B_2 & \cdots & B_{n-1} & B_n \\
p_0 & 0 & 0 & \cdots & 0 & 0 \\
0 & p_1 & 0 & \cdots & 0 & 0 \\
0 & 0 & p_2 & \cdots & 0 & 0 \\
\vdots & \vdots & \vdots & \ddots & \vdots & \vdots \\
0 & 0 & 0 & \cdots & p_{n-1} & 0
\end{bmatrix}
\times
\begin{bmatrix}
t_1 a_0 \\
t_1 a_1 \\
t_1 a_2 \\
t_1 a_3 \\
\vdots \\
t_1 a_n
\end{bmatrix}
=
\begin{bmatrix}
t_2 a_0 \\
t_2 a_1 \\
t_2 a_2 \\
t_2 a_3 \\
\vdots \\
t_2 a_n
\end{bmatrix}
$$

which may alternatively be written as $T \times t_1A = t_2A$. The matrix on the left, T, is called a *transition matrix*, which when multiplied by the vector of ages (A) at t_1 gives the age distribution at t_2. For further information, consider Searle (1966), Caswell (2001), or Gotelli (2001). A numeric example of this approach is given in Appendix 2 in Radosevich et al. (1997). A study by Maxwell et al. (1988) is also described in Chapter 2 in this volume where the matrix model approach was used to simulate population changes of leafy spurge (*Euphorbia esula*) and determine the consequences of several management tactics on patches of that species (Figure 2.9). Note that the Leslie matrix can also be based on an organism's size or growth stage, rather than age, for species in which basic processes of fecundity and mortality are not based strictly on age.

Features of Transition Matrix Models. All populations of plants will eventually reach a stable age or growth stage distribution over time. In other words, the proportion of individuals in each class stays the same or varies little over many time steps. Population growth may still be exponential, as shown in Figure 2.7, but each age or stage class has a constant birth and death rate and the total population increases at constant rate. This rate of growth is a constant, λ, and is the *dominant eigenvalue* of the population, that is, $N_t = \lambda \ N(t)$, which tells how fast a population is growing or receding. Values greater than 1 indicate that the population is expanding, while values less than 1 indicate that it is decreasing and may become extinct. Furthermore, it is expected that populations of plants will return to this equilibrium after a perturbation, such as herbicide application, as shown in Figure 2.9.

It also is possible to calculate an eigenvalue for every age or stage in the transition matrix and to arbitrarily change the numeric value of any stage. By examining how λ changes in relation to large or small modifications in life-cycle stage values, the *sensitivity* of the population at any stage in the life cycle can be determined. For example, if a small change in a stage value causes a large change in λ, the population is sensitive to factors that affect that stage. In contrast, if a large change in a stage value results in a small change in λ, the population is insensitive at that stage in the life cycle. The construction of life tables and Leslie matrix models is considered in substantial depth by Ackakaya et al. (1999) and Caswell (2001). Such models can be invaluable in determining overall expansion rates of weed and invasive plant populations and life stages of the plants that are affected most by management or environmental manipulations (Higgins et al. 1996, Endress et al. in press). It is also possible using such models to determine how the expansion of endangered plants could be affected by the presence or control of adjacent invasive plants (Thompson 2005).

Metapopulations

A *metapopulation* can be thought of as a population of interacting populations, that is, a group of several or many local populations that are linked by immigration and emigration (Hanski 1999, Hanski and Gilpin 1997, Levins 1970, Gotetti

2001). The study of metapopulations requires two important shifts in reference, population *persistence* and *scale*. In the previous discussion of population growth (Chapter 2), the differences between unlimited resources *versus* a finite carrying capacity were explored. We also examined how populations expand or decline in relation to critical points in the life cycle of the individual members. Metapopulation models, in contrast, predict persistence or continuity rather than the size of each population or the number of individuals in a population at equilibrium. In metapopulation analysis, the range of numbers representing population size varies between 0 (local extinction) and 1 (local persistence); no distinction is made among populations that are large or small or those that cycle or remain constant (Gotelli 2001).

In addition, analysis of metapopulations occurs at a much larger spatial scale than that at which individual populations are studied. Metapopulations occur at the regional or landscape scale, which may encompass many connected sites as well as multiple populations. Much as populations are studied by evaluation of fecundity and mortality of individual plants, metapopulation analysis evaluates the presence or extinction of local populations. At this scale, persistence of any particular population is not a focus; rather models describe the fraction of all population sites that are occupied, that is, the extent to which populations fill the landscape (Gotelli 2001).

Risk of Extinction. An important application of metapopulation analysis is the evaluation of persistence or extinction of populations of interest, which is often applied in conservation biology and more recently in invasion biology. At this scale, *local extinction* can occur, where a single population disappears, which contrasts with *regional extinction*, in which all the populations in a region die out. Within a metapopulation, populations are connected through gene flow (immigration via seed or pollen); thus, the risk of regional extinction is usually much less than the risk of local extinction (*extirpation*) for any species. Metapopulation analysis is commonly used to determine the risk of extinction for rare or endangered plant and animal species. In that case, the number of viable populations required for regional persistence is of concern. With invasive plants the opposite is true; the major concern is how many nascent satellite populations are necessary for continued colonization and spread of the species.

It is important to remember that most new plant introductions of even quite invasive species fail. However, once plants are established these areas serve as persistent sources for further invasion, and each new satellite population that forms as a result of long-distance dispersal (Figure 3.2) can become a new source population of that species. The ability of exotic plants to competitively suppress or exclude native plants is often cited to explain exotic species dominance (Keddy 2001, Levine et al. 2003) over a landscape. It also may be, however, that dominance by exotic plants is less due to competition than to interactions with disturbance, land use and cover changes, and reduced dispersal potential of native plants due to habitat fragmentation (Seabloom et al. 2003a,b, Didham et al. 2005, MacDougall and Turkington 2005). The local restoration of exotic-dominated plant communities is particularly problematic in this regard because most common weed control techniques (Chapters 7 and 8) offer only partial or short-term reduction in invasive plant abundance and do little to restore the landscape to an unfragmented state. These invasive "survivors" also could act as a persistent source for new population development.

Metapopulation Dynamics Applied to Invasive Species. The classical model for metapopulation dynamics (Levins 1970, Silvertown and Charlesworth 2001) can be used to express the potential for both local- and landscape-level spread and persistence of exotic species:

$$P_{t+1} = P_t + cP_tV_t - xP_t \qquad (5.2)$$

where P is the number of populated sites, t is time, c is the colonization (immigration) rate, V is the number of vacant sites (equal to $1 - P$), and x is the extinction rate. Therefore, cP_tV_t represents the number of new sites colonized and xP_t is the number of sites where populations have gone extinct or, in this case, where the exotic plant can no longer be found. At equilibrium, $P_{t+1} = P_t$ and thus, from Equation 5.2,

$$V_t = \frac{x}{c} \qquad (5.3)$$

Mathematically, for a species to spread ($P_{t+1} > P_t$), the density of vacant sites (V_t), regardless of cause, must exceed the relative rate of extinction (x/c) (Silvertown and Charlesworth 2001). Similarly, when $V_t < x/c$, the metapopulation of that species will go extinct.

The metapopulation dynamics equations 5.2 and 5.3 above are analogous to the population equations 2.2 (population growth rate in a limited environment) and 5.1 (population growth showing birth and death) and the graph showing logistic population growth (Figure 2.7). At the individual population level, there is a continuous turnover from births and deaths and equilibrium birth and death rates are equivalent. Similarly, at the metapopulation level, there is continuous turnover of individual populations through colonization and extinction (Gotelli 2001). The fraction of sites occupied (P_t) is at equilibrium when the colonization (immigration) rate precisely equals the extinction rate. While these models are useful for predicting metapopulation behavior, in practical terms it may not be possible to determine the location of a colonized site until individuals of the exotic species are already present there. A similar problem is encountered when attempting to predict a safe site for species introduction if the species is not yet present.

The colonization of new sites will depend on biological and physical conditions within a site, such as the amount of unoccupied area and availability of safe sites (with required resources and absence of competition from native plants, predators, and pathogens). The rate of colonization also depends on factors external to the site, that is, the presence of populations at other sites as sources for gene flow (immigration). If many sites are occupied, there should be many individuals migrating, so the probability of colonization would be higher than when only few sites are occupied. Similarly, the extinction rate is zero when none of the sites in the metapopulation are occupied.

DYNAMICS OF WEED AND INVASIVE PLANT SEED

Although it is sometimes convenient to consider separately the two demographic processes related to plant movement, immigration and emigration, both processes can be combined under the general term *dispersal*. Most propagules of weeds or invasive plants are produced on-site from a previous generation and remain there to serve as the primary source of a new population, allowing for the entry of a few immigrants from elsewhere. Thus, immigration is a process of propagule input to an area already inhabited by a species. However, some propagules, especially of invasive plants, always leave the site where they were produced (emigration), starting new colonies often in areas previously unoccupied by that species.

Dispersal means scattering or dissemination. Theoretically, if it is to be successful, dispersal should place a seed in a location that allows a greater likelihood of survival than its location near the parent plant. Seed disperse in *space* and through *time*. Dispersal in space involves the physical movement of seed from one place to another. Harper (1977) indicates that the amount of seed falling on a given unit of area is a function of several factors:

- Height and distance of seed source
- Concentration of seed at the source
- Dispersability of seed (appendages, seed weight, etc.)
- Activity of dispersing agents

Dispersal in time refers to the ability of seed of many species to remain in a dormant condition for some period of time. Thus, the success of a plant species is enhanced by *dormancy* if, at some point in the future, the seedling will be in a microenvironment more favorable for survival than if germination were to proceed immediately.

Seed Dispersal through Space

The opportunity for biological invasion begins with dispersal (Chapters 2 and 3), and many weed and invasive plant species possess appendages to assist in long-distance movement of their seed (Figure 5.5). Because such appendages enhance the ability to move, they markedly increase the likelihood of seed and seedling survival by removing the individual from sources of parental-associated mortality (Figure 3.2). Most theories about colonization have been developed from studies

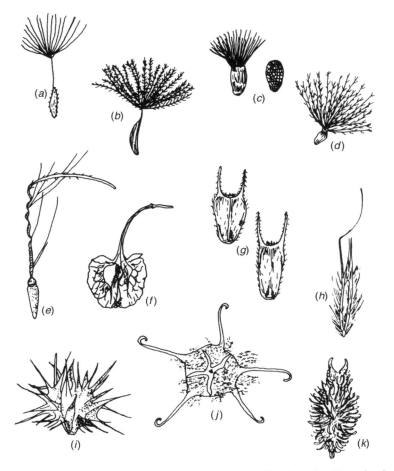

Figure 5.5 Fruits and seed of some weeds showing modifications for dissemination: (*a*) common dandelion (*Taraxacum officinale*); (*b*) meadow salsify (*Tragopogon pratensis*); (*c*) yellow starthistle (*Centaurea solstitialis*); (*d*) Canada thistle (*Cirsium arvense*); (*e*) red-stem filaree (*Erodium cicutarium*); (*f*) curly dock (*Rumex crispus*); (*g*) beggar-ticks (*Bidens frondosa*); (*h*) wild oat (*Avena fatua*); (*i*) sandbur (*Cenchrus pauciflorus*); (*j*) five-hooked bassia (*Bassia hyssopifolia*); (*k*) cocklebur (*Xanthium canadense*). (Compiled from Robbins et al. 1951.) (From Radosevich et al. 1997, *Weed Ecology: Implications for Management*, 2nd ed. Copyright 1997 with permission of John Wiley & Sons Inc.)

of natural ecosystems (Crawley 1987, Mack 1986, Rejmánek and Richardson 1996, Caswell 2001), where areas suitable for occupancy are few and often widely separated. Colonization of natural ecosystems differs from that of agroecosystems because agricultural land is often exposed to very frequent disturbances. As a result, agricultural weeds must adapt to periods of relatively high resource availability and low plant cover (Auld and Coote 1990, Martinez-Ghersa and

Ghersa 2006). Establishment of plant colonizers in range and forest production systems is also subjected to times of disturbance, although such periods are less frequent than on cropped land.

On arable land and other periodically disturbed sites, such as old fields, seed of weeds and invasive plants can disperse both horizontally and vertically. This trait reflects initial dispersal of seed onto the soil and subsequent movement in the soil profile, often with the assistance of implements used for soil tillage or harvest. Thus, most weed seed, even those with special adaptive features for long-distance dispersal, tend to migrate as an advancing front. However, plants from seed that have been widely dispersed tend to colonize as isolated individuals and after high densities are reached begin to spread as fronts (Chapters 2 and 3).

Estimates of Dispersal Distance. Dispersal information is collected in two general ways (Harper 1977). *Displacement* data describe the location of dispersed progeny (e.g., seed) relative to the maternal parent. Such data are obtained by following individual dispersers (e.g., banding and recapture of birds and rodents or mark and recapture of seed, pollen, or insects using color codes, isotopes, or genetic markers and traps). *Density* data describe the number of dispersed individuals observed along linear transects radiating from a point source. These data come from traps (e.g., screens or traps covered with sticky material or pheromones) and give a relative measure of progeny density as a function of distance and direction from the release site. For example, Dwire et al. (2006) quantified seed production and dispersal of sulfur cinquefoil (*Potentilla recta*) in different habitats of northeast Oregon in the United States. Seed dispersal was measured using sticky traps that surrounded individual source plants. Results indicated that 83% of the seed that were produced dispersed within 60 cm from the mother plant, suggesting patch expansion as a front. However, traps were only 1 m away from the dispersing plant in this study, and 60 cm is an insufficient distance to adequately explain the rapid spread of sulfur cinquefoil across that ecoregion. In a study of the effects of habitat on dispersal of the invasive plant artichoke thistle (*Cynara cardunculus*), Marushia and Holt (2006) measured dispersal in a grassland site and a nonvegetated agricultural field. Dispersal in the vegetated site (grassland) followed the expected pattern (Figure 3.2), with the majority of seed falling within 3 m and long-distance dispersal occurring as far as 20 m from the source plant. However, in the nonvegetated site, the majority of seed traveled long distance, landing up to 40 m from the source plants. This study emphasizes the importance of the plant community as well as abiotic factors, such as wind, in weed spread.

The effect of long-distance dispersal on population spread was first noted by Kot et al. (1996), who fit dispersal distances to a data set for displacement of genetically marked fruit fly, *Drosophila* spp. They concluded that the leptokurtic shape (i.e., more peaked about the mode) of the "*dispersal kernel*," especially the length of the tail, determines the probability of long-distance dispersal (Figure 5.6), which is crucial for estimating the rate of spread of invasive plants (Lewis et al. 2005). However, predictions of spread rate differ by an order of

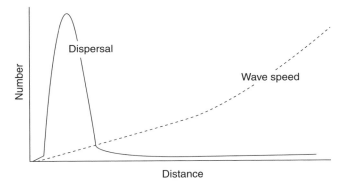

Figure 5.6 Hypothetical relationship of seed numbers to dispersal distance. Note the lep-tokurtic shape of the dispersal kernel, especially the extended tail of the curve that depicts dispersal of few individuals but over a long distance. It is believed that the shape of the dispersal tail is crucial to estimate the spread rate, or wave speed, of invasive plants. (Adapted from Neubert and Caswell 2000. *Ecology* 81:1613–1628.)

magnitude depending on the shape of the tails of related dispersal kernels (Neubert and Caswell 2000). With a few exceptions (Marushia and Holt 2006), the most common sources of information about weed and invasive plant dispersal are anecdotal observations or descriptions of short-distance patterns, which may not represent the maximum dispersal distance of a species. Although some examples exist in the literature where long-distance seed dispersal patterns were measured (Salisbury 1942a, 1961, Sheldon and Burrows 1973, Werner 1975), few are sufficient to plot a dispersal kernel as suggested by Neubert and Caswell (2000). Since mean dispersal distance is an indicator of potential weed spread, it is sometimes used to estimate long-distance dispersal. However, long-distance dispersal should receive greater attention.

Expansion Rate of Invasive Species. Most weeds and invasive plants disperse seed during a narrow window of time or stage of development each year (Harper 1977, Ackakaya et al. 1999). Numerous studies also demonstrate that fecundity and the resulting seed rain and thus soil seed banks of these plants can be massive, suggesting that demographic parameters are important in predicting population spread. Neubert and Caswell (2000) and Caswell et al. (2003) indicate that an invasion is determined by both population growth and spatial dispersal. However, sensitivity analyses in their studies reveal that demography, although important, was less important than dispersal distance in determining overall geo-graphic expansion rate. For example, demography accounted for only 30% of the difference in expansion rate between birds (starling and pied flycatcher) (Caswell et al. 2003). Neubert and Caswell (2000), studying teasel (*Dipsacus sylvestris*), also observed that both demography and dispersal functions were necessary to determine geographic expansion rate, but the dispersal distribution explained far more variation in the rate of geographic expansion than did demography (Caswell et al. 2003).

Radosevich and Wells (unpublished) performed preliminary tests to examine how fluctuations in demographic and dispersal parameters affected the geographic expansion rates of three invasive plants. They found that expansion rate varied by only 25–40%, even when the most sensitive demographic parameter (survival) varied by up to three orders of magnitude in their model. Cannas et al. (2003) developed a simulation model to examine population dynamics and spatial spread of the invasive tree honeylocust (*Gleditsia triacanthos*) in montane regions of Argentina. Their model included life history traits, seed production, and mean dispersal distance of honeylocust and several codominant tree species. They found that the main factors that influenced the velocity of honeylocust spread were dispersal distance and minimum reproductive age. Competition between honeylocust and native trees also contributed to its expansion rate.

Agents of Spatial Seed Dispersal. Wind, water, animals, and humans are the usual agents by which seed are dispersed spatially.

Wind. Seed dispersed by wind can have several distinct forms. They can be dusts (such as orchid seed or fungal spores), winged, or plumed. Seed may be adapted for gliding, such as seed of most conifers, or for rotating, such as seed of maple. The plumed seed characteristic of many species in the family Asteraceae are particularly suited for wind dispersal (Figures 5.5*a–d*), for example, common dandelion (*Taraxacum officinale*), meadow salsify (*Tragopogon pratensis*), yellow starthistle (*Centaurea solstitialis*), and Canada thistle (*Cirsium arvense*). Seed of the plumed type are relatively heavy compared to propagules that occur as dusts or spores and ordinarily would not float in air. However, these species have a specialized featherlike structure, the *pappus*, attached to the seed coat, which allows dispersal by wind. Harper (1977) indicates that in Asteraceae the influence of a pappus on dispersal velocity of a seed is best correlated with the ratio of pappus diameter to achene diameter rather that with the ratio of achene weight to pappus weight. Plants that produce achenes with a high ratio of pappus-to-achene diameter have slower terminal velocities, stay in air longer, and therefore, disperse farther than species with low ratios. Species such as Canada thistle that are well adapted for dispersal in wind have both a small seed and large pappus.

For most systems of wind dispersal, increasing the height of release brings an immediate reward in enhanced dispersal (Harper 1977). For example, the flower stalk of common dandelion exhibits very plastic growth and elongates significantly, especially after flowering. This is interpreted as an effective way to augment the role of the pappus in achene dispersal. The interesting thing here is not only that greater height increases dispersal distance but also that many weeds and invasive plants apparently have evolved mechanisms to place their seed structures higher in the air.

Another effective type of wind dispersal is the rolling action of tumbleweeds, such as Russian thistle (*Salsola iberica*). Stallings et al. (1995) studied the seed-scattering ability and distribution patterns of that species in winter wheat stubble fields of eastern Washington state in the United States. The total amount of seed produced averaged 60,000 seed per plant, and presumably most were dispersed by tumbleweed action rather than by separation from the parent plant. Some plants moved over 4000 m during the six-week study period, always in the direction of the prevailing wind. When compared to stationary plants, wind-blown plants dispersed up to 50% more seed. Other weed species also exhibit this behavior and carry the common name "tumbleweed," for example, tumble pigweed (*Amaranthus albus*) and tumble mustard (*Sisymbrium altissimum*).

Water. Many kinds of weed seed, even those without special modifications, are readily dispersed by water. Irrigation is an important factor in the spread of weeds throughout many agricultural areas of the western United States. Eddington and Robbins (1920) performed the earliest studies of seed dispersal in irrigation water. They found a total of 81 different species of agricultural weeds in 156 catches of irrigation ditches in Colorado. Large numbers of seed can be dispersed in this way.

Weed seed differ in their ability to float in water, although this depends somewhat on the water conditions and manner in which the seed alight upon it (Robbins et al. 1942). Eddington and Robbins (1920) found by dropping 100 seed of 57 different species into water that most species float very well. There are also various adaptations of fruit and seed that aid water dissemination. For example, the fruit of curly dock (*Rumex crispus*) shown in Figure 5.5*f* and arrowhead (*Sagittaria* spp.) have corky "wings" that make them buoyant. Usually weed seed screens are placed in irrigation ditches or water sources to reduce this form of weed seed dissemination.

Animals. Well known are the various forms of hooks and barbs that occur on the outer covering of many fruit and seed of weed species (Figures 5.5*e* and *g–k*). Such appendages are particularly well developed in families such as Asteraceae, Boraginaceae, and some Poaceae (King 1966). Many of these "armed" seed and fruit attach to the fur of animals and are thus dispersed over a potentially large area. Some small seed such as those of crabgrass (*Digitaria sanguinalis*), St. Johnswort (*Hypericum perforatum*), bermudagrass (*Cynodon dactylon*), and sulfur cinquefoil simply lodge temporarily in hair of pasturing animals. Crafts (1975) notes that distribution of medusahead (*Taeniatherum asperum*) and St. Johnswort throughout much of the grazing areas of the Sacramento Valley and Sierra Nevada mountains of California followed cattle and sheep trails.

A less obvious but equally well known method of seed dispersal by animals is in incompletely digested remains of fruit that has passed through the digestive tract. If an animal eats and digests seed, a loss of dispersal results, but if it eats the fruit and passes the seed in feces, a possible gain in dispersal occurs. Harper (1977) and King (1966) present numerous examples of seed dispersal by birds, rodents, and large ruminants. In addition to ingesting seed, the animal simply may move the seed passively from one area to another or collect and store the seed. In this case, dense seedling stands may emerge if a seed cache is buried.

Humans. The role of humans in dispersal of weed seed is especially well developed in agricultural situations. Dispersal of seed is a trait selected against by crop breeders because only the portion of a seed crop that has not fallen to the ground can be harvested. Figure 5.7 is an electron micrograph of two varieties of rice. The caryopsis on the right dislodges and disperses readily, whereas the one on the left can be removed only by physically stripping the panicles. The rice depicted on the left in Figure 5.7 is a commercial rice variety, whereas the one shown on the right was never developed as a crop because of its seed-shattering characteristics. Thus, morphological adaptations that allow seed shatter and dispersal may help ensure success of weed species but are undesirable traits in crops where seed characteristics aiding collection are more desirable.

Many weed species that grow in close association with certain agricultural crops, such as mimics discussed in Chapter 4, have some proportion of seed that shatter and fall. In this way, the site continues to be occupied by succeeding generations. However, some seed also remain with the parent plant and are harvested with the crop. This combination of dispersal mechanisms tends to assure and maintain the crop–weed association since the weed seed is usually replanted with the crop.

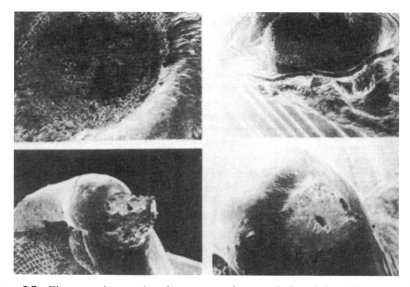

Figure 5.7 Electron micrographs of caryopses of two varieties of rice. The caryopsis on the left is a noncommercial variety of rice in which the caryopsis sits on a "ball-and-socket" arrangement and is easily shattered by harvesting. With the commercial variety of rice on the right, the caryopsis and the ball and socket are fused, an arrangement that must be broken during harvest and makes seed shatter difficult. (Micrographs by F. D. Hess and D. E. Bayer, University of California, Davis.) (From Radosevich et al. 1997, *Weed Ecology: Implications for Management*, 2nd ed. Copyright 1997 with permission of John Wiley & Sons Inc.)

Barnyardgrass (*Echinochloa crus-galli*) is an agricultural weed known to consist of many different varieties that differ in panicle morphology and phenology (see section in Chapter 4 on mimics and Table 4.4). Three distinct varieties have been identified in the rice-growing region of California (Barrett and Seaman 1980). These are *E. crus-galli* var. *crus-galli, Echinochloa crus-galli* var. *oryzicola,* and *E. crus-galli* var. *phyllopogon. Echinochloa crus-galli* var. *crus-galli* and var. *oryzicola* mature early and the caryopses usually have shattered from the panicle at the time of rice harvest. *Echinochloa crus-galli* var. *phyllopogon* apparently germinates later or has a longer life cycle than the other two varieties of barnyardgrass, so that it rarely has shattered completely by the time rice is harvested. Dispersal of this variety, therefore, occurs as the panicles are stripped by the harvester. Because the life cycles of the three varieties overlap to some degree, some seed of each variety occur with the harvested crop. However, most of the seed of *E. crus-galli* var. *crus-galli* and var. *oryzicola* remain in the field, whereas the largest proportion of *E. crus-galli* var. *phyllopogon* remain with the crop seed. It is also interesting that some cultural practices in rice production, such as high water levels and certain herbicides, favor the occurrence of *E. crus-galli* var. *phyllopogon* over other barnyardgrass varieties.

Often elaborate attempts are made to break the weed seed–crop association by various "seed-cleaning" techniques. These methods often take advantage of differential seed coat morphologies between crop and weed. Examples are using weed seed screens to remove barnyardgrass from rice and using various winnowing procedures in cereals. Dodder (*Cuscuta* spp.) seed is removed from alfalfa by mixing iron filings with the seed and exposing them to electromagnets. The filings attach to the rough, reticulate seed coat of the dodder and are thus separated from the smooth alfalfa seed with the magnet. Harper (1977) indicates that corncockle (*Agrostemma githago*), which produces large tuberculate seed, can be removed readily from grain by screening and is no longer considered to be a serious problem in European grain production for that reason.

Another way in which humans have attempted to control the unwanted dispersal of weed and invasive plant seed is through governmental regulation (Chapter 1). Most states and countries maintain seed laws that specify maximum percentages of weed seed contamination allowed in an agricultural crop used for commerce. However, even very little seed dispersed and planted with a crop potentially can infest previously unoccupied fields. For example, as little as 0.25% dodder infestation in alfalfa seed could result in as many as 40,000 dodder seedlings per acre after sowing.

The relationships of human and animal dispersal to invasive plant propagule pressure and, therefore, to invasibility of sites in natural ecosystems were discussed in Chapter 3.

Seed Banks

All viable seed present on and in the soil constitute the soil seed bank. Harper (1977) visualized the soil as a bank or reservoir of seed in which both deposits

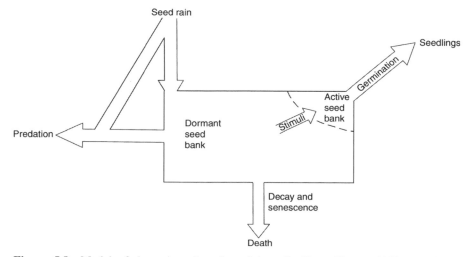

Figure 5.8 Model of dynamics of seed pool in soil. (From Harper 1977, *Population Biology of Plants*. Copyright with permission from Elsevier.)

and withdrawals are made (Figure 5.8). Deposits occur by seed rain from seed production and dispersal, whereas withdrawals occur by germination, senescence and death, and predation. Storage results from the vertical distribution of seed through the soil profile, with most weed seed occurring at shallow depths. Soil seed banks have become a recognized and indispensable part of plant population ecology such that substantial amounts of information are now available about seed bank processes (e.g., Allessio Leck et al. 1989, Baskin and Baskin 1998, Benech Arnold et al. 2000). Moreover, weed scientists and ecologists recognize that information about the dynamics of seed banks allows improved weed management strategies (Altieri and Liebman 1988, Ghersa et al. 1994c, Cavers et al. 1995, Baskin and Baskin 1998, Wiles and Schweizer 2002).

Entry of Seed into Soil. Seed enter soil from several sources but the most common is from plants that mature on an already occupied site. Many assessments of weed seed production are available (Stevens 1954, 1957, Holm et al. 1977, Popay and Ivens 1982, Cavers et al. 1995, Silvertown and Charlesworth 2001, Holmes 2002). In general, the amount of seed produced in various habitats by weeds is astonishingly high (Table 5.1) but can also vary markedly due to plasticity in response to environment. Thus, the actual amount of seed produced per individual plant can vary from nearly nothing to millions, depending on its growing conditions. The importance of high seed production for annual pioneer and exotic invasive plant species was discussed in Chapters 1, 3, and 4.

Holzner et al. (1982) indicate that, generally, agricultural weeds are species that produce large amounts of small seed. However, their seed production is far exceeded by species that colonize natural ecosystems for only a short time

TABLE 5.1 Numbers of Seed and Predominant Species Present in Seed Pools of Various Vegetation Types

Vegetation Type	Location	Seed (m^{-2})	Predominant Species in Soil	Source[a]
Tilled Agricultural Soils				
Arable fields	England	28,700–34,100	Weeds	Brenchley and Warington (1933)
	Canada	5000–23,000	Weeds	Budd, Chepil and Doughty (1954)
	Minnesota, U.S.	1000–40,000	Weeds	Robinson (1949)
	Honduras	7620	Weeds	Kellman (1974b)
Grassland, Heath, and Marsh				
Freshwater marsh	New Jersey, U.S.	6405–32,000	Annuals and perennials representative of the surface vegetation	Leck and Graveline (1979)
Salt marsh	Wales	31–566	Sea rush where abundant in vegetation, grasses	Milton (1939)
Calluna heath	Wales	17,500	*Calluna vulgaris*	Chippendale and Milton (1934)
Perennial hay meadow	Wales	38,000	Dicotyledons	Chippendale and Milton (1934)
Meadow steppe (perennial)	USSR	18,875–19,625	Subsidiary species of vegetation	Golubeva (1962)
Perennial pasture	England	2000–17,000	Annuals and species of vegetation	Champness and Morris (1948)
Prairie grassland	Kansas, U.S.	300–800	Subsidiary species of vegetation, many annuals	Lippert and Hopkins (1950)
Zoysia grassland	Japan	1980	*Zoysia japonica*	Hayashi and Numata (1971)
Miscanthus grassland	Japan	18,780	*Miscanthus sinensis*	Hayashi and Numata (1971)

(*Continued*)

TABLE 5.1 *Continued*

Vegetation Type	Location	Seed (m^{-2})	Predominant Species in Soil	Source[a]
Annual grassland	California, U.S.A.	9000– 54,000	Annual grasses	Major and Pyatt (1966)
Pasture in cleared forest	Venezuela	1250	Grasses and dicot weeds	Uhl and Clark (1983)
		Forests		
Picea abies (100 yr old)	USSR	1200–5000	All earlier successional species	Karpov (1960)
Secondary forest	North Carolina, U.S.	1200– 13,200	Arable weeds and species of early succession	Oosting and Humphries (1940)
Primary subalpine conifer forest	Colorado, U.S.	3–53	Herbs	Whipple (1978)
Subarctic pine/ birch forest	Canada	0	No viable seeds present	Johnson (1975)
Coniferous forest	Canada	1000	Alder (*Alnus rubra*)	Kellman (1970)
Primary conifer forest	Canada	206	Shrubs and herbs	Kellman (1974a)
Primary tropical forest	Thailand	40–182	Pioneer trees and shrubs	Cheke et al. (1979)
	Venezuela	180–200	Pioneer trees and shrubs	Uhl and Clark (1983)
	Costa Rica	742	Pioneer trees and shrubs	Putz (1983)

[a] All references are cited in the original source, Silvertown and Lovett Doust (1993).

Source: Silvertown and Lovett Doust (1993) and Silvertown and Charlesworth (2001), *Introduction to Plant Population Biology* 3rd and 4th eds. Copyright 1993 and 2001 Blackwell Publishing Ltd. (reproduced with permission).

(*ephemerals*), such as *Epilobium angustifolium, Verbascum* spp., or *Solidago* spp. in forest clearings. If weeds/invasive plants are allowed to mature, a substantial amount of seed will continue to enter the soil and become a source for future site occupation. For instance, Wilson (1988) cites an example where weed seed densities in the soil increased in a single year by 370% for grasses and 126% for broadleaved weeds when only cultivation (tillage) was used to control weeds, and presumably the surviving plants matured and dispersed their propagules.

Longevity of Seed in Soil. Weeds vary considerably with respect to the longevity of their seed (Table 5.2), depending upon species, depth of seed burial, soil type, and level of disturbance. Many weed species are noted for the especially long-lived nature of their seed. Information in this area has been collected from two

TABLE 5.2 Percent Viability (%) of Grasses, Legumes, and Weeds after Storage at Three Depths in Mineral Soil[a]

Species	Initial	Stored 1 Year			Stored 4 Years			Stored 20 Years		
		13 cm	26 cm	39 cm	13 cm	26 cm	39 cm	13 cm	26 cm	39 cm
Gramineae (Poaceae)										
Alopecurus myosuroides Huds. blackgrass	44	31	28	5	22	10	53	0	0	0
Holcus lanatus L. velvetgrass	88	68	65	46	17	19	5	0	0	0
Bromus mollis L. soft chess	88	3	11	2	0	0	0	0	0	0
Agrostis tenuis Sibth. common bent	72	92	45	36	28	21	3	0	0	0
Festuca rubra L. red fescue	87	17	14	23	0	0	0	0	0	0
Lolium perenne L. perennial ryegrass										
Early	89	23	30	26	4	T	T	0	0	0
Late	95	86	53	34	22	10	T	0	0	0
Dactylis glomerata L. orchardgrass										
Early	86	48	27	27	0	1	0	0	0	0
Late	94	22	50	49	1	2	11	0	0	0
Phleum pratense L. timothy										
Hexaploid	97	85	96	89	81	32	27	0	0	0
Diploid	89	86	87	61	72	51	39	T	0	0

(*Continued*)

153

TABLE 5.2 *Continued*

Species	Initial	Stored 1 Year			Stored 4 Years			Stored 20 Years		
		13 cm	26 cm	39 cm	13 cm	26 cm	39 cm	13 cm	26 cm	39 cm
Festuca pratensis Huds. meadow fescue										
Early	89	9	T	1	0	0	0	0	0	0
Late	91	4	17	15	T	0	T	0	0	0
Festuca arundinacea Schreb. tall fescue	76	18	21	18	0	0	0	0	0	0
Lolium italicum A. Br. Italian ryegrass	93	29	70	42	3	1	T	0	0	0
Avena fatua L. wild oat	34	19	12	T	3	T	T	0	0	
Legumes (Fabaceae)										
Trifolium repens L. white clover	98	36	11	15	10	6	11	1	6	1
Trifolium pratense L. red clover	93	11	6	6	2	2	1	1	8	0
Late	55	30	26	27	19	7	5	T	2	T
Early	92	14	16	13	19	2	0	0	0	T
Trifolium dubium Sibth. small hop clover	90	13	20	19	9	7	8	1	1	T
Trifolium hybridum L. alsike clover	88	10	8	8	5	7	9	3	2	T

Species										
Medicago lupulina L. black medic	84	5	5	3	1	3	1	0	0	T
Medicago sativa L. alfalfa	95	2	3	3	0	0	0	T	1	T
Weeds (various families)										
Chenopodium album L. common lambsquarters	88	89	89	88	76	83	92	32	22	15
Geranium dissectum L. cutleaf geranium	99	2	15	T	T	0	0	0	0	0
Geranium molle L. clovefoot geranium	99	0	0	0	0	T	0	0	0	0
Lychnis alba Mill. white cockle	95	92	97	100	83	23	83	1	T	2
Matricaria inodora L. scentless chamomile	91	87	82	82	72	53	55	0	0	0
Plantago lanceolata L. buckhorn plantain	98	97	95	60	76	25	42	0	0	0
Polygonum persicaria L. ladysthumb	95	73	98	79	48	27	33	1	2	2
Ranunculus repens L. creeping buttercup	96	97	90	98	64	62	72	51	55	48
Rumex crispus L. curly dock	63	84	52	98	51	53	82	30	26	0

[a]T = trace, less than 1%.

Source: Adapted from Lewis (1973), in Radosevich et al. (1997). *Weed Ecology: Implications for Management,* 2nd ed. Copyright 1997 with permission of John Wiley & Sons Inc.

sources: (1) long-term burial studies and (2) seed collections from soils with a history of no disturbance. Burial studies, such as those initiated by Beal (1911) and Duvel (1903), generally agree with the later observations of Lewis (1973), who showed that seed of grass and crop species succumb early, whereas seed of legumes and weeds remain viable for a long time (Table 5.2). In other studies summarized by Cook (1980), the species composition of seed has been determined in soils that have not been disturbed for a very long time. In all cases, viable seed of weed and pioneer invasive species were found beneath vegetation of substantial age, indicating that they had not been deposited recently.

Perhaps the most interesting reports are those of the longevity of weed seed found in archeological sites. Odum (1965, 1974) reports that annual agricultural weeds predominate in ruderal soils beneath ancient human dwellings. Viable common lambsquarters (*Chenopodium album*) and corn spurry (*Spergula arvensis*) were discovered in soil associated with habitations known to be 1700 years old! Milberg (1990), however, indicates that longevity values provided for those species are unreliable. He believes that values based on archaeological dating for denseflower mullein (*Verbascum densiflorum*) and common mullein (*V. thapsus*), estimated to be 850 and 660 years, respectively, are more convincing. In any event, this information indicates a very long association of these weed species with humans and their endeavors. Table 5.3 lists a number of weed species with seed that clearly have the capacity to remain viable for long periods of time when buried in soil. Information also has been gathered about the longevity of stored

TABLE 5.3 Seed Longevities in Soil and Decay Rates of Populations

Species	Common Name	Longevity[a] (years)	Decay Rate[b] (g)
Chenopodium album	Common lambsquarters	1700	0.105
Thalpsi arvense	Field pennycress	30	0.122
Polygonum aviculare	Prostrate knotweed	400	0.156
Viola arvense	Field violet	400	0.161
Fumaria officinalis	Fumitory	600	0.195
Euphorbia helioscopia	Sun spurge	68	0.206
Poa annua	Annual bluegrass	68	0.237
Capsella bursa-pastoris	Shepherds purse	35	0.244
Stellaria media	Chickweed	600	0.252
Papaver rhoes	Corn poppy	26	0.260
Vicia hirsuta	Tinyvetch	25	0.305
Medicago lupulina	Black medic	26	0.340
Senecio vulgaris	Common groundsel	58	0.340
Spergula arvensis	Corn spurry	1700	0.340
Ranunculus bulbosus	Bulbous buttercup	51	—
Ranunculus repens	Creeping buttercup	600	—

[a]Longevities from Harrington (1972).
[b]Decay rates from Roberts and Feast (1972).

Source: Cook (1980), in Solbrig (Ed.), 1980, *Demography and Evolution in Plant Populations*. Copyright 1980. Blackwell Publishing Ltd., reproduced with permission.

seed *versus* that of seed occurring in the soil. In most cases, storage life is considerably shorter than seed longevity in the soil.

Seed longevity in the soil depends upon the interaction of many factors such as the intrinsic dormancy characteristics of the seed populations, the environmental conditions present in the soil that influence dormancy breaking (e.g., light, temperature, water, and gas environment), and biological interactions (e.g., predation and allelopathy) (Simpson 1990, Fenner 1994, Benech Arnold et al. 2000). The intensity and manner in which these factors interact depend upon seed condition and the location of seed in the soil profile. Seed condition is determined by genotype, environmental factors during plant and seed development, ripening and afterripening requirements, seed morphology (size, shape, coat color, presence or absence of hairy coats, fruit covers, coat roughness, or specific appendages such as awns, etc.), and seed polymorphism.

Density and Composition of Seed Banks. The density and composition of seed in soil vary greatly but are closely linked to the history of the land (Table 5.1). For example, grassland seed banks generally consist of seed associated with non cropped lands, while croplands contain seed of weeds from cultivated fields (Wilson 1988, Smith et al. 2002). When the pattern of seed production, distribution, and storage throughout a successional sequence is studied, the general tendency is for early species to contribute more seed to the seed bank than later ones (Table 5.1). This pattern occurs even though late successional species usually are on the site for a much longer time than are pioneers (Grandin 2001). A significant characteristic of weeds and other pioneer species is the ability to produce a large number of propagules (Cavers and Benoit 1989, Thompson et al. 1998, Silvertown and Charlesworth 2001). This strategy of high reproductive potential when combined with dormancy apparently allows the presence of a large and relatively constant soil seed reserve. In an environment where frequent disturbance is an evolutionary reality, the seed bank must act as a stabilizing factor that assures species survival.

Seed Banks in Agricultural Soils. The studies by Brenchley and Warington (1930, 1933) represent early attempts to estimate weed seed abundance in agricultural soil (Table 5.1). The site examined was on the Rothamsted Experiment Station, England, in a field that had been planted continually to wheat for nearly 90 years. The amount of weed seed found was impressive (28,000–34,000 seed/m^2) and represented 47 species. Almost two-thirds of the seed were species of poppy (*Papaver* spp.). In a subsequent study, Brenchley and Warington (1945) found that different fertility levels (manure) often were associated with different species composition in the soil. However, the overall size of the soil seed reserve remained relatively constant despite the use of various cropping systems, since most species could complete their life cycle between annual tillage operations.

Wilson (1988) indicates that the density of agricultural seed banks can range from zero in newly developed soils to between 4000 and 140,000 seed/m^2 in cropped soil (Table 5.4). He points out that seed densities are influenced by past

TABLE 5.4 Weed Seed in Agricultural Soils

Location	Crop History	Soil Depth (cm)	Average Number of Seeds Collected (seed/m²)	Number of Species	Most Frequently Encountered Species	
					Number	Percent of Total
England	Vegetables	0–15	4,100	76	9	89
Scotland	Potatoes	0–20	16,000	80	6	78
Colorado	Barley–corn–sugarbeets	0–25	137,700	8	2	86
Illinois	Corn–soybeans–corn	0–18	10,200	25	4	85
Nebraska	Corn–field–beans–sugarbeets	0–15	20,400	19	3	85
Washington	Potatoes–wheat	0–30	51,000	23	3	90

Source: Modified from Wilson (1988), in Altieri and Liebman (Eds.), *Weed Management in Agroecosystems: Ecological Approaches.* CRC Press, Boca Raton, FL. Additional references are cited in Wilson (1988).

cropping practices and often vary from field to field. Yet he also notes that seed banks in agricultural fields at different locations often contain the same weed species and share other similarities. Generally, agricultural seed banks are made up of many species but often only a few of these comprise 70–90% of the total seed bank (Table 5.4). These results are confirmed by other weed scientists (e.g., Mohler 2001) with cropland weed species such as velvetleaf (*Abutilon theophrasti*) and giant foxtail (*Setaria faberi*). This predominate set of species may be followed by a smaller subset that comprises 10–20% of the seed reserve. Wilson indicates that a final set, accounting for only a small proportion of the total seed reserve, consists of species that are remnants of past crops (Wilson 1988).

Seed Banks in Soils of Natural Ecosystems. It remains unclear what, if any, role persistent seed banks play in the expansion of exotic invasive plant species. Gratkowski and Lauterbach (1974) observed that native shrubs dominating clearcut sites in the Oregon and Washington Cascade Mountains presumably arose from a long-lasting seed bank that was formed during the previous disturbance (fire) approximately 400 years earlier. Hardseededness of the buried shrub seed was attributed to the presence of the seed bank. Perendes and Jones (2000) found that exotic herbaceous plants such as tansy ragwort (*Senecio jacobaea*) dominated recently clearcut areas in the Cascade Mountains but diminished with canopy dominance of native shrubs or trees. In a 3-year study of invasive potential, Van Clef and Stiles (2001) found no differences in seed bank persistence between exotic and native congeners of the woody perennials *Celastrus* spp. and *Parthenocissus* spp. Higher germination rates of exotic *versus* native congeners of *Polygonum* spp. were observed, although the seed banks for the six species studied lasted only 1 year. Similarly, seed of the invasive bunchgrass kangaroo grass (*Themeda triandra*) persisted less than one year in a semiarid savanna in South Africa (O'Connor 1997). However, plumed achenes of the invasive weed yellow starthistle (*C. solstitialis*) persisted for up to 10 years in rangeland soils in Idaho, although plumeless achenes lasted only 6 years (Callihan et al. 1993). It appears that seed banks for exotic species in natural ecosystem habitats are short lived, but their presence probably depends on site differences that result from disturbance or evolutionary history (Endress et al. 2007) (Chapter 3).

Categories of Seed Banks. Thompson and Grime (1979) measured seasonal variation in seed densities of 10 contrasting habitats and divided the seed in the various communities into four categories (Figure 5.9). Groups I and II consist of *transient* seed that are usually from grasses and forbs and do not persist for longer than one year. Groups III and IV identified by Thompson and Grime consist of the *persistent* seed bank, which is represented by species from a wide range of habitats that persist long enough to become buried in the soil. Similarly, Auld et al. (2000), in a two-year seed burial study in fire-prone plant communities, observed three patterns of seed bank longevity: plants with high seed dormancy, plants with imposed secondary dormancy, and plants without significant dormancy or secondary dormancy imposed (Figure 5.10). Thus, most of the 14 species in this study maintained a significant persistent seed bank for at least two years. King and Buckney (2001) indicate that a persistent seed bank probably accounts for the presence of exotic plants along borders of brushlands near urban areas.

As long as the patterns of disturbance remain unchanged, a seed bank should change little from year to year (Figures 5.9 and 5.10). In other words, if seed of weed or invasive plants are buried in the surface soil layer (0–1 cm), they may

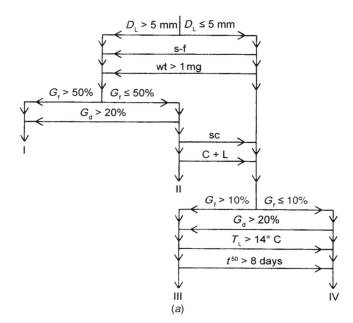

(a)

Figure 5.9 (a) Key to using laboratory characteristics of seed to predict four seed bank types developed in the field (after Grime and Hillier 1981): D_L, length of dispersule; s-f, seed not readily detached from fruit; wt, weight of seed; G_f, maximum percentage germination achieved by fresh seed; G_d, maximum percentage germination achieved by seed stored dry at 20°C for one month; SC, seed requires scarification; C + L, seed requires chilling and light to break dormancy; T_L, lowest temperature at which 50% germination is achieved; t^{50}, time for seed stored dry at 20°C for one month to reach 50% germination. (b) Scheme describing four types of seed banks of common occurrence in temperate regions (after Thompson and Grime 1979). Shaded areas: seed capable of germinating immediately after removal to suitable laboratory conditions. Unshaded area: seed viable but not capable of immediate germination. (I) Annual and perennial grasses of dry or disturbed habitats (e.g., *Hordeum murinum*, *Lolium perenne*, and *Catapodium rigidum*) capable of immediate germination. (II) Annual and perennial herbs, colonizing vegetation gaps in early spring (e.g., *Impatiens glandulifera*, *Anthriscus sylvestris*, and *Heracleum sphondylium*). (III) Annual and perennial herbs, mainly germinating in autumn but maintaining a small seed bank (e.g., *Arenaria serpyllifolia*, *Holcus lanatus*, and *Agrostis tenuis*). (IV) Annual and perennial herbs and shrubs with large persistent seed banks (e.g., *Stellaria media*, *Chenopodium rubrum*, and *Calluna vulgaris*). (From Grime 1989, in Allessio Leck et al. 1989, *Ecology of Soil Seed Banks*. Copyright with permission from Elsevier.)

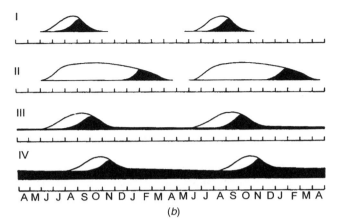

(b)

Figure 5.9 *Continued.*

appear as transient seed banks. However, if seed of the same species are buried deeper (e.g., 10–15 cm), their seed banks appear to be persistent (Chepil 1946, Ballaré et al. 1988, Van Esso and Ghersa 1989, Baskin and Baskin 1998). Thus, a seed bank should be examined in a particular ecological context using demography (e.g., species, seed number, and seed condition) and environmental conditions (e.g., soil depth) to describe it (Bekker et al. 1998).

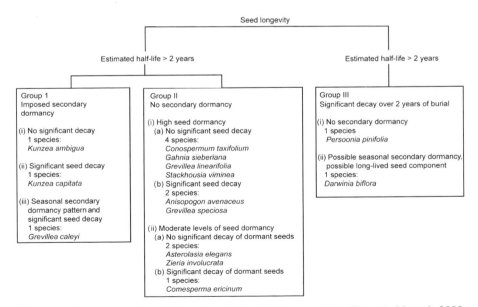

Figure 5.10 Patterns of seed decay in soil for 14 study species. (From Auld et al. 2000, *Aust. J. Bot.* 48:539–548. Reproduced with permission from the Australian Journal of Botany. CSIRO Publishing, Melbourne. http://www.publish.csiro.au/journals/ajb.)

Seed banks also have spatial and temporal dimensions (Wiles and Schweizer 2002) that are not independent of each other. The spatial dimension is given by the vertical and horizontal distribution patterns of seed in the soil while the temporal dimension refers to the life span of seed in the soil. Tillage accounts for most of the changes in the vertical distribution of seed in agricultural soils but many biotic and abiotic factors are involved in the horizontal distribution, as discussed earlier. Soil disturbance also influences the rate at which seed viability is lost (Roberts and Feast 1972, 1973, Lueschen and Anderson 1980, Froud-Williams et al. 1983, Thompson et al. 1998, Kotanen 1996).

Fate of Seed in Soil. Seed is important both for growth and maintenance of existing plant populations and for the initiation of new populations. However, the relative importance of seed recruitment, especially from seed banks, varies among species and among communities (Harper 1977, Louda 1989, Westerman et al. 2003). The dynamics of seed banks are influenced by losses as well as inputs to the soil. Seed lost from the soil seed bank are (1) eaten or digested by rodents, insects, mollusks, or microbes (*predation*), (2) killed by senescence and decay, or (3) removed by germination and emergence (Figure 5.8). Seed predation and senescence/decay are considered here while the processes of dormancy and germination are discussed later in this chapter.

Predation. Predators can influence the input and output of seed to the seed bank at virtually every stage of a plant's life cycle. Louda (1989) indicates that seed predators select seed differentially and, thus, determine the average value of key characteristics of seed that remain in the soil. For example, by finding and using clumped and large seed, predators reinforce other pressures that select for seed traits characteristic of persistent seed banks, including small seed size and hard seed coats. Additionally, fruit and seed consumers change seed distribution by eating, moving, or caching propagules and sometimes increase germinability and recruitment of seed that escape destruction.

Seed predation is believed to influence the dynamics of plant populations that are expanding. Louda (1989) suggests that there are two groups of species defined in the literature with predictable periods of expansion and significant predator impacts (Figure 5.11). First, seed predation appears to change density and relative abundance of dominant species that have annual life histories (e.g., grasses of annual grassland, some agricultural crops) or that have high dependencies on seed recruitment for population maintenance and recovery after disturbance (e.g., mangroves, some trees in temperate forests). Second, seed predation influences recruitment, occurrence, and distribution of moderately large seeded plants with fugitive (plants with changing populations seldom at equilibrium) life histories (Louda 1989). Generally, the risk of predator impact increases as the canopy matures, because a larger canopy provides greater cover.

There is little doubt that seed predation, especially of seed on the soil surface, has a marked impact on seed bank density and composition. For example, Gashwiler (1967) observed that nearly 70% of Douglas fir (*Pseudotsuga*

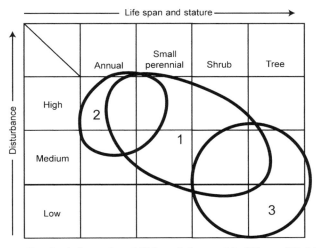

Figure 5.11 Predicted relative vulnerabilities of plants with different life histories to seed predators. Group 1 is composed of fugitive and other species whose life history traits can be considered intermediate between group 2, the ruderals and ephemerals specialized for frequent disturbances or harsh environments, and group 3, long-lived perennials adapted for competitive, stable, low-nutrient environments. Increasing populations of group 1 appear most vulnerable to contemporaneous demographic and distributional effects of seed predation. (From Louda 1989, in Allessio Leck et al. 1989, *Ecology of Soil Seed Banks.* Copyright 1989 with permission from Elsevier.)

douglasii) seed dispersed into recent clearcuts was eaten, presumably by rodents and birds. Similarly, C. M. Ghersa et al. (unpublished) observed that nearly all (99.8%) of the current year's seed rain of barnyardgrass, redroot pigweed (*Amaranthus retroflexus*), and common lambsquarters was eliminated under a canopy of alfalfa. The seed predator in this case was a small field mouse (*Peromyscus* spp.). A several-year rotation of alfalfa is used by farmers in the Pacific northwestern part of the United States as a means to "clean up" extremely weedy fields. Plant residues can also promote the occurrence of generalist weed seed feeders such as beetles, ants, crickets, and small rodents (Liebman and Mohler 2001).

Senescence and Decay. Cook (1980) states that survival of seed stored in soil can be expressed as a negative exponential distribution, whereas shelf storage is best represented by a negative cumulative normal distribution (Figure 5.12). In other words, seed in the soil initially decay faster, then decay of shelf stored seed increases so that the half-life of both types is about the same. Further decay is lower in the soil but quite rapid on the shelf (Roberts 1972b, Cook 1980, Froud-Williams et al. 1983, Ballaré et al. 1988, Van Esso and Ghersa 1989).

Unquestionably, the environmental conditions surrounding the seed during storage, either in soil or on the shelf, affect longevity. Perhaps one reason for these observed differences in longevity is that often seed in soil are maintained in

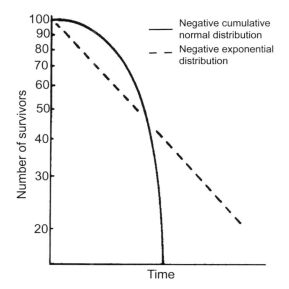

Figure 5.12 Negative cumulative normal distribution of seed viability during shelf storage and negative exponential distribution of survival of seed in soil. (From Cook 1980, in Solbrig 1980, *Demography and Evolution in Plant Populations*. Copyright 1980 Blackwell Publishing Ltd, reproduced with permission.)

a dormant, yet imbibed condition. Most seed in storage, while also dormant, are air dry. Dormant but imbibed seed are capable of many metabolic processes and are therefore able to repair damage to membranes and nuclear DNA as it occurs. It appears that the dormancy mechanisms that prevent imbibed seed from germinating play an important role in the longevity of seed of many weed species in soil. Furthermore, the mortality of weed seed in soil is probably caused by the breakdown of these dormancy mechanisms and, in seed that do not germinate, death by predation or senescence, rather than by viability loss associated with length of storage time as in dry seed.

Wilson (1988) and Thompson et al. (1998) note that longevity of seed tends to increase with depth of burial and to decline as soil disturbance increases. For example, in Duvel's (1903) experiment common lambsquarters seed germinated at a rate of 9% after burial for 39 years in undisturbed soil. In a later study, common lambsquarters seedling emergence was 9% after six years of burial in cultivated soil and 53% after six years of burial in undisturbed soil. Since environmental conditions near the soil surface are usually warmer and dryer than deeper in the soil profile, the inability of seed to remain in a fully imbibed state may account for greater seed losses at shallow depths and after soil disturbance. Similar results have been obtained in numerous other studies using many weed species (Roberts and Feast 1973, Lueschen and Anderson 1980, Froud-Williams et al. 1983, Benech Arnold et al. 2000).

Weed Occurrence in Relation to Seed Banks. For agricultural weeds and most invasive plants, the importance of seed banks is believed to lie primarily in seedling recruitment and subsequent maintenance of high plant densities. A significant characteristic of weeds and other pioneer or invasive plants is production of large numbers of propagules (Tables 5.1 and 5.4). The combined strategy of high reproductive potential and seed persistence allows the presence of a large and relatively constant seed bank. When seed inputs stop, the number of viable seed in the soil declines, usually following the negative exponential model depicted in Figure 5.12. Seed predation, especially at the soil surface, may account for substantial amounts of initial seed loss. For example, when Johnsongrass (*Sorghum halepense*) seed were buried in the field using plastic mesh bags, seed viability declined following the curve depicted in Figure 5.12 and about 20% were viable after 10 years in the soil. However, 80–99% of the seed buried in the same soil without plastic mesh bags for protection were lost during the first year (C. M. Ghersa, unpublished data). Rapid seed bank decline in the absence of inputs has also been observed in many other agricultural and rangeland sites (Chepil 1946, Sarukhan and Gadgil 1974, Scopel et al. 1988, Auld et al. 2000).

However, there is no obvious correlation between the half-life of a species' seed in the soil and the relative abundance of that species above ground (Roberts and Ricketts 1979, D'Antonio et al. 1993, King and Buckney 2001). Many references indicate that some weed seed can remain viable for many years in soil, even when the species that produced it are no longer present above ground. From these data, predictions have been made of long time periods required to completely deplete a soil seed bank of weed seeds (Holzner and Numata 1982, Aldrich 1984, Allessio Leck et al. 1989). Furthermore, after successful introduction, establishment of weeds and invasive plants seems to depend less on the presence of propagules in the soil than on specific environmental characteristics needed for germination and plant growth, because it takes only one seedling to establish a new colony regardless of the amount of seed stored in the soil (Harper 1977, Cavers et al. 1995, King and Buckney 2001).

In order to place weed seed banks into perspective, it is necessary to know from where the seed arrive and the dispersal mechanisms of the source plants. It is also necessary to know about the sources of seed loss and locations of accumulation during the production seasons. It may be that seed banks serve best for short-term dispersal and are of greatest adaptive importance during their first one or two years. For example, if a seed bank provides enough seed to allow several cohorts of weed seedlings in an agricultural cropping season, some individuals should escape the hazards of tillage or other weed control operations. The large number of seed produced by such survivors would be sufficient, due to the reproductive plasticity of most weeds (Chapter 6), to maintain the soil seed reservoir (Ghersa and Roush 1993).

It is possible that large and persistent seed banks of exotic species are an artifact of agriculture, that is, how crops are grown or past land use patterns (Naylor et al. 2005, Endress et al. 2007). Certainly past land uses are known to affect a wide range of ecological processes, which may have strong impacts on future plant abundance and composition (Foster et al. 1998, De Blois et al. 2001). Long-term seed survival is most helpful in explaining how particular plant species remain throughout an entire successional process and how genetic diversity of weeds or invasive plants is maintained in highly selective environments (Putwain and Mortimer 1995). To explain the high densities of weeds that are consistently found in many agricultural crops, adaptive strategies that allow some weed individuals to escape control tactics are important, especially those factors that influence the occurrence of waves of seedling germination. For periodically disturbed natural ecosystems, land use practices that ultimately favor dispersal, recruitment, and survival of invasive plants over time seem equally important.

Dormancy: Dispersal through Time

Dormancy is the temporary failure of viable seed to germinate under external environmental conditions that later evoke germination when the restrictive state has been terminated or released. Dormancy is effectively dispersal through time and is especially critical for annual plants, in contrast to perennials, because the seed of annuals represent the only link between generations of those species.

Descriptions of Seed Dormancy

The extent of both ecological and physiological information on seed dormancy is vast (Roberts 1972a, Taylorson and Hendricks 1977, Baskin and Baskin 1989, 1998, Foley 2001) and sometimes confusing largely because of discrepancies in terminology. Baskin and Baskin (1989) suggest that there are five general types of dormancy exhibited by seed at maturity (Table 5.5). These are distinguished on the basis of the following:

- Permeability or impermeability of the seed coat to water (physical dormancy)

TABLE 5.5   Types, Causes, and Characteristics of Seed Dormancy

Type	Cause(s) of Dormancy	Characteristics of Embryo
Physiological	Physiological inhibiting mechanism of germination in embryo	Fully developed, dormant
Physical	Seed coat impermeable to water	Fully developed, nondormant
Combinational	Impermeable seed coat; physiological inhibiting mechanism of germination in embryo	Fully developed, dormant
Morphological	Underdeveloped embryo	Underdeveloped, nondormant
Morphophysiological	Underdeveloped embryo; physiological inhibiting mechanism of germination in embryo	Underdeveloped, dormant

Source: Baskin and Baskin (1989), in Allessio-Leck et al. (1989). *Ecology of Soil Seed Banks*. Copyright 1989 with permission from Elsevier.

- Whether the embryo is fully developed or underdeveloped, that is, incomplete development of the embryo at seed maturity
- Whether the seed is physiologically dormant or nondormant

Potentially, seed of all three types enter seed banks, but most seed found in seed banks in temperate regions have physiological dormancy, with physical dormancy being second in importance.

Physiological Dormancy. According to Baskin and Baskin (1989, 1998), as seed with physiological dormancy *afterripen* (time from seed maturation to germination), they pass through a series of states known as conditional dormancy before finally becoming nondormant. In the transition from dormancy to nondormancy, seed first gain the ability to germinate over a narrow range of environmental conditions. As afterripening continues, seed become nondormant and can germinate over the widest range of environmental conditions possible for the species (Figure 5.13). However, if environmental conditions (e.g., darkness) prevent germination of nondormant seed, subsequent changes in environmental conditions (e.g., low or high temperatures) cause them to enter secondary dormancy. As seed enter secondary dormancy, the range of conditions over which they can germinate decreases until finally they cannot germinate under any set of environmental conditions (Figure 5.13). Thus, seed exhibit a continuum of changes as they pass from dormancy to nondormancy and from nondormancy to dormancy (Baskin and Baskin 1998).

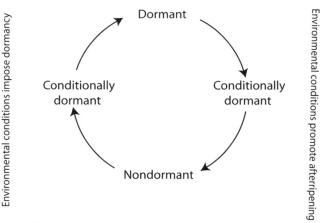

Figure 5.13 Changes in dormancy states of seed with physiological dormancy; seeds are dormant at maturity and go through all possible stages of dormancy cycle. (From Baskin and Baskin 1989, in Allessio-Leck et al. 1989, *Ecology of Soil Seed Banks.* Copyright 1989 with permission from Elsevier.)

Many seed in buried seed banks exhibit annual dormancy–nondormancy cycles as just discussed. These cycles, such as those of obligate winter annuals and spring-germinating summer annuals, are shown in Figure 5.14. From such information on changes in dormancy state (e.g., Figure 5.14), it is possible to predict when exhumed seed, as from tillage or other disturbances that expose soil, will actually germinate. The primary reason that nondormant seed do not germinate while buried is that most of them have a light requirement for germination. This too can be used to manage populations of weeds and invasive plants, as will be seen later (Chapter 7).

Physical Dormancy. The exclusion of necessary environmental factors by certain morphological characteristics, especially of the seed coat, accounts for dormancy of a physical nature. Of particular interest is *somatic polymorphism*—the production of seed of differing morphologies and/or behaviors on the same parental plant. This process is a consequence of divergent cellular differentiation and represents different outcomes of the plant's allocation to seed output. According to Salisbury (1942a) and Harper (1964, 1977), somatic polymorphism is widespread among seed populations of weeds, especially in the families Asteraceae, Brassicaceae, Chenopodiaceae, and Poaceae. Seed polymorphism is generally viewed

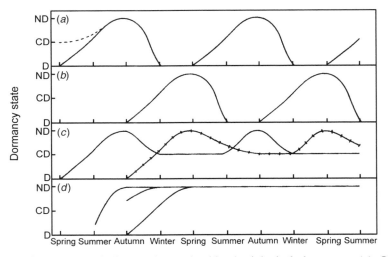

Figure 5.14 Patterns of changes in seed with physiological dormancy. (*a*) Obligate winter annual with annual dormancy/nondormancy cycle. Freshly matured seed in some species are dormant (———) and others are conditionally dormant (•••••). (*b*) Spring-germinating summer annuals with annual dormancy/nondormancy cycle. (*c*) Facultative winter annual (———) and spring- and summer-germinating summer annual (+++++++) with annual conditional dormancy/nondormancy cycles. (*d*) Perennials with no changes in dormancy state after seeds come out of dormancy. (From Baskin and Baskin 1989, in Allessio-Leck et al. 1989, *Ecology of Soil Seed Banks*. Copyright 1989 with permission from Elsevier.)

as a mechanism to enhance species survival when seed are exposed to differing habitats. Some examples illustrate the importance of this process in weed species.

A thoroughly examined case is that of the two seed types of cocklebur (*Xanthium* spp.) which are encased in the fruit and dispersed together. The upper seed, in contrast to the lower one, often fails to germinate when wetted so that at least a year separates the germination of the two types. According to Taylorson and Hendrix (1977), the rates of oxygen diffusion in the two types of seed are similar. Apparently, dormancy in cocklebur involves the presence of a different water-soluble germination inhibitor in each seed type to which the seed coats are impermeable. The presence of oxygen causes the degradation of these two inhibitors and subsequent rupture of the seed coat, but apparently at very different rates in the two seed types. Thus, at least two batches of seed are present in each generation to assure germination in the event that the immediate environment happens to be unsuitable.

Another example of somatic polymorphism is found in common lambsquarters and was studied by Williams and Harper (1965). These researchers found that an individual plant of common lambsquarters produced different types of seed, which are categorized into two groups on the basis of seed color (black or brown), size (small or large), coat characters (thick or thin), proportion of total seed production, and germination requirements. Karssen (1970a,b) observed that the degree of dormancy of common lambsquarters seed was inversely related to size and depended on the thickness of the outer seed coat layer; that is, black seed were small with thick coats and a high level of dormancy. This finding was supported by Williams and Harper (1965), who found large brown seed to have thin coats and low dormancy. Brown seed from common lambsquarters rarely exceed 3% of the total seed produced by a plant and they are the first seed to be produced. The brown seed represent a highly opportunistic strategy, whereas the black seed are more seasonal and predictive in behavior. The combination of germination responses allows the species considerable buffering against sudden selective forces, such as tillage or frosts that might substantially disfavor a single phenotype.

Based on these and other data, Cook (1980) and later Thompson et al. (1998) indicate that seed persistence in soil is favored by a decrease in seed size and relative increase in seed coat thickness, because a proportional decrease in seed size greatly increases the strength of the seed coat relative to the growth force of the embryo during germination. Harper (1977) indicates that the ratio of black to brown seed in common lambsquarters is probably environmentally controlled, as polymorphism is in other genera, since a gradient in proportion of the two seed color morphs exists across Great Britain.

It is likely that seed and dormancy polymorphism are so common among weed species that it is misleading to ascribe to a species any particular germination regime. This seed variability must be considered when attempting to stimulate maximum dormancy breaking of weed seed for subsequent seedling control, since usually only a fraction of the seed germinates even under optimal conditions. Furthermore, earlier or later germinating phenotypes may be favored inadvertently by measures taken to stimulate germination and control of the most typical or abundant polymorph. In determining seed response to control measures, weed scientists should be cautious of possible polymorphic differences within the species they are testing.

Combinations of Physiological and Physical Dormancy. Some seed have a combination of physiological and physical dormancy, for example an impermeable seed coat and a dormant embryo. Clearly dormancy breaking in this situation is a function of the internal conditions of the embryo rather than external environmental conditions. However, the internal causes of dormancy may have been created by severe external constraints. The so-called *hardseededness* (impermeable seed coat) of many legume species that results in response to drought demonstrates this concept. The hilum of legume seed acts as a hygroscopic valve that is activated during dry conditions to allow water loss (Figure 5.15). Hyde (1954) transferred seed of white clover (*Trifolium repens*) to chambers of differing relative humidity and measured the moisture content of the seed following each transfer. He found that under humid conditions water was not able to enter the seed but under dry conditions water vapor could escape. The embryo, therefore, dried to nearly the same water content as the driest environment to which the seed were exposed. Following this treatment, the seed could not imbibe water until the seed coat was broken. By scarifying "hard seed," Hyde found that imbibition would occur (Figure 5.15), thus causing drought-induced dormancy to be broken.

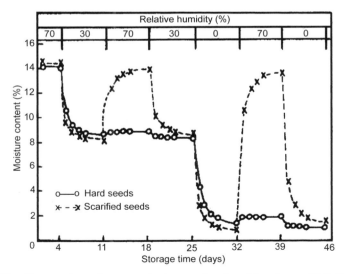

Figure 5.15 Changes in moisture content occurring in white clover (*Trifolium repens*) seeds transferred successively into chambers of different relative humidity. (From Hyde 1954, *Ann. Bot N.S.* 18:241–256. Copyright 2006 Annals of Botany Company, Oxford University Press.)

Seed with Underdeveloped Embryos. Seed typically develop completely while still attached to the parent plant and are mature enough to germinate once they are shed. In some plants, however, the embryo is underdeveloped but not dormant and completion of embryo development occurs after the seed are dispersed from the parent plant. After the embryo becomes fully developed in such species, germination usually proceeds. Baskin and Baskin (1998) indicate that this type of dormancy exists in both tropical (e.g., Magnoliaceae, Degeneriaceae, Winteraceae, Lactoridaceae, Canellaceae, and Annonaceae) and temperate (e.g., Apiaceae and Ranunculaceae) plant families.

Using Seed Dormancy to Manage Weed Populations

With adequate knowledge about the dormancy characteristics of specific weed and invasive plant populations, situations can be identified where very low or no seedling recruitment should occur, even if a high density of seed is present in the soil. For example, weed seed often fail to germinate under dense plant cover or deep in the soil profile. Such locations could be used as weed seed "sinks" (places or conditions where dormant seed remain in the dormant state) by carefully planning logging and grazing systems of harvest or by planning crop sowing, rotation, or tillage operations to reduce seedling germination from existing seed banks. Examples of how dormancy might be used to manage weeds and invasive plants are discussed in Chapter 7.

RECRUITMENT: GERMINATION AND ESTABLISHMENT

Because of the reproductive allocation common to most weeds and invasive plants, a large number of seed and/or vegetative propagules are usually produced and dispersed to the soil. Unquestionably, many of those propagules become seedlings or ramets. *Germination*, the transition from seed to seedling (or bud to ramet), is the most significant way seed are lost from the soil seed bank (Figure 5.8). It is also the most critical (susceptible) phase in the development of a plant. Most pest management strategies against insects or plant pathogens are directed at the most vulnerable stage in the life of the pest organism (Chapter 9). This approach also can be taken with weeds and invasive plants. Following introduction, the timing and amount of weed seed germination undoubtedly influence the spectrum of species within a plant community. In the event that weed control is imposed, germination also may influence the amount of control attained as well as the composition of the native and exotic plant populations afterward.

Seed Germination

The germination of seed involves the initiation of rapid metabolic activity, embryo growth, radicle emergence, and finally emergence of aerial portions of the plant. Radicle emergence is used most often as an indicator that germination has begun. Before shoots emerge from soil considerable underground elongation usually takes place. This growth pattern is important for weeds and invasive plants because seed of many of these species are adapted for shallow or surface germination. The survival of seedlings, therefore, depends on the ability of their primary roots to extract moisture from increasingly lower levels in the soil profile.

Patterns of shoot emergence from the soil are varied, but two principal types are recognized. *Hypogeal* emergence, typical of Fabaceae and Poaceae, occurs when cotyledons remain below the soil, and *epigeal* emergence, for example in Asclepiadaceae and Apiaceae, occurs when cotyledons are carried above the soil surface during emergence. Both methods of emergence are common in weed/ invasive plant species. Monocots, in contrast to dicots, emerge from the soil with the shoot apex encased in a sheath, called a *coleoptile*. The position of the cotyledons in dicots and the degree of mesocotyl and coleoptile extension in monocots can influence the survival of seedlings and are also important for herbicide placement and differential selectivity among weed and crop or other desirable species.

When a seed germinates in natural conditions, the plant essentially "takes a chance" on the soundness of environmental conditions of a site for seedling establishment (Probert 1992). Hence, natural selection probably favors mechanisms that decrease the probability of a seed encountering unfavorable conditions for growth after germination. No doubt these patterns of seedling survival involve dormancy mechanisms, polymorphism, and environmental constraints already discussed. Also obvious with some weed species are flushes of germination, which often occur after tillage or other disturbance such as fire (Popay and Roberts 1970a,b). The scientific literature on seed germination is extensive and demonstrates that germination is influenced by environmental factors including light quality and quantity, temperature (including fluctuations), moisture, and gas ratios (Baskin and Baskin 1998, Benech-Arnold et al. 2000, Mohler 2001) (Figure 5.16). In addition, the age and physiological status of the seed and environmental conditions during afterripening are also important regulators of seed germination.

Light Requirement for Germination. Seed of many plants have a light requirement that must be met before germination can occur. This criterion is especially true for seed of weed and colonizing species. Sauer and Struik (1964) speculated on the ecological significance of this phenomenon after noting the differential emergence of exotic species from soil samples collected at night beneath various stages of old-field succession. Since open habitats produced abundant seedlings, Sauer and Struik suggested that a light-flash mechanism for dormancy breaking may assist pioneer plants in exploiting disturbed environments.

In Wesson and Wareing's classic work (1969a), soil samples were taken at night from beneath an established pasture. After discarding the top 2 cm of soil, they separated the samples by depth and attempted to germinate the seed in them under light and dark regimes. Little germination occurred in the dark, whereas the soil in the light produced many seedlings. They also dug small pits (5, 15, and 30 cm deep) in the same pasture at night and covered some with opaque asbestos, covered others with glass, and left some uncovered, then evaluated the number of seedlings that emerged from each treatment (Figure 5.17). In these experiments,

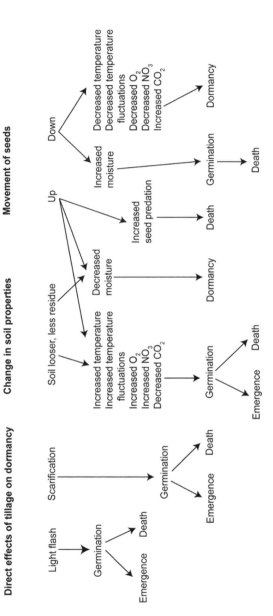

Figure 5.16 Effects of tillage on germination, dormancy, and death of weed seed. (From Mohler 2001, in Liebman et al. 2001, *Ecological Management of Agricultural Weeds.* Copyright 2001 Reprinted with permission of Cambridge University Press.)

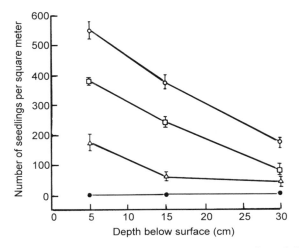

Figure 5.17 Number of seedlings to emerge per square meter from field plots at three different depths. (□) Plots uncovered; recorded after five weeks. (○) Plots covered with glass; recorded after five weeks. (•) Plots covered with asbestos (dark); recorded after five weeks. (△) Asbestos covers removed from treatment 3 and replaced with glass. Emergence recorded for further three weeks. (From Wesson and Wareing 1969a, *J. Exp. Bot.* 20:402–413. Copyright 1969 Oxford University Press.)

germination absolutely required light. In addition, germination was higher in glass-covered than uncovered pits, indicating that germination could be enhanced by increasing temperature but that it was not essential. In further studies using freshly collected seed, Wesson and Wareing (1969b) observed that most of the weed species tested displayed no light requirement when seed were fresh. However, if seed from the same species were buried for 50 weeks, a light requirement was always needed for germination. Furthermore, weed seed that did not germinate within the first few weeks after burial would not germinate until after a light requirement had been fulfilled. Burial actually induces a light requirement in most seed soon after dispersal, which maintains them in a condition of dormancy for as long as they are buried.

It is interesting to speculate on the ecological reasons for the light requirement of weed seed. It is well known that seed of many weed species germinate best at shallow soil depths. Furthermore, a major cause of mortality is deep germination, which prevents seedlings from reaching the soil surface. It seems likely, therefore, that the light requirement may act as a highly predictive indicator of disturbed areas suitable for further colonization. The presence of light might indicate proximity to the soil surface, bare mineral soil, or the absence of an overstory plant canopy.

Phytochrome System. The light-stimulated germination of seed is due to the phytochrome (P) system, which involves two forms of the photoreceptor *phytochrome*. This molecule exists in a biologically active (P_{fr}) and an inactive (P_r) form. The photoconversion of P_r (absorbs wavelengths of light in the red region) to P_{fr} (absorbs wavelengths of light in the far-red region) stimulates germination and other light-responsive processes; that is, red light stimulates $P_r \rightarrow P_{fr}$ whereas far-red light stimulates $P_{fr} \rightarrow P_r$:

$$P_r \underset{\substack{\text{far red light} \\ \text{darkness}}}{\overset{\text{red light}}{\rightleftarrows}} P_{fr} \longrightarrow \text{biological response}$$

Gradually in darkness, such as when seed are buried, P_{fr} declines to a level below that needed for germination; thus an input of red light is required to increase the level of P_{fr} for germination to occur following a prolonged period of burial. In the case of seed, burial appears to induce a dramatic increase in light sensitivity, which is called the very low fluence (VLF) response mechanism (Scopel et al. 1991). The VLF response is mediated by phytochrome and triggered by light exposures that form very small amounts of P_{fr} (i.e., between 10^{-4} and $10^{-2}\%$ of the total phytochrome). This response allows seed to detect microsecond exposures to sunlight when the soil is disturbed, as by tillage (Scopel et al. 1991).

In addition to providing a potential germinant with an indication of its position on or in the soil, the VLF response may reflect the degree of plant community openness or the presence or absence of an overstory plant canopy. Leaves in a canopy transmit considerably more far-red than red light (Figure 5.18). The

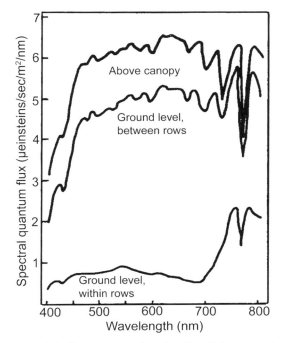

Figure 5.18 Influence of shading upon wavelengths of sunlight present in various regions of sugarbeet field. Within the rows (bottom two curves) there is much less attenuation of far red than the other wavelengths, so shaded plants contain a higher proportion of phytochrome in the P_r form than do unshaded plants. (From Holmes 1975, *Nature* 254:512–514.)

wavelengths of light, including red, intercepted by leaves are either absorbed (most) or reflected (green); light that is transmitted is generally far red and of lower energy. Under those enriched far-red conditions, most seed on the soil surface contain phytochrome in the P_r condition and thus would not germinate. This accounts for the absence of continued weed and invasive plant germination in various crop, grassland, shrub, or tree stands after a canopy begins to close. It also may explain why more weed seedlings are often found between rather than within established crop rows and why invasive plants are often found near trails, roads, and canopy openings (Perendes and Jones 2000).

With most agricultural crop species, in contrast to weeds, a light requirement for germination is not apparent. For example, when planting large-seeded crops, it is advantageous to place seed into the soil where water or other resources are more readily available than on or near the soil surface. It is likely that the light requirement for germination has been bred out of many crop species. Also possible, however, is that once harvested from the parent plant, crop seed never receive enough darkness to induce dormancy fully, or even if such seed are stored in darkness, a sufficient amount of red light is received just prior to planting to induce germination.

Risk of Mortality

Seedlings. As discussed above and in Chapter 3, seed germination is a vulnerable stage in the plant life cycle and seedlings are often exposed to a harsh environment. Therefore, for weed/invasive plant management, it is important to consider all chances for mortality as seedlings develop from seed. The probability of mortality is sometimes called the *critical risk of death*, and it usually determines the initial weed density experienced in a field. As discussed in Chapter 3, each plant species has a certain set of requirements for germination and survival that are adapted to a particular *safe site*. Thus, placement in a suitable safe site minimizes the risk of death. Some of the mortality-avoiding features of weed seed have been discussed already; they include the light requirement for dormancy breaking and its relationship to soil disturbance, polymorphism of seed, and hardseededness. The abundance of *nitrophiles* (species that occur in nitrate-rich habitats) among weeds and some invasive plants also may indicate an adaptation that increases the probability of seedling survival on disturbed sites. There are numerous examples in which potassium nitrate stimulates germination of weed seed (Anderson 1968). This is interpreted as a case of enzyme induction or a signal for germination to proceed rather than as a requirement for nitrogen or potassium fertility.

Certain cultural practices or events, such as annual tillage or frequent disturbance, may be viewed as a significant selective force because microsites (i.e., safe sites) are continually being formed or modified. For example, a certain number of plant seedlings survive each year after an initial disturbance clears and stirs up the soil. However, some species or populations can be favored and others disfavored by the same management practices or disturbance events. Eventually, all the safe sites generated should become filled with surviving individuals of the best adapted genotypes. At that point, seedling saturation of the habitat has occurred and no further recruitment of seedlings from the soil seed bank would be likely. Further germination would most likely result in death of the seedling due to unsuitability of the environment (no safe sites available).

Adopting a cultural or management practice designed to affect the safe sites of a particular component of weed or exotic plant germinants may have considerable utility in reducing densities of such plants below the saturation density. In some cases, an economic weed threshold (Chapter 6) may even be obtained. An experiment by Selman (1970) serves as an example of how safe site modification can influence weed density. Over an 11-year period, Selman compared the levels of wild oat (*Avena fatua*) seedling survival in barley that resulted from either fall or spring tillage practices. During the first seven years of study, he observed that annual tillage in the fall prior to barley planting caused a progressive increase $(1-400 \text{ plants}/\text{m}^2)$ in wild oat density. For the next four years the same field was cultivated in the spring and then planted with barley. The wild oat density decreased to 85 plants/m^2 the first growing season after spring tillage was implemented and eventually stabilized to about 5 plants/m^2 by the end of the study. These results show the value of a timely tillage as a means to reduce weed seedling density and the value of changing a cultural practice to disfavor existing adapted ecotypes. Since a flush of germination of wild oat occurs in both spring and fall, an eventual build-up of the spring germinating ecotype might be expected, similar to what occurred with the fall ecotype. For this reason alternating tillage practices and therefore barley planting dates between fall and spring might maintain the density of both ecotypes of this agricultural weed at a reasonable economic (low) level.

Vegetative Propagules of Perennial Plants. For seedlings of herbaceous perennial plants the risk of mortality should be similar to that of seedlings of annual plants. However, the risk of death for ramets differs considerably. For ramets of perennial plants the most critical time for survival is during active growth immediately following new shoot initiation when new shoots are not completely self-supporting. Consequently, carbohydrate reserves in the roots are lowest at that time. Disturbance, such as repeated tillage at appropriate intervals when carbohydrate reserves are continually being depleted, can have a substantial negative effect on ramet density. Agriculturists have recommended for years to fallow and till repeatedly areas infested with perennial weeds. Similar recommendations are made by foresters to suppress unwanted coppice growth of some tree and shrub species. In the case of agriculture, the interval between disturbance events varies among species but tillage is usually necessary every three weeks during the growing season for maximum carbohydrate depletion and ramet mortality to occur. Too great an interval between disturbances may, in fact, have a detrimental effect on weed control by stimulating ramet production and growth. Interactions between environment and disturbance can enhance ramet mortality as well. For example, considerable density reduction can result from properly timed tillage that takes advantage of both carbohydrate starvation and adverse environmental conditions. In one experiment, nearly 60% reduction in Johnsongrass ramet density resulted from a single properly timed tillage in contrast to an untilled treatment (Radosevich et al. 1975). In that experiment, July tillage allowed sufficient time before fall rains for desiccation of severed rhizomes to occur.

Epidemics of Weeds and Invasive Plants

On agricultural land, soil tillage and herbicides are the main hazards for weed seedlings to overcome. Under these conditions, weed mortality is frequently very high, 80% or more, and occurs during the seedling stage both before and after emergence from the soil (Sagar and Mortimer 1976, Aldrich 1984, Ross and Lembi 1999). In spite of the drastic reductions in seedling density that occur soon after emergence, density-dependent effects can still be observed in the growth and reproduction of weeds. For example, Ballaré et al. (1988) observed that Chinese thornapple (*Datura ferox*) plants originating from high-density patches produced less biomass and seed than those generated in low-density patches. Diseases and predation also may be modified by seedling density during early stages of weed growth (Liebman 2001). These factors of density dependence are generally thought to assure a constant seed rain to the soil.

In most invasive plants, natural dispersal of propagules produces a distribution pattern where the density of seed is very high near the parent plant and dramatically reduced a short distance away from it (Figure 3.2), generating patchy distributions. This pattern may be completely modified under agricultural conditions, however, by tillage, herbicide use, and harvesting operations (Figure 5.19). Such spatial modification in the distribution of seed in agricultural fields reduces patchy distribution of seed and creates an overall reduction in seedling density (Ballaré et al. 1988, Cavers and Benoit 1989, Ghersa et al. 1993, Liebman et al. 2001). Dispersal of seed in the soil profile by tillage also reduces seedling density, changes the rate of seed release from dormancy, and regulates the rate of seedling emergence through time. It is likely that this reduction in seedling density has a dramatic impact on the epidemic nature of weed infestations in cropped fields (Auld and Coote 1990, Maxwell and Ghersa 1992, Ghersa and Roush 1993) because the lack of density-dependent regulation on seed production probably sustains exponential growth of the remaining, "escaped" weed populations, which will produce abundant seed even under high levels of weed suppression (Maxwell et al. 2003, Zavaleta et al. 2001).

Being able to predict the spatial and temporal patterns of seedling weeds and invasive plants is a key problem facing weed scientists and plant ecologists today. This knowledge is needed to adjust control measures to sites and times of development when weeds are likely to be present and to evaluate their risk to crop yield, native communities, or the environment. Such knowledge is also necessary to predict the success of native plant restoration in natural ecosystems following invasive plant removal. Spitters (1989), Benech Arnold and Sanchez (1994), and Ghersa and Holt (1995) review how such predictions might be achieved through models of germination and survival. The goal of such models is to increase the

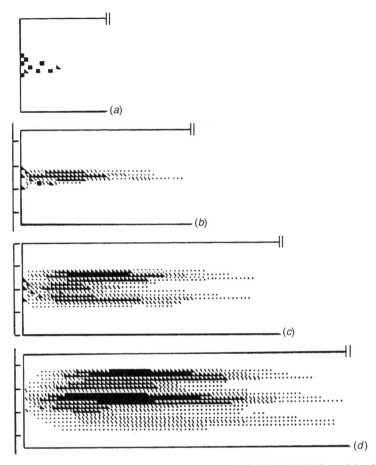

Figure 5.19 Distribution of seed among modules of field. (*a*) Initial seed bank. (*b–d*) Seed bank at end of first, second, and third growing season, respectively. The beginning of the imaginary runs of the combine harvester are indicated by $\square = 0$, $\boxdot = 1$–5, $\boxminus = 6$–15, $\blacksquare = 15$–45, $\blacksquare = 40$–130 seeds per module. (From Ghersa et al. 1993, *Weed Res.* 33:79–88. Copyright 1993 Blackwell Publishing Ltd, reproduced with permission.)

risk of death of weed seedlings or delay growth sufficiently so that weeds are competitively suppressed by crops or other desirable plants.

Predictive Models of Weed Reproduction, Dispersal, and Survival

The value of determining the reservoir of weed seed in the soil and the early fate of seedlings is in the ability to predict potential weed infestations. Using the demographic parameters of seed production and dispersal, seed reserves in the soil, rate of seedling recruitment and expected mortality, it is possible to determine the expected density of weeds likely to occur on a site. Information about the ages of seed that give rise to the majority of new seedlings also contributes to predictions of weed presence.

Weed researchers and population ecologists have developed and are still developing various types of models to describe the demography of plant populations and the competitive interactions among weeds, crops, and plants in natural ecosystems (Sagar and Mortimer 1976, Auld 1984, Spitters 1989, Kropff and Lotz 1992, Maxwell and Ghersa 1992, Silvertown and Charlesworth 2001, Knowe et al. 2005). Such models are excellent tools to understand the components of a species' life cycle, to determine the relationships among these components, and ultimately to predict changes in weed population numbers from one generation to the next as a result of environmental variation, environmental manipulation, or management tactics. As an example of such analyses, we use the classic model developed by Sagar and Mortimer (1976), introduced at the beginning of this chapter (Figure 5.2), for population dynamics of agricultural weeds.

Example: Predictions of Changes in Weed Abundance in Agricultural Fields. In order to demonstrate the value of population models for weed management, Sagar and Mortimer (1976) presented several possible schemes for population regulation in wild oat. Beginning with a hypothetical population of 10 plants and using data from other sources, Sagar and Mortimer considered both best and worst management options for this weed. Figure 5.20*a* represents a situation resulting from effective management in which severe population reduction of wild oat would occur. At each interphase, effective regulation of the weed population is achieved by (a) control of seed output by planting a competitive crop (Chancellor and Peters 1970), (b) harvest before seed is shed and subsequent removal of straw (Thurston 1961), (c) maximum exposure of seed on the soil surface (Wilson 1972), (d) sparse emergence from the soil (Thurston 1961), and (e) high postemergence mortality (Chancellor and Peters 1972). The combination of the interphases in Figure 5.20*a* was predicted by Sagar and Mortimer to cause dramatic deceases in population size (from 10 to 0.018 plants), and contribution to the soil seed bank was only 0.06% of the previously existing seed reserve.

In contrast, Figure 5.20*b* represents a series of poor management options. According to Sagar and Mortimer (1976), the most rapid rate of population expansion arises from (a) minimal crop competition (Chancellor and Peters 1970), (b) poor attempts at seed collection and trash destruction during harvest (Wilson 1972), (c) incorporation of seed into the soil (Marshall and Jain 1967), (d) maximum emergence (Marshall and Jain 1967, Wilson 1972), and (e and f) no subsequent mortality. This interphase combination for the wild oat life cycle (Figure 5.20*b*) resulted in an increase in population size (+1424 plants) and the contribution to the seed bank rose by 80%. By comparing the two life tables for wild oat (Figures 5.20*a,b*), it is possible to see that the greatest control (regulation) occurs (b) at crop harvest, (c) when seed lie exposed on the soil surface, and (d) by failure of seed to emerge from the buried seed reserve. Clearly, the best management is that directed at those particular interphases of the weed's life cycle.

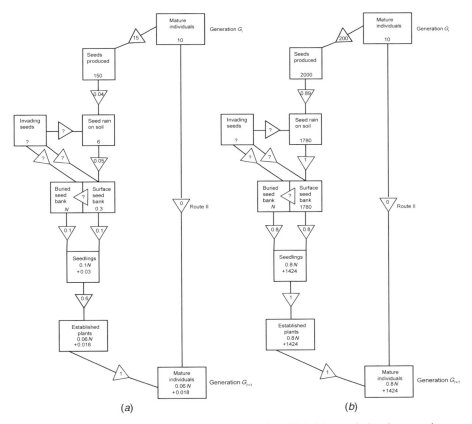

Figure 5.20   Life table of wild oat (*Avena fatua*) in which (*a*) population increase is projected to be low and (*b*) population increase is projected to be high, where *N* represents buried seed bank. The sources of values are acknowledged in the text. (From Sagar and Mortimer 1976, *Ann. Appl. Biol.* 1:1–47. Copyright 1976 Blackwell Publishing Ltd, reproduced with permission.)

SUMMARY

Uncertainty about the density, composition, and population dynamics of weeds and invasive plants often intensifies the perception of risk to agricultural and natural ecosystems and can lead to inappropriate management decisions. Plant population ecology deals with the demographic or numerical aspects of a particular group of individuals and can explain some of the uncertainty inherent in associations of weeds and desirable plants. Demographic approaches can be used to describe the critical life stages of a plant species and examine how weed and invasive plant populations might be expected to change over time, under different environments, or from various management tactics. Natality, mortality, immigration, and emigration are the four fundamental components of plant demography.

However, the complexities of weed populations are best described using diagrammatic or mathematical life tables and, in some cases, difference and transition matrix models. Metapopulation models can be used to describe the persistence of weed and invasive plant populations over a landscape or region.

Seed are important for the maintenance and growth of existing plant populations and for the initiation of new populations. Seed must be both produced and dispersed for a species to be successful. Seed dispersal can be over space, through physical movement, or through time by means of dormancy. Wind, water, animals, and humans are agents of spatial dispersal. The seed bank is a useful concept often applied to the dynamics of seed in or on soil. This concept views the soil as a reservoir of seed to which inputs and withdrawals are made. Most inputs (seed rain) to the seed bank are large for weed species and other colonizing plants. However, the size and longevity of seed banks vary. Although some seed may exist dormant in the soil for a long time, most agricultural seed banks are of short to intermediate duration. The most common forms of withdrawal from the seed bank are germination followed by predation, seed senescence, and decay. There is evidence that if weed seed inputs can be stopped, the number of viable seed stored in soil of cultivated and abandoned fields will decline. Thus, it is important to remove late-germinating cohorts or survivors of weed control tactics to avoid continually reestablishing a reserve of weed propagules in the soil.

Dormancy is the temporary failure of viable seed to germinate under environmental conditions that later allow germination to occur. Seed may be physiologically dormant, maintain physical dormancy, or have underdeveloped embryos. Seed that are dormant do not pose an immediate threat of further infestations unless dormancy is somehow broken. Thus, it may be best to leave dormant seed in that state or to create "sinks" through management activities in which to cache dormant seed.

Most weed/invasive plant seed require light to germinate. This mechanism assures proximity to the soil surface and the absence of other vegetation, since germination is the most critical stage in a plant's development. Many plants have specific physiological or physical adaptations to assure survival in certain locations, called safe sites. Safe sites are usually created for agricultural weeds when crops are planted, but invasive plants must land in an appropriate safe site. Many plants have evolved mechanisms to avoid mortality. It is possible to alter management practices to disfavor weeds previously adapted to the tillage, sowing, or harvesting practices of a cropping system. The ability to predict the spatial and temporal patterns of weed and invasive plant seedling distribution is one of the most pressing problems facing weed scientists today.

Prediction of the abundance and changes in weed and invasive plant populations is possible using demographic models. However, such models usually require large amounts of empirical information about plant life histories in order to be constructed. Nevertheless, such models do exist for many weed–crop associations which demonstrate how improved knowledge of environment and biology can assist in making weed control decisions and either enhance direct weed suppression or eliminate the need for it.

6

PLANT–PLANT ASSOCIATIONS

The interactions that occur among various members of the plant kingdom shape the morphology and life history of individual plants and create the structure and dynamics of plant communities. When an association occurs between a crop and a weed, a loss in crop yield usually results. It is this observation about crop–weed interactions, the potential negative impact of weeds on crops, upon which much of modern weed science is built. Many experiments document crop yield losses from particular weed associations with crops, which creates a need for weed suppression in most crops and many natural resource production systems. Negative plant–plant interactions among invasive plants and native vegetation also cause shifts in species composition and dominance in natural ecosystems (Chapter 3), although some exotic plants may simply be responding to disturbance or other environmental change rather than driving compositional shifts through plant–plant interaction (MacDougall and Turkington 2005).

Experimental methods developed in agroecosystems make it possible to distinguish among the possible types of plant interactions. Most studies of weeds in crops and natural resource systems deal primarily with the negative aspects of the interaction, for example, crop yield loss or negative shifts in native species composition, but there may be neutral or even positive aspects of plant–plant interactions as well. By examining these neutral, negative, and positive aspects of plant–plant associations, it is possible to make management choices about weed levels and cultural practices that optimize production or biodiversity while minimizing costs. It is also possible to correct an environmental deficiency or alter a

Ecology of Weeds and Invasive Plants. By Steven R. Radosevich, Jodie S. Holt, and Claudio M. Ghersa
Copyright © 2007 John Wiley & Sons, Inc.

condition if the cause of an interaction is known, resulting in longer term prevention or restoration than if only containment or control is used to manage invasive plants. This aspect of weed/invasive plant management is discussed in Chapter 9.

NEIGHBORS

Soon after germination a seedling becomes independent of its parental resources in the seed. It begins existence as an individual and extracts from its surroundings the resources necessary for life. Plants, however, rarely exist in isolation. They usually grow with other plants of the same or different species. Therefore, it is important to consider plants as neighbors in the environments in which they grow. There are three types of plant neighbors:

- Parts of the same plant; leaves and branches that extend over other leaves and branches or vegetative shoots from the same parent plant (ramets)
- Individuals of the same species but arising from different seed (genets)
- Individuals of different taxa

Management practices used in agriculture, forestry, rangeland, and urban environments are designed to change plant interactions to improve plant growth or appearance. *Pruning* is a cultural practice used to optimize the interaction among parts of the same plant, while *spacing* or *thinning* is used to optimize the interaction between like individuals. *Weeding* is most often used to change the relationship between neighbors of different taxa.

INTERFERENCE

The general term for interactions among species or populations within a species is *interference*. This is the effect of the presence of a plant on the growth or development of its neighbors. Interference is an alteration in growth rate or form that results from a change in the plant's environment due to the presence of another plant. Burkholder (1952), in his classic paper on interference, categorized the possible interactions that occur among species growing together. He used a scheme where an interaction is symbolically (+ or −) described as an effect on two plant populations. When the two populations are close enough to respond to each other, the interaction is "on," and the interaction is "off" when the two populations are apart. In most cases, an off interaction has no effect, but not always. Table 6.1 lists all the logically possible types of interactions between populations, which may be positive (+), negative (−), or neutral (zero). However, only the relationship between plants in the interaction is identified in Table 6.1. The actual causes of the interactions may include production or consumption of resources,

TABLE 6.1   Complete List of Biologically Possible Types of Interactions[a]

Name of Interaction	On		Off	
	A	B	A	B
Neutralism	0	0	0	0
Competition	−	−	0	0
Mutualism	+	+	−	−
Unnamed	+	+	0	−
Protocooperation	+	+	0	0
Commensalism	+	0	−	0
Unnamed	+	0	0	0
Amensalism	0	−	0	0
Parasitism, predation, herbivory	+	−	−	0
Unnamed	+	−	0	0

[a]When organisms A and B are close enough to participate in the interaction, the interaction is on, otherwise it is off. Stimulation is symbolized as +, no effect as 0, and depression as − .

Source: Burkholder (1952). Cooperation and conflict among primitive organisms. *Am. Sci.* 40:601–631. Reprinted by permission of American Scientist, Journal of Sigma Xi.

production of stimulants or toxins, predation, or protection, which are not described by the scheme.

Of the 10 possible interactions listed in Table 6.1, 3 represent negative effects of interaction. These are competition, amensalism, and parasitism, predation, and herbivory. *Competition* (− − when on and 00 when off) is defined by Barbour et al. (1999) as the mutually adverse effects of organisms (plants) that utilize a resource in short supply. *Amensalism* (0 − and 00) is similar but refers to the interaction in which only one of the plants is depressed whereas the other is unaffected. *Allelopathy*, the inhibition of one plant by another through the release into the environment of selective metabolic by-products, is a form of amensalism. *Parasitism, predation*, and *herbivory* (+ − and − 0) are special forms of negative interference because one plant (or other organism) lives in or on another and thus derives resources directly from its host. Neutral and positive interactions among plants are also possible. These include neutralism, mutualism, protocooperation, and commensalism (Table 6.1), which are described later in this chapter.

Effect and Response

How can interference be positive, negative, or neutral? Vandermeer (1989) describes interference as a double-transformation problem, whereby a plant and the environment in which it exists affect, or transform, each other. In other words, a plant lives according to the dictates of its local environment, yet it is also an important participant in effecting change in that local environment. As stated by Harper (1977):

> Plants may influence their neighbors by changing their environment. The changes may be by addition or by subtraction and there is much controversy about which is

more important. There also may be indirect effects, not acting through resources or toxins but affecting conditions such as temperature or wind velocity, encouraging or discouraging animals and so affecting predation, trampling, etc. (p. 354)

This idea of effect and response is presented diagrammatically in Figure 6.1*a*, in which the environment–organism transformation is shown as a crop–weed interaction. The weed, for example, has an *effect* on the environment. It might remove water or certain nutrients from the soil, leaving it partially depleted, or it could enhance the environment by leaving deposits of nitrogen that it fixed from the air. Both the weed and crop must *respond* to this effect, thus setting up the dichotomy of "effect" and "response" indicated in Figure 6.1*a* (Goldberg and Werner 1983, Vandermeer 1989, Goldberg 1990).

The existence of many forms of interference is based on the observation that interactions among plants occur through an "intermediary," which in Figure 6.1 is collectively called "the environment", but is actually all environmental factors: resources, pollinators, dispersers, herbivores, predators, parasites, or microbial

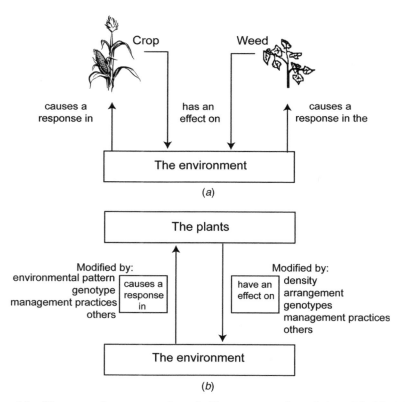

Figure 6.1 Diagrammatic representation of effect–response formulation. (Modified from Vandermeer 1989, *Ecology of Intercropping*. Cambridge University Press, NY.)

TABLE 6.2 Types of Indirect Interactions Among Plants[a]

Types of Interaction	Intermediary	Effect	Response	Net
Exploitation competition	Resources	−	+	−
Apparent competition	Natural enemies	+	−	−
Allelopathy	Toxins	+	−	−
Positive facilitation	Resources	+	+	+
Negative facilitation	Resources	−	−	+
Apparent facilitation	Natural enemies	−	−	+

[a]In this classification, resources of plants include mutualists such as pollinators or dispersers as well as abiotic resources such as light, water, mineral nutrients, and CO_2. In the effect, response, and net columns + and − indicate the effect of plants on abundance of the intermediary, the response of some "target" plant to abundance of the intermediary, and the net effect of plants on the "target" plant, respectively.

Source: Goldberg (1990), in Grace and Tilman (1990), *Perspectives on Plant Competition*. Academic Press Inc.

symbionts (Goldberg 1990). Since all such interactions are therefore indirect, two distinct processes occur. One or both plants in the association has an effect on the abundance of the intermediary and one or both have a response to the changes in the abundance of the same intermediary. The type of interference that occurs depends on the identity of the intermediary and whether effects and responses are positive or negative (Table 6.2). Although the types of interference in Table 6.2 do not conform directly to the interactions described by Burkholder (1952) (Table 6.1), the concept of effect and response suggests a common mechanism through which interactions occur.

Is it Competition?

The scientific literature is inconsistent in use of terminology to describe interference among plants, especially negative interference. For example, competition, according to Burkholder (1952) and Goldberg (1990), describes only one possible type of negative interference (Tables 6.1 and 6.2), and the term implies that the supply of some environmental resource is insufficient for unrestricted growth of the plant in question. A further implication of the term is that the interacting species occupy similar niches so that each is affected by the availability of the limiting resource (Figures 2.1 and 2.7). In this case, the implied cause of the interaction is the differential abilities of the plants to usurp environmental resources.

Unfortunately, competition has been used to describe the negative impact of one species upon another (e.g., weeds on crop yield) without consideration of resource availability or the presence of other organisms. Competition is also used at times to describe an increase in abundance or dominance of exotic invasive plants in natural ecosystems, even though such shifts might occur from positive interactions that facilitated the presence of the exotic species. Thus, it is possible that forms of interference other than competition, for example, amensalism, herbivory, or commensalism, could actually account for the observed changes in plant density, vigor, or yield or species composition. It is also possible that the negative effects of competition in some experiments are confounded with positive effects, since most ecological systems are complex and contain species other than just desirable and undesirable plants.

The interactions in Table 6.2 (Goldberg 1990) are presented and discussed in more depth throughout this chapter on plant–plant associations. However, they are also briefly described here:

- *Exploitation Competition.* Differential ability of plants to acquire and use environmental resources; thus one plant creates a limitation of resource for the growth of another. The environmental resources are light, water, nutrients, and the gases necessary for plants to carry out photosynthesis (CO_2) and respiration (oxygen). These are discussed in depth by Radosevich et al. (1997), Zimdahl (1999), and Booth et al. (2003) and will not be discussed further here. Plant traits that influence exploitation competition include total and above- and below-ground biomass, plant height, and plant canopy area.

- *Apparent Competition.* Alteration of plant size or vigor by differential use of one plant versus. another by a natural enemy (parasite, predator, or herbi-vore). The change in plant size could go unnoticed or influence other inter-actions such as exploitation competition. Louda et al. (1990a) and Connell (1990a) believe that grazers have such a dramatic affect on plants in natural ecosystems that they mask or significantly change the outcome of real competition in many systems.

- *Allelopathy.* Negative influence of one plant on another through the production and release of phytotoxins. This is also an example of *asymmetric competition*.

- *Positive Facilitation.* Favorable response of one or both members of a plant association, often producing an overyield. Resource use can be separated in time or space. Many multicrop associations are believed to create this type of interaction.

- *Negative Facilitation.* Interaction in which neither member of the plant–plant association is damaged as much as expected, creating a net positive response.

- *Apparent Facilitation.* Interaction in which grazing, predation, or parasitism by natural enemies creates a net positive response, but in the absence of the natural enemy both members of the plant association would be damaged by the presence of the other.

MODIFIERS OF INTERFERENCE

Vandermeer (1989) notes that both effects and responses have "modifiers" (Figure 6.1*b*) that influence the direction and extent of positive or negative inter-actions. For example, a plant may affect the environment in a negative way with respect to other plants through nutrient or water extraction or the production of shade or allelochemicals. Thus, a benign environment could become more hostile to other plants. This interaction is generally called *competition*. In another option, the environment may be affected positively by a plant: for example, pollinators attracted to one flowering individual might create a pollinator-filled environment for other individuals of another species (Vandermeer 1989) or the presence of shrubs (e.g., sagebrush in the desert) could collect snow and thus provide more water for grasses beneath them. Vandermeer calls this effect *facilitation*, which refers to all of the positive interactions described in Tables 6.1 and 6.2.

Two principles emerge from this discussion that describe how plant communities develop:

- *Competitive Production Principle.* One species has an effect on the environment, which causes a negative response in the other species. However, there are instances of species existing together where the negative response of one to another is less than would be expected from growing the two competing species together in the absence of other environmental influences.

- *Facilitative Production Principle.* The environment of one species is modified in a positive way by another species such that the growth or development of the first species is enhanced.

These principles are restatements, in a production context, of the competitive exclusion principle and niche theory discussed in Chapter 2.

Figure 6.2 demonstrates how the competitive production and facilitative production principles might act together in plant communities to produce different outcomes. Each situation in Figure 6.2 is modified by plant density. In the first case (*a*) competition dominates while the second (*b*) is dominated by facilitation. In the third instance (*c*) both competition and facilitation dominate but at different plant densities or times.

An abundance of information implicates plant density, proportional relationships among the species in an association, and the spatial arrangement of individual plants as significant factors contributing to interference. These factors are particularly important to consider in competition studies, but they influence the outcome of other forms of interference as well. Because these factors of plant proximity (*density*, *proportion*, and *spatial arrangement*) so dramatically influence the outcome of experiments designed to demonstrate the existence and impact of plant–plant interactions, they are considered here under the general topic of interference.

Space

After germination, plants either exhaust their parental supplies or become independent of them. Further growth then depends on the seedling's ability to extract the resources it needs from its local environment (Chapter 2). The supply of resources may be unlimited in some environments, but limitation is more

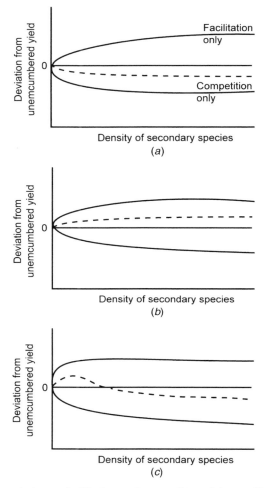

Figure 6.2 Balance between facilitation and competition: (*a*) net effect competitive; (*b*) net effect facilitative; (*c*) net effect competitive or facilitative depending on density of second species. Vertical axis represents yield of primary species. (From Vandermeer 1989, *Ecology of Intercropping*. Copyright 1989. Reprinted with permission of Cambridge University Press, NY.)

common and can be caused by unavailability, poor supply, or proximity to neighboring plants. The presence of neighbors can aggravate an already insufficient resource or create a deficiency where there is ample resource for a single individual. Because resource use is integrated within and among plants, the impact of resources on growth is sometimes considered as a single conceptual unit called *space* (Chapter 2). This concept allows study of the effects and responses of plants without strict regard for the causes of interactions, for example, the actual resource that is limited or the object of competition. Space capture is discussed in Chapter 2 and more thoroughly by Harper (1977) and Radosevich et al. (1997).

Density. *Density* is the number of individuals per unit of area. Typical units of measure are plants per square meter, plants per hectare, or plants per pot. Density is often used to describe the number of plants in a crop, tree, or weed stand. As density increases, a certain level of individuals is reached at which interference occurs among neighboring plants. Plants respond to density stress in two ways: through a plastic response of growth and/or an altered risk of mortality.

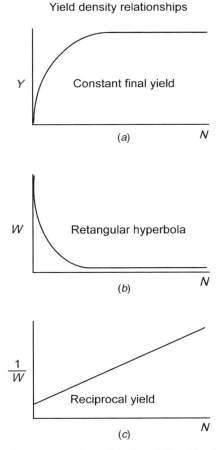

Figure 6.3 Diagrammatic representation of basic relationships between plant yield and plant density: Y = yield of a stand of plants (biomass per area); W = yield of individual plant (biomass per plant); N = plant density. (After Radosevich and Roush 1990, in Radosevich et al. 1997, *Weed Ecology: Implications for Management*, 2nd ed. Copyright 1997 with permission of John Wiley & Sons Inc.)

Effect of Density on Growth. Figure 6.3*a* represents the typical growth response of a plant population to increasing density. Such data are obtained by sowing various densities of a single species and collecting (harvesting) the total plant biomass (yield). With the passage of time plants that grow at high density quickly meet the stress created by the proximity of neighbors, whereas plants at low density do so only as neighboring plants get bigger. Thus, at harvest the total yield per unit of area is independent of density; that is, the yield per unit of area is equivalent over a wide range of either natural or planted densities. This phenomenon occurs because the amount of growth by individual plants decreases as the density increases (Figure 6.3*b*). In its initial phase or at very low densities, the yield of the population is determined by the number of individuals, but eventually the resource supplying power of the environment becomes limiting. This ultimately determines yield, irrespective of the plant density. The relationship between density and plant productivity is reproducible and occurs for a wide range of species and mixtures of species. It is known as the *law of constant final yield.*

Mathematically, the law of constant final yield (Figure 6.3*a*) takes the form

$$Y = w_m N (1 + aN)^{-1} \tag{6.1}$$

where Y is the total yield (biomass) of the population, w_m is the maximum potential biomass per plant in the absence of competitors, N is the density of the population, and a is the area required for w_m (Silvertown and Charlesworth 2001). Figure 6.3*b* depicts the same relationship on an individual plant basis; that is, at high density the total yield is determined by many small plants while at low density it is determined by fewer larger ones. Since individual plant weight w equals Y/N, this relationship is normally written as

$$w = w_m (1 + aN)^{-1} \tag{6.2}$$

This relationship is often linearized by taking the reciprocal of w (Figure 6.3*c*), and the above equation is then written as

$$\frac{1}{w} = B_{i0} + B_{ii} N_i \tag{6.3}$$

where $1/w$ is the reciprocal of individual plant size (weight), $B_{i0} = 1/w_m$, and B_{ii} is a measure of intraspecific competition of species i on itself (Silvertown and Charlesworth 2001). Equation 6.3, which is also depicted in Figure 6.3*c*, is known as the *reciprocal yield law* and is derived from the same parameters as the law of constant final yield. This derivation will become important when interference in mixed stands is examined.

The level of available resource does not alter the relationship of density to yield (Figure 6.3). Either increasing or decreasing the amount of resource determines the ultimate amount of biomass production but does not affect its

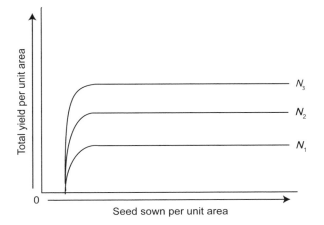

Figure 6.4 Influence of increasing amount of resource on relationship of total yield per unit area to density of seed sown, where N_1, N_2, and N_3 represent increasing increments of a limiting resource, for example, nitrogen fertility. (From Radosevich et al, 1997, *Weed Ecology: Implictions for Management*, 2nd ed. Copyright 1997 with permission of John Wiley & Sons Inc.)

relationship to density (Figure 6.4). Furthermore, under high initial density, final yield may be determined by many small plants or, because of density-determined mortality, fewer larger ones. In either case, the yield per unit area is a constant feature of the environment. This environmental constancy is due to the characteristic nature of plant growth, often referred to as *plasticity*, which is the ability of plants to alter their size, mass, or number in relation to density or other environmental stresses.

Effect of Density on Size Distribution. At any given density of a plant population, a characteristic size distribution of individuals is expected. One way to express this distribution would be to average the size or weight of the population (total weight per number of individuals). An average value is quite misleading, however, because normally very few plants are found that reflect the average size. In most plant populations a size distribution arises in which most are suppressed and small and a few are large and dominant. This distribution was demonstrated by Ogden (1970) with several annual species (Figure 6.5) which had not yet experienced density-dependent mortality. In all cases relatively few plants make up most of the plant biomass.

A similar phenomenon to that of annual plants is observed in stands of trees. The typical stand usually starts with a relatively large number of small trees per unit area, often thousands in natural stands and hundreds if the stand is planted. In both instances, the number of trees decreases over time and the trees that are most vigorous or best adapted to the local environment are most likely to survive. Growth in height is usually the critical factor for tree survival, although taller

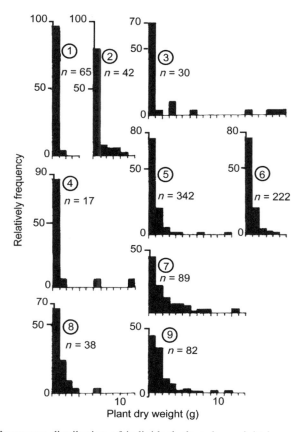

Figure 6.5 Frequency distribution of individual plant dry weight in some mixed annual weed populations in arable field in North Wales, where N = density of individuals per approx. 0.5 m^2: (1) Poaceae—mostly *Poa annua*, (2) *Atriplex patula*, (3) *Polygonum aviculare*, (4) all other species, (5) *Stachys arvensis*, (6) *Stellaria media*, (7) *Spergula arvensis*, (8) *Senecio vulgaris*, (9) *Polygonum persicaria* and *P. lapathifolium*. (From Ogden 1970, *Proc. N.Z. Ecol. Soc.* 17:1–9. Copyright 1970. New Zealand Ecological Society.)

trees are usually largest in other dimensions of growth as well, especially crown size. As weaker (smaller) trees are crowded by their taller neighbors, their crowns become increasingly misshapen and restricted in size (Figure 6.6). Such trees gradually become overtopped and eventually die. Foresters recognize four standard size classes of trees (Smith et al. 1997): dominant, codominant, intermediate, and overtopped (suppressed).

The place that an individual plant occupies within this apparent hierarchy of size classes, regardless of the plant's life cycle, is determined at a very early stage of development. This concept, often called *space capture*, has been recognized as a principle that assists understanding of the many interference processes among neighboring plants.

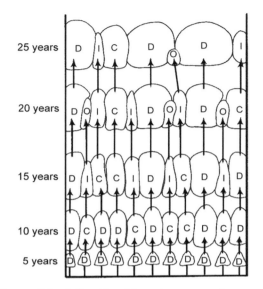

Figure 6.6 Relative spatial relationship of trees in canopy of same pure even-aged stand at succession intervals of age showing differentiation of trees into crown classes. Figure illustrates suppression, as a result of competition, of some trees that were initially dominants: D = dominant, C = codominant, I = intermediate, O = overtopped. (From Smith 1986, *The Practice of Siviculture*. Copyright 1986 with permission of John & Wiley & Sons Inc.)

Effect of Density on Mortality. Plants have an innate capacity for self-thinning as space (implied resources) available to them becomes more and more limited. This phenomenon was first noted by Yoda et al. (1963) and has been termed the $-\frac{3}{2}$ *power law* after the mathematical relationship (slope in Figure 6.7) between plant weight (size) and density that occurs in response to thinning. This relationship expresses a lowered probability of survival as plant size increases. In fact, growth suppression and the occurrence of weak individuals are probably less severe cases of the thinning phenomenon, since death is the most extreme response to stress. Although the power law has been demonstrated repeatedly, Silvertown and Charlesworth (2001) note that the slope of the line is closer to $-\frac{4}{3}$ and that this general relationship between body mass and population density is found in animals as well.

The fact that self-thinning is also an important factor determining yield is reflected in numerous agricultural and forest management studies. Agronomists often recommend certain seeding rates of annual crops, and foresters suggest spacing distances between planted or naturally regenerated trees to avoid self-thinning or growth suppression (Smith et al. 1997). Foresters also empirically determine the relationship between stand density, density-dependent mortality, and individual tree size for every merchantable tree species (e.g., Reineke 1933).

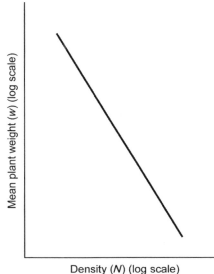

Figure 6.7 Diagrammatic representation of $-\frac{3}{2}$ power law in relation to reciprocal yield law: W = individual plant weight (biomass per plant); N = plant density. Solid line represents reciprocal yield law.

The optimum stand density to obtain a particular size class of tree and the amount of time between thinning or harvest are determined from such relationships.

Increasing the amount of a limiting factor, such as nutrients, often enhances density-dependent mortality because the dominant plants in the density-dependent hierarchy (Figures 6.5 and 6.6) continue to capture most of the resource. Consequently, larger plants become more dominant whereas smaller plants become more suppressed or even die. A notable exception to this generalization is light, because with this environmental resource increased levels of irradiance allow increased survival of plants in all size classes.

Effect of Density on Reproduction. Success of a colonizing species is eventually determined by reproductive output, since the population must be continually maintained through time. The reproductive output of annual plants is especially important since seed are the only link between generations to the inhabited site. Annual weeds and invasive plants apparently regulate and maintain a relatively stable reproductive output through growth plasticity in response to density and mortality.

Palmblad (1968) examined the reproductive response to changing density of eight weed/invasive species: *Bromus tectorum* (downy brome, cheatgrass), *Capsella bursa-pastoris* (shepherd's purse), *Conyza canadensis* (horseweed), *Plantago lanceolata* (buckhorn plantain), and *P. major* (broadleaf plantain), *Senecio sylvaticus* (woodland groundsel), and *S. viscosus* (sticky groundsel), and *Silene angelica* (=*S. gallica*, English catchfly). Seed were sown into bare soil in greenhouse pots at densities ranging from 55 to 11,000 seeds/m^2 (5–200 seeds per pot). Palmblad observed that the amount of seed produced per species was relatively constant and independent of density (Table 6.3). At low density, most of the plants survived to produce an abundance of seed. At high seedling densities, self-thinning generally occurred and the surviving plants were fewer but larger than if there had been no mortality. The combination of more and relatively larger plants at high density usually resulted in similar seed production regardless of planting density. In the case of horseweed, buckhorn plantain, and broadleaf plantain, subsequent reproduction the next growing season by survivors that had remained vegetative enhanced the reproductive stability of those species. Even though seed output varied dramatically from species to species, the total amount of seed produced per species was remarkably uniform across the 200-fold range of densities that was explored.

Although the relationship described above has been demonstrated repeatedly for weeds (reviewed in Canner and Wiles 2002), Watkinson (1985) observed the opposite effect among eight dune and marsh species (Keddy 2001), where reproductive output declined with increasing density of mature individuals depending on the species and habitat conditions in which the plants grew. Others report similar density-dependent reproduction under conditions of pollen (Knapp et al. 2001) or pollinator limitation (Wagenius 2006), enriched CO_2 (He and Bazzaz 2003), or size-dependent flowering (Thrall et al. 1989, Buckley et al. 2001). It appears that theoretically stable reproductive output over a range of densities may not be expressed under suboptimal conditions or in plants of certain life histories.

As described in Chapter 3, propagule availability and pressure are primary factors that influence the abundance and distribution of plants (Flinn and Vellend 2005, Thomsen et al. 2006). Propagule pressure is critical for successful establishment of weeds and invasive species in new areas. The reports by Palmblad (1968) and Canner and Wiles (2002) illustrate how mortality and reproductive plasticity in response to density together ensure reliable seed output for many weeds, a phenomenon that has important implications for weed and invasive plant management. Rarely are all of the weeds or exotic plants in an area removed as a result of any control measure. Plants that escape control have the potential, through survival and subsequent plastic growth, to maintain a relatively constant seed rain. If the seed rain is incorporated into the soil and the seed bank is continually replenished, restoration of the site will be difficult. Thus, because of potentially constant reproductive output, the acceptable density of many weed and invasive plants may be very low. This demonstrates the value of monitoring and "rogueing" fields to prevent establishment of new exotic plants or to remove the survivors that follow weed control. It also suggests that a more realistic approach to weed/ invasive plant management may be to determine optimum weed densities (thresholds) based on desired crop productivity or biodiversity objectives rather than maximum crop output or perfect weed control and to focus control efforts on reducing seed rain and dispersal.

TABLE 6.3 Effect of Density on Mortality and Reproduction in Selected Weed Species[a]

Species \ Density[b]	Percent Mortality Before Flowering					Percentage of Individuals That Produced Fruit					Number of Seed Produced (10^3 Seed)				
	1	5	50	100	200	1	5	50	100	200	1	5	50	100	200
Bromus tectorum	0	—	3	5	12	100	—	81	81	75	2.4	—	2.6	2.6	2.9
Capsella bursa-pastoris	0	0	1	3	8	100	100	82	83	73	23.7	30.5	40.3	37.2	30.1
Conyza canadensis[c]	0	0	1	4	8	100	87	51	42	36	52.6	59.6	40.8	35.3	38.4
Plantago lanceolata[c]	0	0	0	0	1	100	53	8	2	1	2.2	1.4	0.4	0.03	0.06
Plantago major[b]	0	7	6	10	24	100	93	72	52	34	12.0	12.7	8.2	6.6	4.4
Senecio sylvaticus	0	0	3	7	8	100	80	83	49	67	3.8	3.6	4.3	3.7	5.1
Senecio viscosus	0	0	2	2	5	100	100	79	69	68	4.3	11.4	7.7	7.9	7.3
Silene anglica	0	0	0	0	0	100	67	28	25	28	10.3	10.6	18.6	25.4	19.7

[a]Data on percent mortality and fruit production were obtained from Table 5 while numbers of seed produced were obtained from Table 6 in Palmblad (1968). Only first-year data are presented.

[b]Densities are 1, 5, 50, 100 and 200 plants per pot.

[c]At high densities, approximately 8, 50, and 30% of the plants in each pot for *C. canadensis*, *P. lanceolata*, and *P. major*, respectively, remained vegetative and did not flower or produce fruit the first year.

Source: Modified from Palmblad (1968), *Ecology* 49:26–34 in Radosevich et al. 1997, *Weed Ecology: Implications for Management*, 2nd Ed. Copyright 1997 with permission of John Wiley & Sons Inc.

Species Proportion. The density responses of single species have been of concern so far, although all of the relationships discussed above also apply to mixed stands, populations, or communities. In fact, mixtures of species are more common in nature than monocultures of a single species. In agriculture, forestry, and range management, occupation by plants other than ones deemed most desirable is a normal event and species diversity is the rule rather than an exception (National Research Council 1989, Barbour et al. 1999, Keddy 2001). It is only through extreme management measures and expenditures of chemicals, energy, labor, or money that crop monocultures can be attained and maintained (Lewontin 1982, Levins 1986, Levins and Vandermeer 1994, Pimentel et al. 1999, Liebman 2001).

When interactions between at least two species are studied, proportion becomes another factor to consider. Proportion is the relative density or ratio of each species in a stand. Spitters (1983a,b, 1989) demonstrated the importance of species proportion in competition by expanding the reciprocal yield law (Figure 6.3) to include a mixture of two species:

$$\frac{1}{w_i} = B_{i0} + B_{ii}N_i + B_{ji}N_j \tag{6.4}$$

where w_i is the weight of individual plants of species i, B_{i0} is the theoretical mean weight of individual plants of species i under competitor-free growing conditions, B_{ii} is the regression coefficient quantifying the intraspecific effect of density (N_i) of species i on the reciprocal of individual plant weight of species i, and B_{ji} is the regression coefficient quantifying the interspecific effect of density (N_j) of species j on the reciprocal of individual plant weight of species i (Table 6.4). A similar equation can be written for species j or the equation can be expanded to

TABLE 6.4 Statistical Components of Competition and Their Interpretations

Statistic	Component of Competition	Parameter Estimate
Intercept	Maximum potential plant size	B_{j0}, B_{i0}
Regression coefficients	Intensity of intra- and interspecific competitive effects on plant size	B_{jj}, B_{ji} B_{ii}, B_{ij}
Interaction term	Interaction between intra- and interspecific density effects on plant size	B_{pi}, B_{pj}
Ratio of coefficients	Competitive effects of species j density relative to species i density	B_{ii}/B_{ji} B_{ij} / B_{jj}
Model R^2	Overall importance of competition in determining plant size relative to other factors	R^2_{ii}, R^2_{ij} R^2_{jj}, R^2_{ji}
Partial R^2	Importance of competitive effect of each species density	ρR^2_{ii}, ρR^2_{ij} ρR^2_{jj}, ρR^2_{ji}

Source: Modified from Shainsky and Radosevich (1991). *For Sci.* 37:574–592 in Radosevich et al. 1997. *Weed Ecology: Implications for Management*, 2nd ed. Copyright 1997 with permission of John Wiley & Sons Inc.

include more than two species. The regression coefficients describe the effects of intra- (B_{ii}) and interspecific (B_{ji}) competition on individual plant weight. They also indicate that the density of each species relative to that of the other in the stand influences the yield of both species. The values of density (N_i, N_j) in these equations also reflect the need to establish the effects of all species in a mixed stand on the final yield (total outcome) of each species.

The effect of proportion or relative density was often overlooked in early weed–crop competition studies in agriculture but has now become more the norm. It is also an important factor to consider in studies of plant–plant interactions in managed forests and rangelands and exotic plant invasions in natural ecosystems.

Spatial Arrangement. Spatial arrangement is the horizontal pattern of aggregation of plants (Figure 2.5) that reflects dispersal patterns. Fischer and Miles (1973) developed several theoretical stochastic models for interference between crop plants, arranged as a grid of points, and randomly located weeds. They assumed that in the absence of neighbors a plant expands from emergence until it meets another plant, whereupon expansion ceases. Ultimately, each plant establishes a zone of resource exploitation (Figure 6.8) and theoretically many non-overlapping weed and crop domains would occupy an agricultural field, forest, meadow, range, and so on.

Fischer and Miles established that plant arrangement could be an important factor in determining the outcome of weed–crop competition, with weeds gaining least advantage if the crop is planted in square or triangular patterns. In most weed–crop competition studies, spatial arrangement among individual crop plants is held constant, usually in a square or rectangular arrangement, and is assumed to have little effect on the study's outcome. Unfortunately, there are still few experiments to test this assumption.

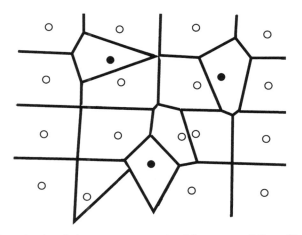

Figure 6.8 Plant size in relation to arrangement and "emergence" time. (From Fischer and Miles 1973, *Math. Biosc.* 18:335–350. Copyright 1973 with permission from Elsevier.)

METHODS TO STUDY INTERFERENCE (COMPETITION)

Many methods have been developed to study plant competition and each considers the factors of density, species proportion, and spatial arrangement to varying degrees (Radosevich 1987). The methods fall into four general types of experimental designs: *additive*, *substitutive*, *systematic*, and *neighborhood*. Although the methods were developed primarily to study competition, some can also be used to explore other kinds of interactions. The interrelationships between negative and positive interactions that occur in plant associations can be revealed or masked by particular experimental designs. In each method, total or individual plant yield, plant growth rate, or plant survival is measured. Each method is a form of bioassay in which the response of one species is used to describe the influence of the others in the mixture.

Additive Designs

Additive designs are perhaps the most common approach used to study plant competition. More than two species can be grown together in additive experiments, but most studies are conducted with only two species, for example, a crop and a weed or a desirable tree and a shrub species. The density of one species, such as the crop, is always held constant while the density of the other is varied, usually by removal or addition. The design is relevant for many agricultural and forestry situations in which at least one species of weed infests an area already occupied by a fixed density of crop or where various weed densities occur from different weed control treatments. In this approach, crop yield usually improves as weed densities diminish until weed levels are reached that do not significantly decrease production further (Figure 6.9).

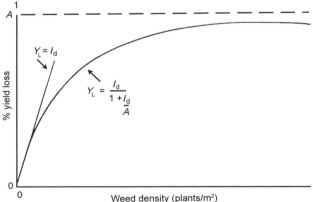

Figure 6.9 Rectangular hyperbolic model for relating yield loss to weed density: $Y_L =$ percentage yield loss; A, $I =$ parameters that determine shape of curve for response of yield loss to weed density. (From Cousens 1985, *Ann. Appl. Biol.* 107:239–252. Copyright 1985. Blackwell Publishing Ltd, reproduced with permission.)

The additive method has been criticized because it does not account adequately for the influence of total density and species proportion on the outcome of competition (Harper 1977, Radosevich 1987, Vila et al. 2004). In the additive approach, the total plant density always varies among treatments and the proportion among species changes simultaneously with total density. Thus, these two factors in the experiment covary, making it impossible to evaluate the effects of either factor alone. Spatial arrangement among plants in additive designs is assumed to be uniform, since the crop is usually planted in a grid pattern and the influence of intraspecific competition is assumed to be constant. However, weed placement is often unreported or unknown in such experiments. This method also does not account for the effects of density on yield of either weeds or crops (discussed above), which may change their spatial arrangement (Hashem et al. 1996, Mohler 1996, 2001).

Production-oriented plant scientists often face a dilemma when accounting for the influence of proximity factors, especially total and relative plant densities, on the outcome of additive experiments. Agricultural and tree crops, for example, are usually grown at a constant density, determined experimentally or intuitively to maximize economic yield, while weeds/invasive plants create conditions where both total and relative plant densities vary. These problems are overcome by more complex experimental designs discussed below.

Substitutive Designs

Many of the criticisms of additive designs can be overcome by the substitutive approach to competition study. There are three general types of substitutive experimental designs: *replacement series*, *Nelder*, and *diallel*. The premise of all substitutive designs is that the yields of mixed stands can be determined by comparison to monoculture yields (Figure 6.10). In a substitutive experiment, the total plant density is held constant while species proportions are varied; thus, the two

Figure 6.10 Replacement scheme with crop plants (**x**) and weed plants (**o**). [From Spitters and Van den Bergh 1982, in Holzner and Numata (Eds.) 1982, *Biology and Ecology of Weeds*. Dr. W. Junk Publishers. Copyright 1982 with permission of Springer Science and Business Media.]

[handwritten: either no interaction (maybe too far apart) or equivalent competitors]

[handwritten: amensalism (0,–) or competition]

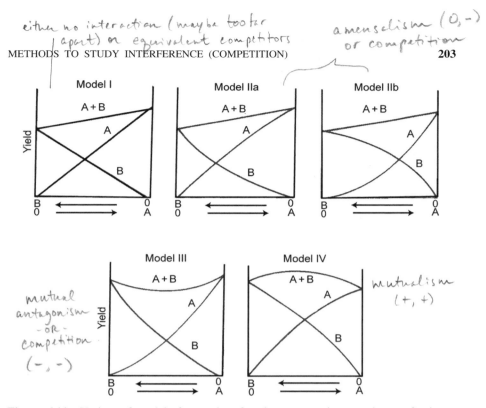

[handwritten next to lower-left: mutual antagonism –OR– competition (–, –)]

[handwritten next to Model IV: mutualism (+, +)]

Figure 6.11 Variety of models for results of replacement series experiments for interference study. The vertical axis indicates some measure of plant yield and the horizontal axis represents the proportion (0–1.0) of two species in mixture. See text for explanation of the models. (Modified from Harper 1977, *Population Biology of Plants*, in Radosevich et al. 1997, *Weed Ecology: Implications for Management.* 2nd Ed. Copyright 1997 with permission of John Wiley & Sons Inc.)

experimental variables are not confounded during the experiment. The spatial arrangement among individual plants in the design is usually nonrandom.

Replacement Series. A replacement series design includes pure stands as well as mixtures in which the proportion of the two species studied is varied while the total plant density is a constant over all treatments (deWit 1960). Figure 6.11 presents the four possible outcomes for the interaction of two species when grown in a replacement series experiment. In Figure 6.11, the vertical axis indicates some level of plant yield and the horizontal axis represents the proportion (0 to 1) of the two species in the mixture.

There are two possible interpretations for the yield-versus-proportion response depicted in model I (Figure 6.11). One interpretation is that the two species are located so far apart that no interaction can occur between them. In order to detect competition (or a positive response), experiments using this approach must be conducted at sufficiently high densities and/or for long enough periods to fall within the range of constant final yield (Figure 6.3). The second interpretation for model I (Figure 6.11) is that the ability of each species to interfere with the other is equivalent; that is, each species contributes to the total yield in direct proportion to its presence in the mixtures.

In some situations, two species make similar demands upon the environment but differ in their response. In models IIa and IIb (Figure 6.11), one species is more aggressive than the other and contributes more than expected to the total yield, while the other contributes less than expected. This is the model for amensalism as suggested by Burkholder (1952) but is also often referred to as competition. In each combination, one curve is always concave while the other is always convex, indicating that the interaction between species is for a common resource(s) and that one species gains more than the other.

In model III (Figure 6.11), neither species contributes its expected share to the total yield. The yield of the two species in any mixture is less than that achieved when either is grown in a pure stand at the same total density. This model represents mutual antagonism or competition (Burkholder 1952) such that maximum productivity results from the monocultures. Mutual benefit is depicted in model IV (Figure 6.11) since both species in the mixtures produce more than is expected from their yields produced in pure stands. Model IV depicts symbiosis, but it also may indicate that each species fails to harm the other as much as expected (Harper 1977, Shainsky and Radosevich 1992, Radosevich et al. 2006). In such situations, each species escapes from some measure of competition with the other (negative facilitation). For example, mutual benefit may occur between certain tree species or between weeds and crops and is important in multicropping situations (positive facilitation).

The value of replacement series designs is their predictiveness. There are four models to interpret neutral, negative, or positive effects between species (Figure 6.11). Predictions of shifts in species composition over time can also be made. In Figure 6.12, species A is more aggressive than species B (model II, Figure 6.11). The dotted line indicates the predicted number of generations for one species to replace the other in a mixed stand, which in this example is about five generations. This type of replacement event is important when determining dominance or species shifts under changing cultural practices or environmental conditions.

It is possible to determine the relative effects of intra- and interspecific interference using the replacement series design, but partitioning the absolute effects cannot be accomplished readily [see Jolliffe (2000) for review]. The replacement series is also limited in that actual and expected monoculture yields, and thus the outcome of any particular experiment, will vary according to the plant density selected for study. Jolliffe et al. (1984) developed a procedure to evaluate quantitatively the results of replacement series experiments. They suggest including several monoculture densities in the design and calculating relative yield responses of the species in the mixture to alleviate this problem. The replacement series design is sometimes criticized for being artificial since a constant density rather than a variable one is used to grow most crops/desirable plants. The method is also cumbersome for studying mixtures of species of different life

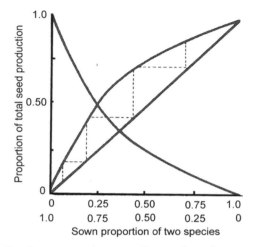

Figure 6.12   Use of replacement series to predict number of generations it will take for one species to displace another as result of interference. The predicted number of generations for the population to change is indicated by the dashed line. The solid diagonal line indicates the yield if both species had identical competitive abilities. (From Radosevich et al. 1997, *Weed Ecology: Implications for Management,* 2nd ed. Copyright 1997 with permission of John Wiley & Sons Inc.)

histories or growth forms. Proportions expressed as ratios of biomass may be more appropriate than ratios based on density when life-forms of the species differ markedly.

Nelder Designs. The Nelder design is usually restricted to the study of competition among individuals of a single species (Nelder 1962). It consists of a grid of plants often planted as an arc or circle (Figure 6.13). The area per plant or the amount of space available to each plant changes in a consistent manner over the different parts of the grid. The influence of another species can be introduced into the Nelder design by overseeding the entire area with a second species. In this case, the effect of intraspecific interference under the constant influence of a "background" species is determined. A qualitative assessment of interspecific competition may be made by comparing arcs with and without the presence of the background species. Interspecific effects may also be examined by alternating the placement of the species along an arc or spoke, so that differing ratios or proportions of the species result (Cole and Newton 1987). Usually every other plant is alternated, giving a 1 : 1 ratio, or species proportion of 0.5. The alternating arrangement of plants can dramatically affect individual plant responses, however, because alternating bigger and smaller plants along a spoke or arc is an artifact of the planting scheme and produces a "wave" of size differences as the experiment proceeds through time.

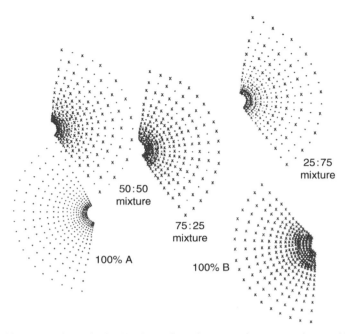

Figure 6.13 Examples of distribution of various species proportions within Nelder design. (From Radosevich 1987, *Weed Technol.* 1:190–198. Copyright 1987. Weed Science Society of America. Reprinted by permission of Alliance Communications Group, a division of Allen Press Inc.)

Another disadvantage of the Nelder design is that only individual plants can be measured, causing difficulty in obtaining the "stand" or population effect from interaction with neighbors that is possible with rectangular designs. The method also does not allow for the partitioning of density and proportion effects on the interaction unless more than one species proportion is used (e.g., Figure 6.13). Nor can the effects of intraspecific competition be separated readily from those of interspecific competition (Radosevich 1987, Radosevich and Roush 1990).

The advantage of the Nelder design is that an array of densities can be studied without changing the pattern of plant arrangement. In addition, only a small area is required to examine the effect of many densities, which is not the case with most square or rectangular designs. This economy of space allows considerable flexibility in dealing with possible environmental gradients in the field. However, arcs are often difficult to plant, especially when spacing along or between crop rows is dictated by equipment or other reasons, but several parallel row arrangements have been proposed (Bleasdale 1967).

Diallel Designs. Plant communities are composed of several to many species. The diallel design combines individuals of each species under study into all possible pairs to examine their interactions (Harper 1977). The experimental design uses only one or two individuals of each species per "treatment." In an experiment involving two species, an individual of each species is grown alone (A and B), two individuals of each species are grown together (AA and BB), and one individual of each species is grown in mixture (AB). This design allows an examination of both intra- (compare A to AA or B to BB) and interspecific (compare AB to either AA or BB) interactions within the framework of a substitutive experiment. The yields of the species mixtures and monocultures are somewhat analogous to the performance of genetic hybrids and inbred lines. Combinations of more than two species also may be examined. The design below uses three species:

A AA AB

B BB BC

C CC CA

The advantage of this approach is the simple design, which can be combined with intensive destructive data collections to determine biomass partitioning among the species under a regime of interference. The researcher, however, is restricted to working with individual plants. Pot or plot size is critical in such experiments, since resources may be relatively unlimited in a system that uses only one or two individual plants. In addition, the influence of density from more than a single neighbor cannot be determined (Radosevich 1987).

Systematic Designs

Because of the joint influences of proximity factors in competition (and interference, more generally), another approach has been used that systematically varies both total and relative plant densities (proportion) (Spitters 1983a,b, Firbank and Watkinson 1985, Cousens 1991). This approach provides a better basis for quantifying competition than either conventional additive or substitutive designs because it provides a broad array of relative densities to examine the consequences of interaction (Roush et al. 1989, Cousens 1991, Cousens and Mortimer 1995, Jolliffe 1997, 2000). Two designs are used to describe density response surfaces systematically: *addition series*—a combination of several replacement series over a range of total densities (Spitters 1983a,b)—and *additive series* or *factorial design*—a combination of additive experiments at different total densities (Rejmánek et al. 1989). The addition series encompasses a triangular portion of a matrix of density combinations (Figure 6.14a), while the additive series includes all possible combinations of several densities of each species (Figure 6.14b). Since both approaches explore a range of total and relative densities systematically, they are considered together in this section.

Addition Series and Additive Series Designs. Spitters (1983a,b) used the reciprocal yield law (Figure 6.3 and Equation 6.4) as the basis for studying plant interactions. As stated earlier, multispecies reciprocal yield (Equation 6.4) describes

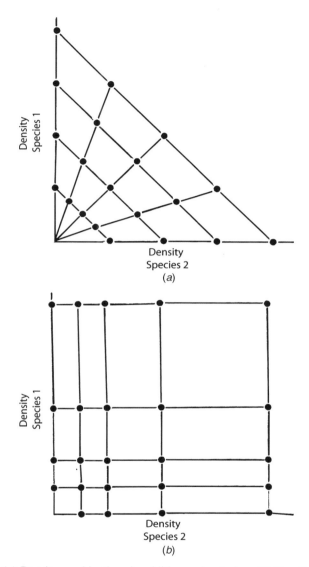

Figure 6.14 (*a*) Density combinations in addition series design. (*b*) Density combinations in additive series or factorial design. (From Cousens 1991, *Weed Technol.* 5:664–673. Copyright 1991. Weed Science Society of America. Reprinted by permission of Alliance Communication Group, a division of Allen Press Inc.)

the relationship between individual plant weight and plant density for more than one species. This equation can be used to determine the yield of a species as a function of the total and relative densities of all other species in a plant mixture. It is possible to describe the yield of a single species (e.g., species i) or to describe the yield of each species as affected by interference from any other species in the plant mixture (e.g., species i, j, or k) using these experimental methods. The addition series and additive series designs represent an approach to examine interactive (positive or negative) relationships using an array of species densities and proportions.

The experimental design for either an addition or an additive series may take the form of a simple matrix of two species. For example, a hypothetical two-way matrix for an addition series is

	0	1A	4A	8A	16A
1B		1/1	4/1	8/1	16/1
4B		1/4	4/4	8/4	16/4
8B		1/8	4/8	8/8	16/8
16B		1/16	4/16	8/16	16/16

where species density per plot in this example ranges from 0 to 16 plants/m^2 in monoculture and from 2 to 32 plants/m^2 in mixture; A and B are the two species and, for example, 1/1 is a mixture with one A and one B per plot. When two species are considered, the addition series is simply a group of replacement series experiments. However, more complex planting arrangements are necessary with more than two species. Figure 6.15 depicts the possible arrangement for an experiment involving three species. The densities of two species ($\times$ and $\bullet$) increase along perpendicular gradients from zero to a high density; then a similar range of densities of another species ($\square$) is superimposed upon species $\times$ and $\bullet$ (Figure 6.15). In this manner, a range of monoculture densities, total densities, and proportions can be varied systematically throughout the experiment. Miller and Werner (1987) and Roush and Radosevich (1985) expanded this method for interactions of four species. The regression coefficients derived from Equation 6.4 or Table 6.4 are then used to determine and separate the effects of intra- and interspecific competition (Carpinelli 2005).

Neither the addition series nor the additive series method accounts for spatial arrangement, so arrangement of the plants must either be constant or be assumed to have a constant effect. Hashem et al. (1996) included several levels of crop rectangularity as another factor in an addition series between wheat and Italian ryegrass (*Lolium multiflorum*) to account for nonregular spatial arrangement.

Neighborhood Designs

While most competition studies concentrate on stand yields, the yield of individual plants may be influenced by nearness to other individuals (Firbank and Watkinson 1987, 1990) and by local variation in the environment (microsites). When individual responses to the proximity of other plants are of primary interest, a neighborhood method may be appropriate to assess interference. In neighborhood designs, performance of a target individual is recorded as a function of the number, biomass, cover, aggregation, or distance of its neighbors. Many

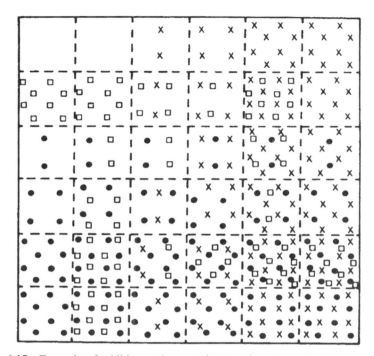

Figure 6.15 Example of addition series experiment using three plant species. Densities range from 0 to 24 plants (symbols per unit of area) and proportions among species vary systematically throughout the design. In this figure, densities of × and ● range from 0 to 8 plants in perpendicular directions. Densities of □ also range from 0 to 8 plants and are superimposed on × and ● in a systematic manner. (From Radosevich 1987, *Weed Technol.* 1:190–198. Copyright 1987. Weed Science Society of America. Reprinted by permission of Alliance Communications Group, a division of Allen Press Inc.)

generalized equations have been developed to represent the relationship of target individual (species) performance to the proximity of neighboring plants (Goldberg and Werner 1983, Silander and Pacala 1990, Wagner and Radosevich 1991, 1998 Tremmel and Bazzaz 1993).

Goldberg and Werner (1983) describe a common experimental approach in which the performance of a single target species is evaluated over a range of densities of a neighboring species. The target species either is grown alone or is surrounded by individuals of the neighboring species (Figures 6.16 and 6.17). The spatial arrangement of individual plants can also vary among target and neighboring species. The effect of the neighboring species on the target species is defined as the slope of the regression of performance (e.g., growth rate, survival, or reproductive output) of target individuals on the amount (e.g., density, biomass, or leaf area) of the neighboring species (Figure 6.18). The relationship is expressed using

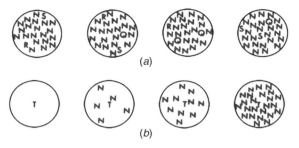

Figure 6.16 Example of experimental design for evaluating competitive effects of one neighbor species (N) on a target (T) with R, Q, and S representing individuals not belonging to neighbor species selected for study: (*a*) initial field condition; (*b*) after treatment. *Only four steps of the neighbor density gradient (after treatment) are shown. The experiment must include a much wider range of densities to estimate accurately the slope of (X_{TN}) of the regression equation $P(T) = Y + X_{TH}[A(N)]$. (From Goldberg and Werner 1983, in *Am. J. Bot.* 70:1098–1104. Copyright 1983. American Journal of Botany.)

a linear regression equation:

$$P(T) = Y - X_n[A(N)] \tag{6.5}$$

where $P(T)$ is the performance of the target individual and $A(N)$ is the "amount" of neighbors. The Y intercept corresponds to the performance of the target species with no neighbors present and X_n is a competition coefficient. A refinement of the neighborhood method is to measure the distance of neighbors from the target

Figure 6.17 Example of neighborhood experiment. Douglas fir planted as grid at 3 m × 3 m spacing are target individuals while various biomass levels of shrub and herbaceous vegetation are neighborhood species. (Photograph by S. R. Radosevich, Oregon State University, Corvallis.)

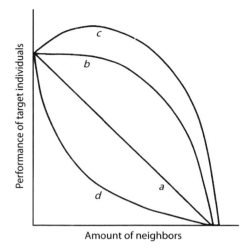

Figure 6.18 Examples of some possible relationships between performance of individuals of target species and abundance of neighbor species. Curve *a* corresponds to equation given in text (linear relationship). Curve *b* represents a relationship in which competitive effects are minimal at low neighbor density but increasingly severe as density increases. Curve *c* represents a quadratic function with a peak at some intermediate neighbor density, indicating beneficial effects at low density but competitive effects at high density. Curve *d* represents a negative exponential function, indicating decreasing per amount competitive effects as density increases. (From Goldberg and Werner 1983, in *Am. J. Bot.* 70:1098–1104. Copyright 1983. American Journal of Botany.)

individuals, so the diminishing effects of more distant neighbors can be incorporated into the regression equations. In that case, the term for amount of neighbors [$A(N)$] can be replaced by $\sum_1 (A_i / d_i^2)$, where A is the biomass or cover of an individual plant of the neighboring species i at a distance (d_i) from the target plant. Several possible interactions can be explored with neighborhood designs (Figure 6.18). Because observations are based on single target individuals, however, many treatments (densities) must be examined to quantify effects of neighbors accurately.

Many experiments have been performed using this general approach to examine plant–plant interactions between a wide array of native and exotic species and in a variety of habitat types (e.g., Putwain and Harper 1970, Fowler 1981, Goldberg 1987, Goldberg and Landa 1991, Shipley et al. 1991, Wagner and Radosevich 1991, 1998, Simard et al. 2006).

Approaches Used to Study Plant Interference (Competition) in Natural and Managed Ecosystems

The methods of studying competition described above were developed and have largely been used in agriculture and other production systems, where all species in the assemblage are planted de novo. Transferring these methods to natural eco-systems is, therefore, problematic if the goal of study is to understand the dynamics of species already present. Nevertheless, competition has also been studied in nonmanaged ecosystems and is hotly debated by plant ecologists as a mechanism for plant interactions and community composition (Goldberg and Barton 1992, Keddy 2001, Gurevitch et al. 2002). Perhaps because of the complexity of natural ecosystems, experiments on competition (interference) focus primarily at the level of individual plants or populations (stands, fields, or sites), but in some cases competition in community assemblages is addressed (Symstad and Tilman 2001, Fargione et al. 2003, Fargione and Tilman 2006).

Regardless of the level of complexity, the study of interactions requires inclusion of experimental controls. This aspect of experimental design can be extremely difficult to achieve without disturbing the system under study and com-promising its "real-world" relevance. However, without appropriate controls, it is difficult to conclude definitively about the role or even occurrence of competition. Control in some cases can be achieved by simple correlations of plant associations in particular environments or, as discussed above, be constructed through veg-etation manipulation. In either case, comparison of competitive and noncompeti-tive "treatments" is the essential ingredient of interference study, and the degree to which experimental control can be attained determines how definitive the exper-iment will be. Keddy (2001) identifies four requirements for competition study:

- Manipulating abundance—increasing (adding) or decreasing (removing) the number or biomass of neighbors is required to establish competition treatments.
- Measuring performance—evaluating plant responses ranging from short-term physiological responses to longer term changes in size, weight, or area provides a measure of the effect of competition.
- Comparing to controls—establishing controls distinguishes scientific experi-ments from all other forms of observation. Controls usually consist of unmanipulated individuals or populations, but the experimental design and controls should also address the modifiers of interference (competition) as well as whether the experimental procedure itself will produce an effect.
- Measuring resource levels—evaluating environmental factors such as resources in both treatments and controls may allow the determination of which resources are affected by the interaction and in turn which resource is changed (reduced) by the presence of neighbors.

Four types of competition studies are routinely conducted by applied and basic plant ecologists, which we call *descriptive*, *retrospective*, *case*, and *gradient* studies.

Descriptive Studies. The first competition experiment, according to Keddy (2001), was conducted by A. G. Tansley and appeared in the *Journal of Ecology* in 1917.

Tansley's experiment was carried out in the Botanic Garden at Cambridge and included two closely related herbaceous species of bedstraw, *Galium saxatile* and *G. sylvestre*, that grow on rocky hillsides and pastures in Great Britain. By using a "common garden" of differing soil types, Tansley observed that G. *saxatile*, was less productive when grown on calcareous soil substrate, whereas *G. sylvestre* did not grow well on an acid peat soil. Thus, in association, *G. saxatile* displaced *G. sylvestre* on acid soil and the opposite occurred on the calcareous substrate. While each species was adapted to a particular soil type, Tansley concluded that competition also regulated their distribution (Gurevitch et al. 2002).

This study creates more questions than it answers, and as Keddy (2001) points out, it demonstrates how current thinking, debate, and even experimental habits concerning competition came about. For example, Tansley's study raises several questions about (1) the role of competition in controlling species distribution, (2) whether competitive outcome is contingent on environment, (3) whether niches are exclusive or inclusive among species, (4) the existence of niche overlap, and (5) dominance and subordinance among species. This study also demonstrates the tendency to use closely related species and the use of common gardens in the study of competition. None of these questions detracts from the descriptions and observations made in the study; rather, they simply open the debate about the process that accounts for the presence of *G. saxatile* and *G. sylvestre* in different places.

Retrospective Studies. Retrospective studies often consider the question of successional dominance, that is, what species will assume a dominant role on a site or landscape over time or as a result of management treatment. Because very few plant scientists or ecologists can begin a succession experiment and wait for it to reach climax condition, a common approach is to examine sites in a region at presumably differing stages of successional development. Data are then analyzed by either correlation or ordination to describe the occurrence of species on the various sites that overlap in species composition and time.

Retrospective studies were the basis by which Connell and Slatyer (1977) derived the three models for succession discussed in Chapter 2. However, as Connell and Slatyer note, it is possible that neutral, positive, or negative interactions could account for these differing models (Grace and Tilman 1990, Barbour et al. 1999, Gurevitch et al. 2002). Furthermore, environment and disturbance may be important in determining the classes of dominants and subordinates in species assemblages across a successional gradient (Keddy 2001). It is possible during retrospective study to quantify species and their abundance in each assemblage or seral stage and to infer community development, but less can be stated about how or why the assemblages occur or change over time without more controlled experimentation, as described above.

Case Studies. The examples of responses of crops or target plants to weed or neighbor density shown in Figures 6.9, 6.11, and 6.18 could be considered case studies, since each of the experiments examines interspecific competition and also applies experimental controls to various degrees. A common type of case study is shown in Figure 6.19, where Wagner et al. (1989) examined the pattern of survival and stem volume growth of ponderosa pine competing with different levels of shrubs. Various weed control tools were used to reduce shrub biomass, which

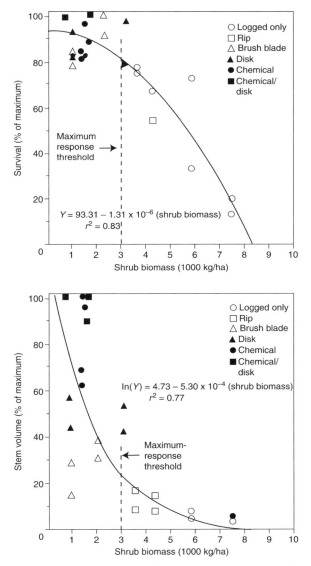

Figure 6.19 Percent of maximum survival (top) and stem volume (bottom) for eight-year-old ponderosa pine seedlings growing with various levels of shrub biomass in south-central Oregon after six site preparation treatments. (From Wagner et al. 1989, *New For.* 3:151–170. Copyright 1989 with permission of Springer Science and Business Media.)

constituted the experimental treatments. A case study of competition is demonstrated by the responses of ponderosa pine survival and stem volume to shrub biomass. However, in this example, the responses are also presented according to the tool employed to remove shrubs. Thus, from Figure 6.19 it is possible to conclude that disking and herbicides are more effective tools for enhancing pine survival and growth than the other mechanical tools used. The quantification of shrub biomass removal with tool use also allows an assessment of interspecific competition and presumably calculation of a competition index (discussed below) for other comparisons. The study is still limited, however, because the influence of tree density or biomass on the outcome of the experiment was not considered. Oliver and Powers (1978) performed a similar experiment in which ponderosa pine density and shrub biomass were varied. They observed that the density or biomass of both species affected the outcome of the pine–shrub interaction. Many such case studies describing weed–crop competition outcomes can be found published in the scientific literature; far fewer actually delve into specific mechanisms that cause the documented responses (Radosevich et al. 1997, Zimdahl, 1999, 2004).

Gradient Studies. Competition may be a fundamental process or property of plant communities, yet most studies of competition are limited to pairwise comparisons. However, the addition of only one more species to any of the methods discussed earlier reveals the limitation of those approaches for examining multispecies interactions. When the research objective is to examine the role of competition within the complexity of natural ecosystems, other approaches must be used. One such approach is the quantification of species occurrence along environmental gradients which exist in most habitats. For example, depressions in grasslands accumulate water and nutrients, whereas ridges are drier and less fertile. Similarly, gradients exist for soil type (and underlying substrate), soil depth, radiation under plant canopies, and even concentration of atmospheric gases. In addition, altitudinal gradients reflect changes in air temperature. An early examination of competition that was influenced by the ability of crop and weed species to differentially exploit the soil (water) resource was reported by Pavlychenko (1940) (described in more detail later in this chapter).

In this context, competition is often studied along gradients of increasing neighbor biomass, representing site productivity, in order to determine the role of competition in community composition (Campbell and Grime 1992, Gurevitch et al. 1992, Keddy et al. 1994, Goldberg et al. 1999, La Peyre et al. 2001). As noted by Gurevitch et al. (2002), however, results from individual studies are contradictory and no general pattern has emerged between competition intensity and site productivity. Further discussion of the use of environmental gradients in competition study is presented in Keddy (2001) and Gurevitch et al. (2002).

INTENSITY AND IMPORTANCE OF COMPETITION

Grace and Tilman (1990) indicate that two distinctions are necessary to fully examine the influence of competition as a possible factor in plant–plant

TABLE 6.5   Organizational Relationships of Intensity of Competition Versus Importance of Competition

	Intensity	Importance
Research focus	Mechanisms	Implications
Levels of organization	Environment, plant population	Plant populations, community
Management implications	Establish yield responses and economic thresholds	Forecast densities and species shifts, implement thresholds, integrate management

Source: Modified from Roush et al. (1989). *Weed Sci.* 37:268–275 in Radosevich et al. 1997, *Weed Ecology: Implications for Management*, 2nd ed. Copyright 1997 with permission of John Wiley & Sons Inc.

interactions: (1) *intensity* of competition and (2) its *importance* (Weldon and Slausen 1986). Intensity integrates physiological and morphological responses of individual plants when in the presence of neighbors of the same or different taxa. Importance, in contrast, describes the role of competition in relation to other processes that may influence the future abundance, density, or species composition of a plant community (Table 6.5). While the distinction between intensity and importance of competition has been the focus of some attention in natural ecosystems (Sammul et al. 2000, Howard and Goldberg 2001, Brooker et al. 2005), it is seldom examined by agricultural and natural resource scientists. For example, most competition experiments and models in agriculture, forestry, and other natural resource systems only consider the degree of desirable plant (crop) yield loss due to competition (intensity), without concern for the role of competition in future weed/invasive plant composition or abundance (Roush et al. 1989, Radosevich and Roush 1990, Vila et al. 2004).

Intensity of Competition

Historically, competition experiments performed in agriculture and forestry have documented levels of crop yield loss rather than the population or community implications of those interactions among crops or trees and weeds. Empirical studies usually have been either additive or substitutive experiments (Vila et al. 2004). Cousens (1985), Hakansson (1988), Jolliffe (1997), and Vila et al. (2004), Zimdahl (2004) have summarized numerous experiments of these types which were conducted over an array of cropping systems and environments. Stewart et al. (1984) and Wagner et al. (2006) provide similar summaries of experiments in young forest plantations. Crop yield response to weed density or weed cover is best described by a rectangular hyperbolic function (Figure 6.9) (Cousens 1985, Auld and Tisdell 1988, Alstrom 1990). A clear law of diminishing returns exists for this relationship between crop yield and weed density. As weed density increases, crop yield diminishes markedly until the density of weeds is reached that does not decrease crop production further.

Competition Intensity Indices. A competition index (CI) reflects intensity of competition and is a useful tool to quantify the competitive effect of weeds in crops and forest plantations and invasive plants in natural ecosystems (Grace 1995, Goldberg et al. 1999, Reynolds 1999, Vila et al. 2004). The CI is also used to compare the competitive effects of weeds or invasive plants under differing cultural practices in agriculture or management regimes in natural ecosystems (Gurevitch et al. 1992, Goldberg et al. 1999) and to compare differences among independent additive experiments. The most commonly used CI is the relative competition index (RCI), which is the proportional decrease in plant performance due to competition, calculated as

$$\text{RCI} = \frac{Y_{\text{no weed}} - Y_{\text{weed}}}{Y_{\text{no weed}}} \tag{6.6}$$

where $Y_{\text{no weed}}$ is the measure of the performance or yield of the crop or other plant of interest when it is growing free of weeds/invasive plants and Y_{weed} its performance or yield when the weed is present. In this way the relative competitive abilities of weeds or invasive plants can be ranked regardless of where or with what crop or other plants they are growing.

Relative Yield. Another useful measure of competition, *relative yield* (RY), is commonly used in agriculture to evaluate weed effects on crops as well as performance of crops grown together in mixed cropping (intercropping) systems. The RY is calculated from controlled experiments, such as those described above, where harvestable yield can be obtained. Both plants of interest are grown in monocultures and in mixtures and total density is held constant, such as in substitutive or systematic experimental designs. Then RY is calculated as

$$\text{RY} = \frac{Y_{\text{mixture}}}{Y_{\text{monoculture}}} \tag{6.7}$$

where Y_{mixture} is the average yield of crop plants when grown with weeds and $Y_{\text{monoculture}}$ is the average yield of the crop when grown without weeds (Harper 1977). When $\text{RY} = 1$, interspecific competition with the weed is not different from intraspecific competition that occurs when the crop is growing alone. If $\text{RY} > 1$, interspecific competition with the weed is less than intraspecific competition of the crop alone, while $\text{RY} < 1$ indicates that interspecific competition with the weed is greater than intraspecific competition of only the crop. Table 6.6 gives the RY of various crops growing in the presences of weeds (Vila et al. 2004). Interspecific competition between crops and weeds in these examples ranged from little (e.g., $\text{RY} = 1.9$) to severe (e.g., $\text{RY} = 0.3$) depending on the weed, crop, and growing conditions. It also is possible to calculate a CI for the species in Table 6.6. However, the two indices (CI and RY) differ in their utility and interpretation since RY is based on a constant experimental plant density, whereas total density can be unknown when CI is determined.

TABLE 6.6 Relative Yield of Several Crops Grown with Exotic Weeds Obtained from Replacement Series Experiments

Reference[a]	Weed	Common Name	Crop	RY[b]
Bridgemohan and McDavid (1993)	*Rottboellia cochinchinensis*	Itchgrass	Corn (maize)	0.6 1.1
Norris (1997)	*Portulaca oleracea*	Common purslane	Common beet	0.8 0.3
Ogg et al. (1993)	*Anthemis cotula*	Mayweed chamomile	Pea	1.9 1.8
Patterson and Highsmith (1989)	*Anoda cristata*	Spurred anoda	Cotton	0.9
	Abutilon theophrasti	Velvetleaf	Cotton	0.8
Wall (1993)	*Setaria viridis*	Green foxtail	Barley	0.7
	Avena fatua	Wild oat	Barley	0.6

[a]All references are cited in Vila et al. (2004).
[b]RY = crop yield in mixture with exotic weeds/crop yield in monoculture.
Source: Modified from Vila et al. (2004). *Biot. Invas.* 6:59–69.

Relative Yield Total. When the yields of both species are of interest and total plant density is held constant, such as in mixed cropping production systems or controlled experiments, it is possible to derive the *relative yield total* (RYT) for the system or experiment. The RYT describes how each species uses resources, that is, space, in relation to the other and is the composite of RY values for both species in the mixture. Thus,

$$\text{RYT} = \frac{Y_{\text{species A mixture}}}{Y_{\text{species A monoculture}}} + \frac{Y_{\text{species B mixture}}}{Y_{\text{species B monoculture}}} \tag{6.8}$$

where $Y_{\text{species A mixture}}$ and $Y_{\text{species B mixture}}$ are the proportional yields of species A and B, respectively, when grown together and $Y_{\text{species A monoculture}}$ and $Y_{\text{species B monoculture}}$ are the yields of species A and B, respectively, when grown separately as pure stands. The RYT values near 1 indicate that the same resource, space, or area is being used by the two competing species, while values less than 1 indicate mutual antagonism (overall loss of space or resources) and values greater than 1 suggest avoidance or symbiosis (overall gain in space or resource use) (Jolliffe 1997).

A similar calculation is *aggressivity* (A)

$$A = \frac{Y_{\text{species A mixture}}}{Y_{\text{species A monoculture}}} - \frac{Y_{\text{species B mixture}}}{Y_{\text{species B monoculture}}} \tag{6.9}$$

This calculation defines the relative success of the two species in using resources and provides a means to evaluate interference among an array of species. For example, Roush and Radosevich (1985) examined the competitiveness of four

annual weed species by combining them as pairs in a replacement series experiment (Figure 6.20). In each combination, one curve is always concave while the
other is always convex, indicating that the species were competing for a common
resource. By calculating both RYT and A values, the following hierarchy
of competitiveness was established among the four weed species: *Echinochloa
crus-galli* (barnyardgrass) > *Amaranthus retroflexus* (redroot pigweed) >
Chenopodium album (common lambsquarters) > *Solanum nodiflorum* (syn.
americanum) (American black nightshade). Similarly, the replacement approach
has been used successfully to assess both perennial weeds (Holt and Orcutt 1991)
and multicropping systems (Trenbath 1976). In Figure 6.21, for example, each
component of the crop mixtures is affected less by interspecific competition than
by intraspecific competition, allowing for overyielding (RYT > 1) when the crops
are grown in combination.

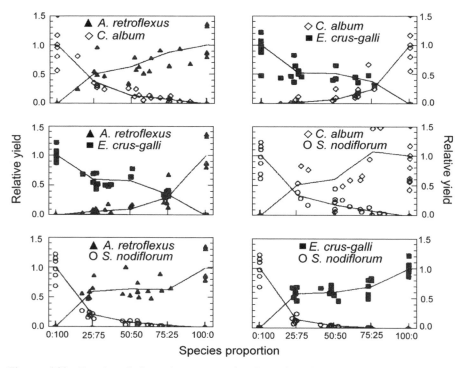

Figure 6.20 Results of six replacement series. In each series relative yields at various
relative proportions are presented for both species. Relative yields represent dry-weight
yields of each species relative to mean dry weight of monoculture treatment (100%) for
that species: (▲) *A. retroflexus*; (◇) *C. album*; (■) *E. crus-galli*; (○) *S. nodiflorum*. (From
Roush and Radosevich 1985, *J. Appl. Ecol.* 22:895–905. Copyright 1985. Blackwell
Publishing Ltd., reproduced with permission.)

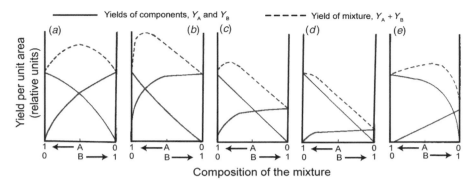

Figure 6.21   Effect of differences in relative aggressiveness and sole crop yields of components on intercrop yield. All yield curves were generated using the deWit (1960) model. In all the 1 : 1 mixtures LER = 1.3, but overyielding by intercrop is found at this proportion only in (a)–(c). Yield components Y_A and Y_B are indicated by solid lines; dashed lines represent the yield of mixtures $Y_A + Y_B$. (From Trenbath 1976, in Papendick et al. (Eds.) 1976, *Multiple Cropping.* Copyright 1976 American Society of Agronomy.)

However, Jolliffe (1997, 2000) points out that RYT may not always be the best method to compare the yields of mixed and monoculture stands of plants. He suggests instead of RYT that *relative land output* (RLO) or *land equivalent ratio* (LER) be calculated. These values are similar to RYT except that the proportion of land devoted to each species in a mixture is compared to an equivalent amount of land growing a pure stand of each species. For example, overyields (RLO > 1) were found from mixtures of red alder and Douglas-fir when RLO was calculated but were not found by calculations of RYT (RYT = 1) from the same experiment (Radosevich et al. 2006).

Intra- versus Interspecific Competition. Advances in experimental designs for quantifying the intensity of competition have come primarily through the use of addition and additive series experiments. Statistical analysis of such studies, using a multispecies form of the reciprocal yield model (Equation 6.4), can describe readily the effects of intra- and interspecific competition as regression coefficients (Spitters 1983a, Roush et al. 1989, Firbank and Watkinson 1990, Shainsky and Radosevich 1992). Table 6.4 summarizes these statistical components of competition. Particularly useful in assessing relative competitive ability is the ratio of competition coefficients, for example, B_{ii}/B_{ji} in Equation 6.4. Generally, intraspecific competition among crop plants is more severe than the interspecific effects of weeds on crop yields. For example, Wilson and Westra (1991) conducted two-year competition experiments with corn and *Panicum miliaceum* (wild-proso millet) using an addition series design. Analysis and coefficients indicated that the influence on corn grain yield of one corn plant was equivalent to that of 14 wild-proso millet plants. Although wild-proso millet is considered to be a serious weed in corn production, these data suggest that corn yields are much more sensitive to corn density than to the presence of wild-proso millet. Similar responses have now been found for associations of many crops and weeds and are summarized by Mohler (2001).

Results from weed–crop experiments suggest that, at the global scale, weeds may be relatively equal in competitiveness and in the intensity of competition with crops or other desirable plants (Vila et al. 2004). However, the existence of competitive hierarchies at local scales seems to be tied closely to environment and other aspects of biology such as physiology, morphology, and carbon allocation (Poorter and Remkes 1990, Cornelissen et al. 1998, Freckleton and Watkinson 2001). Competitive hierarchies, considered a form of *asymmetrical competition*, have been examined in natural ecosystems as well (Howard and Goldberg 2001, Stoll and Prati 2001, Keddy et al. 2002), although seldom in studies of invasive plants. Therefore, predictions of weed–crop or invasive–native plant competitive outcomes will continue to require an understanding of biological and physiological mechanisms of competition examined over a range of environments.

Importance of Competition

The importance of competition in a weed–crop or invasive–native plant community cannot be fully understood from investigations of the intensity of competition. Competition is important in a plant community when it contributes to its organization and dynamics over time, for example, species composition, species shifts, relative fitness among populations, or changes in population densities (Bazzaz 1990, Silander and Pacala 1990, Keddy 2001). Weldon and Slausen (1986) propose that the coefficient of determination (R^2) from regression equations relating plant responses to competition is a suitable measure of the importance of competition (Tables 6.4 and 6.5). They describe equations, derived from neighborhood experiments, in which the slope of the regression quantifies the intensity of competition on plant yield. The R^2 value for those equations suggests how important competition is relative to all other processes that influence plant yield, such as disease, genetics, microclimate, and herbivory (Grace and Tilman 1990).

Roush (1988) and Radosevich and Roush (1990) quantified the role of competition in a community of four annual weed species over a two-year period. Regression models were constructed to relate changes in weed population density during the second year to competition among the weed species during the first year. Results indicated that competition had a significant influence on population growth of the species, especially during the first year; however, the majority of the later variation in population growth was not explained by competition (Roush 1988). She concluded that much of the variation in population dynamics for the species studied is attributable to seed bank and seedling emergence processes. Thus, a more thorough understanding of noncompetitive as well as competitive interactions is necessary to explain and predict the dynamics of plant associations (Chapter 5).

Competition in Mixed Cropping Systems

One of the most inexpensive methods of agricultural weed suppression is to alter crop seeding density and arrangement in monocultures. Mohler (2001) provides an extensive list of crops that respond to changes in their own density in both the presence and the absence of weeds. This response is no doubt due to the high level of intraspecific competition that occurs in most cropping systems (see discussion above).

Another widely practiced means of increasing crop production is by the simultaneous culture of two or more crops on the same piece of land, called *intercropping* (Liebman 1988, Organic Farming Research Organization 2002, Ngouajio et al. 2003, U.S. Department of Agriculture IPM Center 2006). In Figure 6.21, Trenbath (1976) uses a replacement diagram to demonstrate how each component of a crop mixture can be affected by the other, allowing for an overyield when two crops are grown in combination. In each diagram of Figure 6.21, each crop is affected less by interspecific than by intraspecific competition. In each case the RYT (Equation 6.8) is greater than 1 since the crop mixtures failed to penalize each crop as much as expected from their yields in monoculture. A similar observation was made by Shainsky and Radosevich (1992) and Radosevich et al. (2006) when the timber species Douglas-fir is grown together with *Alnus rubra* (red alder) at various proportions.

Liebman (1988) indicates that values of intercrop yield enhancement can, in fact, be quite high (e.g., RYT ranging from 1.38 for a mixture of maize with bean to 3.21 for maize with bean and cassava). These RYT values represent substantial increases in production over what would occur if the crops were grown separately as monocultures. While intercropping techniques are most often used on small farms that employ a minimum of mechanization and other technological inputs, they are not restricted to such situations. In fact, interest in the use of intercrops and cover crops under the fully mechanized cultural practices of temperate zone agriculture is increasing (Horwith 1985, Altieri 1999, Hutchinson and McGiffen 2000, Liebman and Davis 2000).

Weed Suppression in Mixed Planting Systems. Weed suppression is cited as one of the benefits of intercropping (Liebman 1986, 1988, Moody 1988, Altieri 1999, Liebman and Davis 2000, Ngouajio et al. 2003). Vandermeer (1989) believes that the mechanism of weed suppression in mixed cropping systems is competitive impact on weeds by one crop that reduces weed biomass and thereby benefits the second or other crops in the mixture. Perhaps the best known example of this type of weed suppression is the use of *cover crops*, which are solid-grown crops used primarily to protect and cover soil between crop rows or between periods of regular crop production (Aldrich 1984, Zimdahl 1999). Liebman (1986, 1988) reviewed studies of 23 crop and cover crop combinations and found that 20 of them provide significant weed suppression. While these findings with intercrops and cover crops are impressive, Vandermeer (1989) indicates that weed suppression by combinations of two crops can be equivocal. For example, in his literature review Liebman (1986) also found that the suppressive effect on weeds was stronger in intercrops than in the single-cropped components in eight cases, intermediate between single-cropped components in another eight cases, and weaker than all single-cropped components in two cases.

The addition of weeds to an intercrop or cover crop situation creates an interesting ecological system of three or more interconnected competitors. It seems likely that the more complex and well-controlled experimental approaches for competition study, such as the addition series and additive series designs, could help unravel the complexity of interactions that no doubt occur in those systems. Vandermeer (1989) believes that such interactions implicitly involve a positive modification of the environment of one species by another, especially in the case of cover crops. As such, they represent a working example of the facilitation production principle (Figures 6.1 and 6.2). In addition to impacting yield or productivity, changing the planting density or the proportions of plants in agro- and natural ecosystems usually improves the biodiversity of those systems (Liebman and Staver 2001, Radosevich et al. 2006).

COMPETITION THRESHOLDS

Weed management is an essential component of almost every production system because crop yields are affected so markedly by weed presence. In addition, factors in addition to yield such as crop quality, ease of harvest, populations of other pests or beneficial organisms, biodiversity, and ecosystem services are also affected by weeds and invasive plants (Chapter 1). Often the impact of weeds or invasive plants on one of these other factors is so significant that weed control is conducted solely for that purpose. For example, crop quality standards in some vegetable or seed crops are sufficiently high that very few, if any, weeds are tolerated in those crops. Similarly, invasive plants may reduce biodiversity or ecosystem services to such an extent that weed control tactics are introduced for only that purpose. Nonetheless, cost-effective weed management requires that an assessment of potential as well as real damage from weeds and invasive plants be made prior to the introduction of weed control (Chapters 7 and 9). Thus, the concept of competition thresholds for weeds and invasive plants is central to good management of agricultural and natural ecosystems.

Thresholds in Agriculture

Thresholds have many applications in agriculture and are the foundation of *integrated pest management* (IPM). The common IPM thresholds that pertain to weeds are *damage*, *period*, *economic*, and *action* thresholds. Damage thresholds describe the weed population at which negative crop impact is detected, while period thresholds occur during a crop life cycle when weeds are more or less damaging than at other times. Such thresholds are usually expressed in biological terms, such as plant density or weed biomass per unit of area. Glass (1975) and

Coble and Mortensen (1991) define economic threshold as the pest (weed) population density, or damage level, at which control measures should be taken to prevent economic injury to the crop from occurring. This definition of economic threshold or *economic injury level* (EIL) implies that the cost of control should be less than the loss that would have occurred had nothing been done (Norris et al. 2003). The establishment of an *action* threshold, or weed population level at which some action is needed to preclude crop yield loss, necessarily includes predictions of direct effects on crop yield or other forms of economic loss due to weed association with the crop.

As already noted, most of the literature on crop–weed interactions attempts to quantify the negative effects (damage) of weeds on crop yields (Figure 6.9). Economic and action thresholds try to answer the question, "How much will a given amount of weeds reduce both crop yields and profitability?" Damage and period thresholds are discussed in more detail below, while economic and action thresholds for agriculture are discussed further in Chapter 7.

Damage (Density/Biomass) Thresholds. The extent to which crop yields are reduced by weeds depends on many factors, such as crop species and cultivar, weed species present, location or site, and practices used that modify site conditions (Liebman et al. 2001). Differences in weather from year to year also cause annual variations in crop yield, affect weed competitive ability, and confound data interpretation. For these reasons, it seems nearly impossible to determine empirically the yield reductions for even the major crops in association with particular weeds in a region. In addition, experiments on weed–crop competition are rarely conducted at the entire field scale, yet results from small plot experiments may not reflect accurately actual field-level crop responses to weeds. For example, Auld and Tisdell (1988) and Mortensen et al. (1993) demonstrate that weed populations may vary by up to 200% across an agricultural field. These observations indicate that weed densities are extremely variable and that predictions of crop response based on small plot bioassays probably overestimate the value of weed control at the field level.

Nevertheless, many investigations of weed–crop interactions have been conducted over the years, and most demonstrate that weed plants are harmful to agricultural crops, even at low densities (Figure 6.9). The shape of the curve in Figure 6.9 implies that the negative impact of each weed plant on crop yield increases as the weed population declines; that is, even very low weed densities cause substantial losses in most crops with greatest yields occurring where weeds are absent. The relevant question for farmers that is raised by the data that make up Figure 6.9 is whether it is economically reasonable to control weeds to such very low densities. Auld et al. (1987) and Alstrom (1990) suggest that it is not, unless the additional cost of weeding is equal to the value of the weeds' marginal effect on crop yields. In other words, the additional revenue gained from crop yields achieved by weed control must equal the cost of attaining it. Obviously, the economically optimum amount of weeds must be more than zero in this case, unless the crop is infinitely valuable or weed control costs are nothing.

Although the general relationship depicted in Figure 6.9 is also true for forestry situations (Stewart et al. 1984, Wagner et al. 2006), both tree survival and subsequent size are important yield components in young forest plantations. Wagner et al. (1989) examined the patterns of survival and stem volume growth for planted ponderosa pine competing with various levels of woody and herbaceous vegetation. They found that negative hyperbolic curves with opposite concavity described the relationship between the abundance of undesirable plants and tree survival and stem volume (Figure 6.22) of the pine seedlings. From these curves, two types of competition thresholds were identified:

- Maximum-response threshold, a level of competing vegetation abundance at which additional control would not yield an increase in tree performance
- Minimum-response threshold, a level of competing vegetation that must be reached before additional control measures yield an appreciable increase in tree performance (Figure 6.22)

The thresholds for pine stem volume growth occurred at lower competing vegetation abundance than the thresholds for tree survival, indicating that foresters should consider tree survival and tree growth as separate silvicultural objectives when managing competing vegetation in forest plantations. Although the damage threshold approach has not been applied in natural ecosystems, it could be a

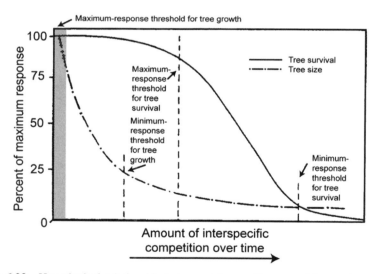

Figure 6.22 Hypothetical relationship between interspecific competition and tree survival and growth. Maximum- and minimum-response thresholds for tree survival and growth occur at different levels of interspecific competition. Maximum-response threshold for tree growth occurs in shaded region under nearly vegetation-free conditions. (From Wagner et al. 1989, *New For.* 3:151–170. Copyright 1989 with permission of Springer Science and Business Media.)

valuable tool for beginning to understand the effects of invasive species on native species survival and abundance (Hughes and Madden 2003, Wiles 2004).

Critical-Period Thresholds. It is the conventional wisdom of many farmers and weed scientists that early weed competition is most detrimental to crop yields and that early weed control is necessary. However, as Zimdahl (1988, 1999) points out, this generalization may not be entirely accurate because it was probably dictated by the method of weed control rather than by biological necessity. For example, hoeing and cultivating are accomplished more easily when both crops and weeds are small. In addition, development of preplant and preemergence herbicides (Chapter 8) could have perpetuated the belief that early weed control was essential.

There is evidence, however, that for certain crops a critical period exists during which weeds should be controlled to prevent yield losses and that for this reason weed control may not be necessary at other times. In *critical-period* studies, crops are kept weed free for varying intervals of time following planting or emergence, and after this period weeds are allowed to grow for the rest of the growing season (Figure 6.23). The resulting data are compared to those of a complementary study in which weeds are allowed to grow for varying intervals of time after crop planting or emergence, with the remainder of the growing season being weed free (Figure 6.23) (Zimdahl 1988, 1999).

As seen in Figure 6.23*a*, the initial weed-free period up to point II results in crop dominance, which diminishes any subsequent competitive effects by weeds on crop yields. If weeds are not controlled, however (Figure 6.23*d*), a period of increasing competition between crop and weed plants follows after emergence, and crop yields are reduced. In Figure 6.23*b*, both weed and crop seedlings are small and far enough apart early in their life cycles so that no interaction occurs, but eventually interference between the species develops. For example, canopies of both weed and crop species would be developing continuously after emergence, but they might not overlap until much later. At this point (I in Figure 6.23*b*) weed control for the remainder of the season prevents crop yield loss. If weeds are not controlled at this point, the canopy of the weed is superimposed upon that of the crop and a loss of crop productivity most likely will result throughout the rest of the season (Figure 6.23*d*). Extrapolation from Figures 6.23*a* and *b* suggests a "critical period" of time (from I to II in Figure 6.23*c*) during which control measures are necessary in order to avoid continuing interference between the crop and weed. Weed removal any time up to the end of the critical period for control, during which crop dominance is being established, would result in no significant crop yield reduction. Further weed control after the critical period most likely is unnecessary to prevent yield loss.

Zimdahl (1988, 1999) points out that the concept of a critical period for weed control has been challenged for a number of crops. In these cases, either the crops are susceptible to weed competition for most of the growing season or a single weeding at an intermediate growth stage is sufficient to avoid yield reduction. In addition, variable cultural practices used in different regions and environmental variation that can occur from growing season to growing season may also affect

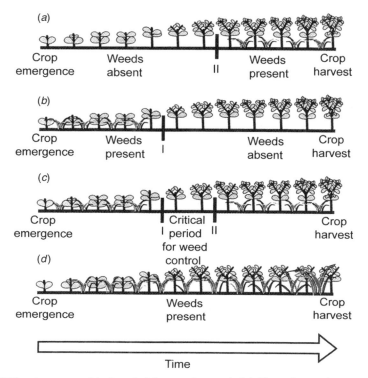

Figure 6.23   Apparent critical period for weed control. (*a*) If weeds are absent up to point II, crop dominance is established and yield losses do not result, even though weeds may be present subsequently. (*b*) If weeds are present for a period of time following crop emergence but are absent for the remainder of the season, yield losses do not result since, presumably, early in the season weeds are too small for competition to occur. (*c*) The combination of results from (*a*) and (*b*) leads to the critical period between points I and II, which is a "window" of time during which weeds must be removed or suppressed to avoid crop yield loss at harvest. (*d*) Situation in which weeds are present throughout the growing season and crop yield loss results. (From Radosevich et al. 1997, *Weed Ecology: Implications for Management*, 2nd ed. Copyright 1997 with permission of John Wiley & Sons Inc.)

germination and growth rate of weeds and crops differentially. Thus, differences in either environmental conditions or the availability of resources from one year to the next could affect the length of the critical period when weeds would need to be controlled.

Thresholds in Natural Ecosystems

The species composition of plant communities changes over time and following disturbance. This ecological process, called secondary succession (Connell and Slatyer 1977, Tilman 1985, 1988, Barbour et al. 1999), is discussed in Chapter 2.

Plant communities are rarely simple in structure, composition, or function, however, and the traditional view of succession has been challenged by other ecologists (e.g., DeAngelis and Waterhouse 1987, Westoby et al. 1989, Pickett et al. 1992, Perry 1997, Briske et al. 2003), who suggest that plant succession is more climate and disturbance driven, especially in rangeland and forest systems, than driven by competition. They argue that many equilibrium states exist among plant communities during succession (the "flux of nature," Chapter 2) and that transitions among these states occur when an *ecological threshold* is crossed.

According to Briske et al. (2003), ecological thresholds are boundaries separating multiple equilibrium states that can be distinguished by changes in community structure and composition or impacts on soil properties that alter site characteristics. Many exotic species that now inhabit native plant communities are believed to disrupt ecosystem function because of their presence (Hobbs and Huenneke 1992, Vitousek et al. 1996, Quigley and Arbelbide 1997, Sheley and Petroff 1999, Harrod 2001, Hobbs et al. 2006) and thus, presumably, drive communities to new equilibrium states. While there is substantial evidence that invasive plants alter the composition of all stages of plant community development, there is still relatively little evidence indicating that they alter significantly the function of the communities they inhabit. However, where such alterations in function occur, the change in the impacted plant community is often dramatic and unlikely to be reversed easily. This aspect of invasive plant impacts and their management are also discussed in Chapters 7 and 9.

The possible impacts of invasive plants on the ecological thresholds of natural ecosystems are only now being realized on a broad scale. The introduction of IPM concepts, which include an integral role for thresholds in informing management (discussed earlier for agriculture and in Chapter 9), has also been slow to emerge in natural ecosystems. Perhaps the threshold most easily recognized by land managers in natural ecosystems is the recently described *early detection and rapid response* (EDRR) threshold. In this case, tactics are employed to eradicate all patches of a new invasive species once it is found in an area (Finnoff et al. 2005). This action implies that any damage by the invader to the plant community is too much. Similarly, there are many examples in agriculture where the rapid eradication of a new weed in a field has resulted in substantial benefits and reduced long-term costs for control measures later when or if the patch expands. According to Hobbs and Humphries (1995), this type of action threshold should be employed in a natural resource production system or natural ecosystem when a resource or area is valuable and the risk to it by weed presence is great (Chapters 2 and 7).

In the later stages of succession, if disturbance to the natural ecosystem is severe or if the presence of invasive plants is ubiquitous, the economic threshold (EIL) could be a better measure of the cost effectiveness of control and restoration. In this case, the damage to a plant community by various weed species and levels of abundance and the costs of control and restoration would be compared to the long-term gain in function or ecosystem services gained by the action (Finnoff et al. 2005) (Chapter 7). In most well-established natural plant communities, species diversity, complexity, and coexistence seem to be the rule rather than simplicity and direct competition (Perry 1997, Barbour et al. 1999). Thus, it is possible in these cases that the threshold for action is higher than zero. It also seems possible that an economic threshold could be lower than anticipated if the long-term benefit of creating and maintaining a self-perpetuating natural production system that is relatively free of invasive plants is great. Relatively intact natural vegetation may not be readily invaded by exotic plants (Parks et al. 2005a), although examples exist that suggest some plant communities are more invasible than others, such as in riparian areas and xeric grassland ecosystems.

Finnoff et al. (2005) suggest that land managers view thresholds for prevention (e.g., EDRR) and control or restoration (EIL) differently. They also suggest a way to interpret the trade-offs in risk between the two management strategies. Unfortunately, it will be difficult to ascertain the thresholds and management trade-offs that occur in natural ecosystems without more complex experiments or at least projections of ecosystem change from the presence and managed reduction of invasive species.

MECHANISMS OF COMPETITION

Most studies of plant competition focus on the demographic relationships (e.g., total and relative densities, arrangements) of the partners that make up the association and on the consequences of competition in terms of plant yield or size (biomass). Few studies are able to examine definitively the environmental, physiological, or morphological mechanisms that underlie competition. Mechanisms of plant competition for environmental resources are demonstrated by the following:

- Resource depletion associated with presence and abundance of neighbors
- Changes in physiological and morphological growth responses that are associated with changes in resources
- Correlations among the presence of neighbors, resource depletion, and growth response

Theories

Mechanisms of plant competition consist of both the effect that plants have on resources and the response of plants to changed resources (Figure 6.1). Several theories have been advanced to explain the relative importance of resource availability, acquisition, and use in relation to characteristics of plants that confer superior competitiveness. Two theories that have received widespread attention are those of Grime (1979) and Tilman (1988). Other theories involve the role of particular traits and growth rates that impart efficient resource use and superior competitiveness among plants.

Theories of Grime and Tilman. Grime explains plant life histories in terms of the processes of disturbance and stress, which select for syndromes of plant characteristics (Chapters 2 and 4). According to Grime, competition is the tendency for neighboring plants to utilize the same environmental resource(s), and success in competition is largely due to the capacity for resource capture (Grime 1979, Grace 1990, 1991). Thus, a good competitor has a high *relative growth rate* (RGR, or relative increase in biomass as a function of existing biomass per unit time) and can use resources rapidly. Tilman, on the other hand, proposes a mechanistic resource-based theory (Figure 6.24) that predicts competitive success as a function of the concentration of limiting resources and ratios of essential-versus-substitutable resources (Tilman 1988, Grace 1990, 1991, Barbour et al. 1999). Thus, competitive success according to this theory is the ability to draw resources down to a low level and tolerate those low levels. A good competitor in this case would be the species with the lowest resource requirement.

Although debate continues about the validity and relevance of these two theories, some of their differences can be explained by the time frame, scale, and associated terminology related to the definitions of competition being used. For example, Grime's *stress tolerator* (Chapters 2 and 4) might be compared to Tilman's *competitor* (Tilman 1988, Grace 1991). Furthermore, while Grime focuses on the role of particular plant traits in competitiveness as well as the role of disturbance, Tilman's theory deals with the dynamics of populations and does not focus on individuals. Both theories help explain the role of resources in competition and how plant traits might confer competitiveness.

Role of Plant Traits. As noted in relation to the theories of Grime and Tilman, plants can be good competitors either by depletion of a resource or by having the ability to grow at depleted resource levels. Unfortunately, relatively few studies examine mechanisms of competition in a quantitative way to determine specific plant traits that correlate with competitiveness. One of the earliest studies on biomass accumulation and competitiveness was conducted by Pavlychenko (1937a) during the third and fourth decades of the last century. He conducted a series of quantitative experiments on biomass patterns, especially roots, of weeds and crops grown under competitive and noncompetitive conditions in the Canadian plains. By using the tedious soil-block washing technique, Pavlychenko (1937a,b) was able to quantify root and shoot distributions of individual crop and weed plants and then relate these measurements to crop yields as a measure of competitiveness (Figure 6.25). His comments concerning root interactions and their influence on shoot development are especially incisive (Pavlychenko, 1940, p. 9):

> Competition begins as soon as the root system of one plant invades a feeding area of another, and usually takes place long before tops are developed sufficiently to exert serious competition for light. Therefore, in dry climates roots actually decide the success or failure in competition between species otherwise equally adapted to a region. The top growth is then developed in proportion to the extent of the root system.

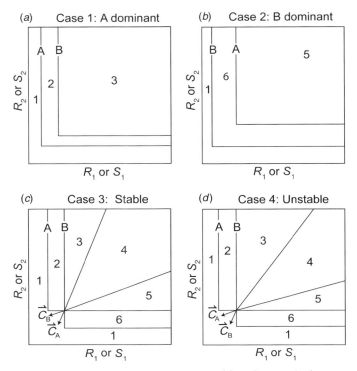

Figure 6.24 Four distinct cases of resource competition. In case 1, the zero net growth isocline (ZNGI, solid lines) for species A is at a lower level of resource supply (R) than that of species B. As a result, A will reduce resource levels to a point below that required for survival of B and competitively exclude B in any habitat that would support its own growth (zones 2 and 3). Case 2 is the converse of case 1. In this case, B wins the competitive interaction and will displace A. In case 3, the ZNGIs cross at a stable two-species equilibrium point. This intersection is stable because each species consumes relatively more of the resource that is more limiting to its growth at equilibrium. In this case, habitats with resource supply ratios in zones 2 and 3 will result in the dominance of species A. Similarly, habitats in zones 5 and 6 will be dominated by B. Habitats with resource supply in zone 4 will result in stable coexistence of the two species. Case 4 is similar to case 3 except that the equilibrium point is unstable because each species uses more of the resource that primarily limits the other species. The outcome of this competitive interaction (except in zone 1 where neither species exists) will be dominated by either species A or B, depending on initial conditions. (From Tilman 1982, *Resource Competition and Community Structure*, Copyright 1982. Princeton University Press. Reprinted by permission of Princeton University Press.)

Gaudet and Keddy (1988) measured the relative competitive ability of 44 herbaceous species and tested whether competitive effect was correlated with simple measurable plant traits. They observed that total and above- and below-ground biomass (size) followed by height and canopy area explained most of the

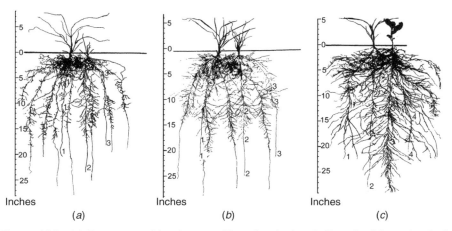

Figure 6.25 (*a*) Root competition between Hannchen barley (left) and wild oat (marked 1, 2, 3) 22 days after emergence. (*b*) Root competition between Marquis wheat (left) and wild oat (marked 1, 2, 3) 22 days after emergence. (*c*) Root competition between Marquis wheat (left, roots marked 1, 2, 3, 4) and wild mustard 22 days after emergence. (Modified from Pavlychenko 1937a, *Ecology* 18:62–79.)

variation in competitiveness of the 44 species in their study against an indicator species. A similar study conducted to evaluate the relative competitive performance of 63 species of terrestrial herbaceous plants also found that plant size (measured as biomass, canopy area, height, and leaf area index) as well as leaf shape best predicted competitiveness (Keddy et al. 2002). Keddy (1992, 2001) notes, however, that morphological traits are only a small subset of the ecological and physiological attributes possessed by plants. As such their role in competitiveness should be viewed cautiously since physiological and morphological traits often interact and compensate strongly (Poorter and Remkes 1990, Radosevich and Roush 1990, Cornelissen et al. 1998, McDowell and Turner 2002).

Plant Growth Rates and Components of Growth. The ability of individual plants to obtain light, water, and nutrients for growth often determines the success of those individuals in resource-rich environments such as agricultural fields or the early stages of succession. In this case successful individuals grow rapidly or large, develop through the various stages of their life cycle, and are eventually replaced by their progeny. On the other hand, the capacity for individual plants to reduce available resources to a low level and then tolerate it is also the mark of successful individuals in competitive environments. The life cycle of unsuccessful individuals is often arrested before its completion. Therefore, plant growth, as well as the developmental stages that accompany it, is fundamental to understanding plant function and the manner of interactions plants undergo with neighbors and with their environment.

Growth Analysis. Beginning with the early works of Blackman (1919) and Kidd and West (1919), techniques have been developed to integrate the effects of environment, development, and size on plant growth. These techniques, collectively called *mathematical growth analysis*, recognize total dry-matter production and leaf area expansion as important processes in determining vegetative growth. The techniques require frequent destructive harvest of plant material throughout a plant's life cycle, which necessitates the collection of copious amounts of data. Use of growth analysis on a large scale expanded greatly with the widespread use of personal computers.

The basic information collected at each harvest includes biomass production of roots, leaves, stems, and reproductive organs and leaf area. From these basic data it is possible to calculate relative growth rates (RGR, R), rates of biomass production per unit of leaf area or net assimilation rate, also called unit leaf rate (NAR, ULR, E), relative leaf expansion rates, and partition coefficients for plant biomass and leaf area, such as leaf area ratio (LAR, F). Thus, the components of plant growth can be compared under a range of environmental conditions and resource limitations. These formulas for calculating growth analysis parameters using several approaches are summarized in Table 6.7. A more complete description of the derived quantities used in plant growth analysis and their mathematical definitions are given in Hunt (1978, 1982) and Chiariello et al. (1991).

Relative growth rate R is considered to be one of the most ecologically significant plant growth indices and can be expressed as

$$R = E \times F \tag{6.10}$$

where R is relative growth rate as defined above, E is the net gain in weight or size per unit of leaf area, and F is the amount of leaf area per total plant biomass—a measure of the relative leafiness of the plant. Thus, R can be expressed in both physiological and morphological terms. The parameter F is the morphological index of plant form, whereas E is a physiological index closely connected with photosynthetic activity of leaves (Hunt 1978, 1982). The splitting of R into its two components is advantageous because it relates biomass increase to the organs most concerned with carbon assimilation, leaves. Splitting of the above growth parameters into other components is also possible, as shown in Table 6.7, for further understanding of plant growth responses to environment.

Relative Growth Rates of Weeds and Invasive Plants. Grime and Hunt (1975) provide one of the largest bodies of comparative data on relative growth rates, 132 species, that is available. The weeds/invasive plants listed in Table 6.8 are most often associated with arable land, grazed meadows and pastures, or other disturbed but productive habitats. Mean (R) and maximum $(R_{\max})$ relative growth rates ranged from 1.0 to over $2.0\,\mathrm{g/g \cdot week}$ and were in sharp contrast to the values of other species in the study from either undisturbed or unproductive environments $(0.5\,\mathrm{g/g \cdot week}$, not shown). Poorter and Remkes (1990) provide similar tables of values for R, E, and F for 24 weeds and invasive plants. Their analysis revealed similar values to those shown in Table 6.8 and a high

TABLE 6.7 Summary of Traditional Growth Analysis Parameters, Units, Instantaneous Values, and Methods of Calculation[a]

Parameter	Units	Instantaneous	Classical (Interval)[b]	Functional	Integral[c]
Relative growth rate (RGR, R, SGR) Specific growth rate	d^{-1} w^{-1}	$\dfrac{1}{W}\dfrac{dw}{dt}$	$\dfrac{\ln W_2 - \ln W_1}{t_2 - t_1}$	$\dfrac{1}{g(t)}\dfrac{df(t)}{dt}$ where $W = f(t)$	$\dfrac{W_2 - W_1}{\mathrm{BMD}}$
Net assimilation rate Unit leaf rate (NAR, ULR, E)	$g\,m^{-2}\,d^{-1}$ $g\,m^{-2}\,w^{-1}$	$\dfrac{1}{A}\dfrac{dW}{dt}$	$\dfrac{1}{t_2 - t_1}\displaystyle\int_{t_1}^{t_2}\dfrac{1}{A}\dfrac{dW}{dt}\,dt$	$\dfrac{1}{g(t)}\dfrac{df(t)}{dt}$ where $A = g(t)$	$\dfrac{W_2 - W_1}{\mathrm{LAD}}$
Leaf area ratio (LAR, F)	$m^2\,g^{-1}$	$\dfrac{A}{W}$	$\dfrac{1}{t_2 - t_1}\displaystyle\int_{t_1}^{t_2}\dfrac{A}{W}\,dt$	$\dfrac{g(t)}{f(t)}$	
Specific leaf area (SLA)	$m^2\,g^{-1}$	$\dfrac{A}{W_L}$	$\dfrac{1}{t_2 - t_1}\displaystyle\int_{t_1}^{t_2}\dfrac{A}{W_L}\,dt$	$\dfrac{g(t)}{h(t)}$ where $W_L = h(t)$	
Leaf weight ratio (LWR)	—	$\dfrac{W_L}{W}$	$\dfrac{1}{t_2 - t_1}\displaystyle\int_{t_1}^{t_2}\dfrac{W_L}{W}\,dt$	$\dfrac{h(t)}{f(t)}$	

[a]Symbols: W = dry biomass (g); t = time, in days (d) or weeks (w); A = assimilatory area (m^2); W_L = dry leaf biomass (g); BMD = biomass duration (g d); LAD = leaf area duration (m^2 d).

[b]Explicit general formulas; see Table 15.2 in Chiarello et al. (1991) for specific formulas for calculating net assimilation rate.

[c]Approximate formulas; formulas for BMD and LAD are given in Section 15.2.1.c of Chiarello et al. (1997).

Source: Chiarello et al. (1991). in Pearcy et al. (1991), *Plant Physiological Ecology. Field Methods and Instrumentation*. Chapman and Hall, NY Copyright 1991, with permission of Springer Science and Business Media.

TABLE 6.8   Maximum Potential Relative Growth Rate (R_{max}), Mean Relative Growth Rate (R), and Plant Species Associated with Arable Land (A), Meadows and Pastures (P), or Manure Heaps (M)

Species	Common Name	Site Association[a]	R_{max} (g/g week)	R (g/g week)
Agropyron repens	Quackgrass	AMP	1.21	1.21
Agrostis stolonifera	Creeping bentgrass	APM	1.48	1.48
Agrostis tenuis	Colonial bentgrass	P	1.36	1.36
Cerastium holosteoides	—	P	1.46	1.46
Chenopodium album	Common lambsquarters	AM	2.12	1.25
Convolvulus arvensis	Field bindweed	A	2.44	1.36
Cynosurus cristatus	Crested dogtailgrass	P	1.54	1.54
Dactylis glomerata	Orchardgrass	PM	1.31	1.31
Festuca rubra	Red fescue	P	1.18	1.18
Holcus lanatus	Velvetgrass	PM	2.01	1.56
Matricaria matricarioides	Pineappleweed	A	1.17	1.17
Plantago lanceolata	Buckhorn plantain	P	1.70	1.40
Poa annua	Annual bluegrass	AM	2.70	1.74
Poa trivialis	Roughstalk bluegrass	PM	1.40	1.40
Polygonum aviculare	Prostrate knotweed	A	1.43	1.43
Polygonum convolvulus	Wild buckwheat	A	1.92	1.35
Ranunculus repens	Creeping buttercup	P	1.39	0.93
Senecio vulgaris	Common groundsel	M	1.63	0.84
Stellaria media	Chickweed	AM	2.43	2.09
Taraxacum officinale	Dandelion	P	1.19	1.19
Trifolium repens	White clover	P	1.26	1.26

[a]Site associations for plant species were determined from Table 1 of Grime and Hunt (1975).

Note: R_{max} is the highest value of R obtained for each species during the periods of observation. R was calculated by Grime and Hunt (1975) from Fisher's (1920) formula ($\log_e W_5 - \log_e W_2)T$, where W_5 and W_2 are whole-plant dry weights at 5 and 2 weeks, respectively, and T is the time interval, 3 weeks.

Source: Grime and Hunt (1975). *J. Ecol.* 63:393–422. Copyright 1975 with permission.

correlation between R and F. Apparently, for these species and under the conditions of the experiment, the more carbon a plant assimilated into leaf area, the faster it could grow and produce new biomass. In a review of 60 publications, R for herbaceous weeds and crops was high and also correlated with F (Poorter and Remkes 1990). Correlations were lower for shrubs, trees, shade plants, and C_4 species, apparently because of the more important role of E (physiology) under conditions where light often limits photosynthesis and in plants possessing the more efficient C_4 pathway of carbon assimilation during photosynthesis.

Many studies have examined life history traits that contribute to invasiveness in nonagricultural ecosystems (Chapter 4; reviewed in Sakai et al. 2001). However, only a few have focused specifically on growth and its components, as was done in the extensive studies of weeds in agricultural and natural production systems described above. Baruch et al. (1989) found higher R in invasive exotic grasses from Africa than in noninvasive native grasses in Venezuela, while Pattison et al. (1998) showed that invasive species in Hawaii have higher R than their native counterparts (Grotkopp et al. 2002). Grotkopp et al. (2002) evaluated 29 pine (*Pinus* spp.) species in order to determine the traits most important in invasiveness. In this study, R was also found to be the most significant factor separating invasive and noninvasive species (Grotkopp et al. 2002). Further analysis revealed that the main component of R that differed between invasive and noninvasive pines was specific leaf area (SLA), a morphological measure of relative leaf production. The consistency of R in contributing to success of weeds and invasive plants across many types of production and natural systems indicates its utility as a predictive tool (Chapter 7).

Relationship of R and Its Components to Competition. Some researchers have examined competition in experiments that describe both individual plant growth and biomass production in mixtures. Simultaneous growth analysis and replacement series experiments performed with pairwise mixtures of four annual weeds revealed that total plant weight, E, and F were positively correlated with competitiveness (Roush and Radosevich 1984). Similar research with perennial weeds and cotton showed that the best predictors of competitive success in mixtures of these species were height, E, R, and initial vegetative propagule weight (Holt and Orcutt 1991). It is not surprising that, for perennial weeds growing with an annual crop, parameters of early establishment (R and initial propagule weight) as well as light utilization (height and E) are important in determining competitiveness. More recent reviews also support the finding that R is positively associated with competitiveness (Keddy et al. 1994). However, some caution is needed in interpreting studies where traits of plants grown alone are used to predict mixture performance, since the same characteristics implicated in competitiveness are often altered by the presence of neighbors (Tilman 1990). Nevertheless, these studies provide a predictive approach to study mechanisms of competition among plants.

OTHER TYPES OF INTERFERENCE THAN COMPETITION

Other types of interference are possible that range from negative to positive interactions (Table 6.1). The negative interactions other than competition include *amensalism*, *parasitism*, *predation*, and *herbivory*, while positive interactions include *commensalism*, *protocooperation*, and *mutualism*.

NEGATIVE INTERFERENCE IN ADDITION TO COMPETITION

Allelopathy

Sometimes the depressive effect of a plant upon its neighbors is so striking that competition for a common resource is not adequate to explain the observation. In this case mortality or a dramatic decrease in biomass is usually evident for one species but not for the other. Such a condition is termed *amensalism* or asymmetric competition. An explanation for such observations is that some plants release into the immediate environment of other plants toxic substances (allelochemicals) that harm or kill them. This phenomenon is called *allelopathy* and is distinguished from other forms of negative plant interference in that the detrimental effect is exerted through release of a chemical by a donor plant. The term allelopathy was coined by Molisch (1937) to describe chemical interactions among plants, including stimulatory as well as inhibitory responses. Many cases of allelopathy also involve the presence of microorganisms in the plant association.

In terms of plant responses, the existence of amensalism, or more specifically allelopathy, has become reasonably well documented over the last several decades (Rice 1984, Putnam and Tang 1986, Rizvi and Rizvi 1992, Inderjit et al. 1995, 1999, Inderjit et al. 1999), and a considerable body of information has accumulated that implicates allelopathy as an important form of plant interference (Table 6.9). Nevertheless, it is difficult experimentally to separate allelopathy from other forms of interference, in particular, competition, in field situations so its existence and role in community and ecosystem functioning remain unclear (Wardle et al. 1998, Gurevitch et al. 2002). Even when allelopathy has been implicated in plant associations, the complexity of the soil rhizosphere makes it challenging to detect the specific chemical involved and show that sufficient quantities are present to cause an effect. Other complications arise when working in field settings, including teasing out the interacting effects of soil chemistry (Inderjit and Weiner 2001), soil microorganisms (Inderjit 2005), herbivory, and other environmental factors (Wardle et al. 1998). Given these difficulties, it is not surprising that most documentation of allelopathy has been done in artificial, highly controlled experiments, such as pots or Petri dishes, where it is often easier to detect allelopathic chemicals and their effects (Stowe 1979, Stowe and Wade 1979, Gurevitch et al. 2002, Inderjit and Callaway 2003). As a result, many ecologists remain skeptical about claims of allelopathy in natural ecosystems.

The one recent, albeit controversial, exception is the case of the knapweeds (*Centaurea* spp.), an invasive species in western North American rangelands. Negative effects of diffuse knapweed (*C. diffusa*) and spotted knapweed (*C. maculosa*) on native grass species were attributed to root exudates (Callaway and Aschehoug 2000, Bais et al. 2003), which were later suggested to be racemic mixtures of catechin (Bais et al. 2002). Other reports question this conclusion, however, and suggest that the effects of knapweeds on other species may be due to alterations in soil pH, nutrients, or microorganisms (Lejeune and Seastedt 2001, Blair et al. 2005). Nevertheless, allelopathy as one possible mechanism of invasiveness in exotic species that did not coevolve with their new neighbors is an intriguing hypothesis that warrants further study (Hierro and Callaway 2003, Callaway and Ridenour 2004).

Despite the controversy surrounding the importance of allelopathy, it is clear that plants produce an abundance of secondary compounds that can have significant adverse impacts on the growth and productivity of other plants as well as on species composition and ecosystem biodiversity in natural ecosystems. As a

TABLE 6.9   Common Weeds and Invasive Plants with Alleged Allelopathic Potential

Scientific Name	Common Name	Reference[a]
Abutilon theophrasti	Velvetleaf	Gressel and Holm (1964)
Agropyron repens	Quackgrass	Kommedahl et al. (1959)
Agrostemma githago	Corn cockle	Gajić and Nikočević (1973)
Allium vineale	Wild garlic	Osvald (1950)
Amaranthus dubius	Amaranth	Altieri and Doll (1978)
Amaranthus retroflexus	Redroot pigweed	Gressel and Holm (1964)
Amaranthus spinosus	Spiny amaranth	VanderVeen (1935)
Ambrosia artemisiifolia	Common ragweed	Jackson and Willemsen (1976)
Ambrosia cumanensis	Cuman ragweed	Anaya and DelAmo (1978)
Ambrosia psilostachya	Western ragweed	Neill and Rice (1971)
Ambrosia trifida	Giant ragweed	Letourneau et al. (1956)
Antennaria microphylla	Pussytoes	Selleck (1972)
Artemisia absinthium	Absinth wormwood	Bode (1940)
Artemisia vulgaris	Mugwort	Mann and Barnes (1945)
Asclepias syriaca	Common milkweed	Rasmussen and Einhellig (1975)
Avena fatua	Wild oat	Tinnin and Muller (1971)
Berteroa incana	Hoary alyssum	Bhowmik and Doll (1979)
Bidens pilosa	Beggarticks	Stevens and Tang (1985)
Boerhavia diffusa	Spiderling	Sen (1976)
Brassica nigra	Black mustard	Muller (1969)
Bromus japonicus	Japanese brome	Rice (1964)
Bromus tectorum	Downy brome	Rice (1964)
Calluna vulgaris	Heather	Salas and Vieitez (1972)
Camelina alyssum	Flax weed	Grummer and Beyer (1960)
Camelina sativa	Largeseed falseflax	Grummer and Beyer (1960)
Celosia argentea	Celosia	Pandya (1975)
Cenchrus biflorus	Sandbur	Sen (1976)
Cenchrus pauciflorus	Field sandbur	Rice (1964)
Centaurea diffusa	Diffuse knapweed	Fletcher and Renney (1963)
Centaurea maculosa	Spotted knotweed	Fletcher and Renney (1963)
Centaurea repens	Russian knotweed	Fletcher and Renney (1963)
Chenopodium album	Common lambsquarters	Caussanel and Kunesch (1979)
Cirsium arvense	Canada thistle	Stachon and Zimdahl (1980)
Cirsium discolor	Tall thistle	Letourneau et al. (1956)
Citrullus colocynthis	Colocynth	Bhandari and Sen (1971)
Citrullus lanatus	Wild watermelon	Bhandari and Sen (1972)
Cucumis callosus	Wild melon	Sen (1976)
Cynodon dactylon	Bermudagrass	VanderVeen (1935)
Cyperus esculentus	Yellow nutsedge	Tames et al. (1973)
Cyperus rotundus	Purple nutsedge	Friedman and Horowitz (1971)
Daboecia polifolia	Heath	Salas and Vieitez (1972)
Digera arvensis	False amaranth	Sarma (1974)
Digitaria sanguinalis	Large crabgrass	Parenti and Rice (1969)
Echinochloa crus-galli	Barnyardgrass	Gressel and Holm (1964)
Eleusine indica	Goosegrass	Altieri and Doll (1978)
Erica scoparia	Heath	Ballester et al. (1977)
Euphorbia corollata	Flowering spurge	Rice (1964)
Euphorbia esula	Leafy spurge	Letourneau and Heggeness (1957)

(Continued)

TABLE 6.9 *Continued*

Scientific Name	Common Name	Reference[a]
Euphorbia supina	Prostrate spurge	Brown (1968)
Galium mollugo	Smooth bedstraw	Kohmuenzer (1965)
Helianthus annuus	Sunflower	Rice (1974)
Helianthus mollis	Ashy sunflower	Anderson et al. (1978)
Hemarthria altissima	Bigalta limpograss	Tang and Young (1982)
Holcus mollis	Velvetgrass	Mann and Barnes (1947)
Imperata cylindrica	Alang-alang (cogongrass)	Eussen (1978)
Indigofera cordifolia	Wild indigo	Sen (1976)
Iva xanthifolia	Marshelder	Letourneau et al. (1956)
Kochia scoparia	Kochia	Wali and Iverson (1978)
Lactuca scariola	Prickly lettuce	Rice (1964)
Lepidium virginicum	Virginia pepperweed	Bieber and Hoveland (1968)
Leptochloa filiformis	Red sprangletop	Altieri and Doll (1978)
Lolium multiflorum	Italian ryegrass	Naqvi and Muller (1975)
Lychnis alba	White cockle	Bhowmik and Doll (1979)
Matricaria inodora	Mayweed	Mann and Barnes (1945)
Nepeta cataria	Catnip	Letourneau et al. (1956)
Oenothera biennis	Evening primrose	Bieber and Hoveland (1968)
Panicum dichotomiflorum	Fall panicum	Bhowmik and Doll (1979)
Parthenium hysterophorus	Parthenium ragweed	Sarma et al. (1976)
Plantago purshii	Wooly plantain	Rice (1964)
Poa pratensis	Bluegrass	Alderman and Middleton (1925)
Polygonum aviculare	Prostrate knotweed	Al Saadawi and Rice (1982)
Polygonum orientale	Princessfeather	Datta and Chatterjee (1978)
Polygonum pensylvanicum	Pennsylvania smartweed	Letourneau et al. (1956)
Polygonum persicaria	Ladysthumb	Martin and Rademacher (1960)
Portulaca oleracea	Common purslane	Letourneau et al. (1956)
Rumex crispus	Dock	Einhellig and Rasmussen (1973)
Saccharum spontaneum	Wild cane	Amritphale and Mall (1978)
Salsola kali	Russian thistle	Lodhi (1979)
Salvadora oleoides	—	Mohnat and Soni (1976)
Schinus molle	California peppertree	Anaya and Gomez-Pompa (1971)
Setaria faberi	Giant foxtail	Schreiber and Williams (1967)
Setaria glauca	Yellow foxtail	Gressel and Holm (1964)
Setaria viridis	Green foxtail	Rice (1964)
Solanum surattense	Surattense nightshade	Sharma and Sen (1971)
Solidago sp.	Goldenrod	Letourneau et al. (1956)
Sorghum halepense	Johnsongrass	Abdul-Wahab and Rice (1967)
Stellaria media	Common chickweed	Mann and Barnes (1950)
Tagetes patula	Wild marigold	Altieri and Doll (1978)
Trichodesma amplexicaule	—	Sen (1976)
Xanthium pensylvanicum	Common cocklebur	Rice (1964)

[a]All references are cited in Putnam and Weston (1986).

Source: Putnam and Weston (1986), in Putnam and Tang (Eds.) (1986). *The Science of Alleopathy*. John Wiley & Sons, NY. Copyright 1997 with permission of John Wiley & Sons Inc.

result, there is considerable interest in exploiting allelopathy for weed control (Rice 1995). Natural plant products are under consideration for the production of herbicides that are relatively more environmentally benign than synthetic herbicides (Duke and Abbas 1995, Inderjit 1999, Duke et al. 2002). Significant progress has been made in both isolation and identification of specific allelochemicals (Dayan et al. 2000, Duke et al. 2000). Bhowmik and Inderjit (2003) review other possible applications of allelopathy in natural weed management, including using allelopathic cover crops or crop residues and using allelopathic crop cultivars to suppress weeds.

Responses of Plants to Allelochemicals. Chemicals with allelopathic potential can be present in virtually every kind of plant tissue, including leaves, flowers, fruits, roots, rhizomes, and seed. However, as noted above, whether the substances can be released into the environment of neighboring plants in sufficient quantities to suppress growth is still a question for many alleged cases of allelopathy (Putnam and Tang 1986, Putnam and Weston 1986, Cheng 1992, Blair et al. 2005). Whether the chemicals once released also persist long enough to suppress succeeding generations of plants also remains an unanswered question. However, allelopathy sometimes provides obvious and startling responses in affected plants. For example, it is well recognized that herbaceous plants growing near black walnut (*Juglans nigra*) may either fail to germinate or suddenly die as a result of juglone, a chemical produced in the leaves and roots of the tree. Dramatic reductions in crop growth have also been attributed to quackgrass (*Agropyron repens*) residues. However, the expression of allelopathy in most cases may be much more subtle than the above dramatic examples and have longer lasting effects (Putnam and Weston 1986, Teasdale 1998, Barbour et al. 1999, Blair et al. 2005).

Toxins from Residues. A primary effect of allelochemicals results from plant association with litter in or on the soil (Liebman and Mohler 2001, Kremer and Li 2003). Numerous organic chemicals are present in plant material and when crop or other plant residues are left on the soil surface after harvest or plowed under, chemicals can be released by rainfall or microbial decomposition (Patrick 1971, Rice 1984, Barnes et al. 1986, Rizvi and Rizvi 1992, Kremer and Li 2003). In the case of plant residues, however, subsequent mortality or growth suppression does not have to be related directly to the release of a toxic organic substance from plant material. Rather, modification of the microenvironment, for example, localized alteration of soil pH or other conditions as a result of litter decomposition, could account for the phytotoxic response (Barbour et al. 1999, Inderjit and Weiner 2001). Also possible is the release of a phytotoxic microbial product that accumulates as residues are degraded (Inderjit et al. 1995, Kremer 1998).

Toxins from Leachates and Exudates of Plants. Another source of allelochemicals is the production and release of toxins (secondary products) by growing plants that ultimately inhibit development of adjacent plants. However, this process does not have as great an effect on yield reduction as litter leaching or decay, perhaps simply due to a concentration effect. Several notable exceptions exist, however. For example, juglone can be leached from living black walnut foliage, as discussed earlier; sesquiterpene lactones found in foliage leachates of plants in the Asteraceae, Apiaceae, and Magnoliaceae can inhibit germination of crop and weed species (Fischer 1986); and caffeine from coffee tree foliage can inhibit the abundance of weeds and young coffee plants in coffee plantations (Waller et al. 1986). Specific evidence by numerous authors of foliage leachates or root exudation by growing plants is summarized in Rizvi and Rizvi (1992) and Inderjit (1999).

Effect of Allelochemicals on Seed. Two functions of endogenous allelochemicals present in seed are the prevention of seed decay and the inhibition of germination. Both of these processes, decay and germination, can account for substantial losses of seed from the soil seed bank (Chapter 5). Similarly, allelochemicals released into the soil from other plants can have an effect on seed in soil. Numerous seed bioassays have been accomplished that demonstrate the effects of allelochemicals in both enhancing and inhibiting germination [see Leather and Einhellig (1986) for references]. For example, the germination of Asiatic witchweed (*Striga asiatica*), an important parasitic weed, is enhanced by the presence of strigol, a substance produced by the roots of susceptible host plants. Putnam and Tang (1986) and Rizvi and Rizvi (1992) indicate that research to either induce or inhibit weed seed germination using allelochemicals would be fruitful areas to pursue for their potential applications in weed management. In addition, allelochemical enhancement of weed seed decay would be a worthwhile effort (Kremer and Li 2003), although little research has been published in this area.

Methods to Study Allelopathy. An array of techniques have been used for the study of allelochemicals (Putnam and Tang 1986, Inderjit 1999, Inderjit and Callaway 2003). For the most part these include specific methods for toxin isolation followed by bioassays to test for phytotoxic activity (Williamson 1990). The methods used for the isolation of toxins range from organic solvent extraction to cold-water infusion. Once a putative toxin is isolated, whether or not it is identified, various bioassay procedures are used in which test plants or seed are exposed to the chemical to evaluate its effect. Typically, bioassays using isolated chemicals are performed in sterile Petri dishes or in pots with soil. A pot method used for many years was the stair-step system (described in Radosevich et al. 1997), in which donor and recipient plants are grown separately in sand solution with the pots alternating in stair-step fashion. The soil solution is circulated from donor to recipient and back again a number of times, sometimes with an exchange column inserted between donor and recipient plants. Thus any substance exuded by roots of the donor can be evaluated for phytotoxicity. Such isolation and bioassay methodology is effective in establishing the existence of naturally occurring toxic substances. More straightforward are studies that use litter or crop residues directly. In these studies, either living or dead plant material is mixed in or placed on the soil for some period of time, after which the soil is bioassayed for allelochemical activity. With either approach, however, results must be interpreted cautiously because the chemical also must be released into the environment and be present at phytotoxic concentrations in order to be considered allelopathic (Williamson 1990, Teasdale 1998).

As noted by Inderjit and Callaway (2003) in their review of methodology for the study of allelopathy, simply isolating chemicals from plants and conducting bioassays do not establish the existence or importance of allelopathy. Designing appropriate experimental protocols for evaluating allelopathy in field settings is particularly difficult, as noted above, due to the confounding effects of other abiotic and biotic factors. Inderjit and Callaway (2003) suggest that evidence for allelopathy in natural ecosystems should include quantification of concentrations released into soil, manipulation of exudates using activated carbon or filtration methods, controlling for effects of resources, and examination of interactions with microbial populations and soil organic and inorganic compounds. Clearly, such investigations are still limited by the lack of proven experimental designs, which are critical for the advancement of understanding of the role of allelopathy in ecosystems (Inderjit and Callaway 2003).

Microbially Produced Phytotoxins

It is well known that microorganisms produce substances detrimental to other organisms, especially other microorganisms. These chemicals may act as repellants, suppressants, inductants, or attractants. Many microbial phytotoxins exist in nature and there has been considerable interest in using either the organisms themselves or the chemicals they produce to suppress weed growth, that is, as naturally produced herbicides (Duke 1986, Duke et al. 2000, Inderjit 2005). Examples of the use of microorganisms for weed control purposes include the soil-borne fungus *Phytophthora palmivora* for control of strangler vine (*Morrenia odorata*), the aerial fungus *Colletotrichum gloeosporoides* ssp. *aeschynomene* for control of northern jointvetch (*Aeschynomene virginica*) in rice and soybean, the wilt fungus *Cephalosporium diospyri* to inhibit Virginia buttonweed (*Diodia virginiana*), and the fungus *Chondrostereum purpureum* applied to cut surfaces of black cherry (*Prunus serotina*) stumps to kill this weed in reforestation areas.

As discussed by Duke (Duke 1986, Duke et al. 2000), in many cases microbially produced substances offer novel chemistries, high efficacy, new and more desirable selectivity, and favorable environmental properties as compared to synthetically produced herbicides. In spite of problems with economical production and in some cases absorption by treated plants, several microbially derived chemicals are now being registered and sold as natural product herbicides. Other substances and microorganisms are being explored as possible weed suppression agents using molecular approaches (Becker et al. 2000, Ruiz et al. 2000). Usually these products are highly specific for the suppression of a single species or group of species.

Parasitism, Predation, and Herbivory

Parasitism, predation, and herbivory are among the most common interactions that occur between plants and other organisms. As described earlier in this chapter, interactions among plants are often indirect since they are caused by direct effects of a biotic intermediary (the parasite, predator, or herbivore) on one member of the interaction. Post et al. (1985) summarizes the relationships among species in higher order interactions in this way: Friends of friends are friends, enemies of enemies are friends, and friends of enemies are enemies. In this sense enemies of friends are also enemies. Thus, it is important to consider how both desirable and undesirable plants are affected by an association to determine if the interaction is beneficial or harmful. Parasitism, predation, and herbivory are usually described as negative interactions because at least one of the partners in the association is impacted negatively (Table 6.1). These interactions, however, are also fundamental for biological control of weeds and invasive plants when the association adversely affects undesirable plant presence or abundance. These three plant associations are considered in more depth below.

Parasitism. A parasite is a plant or animal living in, on, or with another living organism (host) at whose expense it obtains food, shelter, or support. Parasites can be obligate, surviving only in association with the living host, or nonobligate, living either saprophytically or on a living host. In addition, some parasitic flowering plants are hemiparasites, that is, plants with chlorophyll that depend on the host for water and mineral nutrition.

Most parasitic flowering plants occur in about 10 families, but only four families contain the most troublesome plant parasites in agricultural (Figure 6.26) and forestry systems (Figure 6.27). These are Convolvulaceae (*Cuscuta*, dodders), Loranthaceae (*Arceuthobium*, *Phoradendron*, *Viscum*; mistletoes), Orobanchaceae (*Orobanche*, broomrapes), and Scrophulariaceae (*Striga*, witchweeds). Each genus is represented in North America except *Viscum*, although *Striga* is found only in limited areas of North and South Carolina. Parasitic plants are also of major importance in tropical agriculture (Akobundu 1987). In general, parasitic weeds are grouped into root parasites, such as the witchweeds and broomrapes, and stem parasites, such as the dodders, mistletoes (*Loranthus* spp.), and dwarf mistletoes (*Arceuthobium* spp.). The characteristics and economic importance of the various species within each genus have been described by other authors (King 1966, Kuijt 1969, Musselman 1987, Zimdahl 1999). For this reason this section concentrates on the features of parasitic plants that make them unique among weeds.

Adaptations of Parasitic Weeds for Dispersal and Germination. In order to survive, seedlings of parasitic plants must quickly find a suitable host. There are three methods through which parasitic plants increase the probability of successful contact with their host. In a number of parasitic species, for example, dodders, the seed are relatively large, which allows for a period of radicle extension until a host plant can be found. Another mechanism relies on birds for dispersal. Kuijt (1969) indicates that a precise mode of dispersal has evolved in these cases, which relies on birds to deposit the seed of the parasite, for example, mistletoe (*Phoradendron* spp.), on host branches. A similar mechanism of dispersal that assures the same end is the propulsion of seed from the fruit of dwarf mistletoe

Figure 6.26 Dodder (*Cuscuta* sp.) attached to alfalfa. (Courtesy of A. P. Appleby, Oregon State University.)

into the branches of the same or a nearby tree. The third adaptation for host location requires a biochemical exudate, which is produced by the root of the host plant, in order to initiate germination of the parasite seed. This requirement for germination is most pronounced in broomrape seed.

A further adaptation for host location following germination is demonstrated by both the broomrapes and the witchweeds, where chemotropic growth of the radicle occurs toward the root of their host plants. Although this feature of germination is probably highly evolved and acts to enhance seedling survival, it can also be used to obtain some control of the weed species. For example, species of witchweed can be induced to germinate by species that are not preferred hosts,

Figure 6.27 Dwarf mistletoe (*Arceuthobium americanum*) on lodgepole pine. (Photograph courtesy of W. Theis, U.S. Forest Service, Corvallis, OR.)

that is, "trap" crops. Since the trap crop cannot support the growth of the parasite, the abundance of witchweed seed in the soil is reduced through seedling mortality. A review of the use of trap crops in management of parasitic weeds is provided by Chittapur et al. (2001).

Physiological Characteristics of Parasitic Weeds. Although many parasitic weeds contain at least some chlorophyll, others do not. Some species that have chlorophyll apparently photosynthesize to only a limited degree, for example, dodders and dwarf mistletoes, whereas others fix carbon nearly as well as other nonparasitic members of their families. Experiments that utilize radioactively labeled elements and substances demonstrate the passage of organic material, minerals, and water from host to parasite. However, the degree of dependence on the host plant often varies with age and species of parasitic plant. For example, witchweed attaches to the roots of a host plant soon after germination but does not emerge from the soil for several weeks. During this time witchweed is dependent upon the host plant, but once seedlings emerge, witchweed plants produce chlorophyll and begin to generate their own assimilates. Water and mineral nutrients still must be obtained from the host plant, however. Broomrape, on the other hand, is a root parasite that lacks chlorophyll and depends on its host plant for its total sustenance.

The major organ of parasitic weeds for attachment and penetration of host tissue is the *haustorium*. Haustoria vary in structure according to species but all have a similar function, which is attachment and subsequent transport of materials from host to parasite. Figure 6.28 depicts penetration of the haustorium of field dodder (*Cuscuta campestris*) into a species of *Impatiens*. Since the hyphae of the haustoria contact both xylem and phloem of the host plant, transport of water, minerals, and hormones, as well as carbon compounds occurs (Musselman 1987).

It is apparent that parasitic weeds often have a significant detrimental effect on their hosts. However, in many cases a fine line exists between parasitism and hemiparasitism, which may have much more subtle effects. It is likely that parasitic plants have evolved closely with their host species, and perhaps study of this close evolutionary association can lead to better management of these species. Detailed studies concerning responses to chemical signals and their mechanisms of production and transport, which seem almost universal among parasitic plants, may aid in understanding the molecular regulation of parasitism and lead to strategies for control. For example, the haustoria of parasitic weeds apparently do not form if the plants are not grown in association with host plants but are rapidly induced in the presence of host roots or host root exudates.

Predation. Predation of weeds and invasive plants involves an association with some other type of organism, such as insects, rodents, or birds, or in some cases fungi or bacteria. As seen in Chapter 5, seed predation, or *granivory*, is one of the most important processes that regulate the occurrence of seed in seed banks, although seed predators can also be effective at seed dissemination in some circumstances (Louda 1989). Demographic models in which soil seed losses include death only by seed aging and germination usually give outputs where the seed

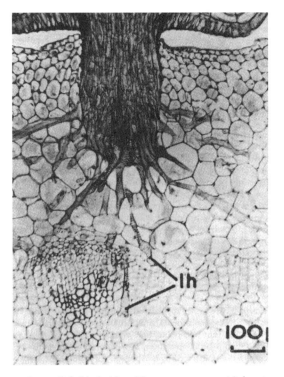

Figure 6.28 Haustorium of field dodder (*Cuscuta campestris*) in stem of *Impatiens* sp. (From MacLeod 1962, The Botanical Journal of Scotland, *Trans. Bot. Soc. Edin.* 39:302–315. Copyright 1962.)

population grows exponentially or reaches unrealistically high values because they do not account for seed losses from predation or dissemination to other areas. Sagar and Mortimer (1976) briefly explored the potential for seed predation to influence subsequent weed populations of wild oat under worst case/best case management scenarios (Figure 5.20). In that scenario, exposure of seed on the soil surface resulted in significant losses due to predation. Using a simulation model to estimate overall seed losses due to predation in various crops and cropping systems, Westerman et al. (2006) also showed that weed seed predation was important in reducing weed seed in crop fields and could serve as a tool in ecological weed management.

While there are many published accounts of weed seed predation by a variety of taxa, fewer studies quantify the interactions among predation of the weed/invasive plant seed rain, seed bank dynamics, and the subsequent distribution and composition of weed populations on agricultural, forest plantation or rangeland sites (Crawley 1992, Cromar et al. 1999, Liebman and Mohler 2001, Kremer and Li 2003). Several recent studies have examined seed predation at the community to landscape scale. Menalled et al. (2000) compared seed predation in complex and simple agricultural landscapes. In both landscape types there was significant postdispersal weed seed predation with a tendency toward higher removal rates in the complex landscape. However, seed predation showed a high degree of variability within and among agricultural fields (Menalled et al. 2000). In a study of agricultural fields with five contrasting management histories, Davis et al. (2006) showed that management history, microbial community composition (fungi and bacteria), and weed seed mortality were correlated. It is clear that habitat management is an important factor in maintaining stable insect and natural enemy populations in agricultural systems and may have a similar function in increasing weed seed predation (Landis et al. 2005). Thus, weed management techniques that maintain habitat for seed predators, such as using cover crops and mulches, herbaceous filter strips, or fallow periods, as well as low-input cropping systems may enhance weed control through indirect effects on seed predation (Chapters 7 and 9).

Herbivory. Herbivory by animals can decrease growth and fecundity, stimulate compensatory growth, or in severe cases cause mortality of plants. Defoliation by insect, rodent, or ungulate grazing is perhaps the best known example of herbivory. When grazing occurs on a crop or other desirable plant, weeds or invasive plants usually increase in abundance, as seen in Figure 6.29. In this example, cutworm grazing on alfalfa decreases alfalfa cover and allows weeds also present in the field to increase.

Empirical studies also suggest that herbivory contributes to long-term changes in native plant community composition and structure, trajectories of succession, and competition among plant species (Molvar et al. 1993, Hobbs 1996, Augustine and McNaughton 1998). Grazing animals in these cases function as a chronic disturbance that exerts a continuous influence or pressure on the ecosystem for a long period, as opposed to the episodic disturbances of fire, some insect defoliation, or timber harvests (Kie et al. 2003). Herbivory also has the potential to contribute to the establishment, spread, and persistence of exotic invasive plants in such ecosystems. For example, Belsky and Gelbard (2000) believe that livestock grazing (cattle, horses, sheep) has been a major contributor to exotic plant invasions into rangelands throughout the western United States. However, the effects of herbivore grazing on plant community structure and composition remain poorly understood because most studies have only been correlative or quasi-experimental, relying on animal exclosures with limited replication, size, or experimental control (e.g., only presence/absence) over possible herbivores.

Parks et al. (2005b) indicate that both domestic and wild ungulates contribute to exotic plant invasions of forest and rangelands by the following

- Selective grazing and reduction or elimination of native plants that favors an increase in invasive plants (Hobbs 1996, Augustine and McNaughton 1998, Vesk and Westoby 2001)
- Transport of invasive plant seed into uninfested areas, which occurs when animals consume seed in one area and later regurgitate or defecate them in another (endozoochory) (Janzen 1984, Mack 1991, Sheley and Petroff 1999)

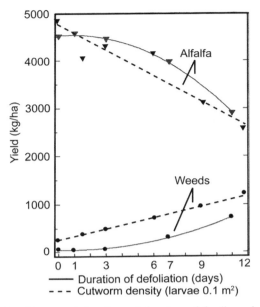

Figure 6.29 Effects of increased density and duration of feeding of variegated cutworm on alfalfa yield and on growth of competing weeds (Redrawn from Buntin and Pedigo 1986: in Norris et al. 2003, *Concepts in Integrated Pest Management*, Pearson Education, Upper Saddle River, NJ.)

- Transport of seed attached to skin, fur, or hooves to another location where it detaches (epizoochory) (De Clerck-Floate 1997, Olson et al 1997)
- Disturbance or alteration of soil by trampling vegetation and creating bare-soil conditions (Augustine and McNaughton 1998, Lonsdale 1999).

Undoubtedly livestock and other herbivores are significant factors that led to replacement of perennial bunchgrass species by cheatgrass and other exotic invasive plants in Great Basin and the United States (Mack 1981, D'Antonio and Vitousek 1992, Parks et al. 2005a).

Apparent Competition. Crawley (1983) and Louda et al. (1990a) believe that herbivory is so important in natural systems that it often masks or changes significantly the outcome of competition had the plants not been grazed, a phenomenon Louda et al. (1990a) and Connell (1990a) call *apparent competition*. In this case, the impacts of herbivory on plant species richness (composition) probably do not occur directly through the animals eating plant populations to local extinction, as suggested by Parks et al. (2005b), but through modifications in the plants' competitive abilities following feeding activities. If this hypothesis is correct, farmers, rangeland managers, and foresters may be able to modify the grazing activities of domestic animals, wildlife species such as deer and elk, and even "pest" organisms such as insects, slugs, and snails to limit the dispersal and growth potential of weeds and invasive plants.

Habitat Modification in Cropping Systems. The presence of certain weeds in and around agricultural fields often reduces specific insect pest populations. For example, Andow (1983, 1988) observed that 105 species of insect herbivores were more abundant in crops grown as monocultures than in weedy or intercropped systems. Certain weeds may also be preferred hosts, or decoys, for crop diseases and insects, thereby reducing crop damage from them. In some cases, the decoy plants may even stimulate germination of soil-borne pathogens, which subsequently die due to unavailability of the suitable host (Charudattan and DeLoach 1988).

Liebman (2001) and Norris et al. (2003) identify several basic principles that underlie weed–insect/pathogen interactions. They suggest that habitat diversification in crops leads to increased stability of insect populations, including pests, their predators, and other beneficial organisms. This argument is also supported by Swift and Anderson (1993), who indicate that both ecosystem stability and productivity increase as a function of enhanced plant biodiversity. In many cases, weed removal has been shown to increase the incidence of pest attacks on crops, presumably due to a reduction in the diversity of plant resources available for the pests. However, it is also likely that at least an equal number of examples exist where the presence of weeds increases crop pest populations (Norris et al. 2003). There are enough instances, however, where alterations of crop and weed densities have proven beneficial to crops to warrant a reevaluation of present weed control practices (Liebman 2001) (Chapter 7) and to reconsider the definition of a weed (Table 1.1).

POSITIVE INTERFERENCE

Facilitation

The process of *facilitation* is presented here as being complementary to competition in order to account for the many cases in agricultural and natural ecosystems in which one species provides some sort of benefit for another. On one hand, the plant species (e.g., weeds and crops or exotic and native plants) use different components of an ecosystem or use the same component in different ways. Within this general mechanism, both or all partners in the association might benefit simply from reduced competition. On the other hand, the species may alter the environment of one another positively, which is called facilitation or the facilitative production principle, according to Vandermeer (1989) (discussed earlier in this chapter).

Consider a hypothetical situation in which two plant species compete intensely for a commonly shared resource. However, they are only able to do so because of the protection provided by one species from a critical herbivore, while the other species fixes nitrogen from the air for use by both species. In this situation, the beneficial (facilitative) effects could be strong, but the competitive response is sufficiently intense to offset them, leading to an RYT of less than 1. Vandermeer (1989) points out that a standard experimental procedure would not suggest the operation of any other process than competition. Thus, when developing experimental approaches to study plant interactions, it is important to develop a framework that recognizes the importance of facilitative components of plant–plant interactions as well as competitive ones (Tables 6.1 and 6.2). Such complex interactions could occur not only in intercropping situations but also in monoculture crops that are invaded by more than a single species of weed or in which crop-planting densities are varied (Liebman et al. 2001).

In this chapter, only associations among plants are explored, especially crop–weed and native–exotic plant interactions. It is certain, however, that the interactions listed in Tables 6.1 and 6.2 are not restricted only to plants. In fact, many of the most biologically interesting and economically relevant examples of interaction occur among plants and other organisms, such as *protocooperation* and *mutualism*.

Commensalism. *Commensalism* (Table 6.1) is a one-way relationship between two organisms. In this type of association, only one organism is stimulated by the presence of the other and inhibited by its absence, whereas the other, or host, is unaffected. Common examples of commensalism are those in which the host organism serves as a surface for attachment and support or a means of shelter for the other organism without itself being affected (Whittaker 1975, Barbour et al. 1999). The organism that benefits gains physical anchorage or protection from the environment.

Commensalism between plants is demonstrated by epiphytes, which are plants that grow on other plants. The epiphyte uses the host for mechanical support rather than as a source of nutrients and water that are supplied by humid air or rainwater. Many organisms such as algae and lichens or nonvascular plants such as mosses are commonly found growing on the bark of trees. Examples of epiphytes among vascular plants include herbaceous perennial plants such as some ferns, bromeliads, orchids, and cacti.

The nurse plant syndrome is another type of commensal relationship between plants. In these cases, one plant is generally found growing in the shade or shelter of another, or nurse, plant. The nurse plant host, usually a shrub or perhaps an adult form of a seedling plant being benefited, provides shade, protection from high temperatures, soil drying, and frost, and sometimes protection from herbivores (Holmgren et al. 1997, Barbour et al. 1999). This pattern of growth is exhibited by the saguaro cactus as well as several species of desert annual plants that are commonly found growing under shrubs (Walker et al. 2001). In all these examples, the species that are positively associated with a host plant show no host specificity, nor is the host affected by the presence of the associated plant. In the case of desert annuals, in addition to providing protection, the shrub hosts act as a barrier to trap organic debris and therefore provide a suitable substrate for growth. While the nurse plant syndrome is difficult to prove and thus not well understood, it appears to be most common in harsh environments, particularly arid ones, where it can influence community structure and diversity (Haase 2001,

Tewksbury and Lloyd 2001, Flores and Jurado 2003). Interestingly, facilitation by nurse plants is being explored as a means of improving restoration success in certain environments, particularly the Mediterranean Basin (Castro et al. 2002, Gomez-Aparicio et al. 2004).

Protocooperation. The above examples are rather specialized in nature, and no proven examples are found in the literature of commensalism occurring among weeds and crops. It is much more common that two plants that interact affect each other reciprocally. This is the case with *protocooperation*, where both organisms are stimulated by the association but unaffected by its absence.

Root Grafts and Mycorrhizae. An example of protocooperation that occurs among species in many different habitats is natural root grafts. Generally occurring between trees, these grafts allow a mutual exchange of photosynthate and other materials to occur (Barbour et al. 1999). In many instances fungal hyphae (mycorrhizae), which dramatically enhance soil nutrient uptake, may also link two or more plants together, thus facilitating the protocooperation (Krishna 2005). Mycorrhizal associations of fungi with plants are also discussed later under mutualism.

Exudates. Another type of plant interaction frequently associated with negative effects (allelopathy) is the exudation of materials from the roots of one plant and subsequent absorption by the roots of another. This type of transfer, involving the one- or two-way movement of organic and inorganic metabolites, may be beneficial in some cases. For example, Neill and Rice (1971) observed that soil from the rhizosphere of western ragweed (*Ambrosia psilostachya*) markedly stimulated the growth of several species of plants growing in the same field. Similarly, Gajić and her associates (1986) demonstrated that wheat grain yields were increased appreciably when grown in mixed stands with corncockle (*Agrostemma githago*) as compared with pure stands of wheat. Seedlings of corncockle stimulated the growth of seedlings of wheat when both were grown on agar in sterile culture, indicating the presence of a root exudate. Subsequent isolation of the exudate, agrostemmin, and application to a wheat field (1.2 g/ha) increased grain yields of both nitrogen-fertilized and unfertilized stands. Grodzinski (1992) also observed positive benefits as well as negative ones from exudates of mustards (*Brassica* spp.) on a variety of crop seed and seedlings.

Leaching of metabolites from above-ground plant parts by rain, dew, and mist is also a source of movement of beneficial substances between plants. Calcium, phosphorus, magnesium, and other inorganic nutrients as well as carbohydrates, amino acids, and organic acids can be transferred in this manner. Plants for which beneficial leaching of nutrients was demonstrated include many vegetable crops, grains, grasses, cotton, and tobacco. Residues from water and methanol extracts of over 90 weed and crop species were found to have either stimulatory (6 species) or allelopathic (18 species) effects on the test plant, purple top turnip, at concentrations ranging from 3 to 300 ppm (Nicollier et al. 1985). It is important to realize the potential role of root exudation and foliage leaching in maintaining soil fertility and perhaps influencing plant succession.

Intercropping. Probably the most well known form of protocooperation among plants occurs in polycropping or intercropping systems, discussed in detail above in the section on competition. The practice of simultaneously planting two crops is in widespread use throughout the world, particularly in the tropics. Depending on the particular crops in the intercrop mixture and the environment in which they are grown, much evidence has accumulated that confirms that growth and yield are sometimes greater when plants are grown simultaneously than if they are grown alone as monocultures (Altieri and Lieberman 1986, 1988, Vandermeer 1989, Fukai and Trenbath 1993, Radosevich et al. 2006). However, such factors as plant density, spatial arrangement, time of planting, rate of development, and soil fertility must be considered in planning and evaluating results of intercropping experiments in order to optimize any potential yield advantage. Furthermore, a clear understanding of the shared resources among component crops will help develop appropriate management practices and choose the best cultivars for intercrops (Fukai and Trenbath 1993).

Mutualism. In contrast to the other types of positive interactions, *mutualism* is an obligatory relationship. The benefits gained by each partner in the association link them into mutual, physiological interdependence. In the event that one partner is absent, all suffer or in some cases cannot even exist as free-living organisms. Mutualism is distinguishable from protocooperation because the yield advantage in protocooperation (e.g., intercropping) often occurs because the plants fail to suffer in the presence of each other rather than because of benefits they afford each other. *Symbiosis* is another term used for positive interactions. Symbiosis is defined as the permanent, intimate association of two dissimilar organisms (Whittaker 1975, Leake et al. 2004). Because it is generally used to refer to mutually beneficial relationships, symbiosis is often equated with mutualism.

Lichens. One of the most well known mutualistic relationships in nature is lichens. A lichen is a symbiotic association between an alga and a fungus. The hyphae of the fungus surround the algal cells to afford protection from the environment and a favorable environment for growth. Photosynthates and often nitrogen compounds are supplied by the alga to the fungus.

Mycorrhizae. Many fungi are also found in symbiotic or mutualistic associations with roots of higher plants in the form of mycorrhizae (Allen 1991, Bever 2003, Leake et al. 2004, Krishna 2005). The fungal hyphae in these associations form a mantle over the root surface and penetrate inter- and/or intracellular regions of the root cortex. Many benefits to the host plant due to mycorrhizae have been found, such as serving as root hairs and thus increasing the root absorptive surface; increasing supplies of nitrogen, phosphorus, other nutrients, and water to the root; and increasing the decomposition of organic matter in the vicinity of the plant roots that is then available for plant uptake (Allen et al. 2003). The plant in turn supplies carbohydrates and other metabolites that benefit the fungus. Mycorrhizae occur in many diverse habitats and on a variety of plants, including grasses, shrubs, and trees. Through this association the host plant is often able to grow in nutrient-poor soil that otherwise would be unsuitable for normal growth and development.

As noted by Allen et al. (2003), although much is known about how mycorrhizae work, the functional role of mycorrhizae in ecosystems remains unclear. Recent evidence suggests that in some systems plant invasions may alter mycorrhizal associations with native species. For example, Hawkes et al. (2006), using polymerase chain reaction (PCR) amplification, examined how plant invasions alter the arbuscular mycorrhizal fungi associated with the roots of native plants in grasslands of California and Utah. They found a significant effect of exotic grasses on the diversity of mycorrhizal fungi colonizing native plant roots. Mycorrhizal fungi decreased on the roots of the native grasses *Hilaria jamesii* and *Stipa hymenoides*, but the number of mycorrhizal associates increased on the roots of two native forbs in the study. They also found that the presence of exotic grasses resulted in a compositional shift among fungi, with mycorrhizal *Glomus* spp. being replaced by nonmycorrhizal fungi. These scientists believe that alteration of the soil microbial community by invasive plants could both provide a mechanism for successful invasion and account for resulting effects of invaders on the ecosystem (Hawkes et al. 2006).

Nitrogen Fixation. Another type of mutuality involving higher plants is symbiotic nitrogen fixation. This process involves the conversion of nitrogen gas to organic ammonia, which only certain prokaryotic organisms can accomplish. Species of the nitrogen-fixing blue-green algae *Nostoc* and *Anabena* are found in symbiotic association with certain liverworts, cycads, and *Azolla*, a genus of small aquatic ferns. Many legumes are associated with the nitrogen-fixing bacterium *Rhizobium*, and certain shrub and tree species, such as *Purshia tridentata* and *Alnus rubra*, are associated with nitrogen-fixing bacteria as well. Generally, the nitrogen-fixing symbiont is localized in morphologically specialized structures on the host plant, for example, in root nodules of legumes. Frequently the symbiont is morphologically and physiologically distinct from the free-living species of the same genus.

The *Azolla–Anabena* association is among the most agronomically important of the mutualistic relationships so far described. *Anabena* supplies *Azolla* with its total nitrogen requirement and the fern supplies carbohydrates and metabolites to the alga. This particular association is used around the world as a nitrogen source in the culture of rice. *Azolla* either is grown concurrently with rice or may be used as a green manure after harvest of the rice plants. It has been estimated that three-fourths of all the nitrogen required by rice in California can be met by cultivating *Azolla* in rice fields.

The symbiotic nitrogen-fixing association between legumes and *Rhizobium* has also received considerable attention for many years. It was first discussed by Wilson (1940) and has been the subject of many reviews (Dilworth and Glenn

1991, Stacey et al. 1992, Postgate 1998, Werner and Newton 2005). By 1940, cropping mixtures involving legumes and nonlegume combinations of plants were in widespread use. Such mixtures as clovers with grasses, peas with oats, cowpeas or soybeans with maize, and winter vetch with rye or wheat are still frequently encountered. Many benefits are attributed to these mixtures, mostly due to the symbiotic nitrogen-fixing association of the legume and bacterium. These benefits include excretion or exudation of nitrogen by the legume for use by the nonlegume, stimulation of soil microorganisms, and return of nitrogen to the soil by sloughing off of roots and nodules. These practices are in use today either with legume–nonlegume cropping mixtures or legume–nonlegume rotations.

According to the diagrams of plant–plant interactions depicted in Figures 6.1 and 6.2, the nitrogen-fixing associations discussed above improve (have an effect on) the environment of the nonsymbiotic species (e.g., rice or other nonlegume) resulting in a dramatic improvement in yield (response) of the entire mixture. The long-term, evolutionary advantages of mutualism in the examples discussed above are long-lived absorbing organs and tight nutrient cycling. Thus, both partners can tolerate low levels of available nutrients due to increased efficiency of extracting essential minerals. Consequently, increased ecological amplitude is gained by the partners in association. Unfortunately, at least one example exists where the introduction of an exotic nitrogen-fixing tree was detrimental because it replaced native trees in Hawaii that had evolved in a nutrient-poor environment.

SUMMARY

Associations among members of the plant kingdom are common. Often when such associations occur between crops and weeds, crop yields are reduced. Most research deals with quantifying this negative aspect of plant association, although plant–plant interactions have more complex effects than simply yield reduction. Interference is the general term used for interaction among plants, that is, the effect that the presence of one plant has on another. There are many forms of interference ranging from negative to neutral to positive.

Density is the most common plant factor that modifies interference. Plants respond to density stress in two ways, by plastic growth and an altered risk of mortality. Species proportion and the spatial arrangement among individuals in a population or plant stand are other factors influencing interference. There are many different methods to study plant interference, particularly competition. These methods are based on the law of constant final yield and consider plant density, proportion, and arrangement to varying degrees. Experimental designs for studying interference include additive, substitutive (replacement series, Nelder, and diallel designs), systematic (addition series and additive series), and neighborhood experiments. Most experiments document well the levels of yield loss or changes in plant composition from association with weeds or invasive plants (intensity of competition). However, the importance of competition on the long-term dynamics of plant community structure and composition is less well understood. The experiments most often conducted in natural ecosystems are descriptive, retrospective, case studies, or gradient studies.

There are several types of thresholds relevant to weed/invasive plant management. In addition to ecological thresholds, these include damage, period, economic, and action thresholds that are based on IPM concepts. Thresholds are also used to evaluate the cost effectiveness of weed control tools and tactics. Several theories attempt to account for mechanisms of plant competition. One prevalent theory explains competitiveness on the basis of plant traits, particularly relative growth rate, which enables a plant to use resources rapidly. Another theory explains competitiveness on the basis of responses to disturbance and nutrient fluxes and equates competitiveness with low resource requirement.

Other forms of interference than competition are possible among plants. These include amensalism, parasitism, predation, and herbivory, which are negative interactions, and commensalism, protocooperation, and mutualism, which are positive interactions. The most common form of amensalism is allelopathy, the detrimental effect exerted on a plant from the release of a chemical by another plant. The primary effect of allelopathy on crop production results from an association with plant residues, although another source of allelochemicals can be "secondary products" leached or exuded from foliage or roots. Seed can also be an important source of allelochemicals. Recently, there has been considerable effort to isolate and produce specific allelochemicals or the microorganisms that produce them for use as naturally produced herbicides. Parasitic weeds are plants that live on or in another plant at whose expense they obtain food, shelter, or support. Parasitism is an obligate relationship for the parasite because without the host plant it will die. Parasitic flowering plants can be important weeds in cropland and forests of both temperate and tropical areas. These weeds usually have specific, highly evolved adaptations that assure finding and attaching to the host plant. Predation is most often associated with seed loss in or on the soil. Herbivory accounts for a process known as apparent competition. Herbivory is also one of the major causes for replacement of native grasses in prairies, grasslands, and savannahs.

Commensalism is a noninjurious relationship between two plants in which one is benefited without any harm occurring to the other. Commonly, the host plant serves as a source of support, anchorage, or protection for the other plant without itself being affected. The nurse plant syndrome is probably the best known form of commensalism. During protocooperation the two plants in the association are stimulated by the interaction but unaffected by its absence. Root grafts, mycorrhizal associations, and chemical exudates are all ways in which protocooperation can occur. Intercropping and polycropping are much utilized forms of protocooperation that occur among plants. In intercrop situations two crops are grown together, usually simultaneously, resulting in more total crop production than if the two crops had been grown separately. Weeds in an intercrop situation add significant complexity to the association and to its experimental analysis. Mutualism is an obligatory relationship between two organisms in which both partners benefit from the association. However, both suffer if one of the partners is absent. Mutualistic relationships among plants are rare but are quite common between plants and microorganisms. Lichens and symbiotic nitrogen fixation are well-known forms of such symbioses. Often nitrogen-fixing and non-nitrogen-fixing species are mixed to create a positive, highly productive plant association, but this can be detrimental if plants in natural ecosystems evolved in nutrient-poor conditions.

7

WEED AND INVASIVE PLANT MANAGEMENT APPROACHES, METHODS, AND TOOLS

Production of crops or other desirable plants or protecting natural habitats usually requires that the competitive impact of weeds and invasive plants be minimized. This is accomplished using various tools for weed control which, obviously, must not injure the crop or desired vegetation. Weed control is a component or tactic of the more general strategy of vegetation management, which includes fostering beneficial vegetation as well as suppressing undesirable plants. Successful vegetation management depends on knowledge about plant identification, life history, biology, associations with other organisms, and the selection of the proper weed control method.

In the preceding chapters, characteristics that make weeds and invasive plants successful organisms in agricultural, forest, and rangeland habitats were examined. The following chapters describe the approaches, methods, and tools used to control undesirable vegetation as well as general principles of weed/invasive plant management. Implicit in this discussion is the premise that better management—greater effectiveness of tools and tactics for prevention, suppression, or containment of weeds—results from increased knowledge of weed/invasive plant biology and the interrelationships of plants with each other and the environment they all share.

PREVENTION, ERADICATION, AND CONTROL

Management of weeds and invasive plants is a general strategy that encompasses the approaches of prevention, eradication, and control. *Prevention* involves procedures

Ecology of Weeds and Invasive Plants. By Steven R. Radosevich, Jodie S. Holt, and Claudio M. Ghersa
Copyright © 2007 John Wiley & Sons, Inc.

that inhibit or delay establishment of weeds in areas that are not already inhabited by them. These practices restrict the introduction, propagation, and spread of weeds on a local or regional level. Preventive measures include cultural practices, such as using clean crop seed to reduce weed dissemination into agricultural fields or using weed-free feed for stock animals that graze on rangeland or travel through forests and parkland. Prevention also includes the use of quarantines and weed laws. Surveys and monitoring are the first step in prevention of invasive species in natural ecosystems or natural resource production systems other than agriculture. These methods of weed management are discussed later in this chapter. *Eradication* is the total elimination of a weed species from a field, specific area, or entire region. It requires the complete suppression or removal of seed and vegetative parts of a species from a defined area. Although several regional eradication projects have been attempted in the United States, this goal is rarely, if ever, achieved without monumental effort. Eradication is usually attempted only in small areas or those with high-value crops or land use because of the difficulty and high costs associated with these practices. *Control* practices reduce or suppress weeds in a defined area but do not necessarily result in the elimination of any particular species. Similar to control, *containment* is often a goal of management of invasive plants, where the infestation is held to a defined geographic area and not allowed to spread. Weed control, therefore, is a matter of degree that depends upon the goals of the people involved, effectiveness of the weed control tool or tactic used, and the abundance and competitiveness of the weed species present. There are four general methods of weed control—*physical, cultural, biological, and chemical*—which are described later in this chapter.

WEED MANAGEMENT IN AGROECOSYSTEMS

Weed management options in agriculture should be considered well before crop planting because many weed problems are created or enforced by the manner that humans choose to grow crops, harvest forests, or graze rangeland. For example, once weeds become established in a field, they rarely can be eradicated from it. Ballaré et al. (1987) observed that weed control efficiencies up to 95%, a common level used by weed scientists to rate herbicide performance as excellent, cannot prevent weed populations from increasing year by year. Norris (1992) calculated that weed reductions must exceed 99.9% to reduce seed inputs to the soil enough to maintain a stable seed bank that does not increase. Liebman et al. (2001) believe that further improvements in current weed containment procedures will be brought about only by a thorough understanding of weed biology in the cropping environment, coupled with a management system that utilizes all suitable techniques to reduce weed populations and maintain them below levels that cause increases in the seed bank.

Economics and Biology of Weed Control: Whether to Control Weeds

Most decisions about weed/invasive plant management are based on three elements: (1) weed responses to control methods, (2) the opportunity to improve productivity, and (3) profitability. Each of these elements has been studied separately; however, they do not act independently because each one influences the relative importance of the other two elements in an iterative, interactive manner. Radosevich and Shula (1994) integrate these three elements of a weed management decision into a simple model based on empirically derived functions from the weed science literature. As described below, this model demonstrates the importance of economic thresholds for weed management and questions the value of high input costs to achieve very high levels of weed control (Radosevich and Shula 1994).

Weed Response to Control. The search for cost-effective methods to control weeds has usually focused on physical soil disturbance by tillage, fire, grazing, or herbicides as a means to prepare land for planting or to suppress vegetation during crop, tree, or forage growth. Experiments on these methods usually describe the degree of disturbance, levels and combinations of herbicide doses, or times of treatment or herbicide application required to kill or suppress weeds (Zimdahl 1980, Stewart et al. 1984, Streibig 1988). Such experiments are bioassays (Streibig et al. 1993) in which target or test plant species are subjected to various intensities, such as levels, doses, timings, combinations, or forms of tools to inhibit weed growth. Injury to the target (weed) species, usually measured as reductions in occupancy, cover, height, or biomass, is then compared to the condition of untreated plants.

It should be noted, however, as a point of caution, that experiments to compare tools or their intensities usually assume the context of an existing production system. This assumption can mask the importance of site and abiotic and biotic environmental factors on tool performance. Therefore, adopting new technologies exclusively from such experiments can result in higher than optimum herbicide rates, number of applications, or levels of soil disturbance because higher intensity of tool use usually compensates for a more thorough understanding of environment and biology. In contrast, lower herbicide rates, levels of disturbance, and so on, are usually possible when local information about environmental variation or species responses is incorporated into such experiments.

Both logistic and sigmoidal functions are used to describe plant responses to tool intensity and herbicide dose (Figure 7.1, top). Input costs for weed control are also reflected by such curves because costs generally increase proportionally to the intensity of the control measure used (Figure 7.1, bottom). Both relationships in Figure 7.1 show that beyond a certain point a diminishing increment of return (weed control) is achieved with each increasing increment of herbicide or cost.

Opportunity to Improve Productivity: Crop Response to Weeds. As already discussed in Chapter 6, crop growth can be seriously affected by the presence of weeds. Many experiments which collectively examine a wide range of crops, weed species, locations, environments, and control tactics demonstrate that the amount of increased crop response is generally related to the degree of vegetation suppression. Maximum crop production always results when crops grow without weeds and lower yields occur if weeds are abundant. Cousens et al. (1984) and

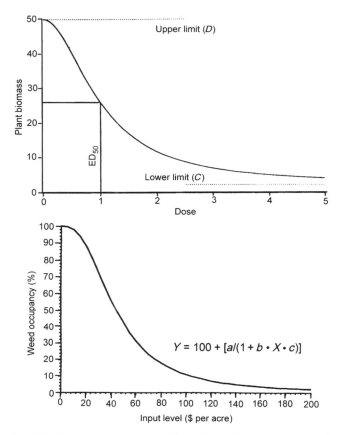

Figure 7.1 Logistic dose–response curve. *Top*: Plant biomass plotted against herbicide dose (untransformed) (Streibig et al. 1993). *Bottom*: Weed occupancy plotted against input level (cost per acre). (From Radosevich and Shula 1994, in Radosevich et al. 1997. *Weed Ecology: Implications for Management*, 2nd ed. Copyright 1997 with permission of John Wiley & Sons Inc.)

Alstrom (1990) indicate that results of these crop–weed experiments all conform to the same mathematical function, shown in Figure 7.2. This negative hyperbolic response between crop yield and weed density or biomass is consistent with the reciprocal yield law (Shinozaki and Kira 1956), which also has been demonstrated repeatedly for a wide range of species and environmental conditions (Figure 6.3). Interspecific competition for environmental resources is believed to be the major process responsible for this negative interaction among plants (Chapter 6).

Profitability: Value of Weed Control. The economics of weed control depend not only on the gain in crop yield from the absence of weeds (Figure 7.2) but also on the monetary value of the extra yield and the costs of the weed reduction (Alstrom 1990). This subject is not straightforward, although it has been covered

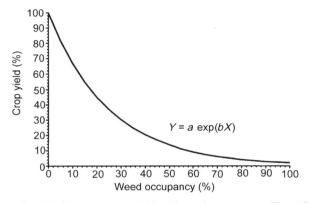

Figure 7.2 Relationship between crop yield and weed occupancy. (From Radosevich and Shula 1994, in Radosevich et al. 1997, *Weed Ecology: Implications for Management*, 2nd ed. Copyright 1997 with permission of John Wiley & Sons Inc.)

comprehensively by Auld et al. (1987). These authors indicate that three alternatives are possible to describe how revenues are obtained from weed control. They may (a) follow the law of supply and demand, that is, increase initially and then slow as more of the product is produced, (b) increase as the amount of product increases, or (c) increase initially and then decline as social or environmental costs become evident (Figure 7.3).

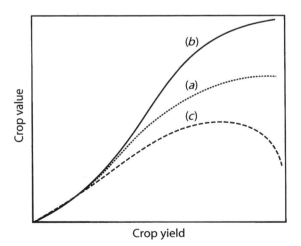

Figure 7.3 Relationship of crop value to crop yield. (*a*) Value of crop increases initially with increasing yield, then remains constant according to supply and demand. (*b*) Value of crop increases with increasing yield. (*c*) Value of crop increases initially as in (*a*) and (*b*), but eventually declines as a result of "external" environmental and social costs. (From Radosevich and Shula 1994, in Radosevich et al. 1997, *Weed Ecology: Implications for Management*, 2nd ed. Copyright 1997 with permission of John Wiley & Sons Inc.)

The second alternative (Figure 7.3*b*) is generally regarded as most appropriate for agricultural systems of production, whereas the alternative in Figure 7.3*a* is also appropriate for forestry, where, for example, a premium price is attained for large or high-quality logs. Under both of these alternatives, the value of further weed control generally rises as the price of the commodity rises because, to a limit, a greater amount of a more valuable commodity is produced. The alternative in Figure 7.3*c* has now become more evident as social and environmental costs for increased pesticide regulation and registration, environmental impact reports, mitigation or restrictions resulting from pesticide residues, and reforestation and other restoration efforts are being expressed [See Radosevich et al. (1997) or Liebman et al. (2001) for more discussion about these costs.]

Using the economic alternative shown in Figure 7.3*a*, each element of a weed control decision was combined to construct the relationship shown in Figure 7.4, which demonstrates the profitability arising from various hypothetical treatment intensities or input costs (Radosevich and Shula 1994). Efficiency, or inputs *versus* outputs, is shown as the diagonal line in Figure 7.4 and can be expressed as any common currency (e.g., money, energy, or time). Presumably farmers, foresters, and land managers will not make inputs that cannot be recovered as outputs. Thus, management tactics that result in greater benefits than the input costs to accomplish the task are "profitable" and probably would be continued. Strong motivation would be expected against management practices for which output benefits did not at least equal input costs (Figure 7.4) (Radosevich and Shula 1994).

An evaluation of gross output (revenue) in this model (Figure 7.4) reveals that most intensities of weed control fall within the profitable range. However, when

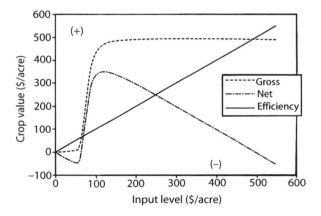

Figure 7.4   Relationship of crop value to input level using hypothetical costs of inputs and crop values. The diagonal line represents efficiency (i.e., costs of inputs equal value of outputs); values above diagonal line are "profitable," while those below line are "unprofitable." (From Radosevich and Shula 1994, in Radosevich et al. 1997, *Weed Ecology: Implications for Management*, 2nd ed. Copyright 1997 with permission of John Wiley & Sons Inc.)

net benefit (revenue − cost) is plotted against input costs, many intensities of weed control result in a loss (Figure 7.4). When control measures are so low and, therefore, weed abundance is so high that substantial crop losses occur, losses in net revenue obviously result, but even greater losses in net revenue can happen if input costs are too high (Figure 7.4) (Auld et al. 1987). In this analysis, the threshold level of input costs is achieved at about 80% weed control, with greatest benefit arising when weed control is between 80 and 90% (Figures 7.1 and 7.4). These rather high levels of weed control also were achieved in this analysis at relatively low projections of input cost. Weed control levels beyond 90% hypothetically would result in losses in net revenue even though the practices may still be within the profitable range.

This analysis questions the value of high input costs to achieve very low weed occupancy. It is technologically feasible to create nearly vegetation free and, therefore, highly productive agricultural fields and forest plantations. However, the economic and environmental desirability of doing so requires careful consideration. Farmers and foresters cannot afford to risk continual erosion of the financial, environmental, and biological ingredients necessary for long-term productivity of their cropping systems. Management for maximum (crop) or short-term (forest plantation) gains in production (Figure 7.2) probably requires greater investments of money, energy, and time and provides less net return than managing weeds at more optimal levels (Figure 7.4). However, attaining optimal levels of weed control requires much greater knowledge and understanding about the environment and biology of weed and crop, tree, or rangeland species. Unfortunately, inputs are often substituted for this greater understanding.

Influence of Weed Control on Agricultural Crops and Weed Associations

Weed control practices influence plant communities (crops and weeds) in two major ways—by reducing plant density and altering species composition in the area being treated.

Reduction in Weed Density. The effects of weed control on total plant density are usually obvious since most tools effectively reduce the abundance of weeds in a crop. Agriculturalists still rely heavily on mechanical forms of soil disturbance from plows, disks, harrows, and so on, and on herbicides for weed control both prior to and after planting. These tools reduce plant (weed) density very well. This reduction in total plant density generally results in favorable crop yield responses from weed control (Figures 6.9 and 7.2).

Alteration in Species Composition. Weed control practices also influence markedly the plant species composition of a field. However, this impact often is not recognized as an important consequence of weed control. Consider the rolling cultivator in Figure 7.5. This implement reduces weeds between the crop rows effectively but leaves both crop and weeds within the rows untouched. Thus, the cultivator affects both total plant density and the proportion and composition of

Figure 7.5   Rolling cultivator being used to control weeds in sugarbeet crop. (Photograph courtesy of R. F. Norris, University of California, Davis.)

species remaining in the field. Few studies exist that explore the effects of weed control or other cultural practices on subsequent weed stands and crop production. One study is that of Haas and Streibig (1982), in which some weed species increased while others decreased depending on method of cultivation and practices such as use of herbicides or combine harvesting that increased soil compaction, lime, and nitrogen. Compositional changes in plant species are also quite dramatic during forest regeneration (Walstad and Kuch 1987, Balandier et al. 2006, Wagner et al. 2006) and rangeland restoration (Sheley and Petroff 1999). In these cases, the total density of plants is temporarily reduced by weed control, but the composition of the tree, grass, or shrub stand is often affected for decades (Radosevich et al. 2006, Wagner et al. 2006).

Changes in composition within a weed species may also occur as a result of weed control practices, as discussed in Chapter 4. For example, Price et al. (1984) collected samples of wild oat (*Avena fatua*) along transects encompassing rangeland, ditch banks, and agricultural fields. Using electrophoretic enzyme analysis, these authors determined that cultural practices for cereal production apparently selected specialized wild oat ecotypes, most likely in response to the frequent tillage necessary to grow cereal crops. The topic of plant species adaptation to methods of crop production and weed control was discussed in depth in Chapter 4 and by Mohler (2001).

Herbicides (chemical control) effectively reduce weed density and also have a marked and rapid effect on weed species composition. This fact is the basis for selective weed control, which will be discussed in Chapter 8. *Selectivity* means that some plants in the crop–weed association are killed while others are not. Thus, when herbicides are used in the same location over time, a species compositional change toward more tolerant weed species is often observed. For example, a tendency for grass weeds to be favored over broadleaved (dicot) weeds usually follows repeated annual applications of the herbicide 2,4-D (2,4-dichlorophenoxy acetic acid) and other similar herbicides that act selectively against dicot plants. A similar phenomenon occurs when the crop and weed species in a field are taxonomically related. Plants in the same family or genus may share physiological or morphological traits that result in similar responses to particular herbicides. Thus, herbicides that do not readily control plants in the nightshade family (Solanaceae) are often used in crops such as tomato and potato that are in that family, but over time a shift in weed species composition toward weedy nightshades (*Solanum* spp.) may occur.

An extreme case of selection within weed species due to herbicide use is *herbicide resistance*, the naturally occurring inheritable ability of some weed phenotypes within a species to survive an herbicide treatment that should, under normal-use conditions, effectively control them. Since the 1970s, over 300 cases of resistance have been documented worldwide in at least 180 different weed species (Heap 2006). Over the last 35 years is has been well documented that repeated applications of herbicides with similar modes of action on the same field site impose sufficient selection pressure to increase resistance within formerly susceptible species (LeBaron and Gressel 1982, Holt and LeBaron 1990, Holt 1992, Powles and Holtum 1994). This topic has been reviewed extensively (Holt et al. 1993, Powles and Shaner 2001, Moss 2002) and cataloging the documented cases worldwide is the basis of an active website (Heap 2006, www.weedscience.org). In spite of the large number of cases, herbicide resistance has become a limiting factor for crop production only in a few local cases within a country or region (e.g., resistant annual ryegrass, *Lolium rigidum*, in Australian cereal production). Nevertheless, this phenomenon provides a dramatic example of the effects of control practices on weed population and species composition, which must be considered when designing management strategies. Particularly in instances where herbicide-resistant crops are grown, management to avoid selection for resistant weeds is critical (Shaner 2000).

Influence of Weed Control on Other Organisms

Regardless of the method used, weed control suppresses, removes, or destroys vegetation, which results in modification of the environment and habitat of other organisms (crop pests, non pests, and beneficial insects). Because weeds, like crops, are primary producers (Chapter 2), weed control must be recognized as a component of management programs that target crop pests but can affect other organisms as well (Figure 7.6).

There is little reason to maintain a level of plant species diversity in a crop field if the objective is only to control weeds. In this case, the presence of other plant species will decrease the level (efficiency) of weed control observed and could also reduce crop yield. However, if insect or disease management is also a consideration, the presence or absence of certain plant species can be a significant factor in the success of those programs (Altieri 1999, Schlapfer et al. 1999, Sheley and Kruger-Mangold 2003). Altieri et al. (1977) was one of the first

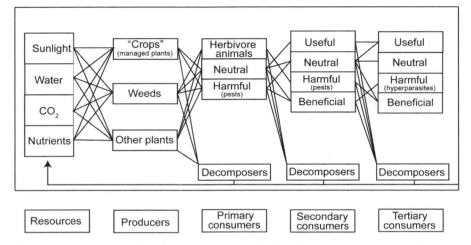

Figure 7.6   Summary of food web connections for hypothetical agroecosystem. Note that all resource and energy flow is from left to right, except for nutrient recycling. (Modified from Norris et al. 2003, *Concepts in Integrated Pest Management.* Pearson Education, Upper Saddle River, NJ.)

agricultural scientists to observe that a diversity of plant species reduced the magnitude of insect attacks in cropping systems. The increased level of species diversity in those experiments either decreased the incidence of phytophagus insects or increased the numbers of beneficial insects, which then lowered the pest insect population. Such experiments have since been repeated for many cropping systems (Liebman 2001) and the benefits of maintaining field-level biodiversity for crop production are becoming well documented (Liebman et al. 2001, Norris et al. 2003). However, vegetation diversity can also lead to increased insect attack in some cases, since weeds may provide a food source or habitat for insect pests that damage crops (Norris et al. 2003). Weed control can decrease the incidence of insect pests in numerous crops. Depending upon the situation, weed species can act directly as alternate food sources for phytophagus insects, provide a food source for phytophagus insects on which beneficial insects feed, or modify the microenvironment, allowing survival of beneficial natural enemies during adverse conditions (Norris 1982, Norris et al. 2003).

The role of weed species diversity on pathogen and nematode populations is poorly understood. It is known that maintaining large areas of a single crop or tree species can result in widespread disease. In some cases weed control leads to fewer disease and nematode problems. Within a field, evidence also suggests that some weeds can serve as alternative hosts of nematodes or protect nematodes from pesticides and the environment (Thomas et al. 2005). However, certain weeds can also provide nematode suppression through antagonism, contribute to changes in nematode biotic potential, or exert indirect effects on nematodes through competition with crops (Thomas et al. 2005). Schroeder and colleagues hypothesize that weed communities in many agricultural fields are dominated by plant species that are either tolerant or resistant to the endemic pest complex, particularly the soil pest complex, because of constant selection pressure from these pests (Schroeder et al. 2005). Unfortunately, little empirical data are available to test this hypothesis or conclude whether changes in abundance of diseases or nematodes result from either maintaining weed diversity or practicing weed control. Because of the lack of clear answers about the role of weed diversity in overall pest management, land managers and farmers are faced with a potential dilemma (Mohler 2001). Should all weeds be controlled or should some of them be left to assist in the management of other pest organisms? To answer this question, the costs and benefits of weed control must be weighed against potential costs and benefits derived from alternative strategies to control other pest populations (Figures 7.4 and 7.6). It is probably possible for weeds to be managed in such a way as to maintain crop yields and sustain beneficial populations of other organisms. However, such management tactics must be examined carefully for each cropping system, since any increase in plant diversity most likely complicates weed control efforts.

MANAGEMENT OF INVASIVE PLANTS IN NATURAL ECOSYSTEMS

The objectives of invasive plant management in natural ecosystems are similar to those of weed management in agroecosystems, although the specific considerations in evaluating whether to control invasive plants differ profoundly from those pertaining to agricultural weed control (Chapter 2). Economic cost–benefit analysis, as described above for agricultural weeds, is more elusive in ecosystems where no harvestable commodity is produced and the value of the land or services it provides is intangible (Pimentel et al. 1997, 1999). Chapter 2 outlined principles for setting priorities for management of invasive plants based on the risk of invasion and the value of land (Hobbs and Humphries 1995). Below, the practical considerations in invasive plant management are discussed.

Approaches to Prioritize Management

Rejmánek (2000) reviews approaches that are used to achieve the goals of prevention, eradication, or control of invasive plants. These approaches are *stochastic*, *empirical-taxon specific*, *evaluation of biological characters*, *evaluation of habitat compatibility*, and *experimentation*.

The *stochastic* or predictive approach focuses on initial population size and number and timing of introductions as factors that increase the probability of invasion success. For example, Mulvaney (unpublished data cited by Rejmánek 2000) observed a strong correlation between the extent of planting and the probability that a woody plant species would become naturalized in southeastern

Australia. In Venezuela, a significant correlation was found between the total number of known localities of an invader and the years since the first observation of the species was recorded (Rejmánek 2000). While this approach is useful for making probabilistic predictions about potential establishment and spread, S. J. Owen et al. (1997) found a similar correlation on New Zealand conservation lands, where the most widespread invaders are those that were introduced early and have the longest residence time. The shortcoming of this approach is that it tells relatively little about the potential impact of the species or the environmental and biological factors necessary to manage it (Mashhadi and Radosevich 2003).

An *empirical-taxon specific* management approach is based on information about the invasiveness of a species elsewhere. Information about the population growth of a particular species in a specific habitat can help land managers when confronted with a new introduction. Knowing the experiences of others also can help land managers make decisions about control and/or eradication of invasive plants. For example, Williams et al. (2001) indicate that 80% of the exotic weed species in New Zealand are also described as invasive outside that country. Rapoport (1991) *evaluated the biological characters* of invasive plants. He reports that 10% of the estimated 260,000 vascular plant species on earth are potential invaders, but only 15% of these potential invaders have actually invaded an area outside their native range. The traits that make some plants more weedy or invasive than others have been studied extensively since Baker conceived the concept of an "ideal weed" (Table 1.2 and Chapters 4–6) (Rejmánek 2000, Sakai et al. 2001). Understanding how and why certain biological characters promote invasiveness in a species is an important management tool (Chapter 6). Statistical tools like discriminate analysis, multiple logistic regression, path analysis and classification, and regression trees can be used to assess biological characters responsible for invasiveness.

The fourth approach to management of invasive plants involves *evaluation of habitat compatibility*, also referred to as *invasibility*, to determine whether a particular species can invade a particular habitat or habitat type. This approach is based on the assumption that climate is the overriding factor that determines the suitability of a site for an invasive species (Woodward 1987, Panetta and Mitchell 1991, Mack 1996). Several models have been developed that predict the potential new range of an invader based on identifying regions that are climatically similar to the species' native range (Sutherst and Maywald 1985) (Chapter 2). This approach is most powerful when combined with analysis of other factors, such as soil type, that can influence weed distribution. An extension of this approach is *niche modeling*, reviewed by Peterson and Vieglais (2001), Peterson (2003), and Thuiller et al. (2005), which models the ecological niche of a species in its native range and projects it onto other potentially habitable areas (Chapter 2). Finally, *experiments* can be conducted to test empirically the predictions made from the four approaches described above (Rejmánek 2000). However, the time lag that is inherent in most invasive episodes (Kowarik 1995) usually makes such experiments unappealing for land managers (Radosevich et al. 2005).

Documenting Invasions

Which species should be managed to limit their impact in an area? Which ones should be ignored? When is eradication a reasonable option; when is it not? To answer such questions land managers need to know the location and patch sizes of invasive plants in the areas they manage. Unfortunately, it is often difficult to detect invasive plants over large and remote areas in an efficient and cost-effective manner. Thus, tools and techniques used to locate invasive plants, measure population sizes, and store data are of considerable interest to land managers. Several recent chapters and books review approaches and methods for documenting invasions (Jackson 1997, McPherson and DeStefano 2003, Mack 2005). These include conducting ground-based observations, evaluating herbarium specimens and floras, and using aerial photography or satellite imagery (Mack 2005).

In ground-based detection, the most convenient estimate of weed population size is often plant canopy cover. Some land managers rely on visual sightings and estimates of cover, while others document cover by measuring patch diameter, radius, or plant density. Standard vegetation sampling methods are typically used, such as point or transect sampling (Mueller-Dombois and Ellenberg 1974, Barbour et al. 1999) combined with statistical analysis to assist interpretation. Other biological and environmental information is often collected to describe habitats invaded by the species in question. The specific type of data collected depends on how the information will be used. For example, risk assessment requires information on ecological and environmental characteristics of a site, biological characteristics and impacts of the invader, and locations of invasive plants from inventories and surveys. Managers conducting inventories or surveys also use specialized equipment such as global positioning systems (GPSs) to locate invasive plants and geographic information systems (GISs) to store biological and environmental information about the invasive species. This equipment is now readily available but was only experimental as little as a decade ago.

Another critical question central to invasive plant management is how to accurately and cost effectively inventory and monitor invasive plants across large landscapes. This question is particularly important since many invasive plants are easiest to control when populations are small yet control may be difficult to perform because of financial or personnel constraints. A wide range of remote sensing techniques have been used to detect invasive species, from coarse-scale satellite imagery to fine-scale aerial photography taken from fixed-wing aircraft. Despite the advocacy of remote sensing as a key to early detection of plant invasions, the results have generally been mixed (Carneggie et al. 1983, Johnson 1999, Everitt et al. 2001, Lass et al. 2002, Ramsey et al. 2002), and some form of on-the-ground survey is usually necessary to validate the results. Rew and Pokorny (2006) provide a description of both on-the-ground and remote sensing approaches used in western North America to assess invasive plant occurrence.

Terms Used by Land Managers. Terms like *survey*, *inventory*, *monitoring*, and *mapping* are often used by land managers to describe how they find and document the presence of exotic species. Sometimes these terms are used interchangeably, which can cause mistakes in how data are used to make management decisions (Rew and Pokorny 2006). Since comprehensive inventories of invasive plants are often not practical, land managers usually rely on strategic sampling, that is, surveys, to find where invasive plants are located.

Surveys and *inventories* are single point-in-time observations to determine the location and abundance of one or more exotic species in a particular management area (NAWMA 2002). An invasive plant survey samples a representative portion of a greater weed population, whereas an inventory determines the extent of the entire population of a targeted species within the defined geographical area (Valladares and Pugnaire 1999). The units of a survey are determined from either random points or transects. *Monitoring*, on the other hand, requires returning over time to the area being assessed to determine the degree of invasiveness (expansion) of the species, the impact of the species on the function of the ecosystem inhabited, or the effects of treatment. *Mapping* is a way in which the data are visually represented.

Rew et al. (2005) indicate that, historically, inventories and surveys have been linked to management, while monitoring may not occur until after the management activity has occurred, if at all. In addition, some exotic species may be more destructive than other species in a particular area, and some environments may be more susceptible than others to invasive plants. Thus, Rew et al. believe that monitoring should be included in all phases of management, even the earliest, to determine the risk of invasive plant expansion in the communities and habitats they occupy.

Incorporating Risk Assessment into Invasive Plant Management

Risk assessment differs according to the identity and scale of what is at risk (Figure 2.12). Prather (2006) indicates that assessments can be made on the basis of (1) individual species or (2) plant communities or habitats being occupied.

Individual Species Approach. A species-based approach can be used to determine which plants from a survey are likely to become invasive and, thus, help prioritize which species should be targeted for prevention or eradication programs. Reichard and Hamilton (1997), Daehler and Carino (2000), and Daehler et al. (2004) indicate that the single best predictor of invasiveness is when a species is invasive elsewhere. This characteristic alone accounts for 50% of the invasive plant species in Hawaii. By combining specific traits (characteristics of invasiveness) and prior invasive history, Pheloung (2001) created a screening method that correctly identified about 80% of the invasive plants species in Australia (Figure 7.7). The non-profit organization NatureServe (http://www.natureserve.org/getData/plantData.jsp) has also developed a screening approach to determine plant invasiveness. This approach combines species traits with ecosystem responses and the difficulty of managing the species of concern.

Questions forming the basis of the weed risk assessment model.

		History/Biogeography	
A	**1** Domestication/ cultivation	1.01 Is the species highly domesticated? if answer is "no," go to question 2.01	
C		1.02 Has the species become naturalized where grown?	
C		1.03 Does the species have weedy races?	
	2 Climate and distribution	2.01 Species suited for Australian climates (0 low; 1—intermediate; 2—high)	
		2.02 Quality of climate match data (0—low; 1—intermediate; 2—high)	
C		2.03 Broad climate suitability (environmental versatility)	
C		2.04 Native or naturalized in regions with extended dry periods	
		2.05 Does the species have a history of repeated introductions outside its natural range?	
C	**3** Weed elsewhere	3.01 Naturalized beyond native range	
E		3.02 Garden/amenity/disturbance weed	
A		3.03 Weed of agriculture/horticulture/forestry	
E		3.04 Environmental weed	
		3.05 Congeneric weed	
		Biology/Ecology	
A	**4** Undesirable traits	4.01 Produces spines, thorns, or burrs	
C		4.02 Allelopathic	
C		4.03 Parasitic	
A		4.04 Unpalatable to grazing animals	
C		4.05 Toxic to animals	
C		4.06 Host for recognized pests and pathogens	
E		4.07 Causes allergies or is otherwise toxic to humans	
E		4.08 Creates a fire hazard in natural ecosystems	
E		4.09 Is a shade-tolerant plant at some stage of its life cycle	
E		4.10 Grows on infertile soils	
E		4.11 Climbing or smothering growth habit	
		4.12 Forms dense thickets	
E	**5** Plant type	5.01 Aquatic	
C		5.02 Grass	
E		5.03 Nitrogen-fixing woody plant	
C		5.04 Geophyte	

Figure 7.7 Questions forming basis of weed risk assessment model. (From Pheloung et al. 1999, *J. Environ. Manag.* 57:239–251. Copyright 1999 with permission of Elsevier.)

Plant Community or Habitat Approach. The invasibility of plant communities and habitats varies and can be modified by human activities (Chapter 3). Rew et al. (2005) indicate that prior knowledge should be used to develop any survey scheme in order to maximize finding invasive plants. For example, since invasive plants are most often introduced by humans, the vectors of human transport such as roads and trails should be examined. Roads and trails are also places of high

C	**6 Reproduction**	6.01 Evidence of substantial reproductive failure in native habitat	
C		6.02 Produces viable seed	
C		6.03 Hybridizes naturally	
C		6.04 Self-fertilization	
C		6.05 Requires specialist pollinators	
C		6.06 Reproduction by vegetative propagation	
C		6.07 Minimum generative time (years)	
A	**7 Dispersal**	7.01 Propagules likely to be dispersed unintentionally	
C	**mechanisms**	7.02 Propagules dispersed intentionally by people	
A		7.03 Propagules likely to disperse as a produce contaminant	
C		7.04 Propagules adapted to wind dispersal	
E		7.05 Propagules buoyant	
E		7.06 Propagules bird dispersed	
C		7.07 Propagules dispersed by other animals (externally)	
C		7.08 Propagules dispersed by other animals (internally)	
C	**8 Persistence**	8.01 Prolific seed production	
A	**attributes**	8.02 Evidence that a persistent propagule bank is formed (>1 year)	
A		8.03 Well controlled by herbicides	
C		8.04 Tolerates or benefits from mutilation, cultivation, or fire	
E		8.05 Effective natural enemies present in Australia	

A = agricultural, E = environmental, C = combined

Figure 7.7 (*Continued*).

and repeated disturbance, which increases the likelihood of invasive plant introduction (Chapter 3). In addition, occupancy by invasive plants usually declines as plant cover increases such that a gradient of decreasing occupancy from areas of low to high cover should be expected. Some invasive plants also may be restricted to certain plant communities (Zouhar et al. 2007), while other communities are unlikely to be occupied by exotic plants (Parks et al. 2005a). Finally, Rew et al. (2005) point out that the distribution of invasive plants has a very high component of randomness and that any sampling strategy should be structured to account for this variability.

Radosevich et al. (2005) provide an example of how to construct a risk assessment model based on the susceptibility of native plant communities to invasion, disturbance history of sites, and proximity to current infestations. The output of such a model, presented as a map, is shown in Figure 7.8. Although risk assessment can be a valuable tool for land managers, it requires good information on species biology, site characteristics, and reliable position coordinates for an area, watershed, or region. Heger and Trepl (2003) and many others (e.g., Prather 2006) also provide examples and describe computer programs designed to predict the occurrence or assess the spread of invasive plants using a community or habitat approach.

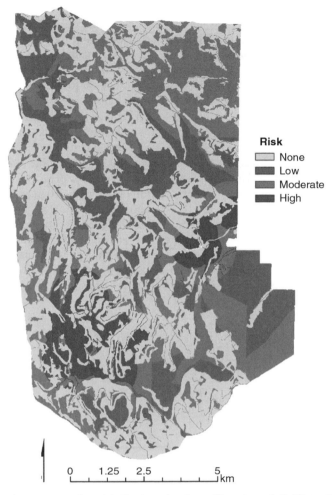

Figure 7.8   Areas at varying risk for invasion by sulfur cinquefoil (*Potentilla recta*) in Starkey Experimental Forest and Range Research Facility, Oregon Department of Fish and Wildlife, La Grande, OR. Risk depends on integration of habitat susceptibility, disturbance, and proximity to current infestations. [From Rew and Pokorny (Eds.) 2006, *Inventory and Survey Methods for Nonindiginous Plant Species.* Montana State University Extension Service, Bozeman, MT. Copyright 2006 with permission of L. Rew, Montana State University, Bozeman.]

Risks Associated with Action and Inaction

As described in Chapter 1, the language used to describe weeds and invasive plants is often emotive (Table 1.1). Even the mere presence of an exotic plant is sometimes enough to cause a farmer or land manager to initiate weed control or invasive plant management procedures. This approach is intuitively appealing particularly if an exotic plant is known to be highly competitive, disrupt ecosystem processes, or be associated with declines in native species (Hobbs and Mooney 1991, Blossey 1999, Davis et al. 2000, Mack et al. 2000). However, many of the cases described in these reports document only suspected impacts (Blossey 1999) or lack rigorous quantitative methods to assess impacts (Mack et al. 2000, Rew et al. 2005).

In addition to the risk associated with invasive plant presence, uncertainty can arise from treatments or tools used to prevent, eradicate, or contain these plants, which are also not without potential impact (see below and Chapter 8). However, taking no action (Rejmánek and Pitcairn 2002) or delaying action (Higgins et al. 2000) also can have an impact by increasing later costs of control or decreasing the chances of success. Thus, proactive management requires identification and eradication of small patches of potentially invasive plants before they become widespread. Nevertheless, just as in agroecosystems, the routine spraying of natural ecosystems rarely results in the long-term diminution of large populations of invasive plants even when the locations of such populations are known. Rew et al. (2005) offer an approach (Figure 7.9) where monitoring is used to improve the reliability of management tactics on areas inhabited by invasive plants. They suggest that land managers consider the following points when developing a weed management plan:

- Develop monitoring plans in the context of the overall objectives of land management.

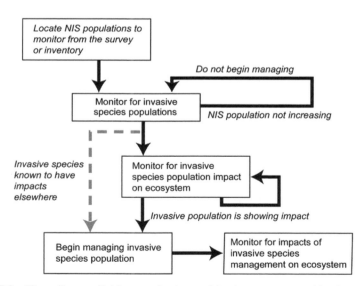

Figure 7.9 Flow diagram linking monitoring and land management objectives for exotic invasive plants. (From Maxwell 2005, *CIPM Online Textbook*, with permission of B. D. Maxwell, Montana State University, Bozeman, MT.)

- Select methods that will quickly meet monitoring objectives.
- Clearly link monitoring output to management decisions, which may include not to manage as well as to manage invasive plants.

Framework to Combine Research and Management of Invasive Plants

Preventing, reducing, or eliminating undesirable impacts of invasive plants is a challenge facing public and private land managers around the world. Hobbs and Humphries (1995) suggest an approach to set priorities for management of invasive plants based on land value and the degree of disturbance or risk of invasion (Figure 2.12). However, actual management of invasive plants is usually more difficult than depicted in Figure 2.12 and cannot be accomplished merely by applying the methods and tools used in agroecosystems. Because of the potential seriousness of an invasive plant problem, there is also a need to develop research programs that facilitate relationships between scientists and land managers, similar to the research–extension–outreach continuum typical of land grant universities in the United States. Research in natural systems, however, sometimes takes too long to accomplish and report, especially when managers must act quickly to prevent or contain the spread of an emergent invasive plant problem (Jasieniuk 2004).

Radosevich et al. (2005), using the invasive plant sulfur cinquefoil (*Potentilla recta*), suggest a framework in which empirical experiments, risk assessments of invasive species, and projections of species introduction and spread across susceptible landscapes are used to help land managers conduct and evaluate management activities (Figure 7.10). In this scheme, managers indicate concerns and provide information about day-to-day management activities for invasive plants.

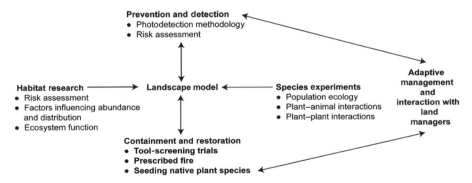

Figure 7.10   Regional approach for sulfur cinquefoil (*Potentilla recta*) research and management in Blue Mountains Ecoregion of Pacific Northwest, United States. The research approach was developed for management of a single species and may become more complex as additional invasive plants are added to framework. [From Radosevich et al. 2005, in Inderjit (Ed.) *Ecological and Agricultural Aspects of Invasive Plants.* Copyright 2005 with permission of Birkhäuser-Verlag, Basel.]

This often results in increased applicability of the research (Randall 1996, Byers et al. 2002). The approach suggested by Radosevich et al. (2005) incorporates habitat-level (Werner and Soule 1976, Zouhar 2003) and species-level (Sheley and Petroff 1999, DiTomaso 2000) experiments on age structure, population dynamics, competitive ability, dispersal, disturbance, and herbivory into a landscape-level model (Neubert and Caswell 2000, Radosevich and Wells unpublished). When these activities are combined with a GIS-based risk assessment (Figure 7.8), it is possible to project expansion of the species over time, which is helpful to land managers because the consequences of either management or no action can be determined and policies can be justified.

In areas where invasive species are already well established, tools for effective containment followed by restoration are needed to prevent reinvasion of a site (Allen et al. 1997). Invasive species containment and restoration experiments are often critical when promoting regional invasive plant management programs. There are, however, significant conceptual and logistical challenges to implementing scientifically sound restoration research. Michener (1997) and Suding et al. (2004) review these constraints and discuss approaches and analytical tools for ecological restoration research. Typically, control strategies for invasive plants (see below) that focus solely on the reduction of undesirable species by herbicide or fire often fail as other weeds quickly colonize the area (Bussan and Dyer 1999, Sheley and Petroff 1999, Sheley and Krueger-Mangold 2003). For this reason, and since little information exists on the methods of controlling sulfur cinquefoil (Lesica and Martin 2003), Parks and colleagues (in Radosevich et al. 2005) established an experiment to determine which herbicides, rates, and timing of application are most effective on the weed while minimizing their impacts on native plants. A further objective of the research was to determine if postherbicide reseeding facilitates native plant establishment. Thus, another experiment was established to examine the importance of prescribed fire, herbicide application, and native grass seeding on grassland restoration. This information is helpful to land managers because it is unclear whether fire, an increasingly popular vegetation management tool, facilitates, inhibits, or has no effect on the spread of sulfur cinquefoil. Basic information on species and community responses to tools for invasive plant management, such as that examined in this research, is necessary for resource managers to assess ecological responses to treatments and for the development of postmanagement strategies that enhance plant community restoration, wildlife habitat, or other components of biodiversity (Figure 7.10).

Results for such an approach (Figure 7.10) provide land managers, especially those interested in promoting native plant communities, guidance in integrating management tools (e.g., herbicides, fire, and native plant seeding) and techniques with information on plant invasiveness and community invasibility into an overall invasive plant management program. The framework discussed by Radosevich et al. (2005) (Figure 7.10) was designed from the onset to seek land manager input, and each experiment was conceived and implemented with a range of agency and private land managers involved. During the course of data collection, field tours to view experiments and field training sessions were conducted as an integral part of the program.

METHODS AND TOOLS TO CONTROL WEEDS AND INVASIVE PLANTS

Many of the methods and tools used to control, contain, or eradicate weeds and invasive plants are centuries old, while others have only been developed during the last several decades. In general, the approaches (methods and tools) used to reduce the abundance or vigor of unwanted vegetation vary only slightly according to habitat, that is, whether they are used in agriculture, other natural resource production systems like rangeland or forest plantations, or natural ecosystems. The tools used for weed control are categorized as physical, cultural, biological, or chemical tools. Although some approaches are more appropriately used in agriculture than other ecosystems, the methods and tools used to control or contain both agricultural weeds and invasive plants are discussed together in this section.

Physical Methods of Weed Control

Physical methods of weed control include any technique that uproots, buries, cuts, smothers, or burns vegetation. These methods consist of hand pulling and hoeing, fire, flame, tillage, mowing and shredding, chaining and dredging, flooding, and mulches and solarization.

Hand Pulling and Hoeing. Hand pulling and hoeing are the oldest and most primitive forms of weed control. However, it is estimated that over 70% of the world's farmers, mostly in developing countries, still use hoes or other manual implements to cultivate their cropland. A variety of hand implements have been developed for the removal of weedy vegetation. These tools range from rather primitive devices to sophisticated implements. Hoeing and pulling are most effective on annual or simple perennial plants that are not able to sprout from roots or other vegetative organs.

Even though manual methods of weed control have declined in developed nations, hand hoeing is still practiced in certain high-value crops or when other types of weed suppression are not possible. For example, hoeing is still practiced in crops where selective herbicides have not been developed or the area to be weeded is too small for most equipment. Hand hoeing and pulling also can be effective and economical for "rouging" the few individual weeds that escape other control measures or that infest a field for the first time. These few individual plants, if left unattended, have the potential to replenish the soil seed reservoir of a field (Chapter 5). Similarly, the earliest stages of colonization by an invasive plant can sometimes be eradicated by hand pulling in areas where financial or environmental constraints prohibit use of other methods.

Figure 7.11   Using prescribed fire to prepare a forest site for tree planting. The helicopter is used to rapidly ignite the perimeter of the site. (Photograph by E. Cole, Oregon State University.)

Fire. Fire is another tool that has been available to humans for centuries for the manipulation of vegetation. It is still used extensively to remove vegetation and crop residues in agriculture and to prepare forest lands for regeneration (Figure 7.11) after clearcut logging. The burning of agricultural fields to remove residues of the previous crop can suppress pathogens, insects, or weeds that might occur in the new crop. Fire is also sometimes used to remove weeds and other residue from along roadsides, canal banks, ditches, and vacant areas. In addition, fire has been used to manage fuel breaks in shrublands that are prone to wildfire (Figure 7.12). Broadcast burning is also an accepted and effective method to

Figure 7.12   Controlled burn used to convert shrubland to grassland. The technique is also used to maintain fuel breaks in dense shrubs or chaparral. (Photograph by S. R. Radosevich, Oregon State University.)

periodically increase rangeland productivity by stimulating growth of certain fire-adapted grass species.

Muyt (2001) notes that fire can be an effective weed management tool in certain natural ecosystems when applied carefully and selectively. He indicates that fire is used for various management purposes, including:

- Destroying mature plants and eliminating seedlings
- Exhausting weed seed banks
- Improving access for follow-up treatments and encouraging regrowth
- Removing dried weed material
- Stimulating germination, growth, and spread of indigenous vegetation

Fire also favors some weeds and invasive plants and can have a detrimental effect on natural ecosystems (Brown and Kaplar-Smith 2000, Muyt 2001, Rice 2004), so it is critical to assess whether fire is an appropriate weed/invasive plant management technique before conducting any controlled burn. The use of pre-scribed fire to control invasive plants is reviewed in DiTomaso et al. (2006) and DiTomaso and Johnson (2006).

Flame. For selective weed control in certain agricultural crops, heat can be applied with a tractor-drawn propane flamer. The principle of selective flaming is to direct heat toward the ground and avoid injuring the crop. Flaming is used either in crops in which the growing points (meristems) are beneath the soil and protected or if the crop is relatively tall and woody. In either case, the crop plants withstand the heat of the burner whereas small succulent weeds do not. The principle behind flaming is to control the intensity and exposure of the heat, causing cell rupture rather than com-bustion of plant material. The effects of flaming may not be apparent for several hours after treatment. This weed control technique has been used successfully in alfalfa, cotton, sugarcane, soybean, and peppermint.

Tillage (Cultivation)/Disturbance. *Tillage* or *cultivation* is disturbance of the soil. A major benefit of tillage is prevention and suppression of weeds. Tillage suppresses weeds by breaking, cutting, or tearing them from the soil, thus expos-ing the vegetation to desiccation, and by smothering them with soil. Repeated tillage may deplete weeds from fields by diminishing seed or vegetative propa-gules in the soil, providing that "escaped" plants are not allowed to reproduce. Repeated tillage may also exhaust the carbohydrate reserves of perennial weeds, thus suppressing them.

Tillage is also important to crop production for reasons other than weed control. Some of these other reasons for tillage are as follows:

- Seed bed preparation
- Burial of crop residues
- Control of plant pathogens, insects, and rodents

- Temporary improvement of soil physical conditions
- Improvement of surface conditions for rainfall reception
- Altered surface roughness
- Incorporation of fertilizers, herbicides, or soil amendments
- Moderation of soil surface microenvironment

Weed seed near the soil surface are usually not injured by tillage, but effective control of weed seedlings can result from a properly timed cultivation. Weed mortality is most likely to result from tillage when plants are small. Annual, biennial, and simple perennial weeds are most susceptible to tillage practices, which sever shoots from the root or uproot the entire plant. Only the shoots of creeping and woody perennials are killed by tillage. Best results from cultivation (tillage) occur when the soil is dry and the weather is hot. These conditions allow the optimum opportunity for weeds to desiccate and die following soil disturbance. Substantial control of perennial vegetative reproductive organs (rhizomes, stolons, etc.) can result when tillage is performed under hot, dry conditions.

Suppression of Seedling Weeds by Tillage. Covering small weeds with soil smothers them by disrupting the reception of light necessary for photosynthesis. This is an especially useful technique to control small weeds that occur within rows of crop plants. Covering small seedling weeds with soil is accomplished by a variety of manual and mechanical tools, such as rolling cultivators that are designed for that purpose.

Repeated tillage can also reduce the level of weed seed present in the soil because a portion of the dormant seed in the soil will usually germinate when the proper environmental stimuli (light, moisture, temperature) are provided. Tillage brings weed seed to the soil surface where germination is most likely and subsequent tillage kills them. A seed bank can be depleted in this manner, providing the germinants are not allowed to mature and reproduce. However, if tillage is not frequent enough to prevent weeds from producing seed, annual weeds can respond to the reduced densities caused by tillage and increase both size and seed output (Chapters 5 and 6). Frequent cultivations tend to disfavor the long-term presence of perennial weed species.

Light Requirement for Germination and Night Tillage. Field experiments demonstrate that 40–90% increases in germination of many dicotyledonous and grass weeds result from tillage operations performed during the day as compared to identical tillage performed at night (Figure 7.13; see Chapter 5 for discussion). Scopel et al. (1994) found that light acting through the photoreceptor phytochrome is the environmental signal that allows seed to detect such disturbances to the soil. The enhancement of weed seed germination during daylight (Figure 7.13) is due to light that penetrates into the soil during the actual tillage operation. These experiments indicate that if tillage operations are performed at night or with light-proof implements substantial reductions in weed germination and abundance result.

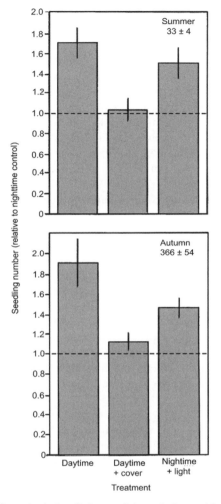

Figure 7.13 Effect of manipulating light conditions during cultivation on emergence of dicot weed seedlings compared with nighttime (no light) control. Absolute densities after nighttime cultivations are indicated at top of each panel in plants per square meter. Seedling counts were performed three weeks after cultivations. Species were (summer) *Amaranthus retroflexus*, *Solanum nigrum*, and *S. sarrachoides*; (autumn) *Lamium purpureum*, *Cerastium vulgatum*, *Veronica* spp., and *Stellaria media*. (From Scopel et al. 1994, *New Phytol.* 126:145–152. Copyright 1994. Reproduced with permission of New Phytologist Trust.)

A light requirement for seed to be released from dormancy is a widespread mechanism found in almost all plant species except crops. This requirement is satisfied even after a brief exposure to light, such as that resulting from soil tillage or other disturbances that expose soil. Thus, germination of species whose seed require light should be impeded when a no-tillage system (see alternatives to conventional tillage, below) is implemented for crop production. The light-requiring seed of such weed species would remain dormant, in the soil and eventually be lost due to predation, senescence, or decay (Chapter 5). Changes in composition of the weed flora favoring species that produce seed without a light requirement for germination have been observed following the implementation of no-tillage systems (Tuesca et al. 1995).

The requirement of light for germination of weeds and invasive plants in nonagricultural and natural ecosystems may also account for the widespread occurrence of these species along roads, trails, and forest clearcuts. Such areas experience soil disturbance, which favors germination. However, these areas are also low in vegetative cover and full plant canopies can be slow to recover during secondary succession. Thus, disturbance of resident vegetation along roads and in other open areas provides an environment with little competition, which also favors invasive plants.

Suppression of Perennial Weeds by Tillage. Perennial and biennial weeds can be suppressed by repeated tillage of fields that they infest. Repeated tillage results in a process called *carbohydrate starvation.* Following each tillage operation, new shoots emerge from the root or rhizome system of perennial weeds, which requires energy in the form of stored carbohydrates. The objective of repeated tillage is to deplete the carbohydrate reserves in the underground storage organs, eventually causing starvation and death of the plants. Cultivation every 10–14 days after emergence of new vegetative growth for at least one growing season is necessary for maximum carbohydrate starvation of most herbaceous perennial weeds (Ashton and Monaco 1991). Perennial weeds may actually be stimulated by cultivation if it is not frequent enough to cause carbohydrate starvation of the underground storage system.

Equipment Used for Tillage. In modern agriculture, mechanical equipment is used before and after crop planting to control weeds. The tools used for mechanical tillage are either *primary* or *secondary* tillage implements. Primary tillage implements are used to break and loosen the soil from 10 to 90 cm in depth. Usually, primary tillage is done only at the beginning of each growing season. Secondary tillage equipment is used to disturb the soil to 10 cm or less. These implements are used more often than those for primary tillage operations. Often several cultivations are necessary during a growing season to prepare the soil for planting, ensure optimum crop development, and control weeds. The type and size of tillage implement depend on the crop, soil, acreage, and power available to operate the equipment. In all cases the power to drive the implements is provided by animals or engines, rather than by humans. Excellent discussions of the methods and tools used for mechanical tillage are provided by Aldrich (1984), Ross and Lembi (1985, 1999), and Radosevich et al. (1997). Thus, the reader is referred to these texts for a more comprehensive review of this subject.

Problems with Tillage. The influence of tillage on weed seed in the soil is often unpredictable. In some instances tillage may decrease weed seed abundance while it may maintain seed abundance in other situations. Stirring weed seed in the soil profile often enhances germination and assists in depleting the reservoir of dormant seed, provided that control measures occur before weeds reproduce. However, seed burial by tillage may prolong the time that seed of some weed species exist in the soil, increasing their responsiveness to light stimuli and susceptibility to predation and decay (Chapter 5). It is also possible that repeated tillage favors some weed species or phenotypes through altered selection pressure (Chapter 4).

Tillage assists the dispersal of some weeds. Roots and rhizomes of herbaceous perennial weeds can be spread from field to field with tillage implements. In addition, infrequent tillage enhances perennial weeds by stimulating sprouting from severed vegetative organs. Tillage also can damage crop roots. However, such injury can be minimized if late season and deep cultivations are avoided. Tillage exposes soil to weathering, which can make wind and water erosion possible. However, equipment has been developed and practices can be used that minimize soil erosion. Compacted soils have also been attributed to mechanical tillage, especially when equipment operations are performed on wet soils.

Alternatives to Conventional Tillage. Several alternatives to conventional tillage have been developed to aid in soil conservation, reduce energy requirements, and decrease expenses. These systems are described by Ross and Lembi (1999):

- *Conservation Tillage.* Any tillage system that reduces the loss of soil or water when compared to a nonridged clean tillage system. This tillage system generally leaves a layer of crop residue on the soil surface that slows soil erosion and conserves water.
- *Minimum Tillage.* The minimum amount of tillage required for crop production or for meeting the tillage requirements under existing soil and climatic conditions. This system means eliminating excess tillage operations.
- *Reduced Tillage.* Use of primary tillage in conjunction with special planting techniques to reduce or eliminate secondary tillage operations.
- *No Tillage.* A 2–5-cm-wide seedbed is prepared with a special coulter or disc attached ahead of the planter. The crop seed are planted directly through the crop residue and no further tillage is required. In no-till systems, weed control is usually accomplished by chemical methods.

Ridge tillage is a form of conservation tillage that appears to overcome many of the soil microenvironmental, soil compaction, and weed control problems associated with other conventional and untilled systems. In the spring the tops of the ridges are tilled for planting (Figure 7.14). This removes crop residues from the ridge tops and disturbs the soil enough to create a seedbed. Soil on the ridges is also generally warmer than that between the ridges or in fields without ridges, which facilitates crop emergence. Tilling only the tops of the ridges disturbs fewer weed seed, reducing weed germination. Erosion is slowed because soil and

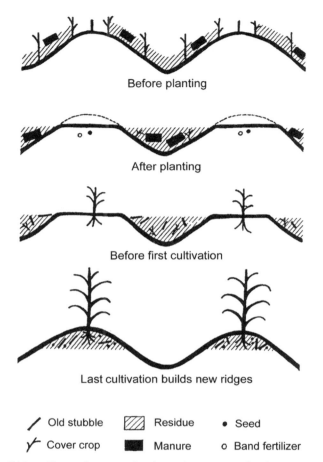

Before planting

After planting

Before first cultivation

Last cultivation builds new ridges

/ Old stubble   ▨ Residue   • Seed

ʏ̶ Cover crop   ▰ Manure   ○ Band fertilizer

Figure 7.14   Ridge tillage advantages in alternative crop production systems. The planter tills 5–10 cm of soil in a 15-cm band on top of ridges. Seed are planted on top of ridges and soil from ridges is mixed with crop residue between ridges. Soil on ridges is generally warmer than soil in flat fields or between ridges. Warm soil facilitates crop germination. Presence of crop residues between ridges also reduces soil erosion and increases moisture retention. Mechanical cultivation during growing season helps control weeds, reduces need for herbicides, and rebuilds ridges for next season. (From National Research Council 1989, Copyright 1989. National Academy Press, with permission.)

crop residues between the rows are not disturbed. Weeds that emerge later in the growing season tend to be between the ridges. Cultivations can control these weeds and less soil compaction occurs in the crop rows, thereby further enhancing crop growth and water infiltration [National Research Council (NRC) 1989].

Mowing and Shredding.   Mowing is used to control weeds by cutting or shredding their foliage. Mowing is usually accomplished to facilitate other management activities that require removal of vegetation. For example, it is often used to temporarily suppress weeds and cover crops in orchards and vineyards to assist harvesting operations. Mowing can also reduce water use in orchard and vine crops, since plants without foliage cannot use soil moisture rapidly. Other crops, such as alfalfa, pastures, and turf, are mowed as a normal production practice, which also suppresses some weeds.

In some situations, it is desirable to decrease the amount of vegetation that is already present without killing it. Power line rights-of-way, some roadsides, ditch banks, abandoned cropland, or vacant lots are areas where weeds are mowed periodically. Mowing of herbaceous vegetation should occur before the plants set seed in order to discourage weed seed dissemination. Since mowed plants, even annuals, usually regenerate new shoots, frequent mowing is often required to prevent seed production.

Mowing can suppress some perennial weeds through carbohydrate starvation, as discussed above. Similar to tillage, frequent mowing stimulates new shoot development, which eventually depletes the plants' carbohydrate reserves if done often enough. Mowing every few weeks for at least one to two growing seasons is usually necessary to satisfactorily suppress herbaceous perennial vegetation in this way. At no time during the growing season can the weeds be allowed to replenish their underground supply of carbohydrates for this system of weed control to be effective.

Mowing alters the stature of some weeds. A common morphological response of plants that have been mowed is to regenerate several new shoots from below the cut. Therefore, repeated mowing can change the appearance of a weed from a single-stemmed, tall, upright form to a plant with multiple shoots that are relatively prostrate. Weeds that occur in frequently mowed crops, such as alfalfa or turf, often assume this prostrate appearance due to either plasticity or selection of better adapted ecotypes (Chapter 4). Even low-growing prostrate weeds produce seed, however, so weed seed abundance will probably increase in the soil when mowing is practiced for weed control.

Equipment Used for Weed Mowing. There are three basic types of mowers: the *sickle bar mower, rotary mower*, and *reel mower* (Ross and Lembi 1985, 1999). The *sickle bar mower* is a tool used for cutting alfalfa and other types of hay. It consists of a cutting bar with attached guards and knife (sickle) that is driven back and forth in a horizontal direction. The motion of the sickle through the guards cuts the plants (Ross and Lembi 1985, 1999). This tool is also used to cut tall weeds in pastures and rangelands and along roadsides. The *rotary mower* or *flail* is an implement that consists of a horizontal blade attached to a perpendicular revolving shaft. It is used for mowing weeds in orchards, along roadsides, or in vacant areas. A heavy-duty machine (Figure 7.15) has also been developed for cutting brush and trees from under power lines, along other rights-of-way, and in certain forest situations. *Reel mowers* are used for management of turf. The tool consists of several spirallike blades that rotate against a stationary cutting blade. Although an important tool for maintaining healthy turf, the reel mower is used sparingly for weed control.

Figure 7.15 Rotary mower used for vegetation control in forestry. (Photograph by S. R. Radosevich, Oregon State University.)

Tools Used for Cutting and Mowing Weeds by Hand. Various kinds of *hand sickles*, *scythes*, and *machetes* are used to cut weeds, invasive plants, and other vegetation. These tools are still in common use in many areas of the world for the suppression of herbaceous vegetation. Often manual methods are the only practical means to manage vegetation in small close-cornered areas, where economics or topography does not allow more elaborate tools or where other methods of weed control have been restricted. Manual cutting is an important way to alter the composition of forest stands by removal of cull, unmerchantable, or weed trees. This practice can also be used to remove shrubs that are in proximity to commercially desirable trees. The gasoline-powered chain saw is the major tool for accomplishing these activities (Figure 7.16). Most hardwood tree species and

Figure 7.16 Brush cutting using chain saw. (Photograph by R. G. Wagner, Oregon State University.)

some conifers sprout after the stem and foliage have been removed. However, the amount of sprouting may be diminished significantly by cutting trees at certain times of the growing season. For example, the sprouting ability of red alder (*Alnus rubra*) is reduced when cutting is performed in July and August rather than at other times of the year.

Chaining and Dredging. *Chaining* involves use of a heavy chain, similar to that used to anchor ships, which is dragged between two tractors. In some cases, a metal blade is welded across each link of the chain. As the chain "rolls" between the two tractors, shrub stems are crushed and some plants are uprooted. This procedure is used to prepare shrublands or chaparral for rangeland improvement. *Dredges* are used primarily to remove submerged and immersed aquatic weeds from canals and rivers. Chaining is also used for aquatic weed control by "dragging" canals to tear loose aquatic plants growing there.

Flooding. *Flooding* is used in some regions to control established herbaceous perennial weeds. It has been used successfully to control Johnsongrass (*Sorghum halepense*), Russian knapweed (*Centaurea repens*), hoary cress (*Cardaria draba*), and silverleaf nightshade (*Solanum elaeagnifolium*). Complete submergence for one to two months during the summer is necessary to kill these species. Water depths of 15–25 cm are necessary so that the weeds cannot extend their foliage above the water surface. Flooding for only a few weeks rarely has an adverse effect on vegetative reproductive organs of weeds or weed seed buried in the soil.

In areas suitable for rice production, rotation to this crop permits both crop production and control of some perennial weeds. However, annual weeds such as barnyardgrass (*Echinochloa crus-galli*) and sprangletop (*Leptochloa* spp.) are associates of rice culture and not controlled well by flooding. Both weed species have been suppressed in rice by maintaining high water levels (15–20 cm). However, this practice is not desirable because deep water is also detrimental to rice at certain stages of development.

Mulching and Solarization. The purpose of *mulches* used for weed control is to exclude light from germinating plants. The exclusion of light inhibits photosynthesis, causing the plants to die. Commonly used mulches are straw, manure, grass clippings, sawdust, rice hulls or other crop residues, paper, and plastic. Artificial mulches made of woven plastics are also available for use. These mulches exclude particular wavelengths of light, usually in the photosynthetically active region of the light spectrum, but allow water, nutrients, and air to penetrate into the soil. Mulches are most effective for controlling small annual weeds but larger plants and some perennials can be suppressed by using mulches. Crops in which mulches are used are strawberries, pineapple, sugarcane, and some vegetable crops; mulches are also used in home gardens and landscapes. Organic food growers often rely heavily on mulches for weed control. Additional benefits arise from using mulches, including protection of low growing plants from frost in northern regions and reduction of soil temperature in warmer climates.

Soil solarization involves covering moist tilled soil with clear plastic to kill imbibed weed seed. (Note that black plastic used to cover soil is considered mulch, not a solarization technique.) The plastic sheets are left covering the soil surface for about four weeks. Long periods of high-intensity solar radiation that elevate temperature in moist soil are needed for best results. It is uncertain whether high temperatures or other factors increase mortality of weed seed. In some cases, high temperature stimulates germination but seedlings cannot survive under the plastic. Solarization was initially developed for control of soil micro-flora and microfauna, which can also be injured or killed by this practice. Solarization has recently been tested for use in restoration of abandoned farmland to reduce abundance of resident exotic weeds prior to planting native grassland species (E. A. Allen, unpublished).

Cultural Methods of Weed Control

Cultural methods of weed suppression often occur during the normal process of crop production. These practices include *weed prevention, crop rotation, crop competition, smother crops, living mulches and cover crops,* and *harvesting* operations.

Weed Prevention. The prevention or quarantine of a weed problem is usually easier and less costly than control or eradication attempts that follow weed intro-ductions because weeds are most tenacious and difficult to control after they become established (Figure 2.15). If weeds are allowed to develop a reservoir of seed or buds, they usually will be present in that location for many years, even decades. The following measures prevent the introduction of weeds into noninhabited areas:

- Use "clean" (weed-seed-free) crop seed for planting.
- Use manure only after thorough fermentation to kill weed seed.
- Clean harvesters and tillage and road building/maintaining implements before moving to non-weed-infested areas.
- Avoid transportation and use of soil or gravel from weed-infested areas.
- Inspect nursery stock or transplants for seed and vegetative propagules of weeds.
- Avoid planting exotic or invasive plants around homesites.
- Remove weeds from near irrigation ditches, fence rows, rights-of-way, and other noncrop land.
- Prevent reproduction of weeds.
- Use weed seed screens to filter irrigation water.
- Restrict livestock movement into non-weed-infested areas.

Other practices used to prevent and avoid potential weed problems at the state, regional, or national level are *weed laws, seed laws,* and *quarantines.*

The *Federal Noxious Weed Act* was enacted in the United States in 1975. This law prohibits entry of weeds into the United States by providing crop inspection for weed seed at ports of entry. The law also allows establishment of quarantines and provides for the control or eradication of weeds that are new or restricted in distribution. Other local, county, and state *weed laws* have been enacted so that property owners or public agencies must maintain a program of weed prevention or control on their lands. These weed laws permit the formation of *weed control districts* at the local level. It is the obligation of the district, through the activities of a superintendent, to diminish or restrict the occurrence of certain noxious weeds within its jurisdiction. The success of such laws depends upon the level of funding available, the knowledge of the superintendent about weed control measures, and the cooperation of public and private land owners to see that weed suppression programs are accomplished.

Seed laws are used primarily to assure the purity of crop seed and to restrict the dissemination of weed seed across political boundaries. In the United States, every state has a seed law that generally conforms to the statutes of the *Federal Seed Act*. Most states define two categories of noxious weeds, usually primary and secondary, which must be excluded from crop seed. No crop seed containing any primary noxious weed seed may be sold under this law, while only a small percentage (approximately 0.25%) of secondary noxious weeds may be present in crop seed.

Quarantines are enacted to isolate and prevent the dissemination of noxious weeds within a defined area or region. However, few quarantines have been enacted against weed species. A notable example of a successful quarantine is one established for witchweed (S*triga asiatica*) containment in portions of North and South Carolina. The strict regulation of farm material (farm products, residue material, etc.) in combination with other weed control methods has restricted this weed and significantly reduced its abundance in the quarantine area.

Crop Rotation. *Crop rotation* is the practice of growing different crops on the same land from year to year. It is a predominant method of weed control in many annual and short-lived perennial crops. Certain weed species are often associated with particular crops [e.g., barnyardgrass in rice, mustard (*Brassica* spp.) in cereals, dodder (*Cuscuta* spp.) in alfalfa, foxtail (*Setaria* spp.) in corn]. Therefore, populations of such weeds will usually increase whenever the crop is grown on the same ground continuously for several years. This increase in weed abundance happens because the same environmental or cultural conditions that favor crop production also favor the weeds (Chapter 4).

Weed associations with crops may be discouraged by growing crops in sequences that have sharply contrasting growth and cultural requirements. This practice discourages the development or evolution of weed populations that are well adapted to the growing conditions of any particular crop. Crop rotation also permits the use of different herbicides or other tools to select against weed populations that might be herbicide resistant.

The rotation of a solid-seeded crop, like alfalfa or cereals, to a row crop, like tomato, cotton, corn, or soybean, often allows a concomitant shift in weed control practices because of the differing cultural techniques necessary to grow crops in narrow *versus* wide rows. The rotation from crop production to *fallow* (no crop) also permits weed control measures that are not possible when a crop is always present. During the fallow period, use of a different form of tillage, implement, or control method may be possible to reduce weed abundance. Liebman and Staver (2001) note that crop rotations, in addition to providing weed control, often improve crop yields and quality by enhancing soil conditions and disease or insect control.

Competition. Crop yields often depend on the amount, size, and proximity of weeds present after crop emergence (Chapter 6). Weed vigor is similarly influenced by crop abundance, size, and proximity (Mohler 2001). Therefore, cultural practices that shift the balance of competition toward the crop will usually disfavor weed occurrence and improve crop yields. Factors that improve *crop competitiveness* include the following:

- Selection of well-adapted crop varieties
- Optimum planting date
- Optimum planting arrangement (row spacing)
- Soil amendments, such as fertility and lime
- Proper water management
- Use of "smother" crops

Practices that provide vigorous uniform crop establishment usually assist in reducing weed prevalence. Numerous examples exist in which poor crop development allowed increased weed growth. Poor performance of the crop may be genetic in origin (selection of the wrong variety) or caused by an array of cultural and environmental factors. The choice of planting date also influences the level of crop competition and necessity for weed control. For example, alfalfa in California may be planted in either autumn or spring. When seeded in the fall, seedling alfalfa plants are exposed to several months of cool, wet weather that slows their growth. In contrast, winter annual weeds grow vigorously under those environmental conditions, making chemical weed control necessary in fall-planted alfalfa stands. Alfalfa planted in the spring grows more quickly than that planted in the fall and thus can be more competitive with weeds.

Cultural practices that provide adequate soil fertility and water availability are necessary to ensure good vigorous crop growth. Poor irrigation or fertility practices create crop stress, which may favor weed occupation, abundance, or competitiveness. Excess water or fertility can also disfavor crop growth and favor the occurrence of flood-tolerant or nitrophilous weeds.

Row Width. Cultural practices such as manipulating row width can improve crop competitiveness. Mohler (2001) cites nearly 60 studies where both agricultural weeds and crops respond to changes in crop row spacing. Weeds are most often disfavored by narrower crop rows. For example, Rodgers et al. (1976) observed that if cotton was grown in rows 105 cm apart, 14 weeks of weed control were required to prevent yield losses. However, the weed control period was reduced to 10 and 6 weeks when the row widths were decreased to 77 and 52 cm, respectively. These required "weed-free" periods corresponded to the time required for cotton to develop a closed canopy. As crop plants grow and increase in leaf area throughout the growing season, many of them become highly competitive to weed seedlings growing beneath them. Thus, control measures that inhibit weeds soon after crop emergence can provide apparent season-long weed suppression. It is this observation that suggests the existence of critical time periods or temporal thresholds for weed control following crop emergence (Chapter 6).

Manipulation of Plant Canopies. Plant canopy cover can be used to reduce weed abundance by manipulating the conditions weed seed require to either germinate or remain dormant (Figure 5.18). As already discussed, almost all weed seed have a light requirement for dormancy to break and germination to proceed (Chapter 5). Plant canopies can be manipulated to suppress weeds by the density and arrangement of the desirable plants, whether they are agricultural crops, forest trees, or native species planted for restoration of an invaded site (Quinn 2006). It is also possible to manipulate plant canopies using cover crops and by scheduling the timing of emergence of the crop or other desirable plants. Ghersa et al. (1994b) provide an example of this latter process.

In Oregon, wheat is sown from the end of September through October. Within a month after crop sowing, Italian ryegrass (*Lolium multiflorum*) seedling recruitment occurs. However, Italian ryegrass seed have particular thermal and light requirements for germination, which can be disrupted by the presence of a plant canopy. Ghersa et al. suggest that a fast-growing summer annual crop such as Italian millet or sorghum could be sown in Oregon in late August or early September to produce a dense initial plant canopy. Wheat could then be sown into this canopy using a no-tillage system. Within a month or so after wheat sowing, frost, or in some cases herbicide, would kill the summer annual cover crop. By the time the summer annual species dies, wheat should have sufficient green canopy to filter the incoming sunlight, thus preventing germination of Italian ryegrass and other winter annual weeds. In an experiment carried out to study the effect of short-duration plant shading on biomass production of Italian ryegrass and wheat, Ghersa et al. (1994b) observed that over 50% reduction of Italian ryegrass vegetative biomass was obtained where winter wheat was sown under very sparse canopies of barnyardgrass as the cover crop.

Smother Crops. Some crops can suppress weed growth significantly through an ability to grow fast or because they are planted at high density. Crops with these characteristics are called *smother crops*. They include foxtail millet, buckwheat, rye, sorghum, sudangrass, sweetclover, sunflower, barley, soybean for hay, cowpea, clover, and silage corn (Ross and Lembi 1999). Often, these crops are solid seeded or planted in very closely spaced rows. They also may be used in rotations or mixtures with other crops.

Living Mulches and Cover Crops. In *living mulch* or *cover crop* systems, two crops are grown simultaneously, but usually one of them is more economically important than the other. Although the term "living mulch" has a recent origin, the concept does not. For example, intercropping of corn and legumes was studied in the 1930s and 1940s as a means to improve production of both crops. The current living mulch concept, however, involves growing row crops in an established plant cover provided by a cover crop or "green mulch." In this way, the occurrence of bare soil is minimized and weed seed germination is reduced (Teasdale 1998). Cropping systems presently under study with another crop species used as a living mulch or cover crop include alfalfa, clover, corn, soybean, several vegetable crops, and dry beans. Species sometimes used as cover crops are sun hemp, marigold, alfalfa, white clover, and cowpea, although other species are also under study.

The benefits of living mulch production systems are reduced soil erosion, stabilization of soil organic matter layers, improved soil structure, reduced weed abundance and competition, control of insect herbivores and plant pathogens, and diminished soil compaction (Altieri et al. 1977, Altieri 1999, Enache and Ilnicki 1990, Worsham et al. 1995, Yenish et al. 1996, Ngouajio et al. 2003). For example, a crop variety of *Brassica napus* (mustard or rape) has been used effectively as a natural fumigant against soil-borne pathogens and nematodes when its foliage is incorporated into the soil and adequate time allowed for its residue to decompose. Legume mulches supply nitrogen to the associated crop. However, substantial losses in crop yield can result from living mulches if competition between the "mulch" and crop plants is allowed to develop. Minimal competition has been found in soybean–winter rye, cabbage–fescue, and vegetable–marigold or sunhemp crop–mulch systems, but nearly all other living mulch systems require some form of cover crop suppression by either chemical or mechanical means.

Harvesting. Although not considered a method of weed control, harvesting can provide a certain level of weed removal or suppression, especially in short-statured herbaceous perennial crops. For example, it is common to harvest alfalfa several times during a growing season. The timing of harvest operations, relative to water availability and germination characteristics of weed species, can substantially improve weed control throughout the entire growing season. Summer annual weeds present in alfalfa at the time of first cutting can be reduced by timing harvests to shade and thereby suppress seedlings that germinate as the crop canopy develops. The frequency of grazing on rangelands and pastures has a similar effect on weed/invasive plant abundance and distribution in those production systems.

The use of combine harvesters as an artificial weed seed "predator" rather than as a disseminator of weed seed (Figures 5.19 and 5.20, respectively) can also be used to reduce the abundance of weed seed in agricultural fields.

Biological Control: Using Natural Enemies to Suppress Weeds

Both agroecosystems and natural ecosystems contain many organisms (Figure 7.6), including crops or desirable plants, weeds or invasive plants, pathogens, insects, and animals. Some of these organisms utilize specific plant species as a food source or host organism. Many weeds and invasive plants were introduced into new regions without their associated natural enemies. For this reason, weeds and invasive plants often grow as solid extensive stands in new areas of introduction, whereas in their native area they may exist as scattered patches or clumps due at least in part to negative interactions with other organisms. The lack of natural enemies following weed/invasive plant introduction can allow the weed population to increase rapidly to levels that eventually conflict with human interests.

Biological weed control is the use of living organisms to lower the population level or competitive ability of a species so it is no longer an economic problem. Spencer (2000) and Coombs et al. (2004) provide more complete reviews of this method of weed and invasive plant control and the process of plant suppression by biological control agents. Figure 7.17 depicts this process and compares it to weed responses from herbicide application. Note that biological control can have long-lasting effects on weed populations, in contrast to most herbicidal control. Not shown in the Figure 7.17, however, is the longer time frame required for biological control agents to become established and exert their effect. Cruttwell McFadyen (2000) provides numerous examples of successful biological weed control that utilizes one or more natural enemies (Table 7.1). Some other successful biological control agents are shown in Figure 7.18. Cruttwell McFadyen also indicates that although some biological control programs have saved millions of dollars, successes are generally not well recognized. She attributed failures of biological control to poorly resourced programs, long time lags for full success (often greater than 20 years), and failure to record the full extent of the prebiological control weed infestation.

Biological control differs from other methods of weed control in several ways (Rosenthal et al. 1989):

- It does not necessarily kill the weed outright; instead, often only the competitive or reproductive ability of the affected plant is reduced.
- It may be slow acting, often requiring years to achieve acceptable control levels. Thus, biological control should not be attempted when the destruction of the weed is needed immediately.
- It is relatively inexpensive, especially in contrast to the high costs of development (and frequent use) of herbicides.

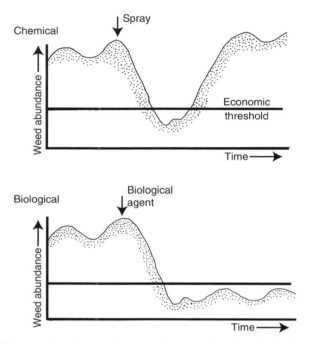

Figure 7.17 Diagram showing weed suppression over time from biological and chemical control methods. Note that successful biological control results in equilibrium of weed abundance below an economic threshold through time. (From Adkins 1995, *Weed Science Lecture Guide*, with permission of S. Adkins, University of Queensland, Brisbane, Australia.)

- Biological control is selective. Control agents must be host specific for the weed that they affect to prevent damage to crops or native plants. This selectivity is an advantage when control is directed at only one weed species. However, it is a disadvantage when several weed species must be suppressed at the same time.
- Because of the high host specificity, biological control agents should not cause harmful side effects.
- Biological weed control is often permanent. The object of biological control is not to eradicate the weed. Ideally, some of the weed population should always be present to maintain a population of the natural enemy. Such control is permanent once the weed and natural enemy populations are in equilibrium (Figure 7.18).

Procedures for Developing Biological Control. The first step in developing a biological control program is to determine the suitability of the weed or invasive plant problem for biological control measures. Like any form of weed management (Figures 7.1–7.4), economic losses caused by the weed or invasive plant

TABLE 7.1   Examples of Successful Biological Control of Weeds with Introduced Insects and Pathogens

Weed Species	Agents Introduced	Contributed to Success	Countries
Ageratina adenophora	2	2	Hawaii
Ageratina riparia	3	3	Hawaii
Carduus acanthoides	3	2	United States
Carduus nutans	5	2	Canada, United States
Carduus tenuiflorus	4	1	United States
Centaurea diffusa	14	5	Canada, United States
Centaurea maculosa	13	4	Canada, United States
Chondrilla juncea	4	1	Australia, United States
Chromolaena odorata	3	2	Guam, Ghana, Indonesia, Marianas
Senecio jacobaea	6	4	Australia, Canada, New Zealand, United States
Xanthium strumarium (occidentale)	5	2	Australia
Harrisia martinii	2	1	Australia, South Africa
Opuntia aurantiaca	6	1	Australia, South Africa
Opuntia elatior	1	1	India, Indonesia
Opuntia ficus-indica	9	3	Hawaii, South Africa
Opuntia imbricata	1	1	Australia, South Africa
Opuntia leptocaulis	1	1	South Africa
Opuntia littoralis	1	1	United States
Opuntia oricola	1	1	United States
Opuntia streptacantha	6	2	Australia
Opuntia stricta	9	2	Australia, India, New Caledonia, Sri Lanka
Opuntia triacantha	3	2	West Indies
Opuntia tuna	3	2	Mauritius
Opuntia vulgaris	4	2	Australia, India, Mauritius, South Africa, Sri Lanka
Hypercium perforatum	11	2	Canada, Chile, Hawaii, South Africa, United States
Cordia curassavica	3	2	Malaysia, Mauritius, Sri Lanka
Euphorbia esula	18	3	Canada, United States
Sesbania punicea	2	2	South Africa
Hydrilla verticillata	4	1	United States
Lythrum salicaria	4	2	United States
Sida acuta	2	1	Australia
Clidemia hirta	7	1	Fiji, Hawaii, Palau
Acacia saligna	1	1	South Africa
Mimosa invisa	2	1	Australia, Cook Islands, Micronesia, Papua New Guinea (PNG)
Emex australis	4	1	Hawaii

(Continued)

TABLE 7.1 *Continued*

Weed Species	Agents Introduced	Contributed to Success	Countries
Tribulus cistoides	2	2	Hawaii, PNG, West Indies
Tribulus terrestris	2	2	Hawaii, United States
Alternanthera philoxeroides	3	1	Australia, China, New Zealand, United States
Pistia stratiotes	1	1	Australia, Botswana, Ghana, PNG, South Africa, United States, Zambia, Zimbabwe
Salvinia molesta	1	1	Australia, Fiji, Ghana, India, Kenya, Malaysia, Namibia, PNG, South Africa, Sri Lanka, Zambia, Zimbabwe
Eichornia crassipes	7	2	Australia, Benin, India, Indonesia, Nigeria, PNG, South Africa, Thailand, Uganda, United States, Zimbabwe

Source: Modified from Spencer (2000); data from Julien and Griffiths (1998), Olckers et al. (1998), and Briese (2000). *International Symposium on Biological Control of Weeds*, Bozeman, MT.

plus the costs of other control measures should exceed the cost of the biological control project. Generally, biological control has been most useful when (1) current control measures are inadequate or expensive, (2) land values are low, (3) infested areas are vast and/or target plants are widely dispersed, and (4) no closely related crops or other plants of economic or ecological importance are present within the region of infestation.

Next, natural enemies of the weed or invasive plant species must be surveyed in both its native and its naturalized locations. The biological control agent should be damaging to the weed and be able to survive in the area of introduction. Thus, careful study of the distribution and feeding behavior of the potential biological control agent is needed. Host specificity in feeding, development, and reproduction must be demonstrated by the potential biological control agent. When selecting target weeds/invasive plants and potential organisms for host specificity testing in the United States, the Federal Working Group for Biological Control of Weeds (WGBCW) is consulted. This organization is composed of representatives of the U.S. Department of Agriculture (USDA), Department of Interior, Environmental Protection Agency (EPA), and Army Corps of Engineers. The recommendations of the WGBCW are made to the Animal and Plant Health Inspection Service (APHIS) of the USDA. The APHIS gives the final approval for importation, testing, and release of biological control agents in the United States. Testing of biological control agents involves experiments for host specificity in quarantine facilities and eventual release in the field.

(a)

(b)

(c)

Figure 7.18 Examples of biological control using (a) *Chrysolina* spp. beetles for St. Johnswort (*Hypericum perforatum*) control, (b) cinnabar moth larvae on tansy ragwort (*Senecio jacobaea*), and (c) "feeder" geese for grass suppression. (Photographs courtesy of W. B. McHenry and C. L. Elmore, University of California, Davis.)

Methods to Implement Biological Control Programs for Weeds. The implementation method used most for the biological control of weeds/invasive plants is the "*classical*" approach. This method relies on the utilization of exotic herbivores or pathogens with sufficiently narrow host ranges and is the procedure described above. The natural enemies are usually sought in areas where the weed–herbivore or weed–pathogen association coevolved. The classical approach has been most effective against naturalized invasive plants in natural resource production systems like rangelands and forests and in riparian areas. Another useful method of biological control is *augmentation*. With augmentation, the biological control agent is collected or mass reared, then periodically released to control a weed or invasive species. This approach is particularly suitable for areas in which the natural enemy is unable to survive adverse climatic conditions or when its population is insufficient to maintain acceptable control. Augmentation depends upon the ease of collection, rearing, transport, and distribution of the biological control agent.

Hazards of Biological Control. A significant hazard with the introduction of biological control agents is their accidental feeding on desirable plants that are closely associated with the target weed species. Although biological control of weeds has reduced the impact of certain target species (Table 7.1), Louda et al. (2003) note that some rare plants have declined after feeding by biological control agents. Pemberton (2000) also found, for example, that 15 insect species introduced for biological control fed on 41 different native plants species (Prather 2006). Louda (2000) indicates that past uses of natural enemies for weed or invasive plant control provide insight for the following eight lessons that she believes should be incorporated into biological control programs:

1. Better a priori quantification of the occurrence and detrimental effects of weeds
2. Improved incorporation of ecological criteria to supplement the phylogenic information used to select plants for prerelease testing
3. Increased assessment of plausible direct and indirect ecological interactions when an agent looks promising but feeding tests suggest it is not strictly monophagus, including factors determining host use and limiting population growth
4. Quantitative evaluation of the efficacy of the proposed biological solution, including evidence that the agent can reduce persistence and regeneration of the weed
5. More evidence on efficacy and cost of alternative control methods
6. Expanded review both prior to release and periodically afterward
7. Addition of postrelease evaluations and redistribution control
8. A rethinking of the situations that qualify for the use of biological control releases

The use of randomly amplified polymorphic DNA polymerase chain reaction (RAPD-PCR) may provide a means to better match natural enemies to their hosts. This approach has been used successfully by Ruiz et al. (2000) to map locations of 71 Russian knapweed (*Acroptilon repens*, synonym *C. repens*) populations in the western United States and Turkey, Kazakhstan, and Uzbekistan.

Grazing. *Grazing* is perhaps the oldest and most common form of biological weed control. It can be accomplished using a wide array of animals that eat vegetation, including large ruminants and ungulates, birds, insects, and fish (Figure 7.18). For example, geese are sometimes used to remove grass weeds from peppermint and orchards. Sheep are used at times to suppress herbaceous weeds in fast-growing established alfalfa stands, while blackberry can be controlled effectively by goat grazing. Certain species of fish have also been used to suppress aquatic weeds in canals and lakes.

Grazing, however, can be an agent of weed propagule dissemination and can suppress native plant populations (Chapter 5). The use, timing, and rotation of grazing animals for weed and invasive plant suppression should, therefore, be done with care to minimize their negative impacts on native vegetation.

Mycoherbicides. Plant pathogens have been used effectively to control weeds in augmentative-type biological control programs because plant pathogens are easily and cheaply cultured on artificial media, whereas insects and other biological control agents are not. Furthermore, pathogens may be applied to field situations using the same techniques and devices as used for herbicide application. An organism used in this manner has been termed a *bioherbicide* or if the organism is a fungus a *mycoherbicide*. Cruttwell McFadyen (2000) indicates that in addition to the list of successful biological control agents in Table 7.1, three invasive plants, northern joint vetch (*Aeschynomene virginica*), milkweed vine (*Morrenia odorata*), and broomrape (*Orobanche ramosa*), have been controlled successfully by fungi applied as mycoherbicides.

Candidates for mycoherbicides must produce large amounts of easily collected inoculum and be:

- Easily cultured in the laboratory
- Highly virulent to the weed or invasive plant
- Selective to desirable plant species
- Safe to humans and animals

Allelopathy. *Allelopathy* is any direct or indirect harmful effect of one plant on another through the production and release of secondary chemicals into the soil rhizosphere. Allelopathy is a form of negative interference, *amensalism*, and can be confused with competition when not studied carefully using proper controls (Chapter 6). Allelopathy has emerged as an intriguing method of using plants or plant residues to control weeds. Many smother crops and living mulches (cover crops) may be allelopathic to other plants or to themselves (*autotoxic*). Allelopathic crops offer potential for the development of cultivars or the extraction or manufacture of allelochemicals they produce for suppression of weeds in agroecosystems. The formation of toxic substances by microbial decay of plant residues may be an important mechanism for controlling seedling weeds with natural mulches. Recent investigations in rangeland systems suggest that some native

TABLE 7.2   Factors that Influence the Outcome of Bioassays for Allelopathy

Parameters
Assay species
Significance of an appropriate control
Seed germination
Seed number in relation to solution volume
Root–oven-dry and fresh weight, length
Physical state of allelopathic materials
Solvent for extracting allelochemicals
Concentration-dependent bioassays
Joint action of allelochemicals
Selection of allelopathic candidate
Sorption of allelochemicals in soil
Soil texture
Soil chemistry
Soil microbial ecology

Source: Modified from Inderjit and Nilsen (2003), *Crit. Rev. Plant Sci.* 22:221–238.

species may suppress invasive spotted knapweed (*Centaurea maculosa*) through the release of allelochemicals (Kulmatiski et al. 2004). Thus, the use of allelopathy to control weeds may also be considered a form of biological control, only using plants.

Allelochemicals are present in all types of plants and tissues and are released by a variety of mechanisms, including decomposition of plant residues (shoot or root), root exudation, and volatilization (Weston 2005). As noted in Table 7.2, approaches for documenting allelopathy unequivocally as well as identifying the active chemical remain problematic. Weston (2005) indicates that additional information is needed on allelochemical mechanisms of release, selectivity and persistence, mode of action, and genetic regulation. With such information, the use of allelopathic plants or substances isolated from them and identified or produced transgenically (Becker et al. 2000, Shaw and Milne 2000) may become an important form of weed control in the future.

Chemical Control

Chemicals, like other methods of weed control, have been used for centuries to suppress or remove weeds. Crafts (1975) indicates that solutions of sodium nitrate, ammonium sulfate, iron sulfate, and sulfuric acid were all effective treatments for weed control in cereal crops by 1900. Early nonselective uses of other chemicals, often industrial by-products or salt, had been used prior to that time to kill weeds in noncropland areas. The use of chemicals for weed control expanded rapidly during the middle portion of the twentieth century, following the discovery of several synthetic organic substances that killed or suppressed vegetation. Herbicidal oils and dinitrophenolic compounds were among the first organic chemicals to be used for weed control. However, it was the discovery of the plant-growth-suppressing ability of 2,4-D shortly after World War II that led to the expansive growth of chemical tools for weed control.

Herbicides. *Herbicides* are organic, synthetic chemicals used to kill or suppress unwanted vegetation. Herbicides now lead all other pesticide groups in total acreage treated, amount produced, and total value from sales. Ross and Lembi (1999) list the following reasons for the overwhelming success of herbicides as tools for weed control in agriculture:

- Herbicides allow the control of weeds where cultivation is difficult, for example, within and between narrowly spaced crop rows.
- Herbicides reduce the number of tillage operations needed for crop establishment. The amount of tillage reduction may be only a few operations or entire reliance on chemical weed suppression, such as in no-tillage systems.
- Controlling weeds with chemicals often permits earlier planting, since some tillage operations can be eliminated.
- Herbicides have reduced the amount of human effort expended for hand and mechanical weeding. In crops where herbicides are available, the costs associated with weeding often can be reduced substantially. These cost reductions may be directly or indirectly associated with reduced managerial requirements related to the hiring, overseeing, or housing of labor.
- Herbicides allow greater flexibility in the choice of management systems. Less reliance on crop rotation patterns, tillage implements and timings, and fallow periods allows greater selection of crops and management options.

Problems with Herbicides. It should be realized, however, that herbicides are only one type of tool available for weed or invasive plant control. Because of their effectiveness, there is sometimes a tendency among growers, land managers, and their advisors to expect that any vegetation management problem can be controlled effectively by chemical means. Such an attitude sometimes leads to more expensive or less effective vegetation management because other proven methods such as prevention and sanitation, tillage, crop competition, and rotation are overlooked as viable options.

Other potential problems associated with herbicide use include (1) injury to nontarget (e.g., native and/or rare plants) vegetation, (2) crop injury, (3) residues in soil or water, (4) toxicity to other nontarget organisms, and (5) concerns for human health and safety. The increased legal and regulatory requirements for herbicide application and worker safety are other concerns associated with the use of herbicides. In many cases, these problems or disadvantages can be overcome by proper selection, storage, handling, transportation, and application of the chemical.

Timing and Uses of Herbicides. Herbicides are often applied at various times during a growing season depending upon differences in development or stage of growth between the weed and the crop or other desirable species. Specific terms are used to describe these differences in herbicide application time (Figure 7.19). This figure also demonstrates how the sequence of herbicide applications fits with cultural practices and stage of growth of weeds and crop plants. These terms also explain how herbicides are used and categorized.

Preplant applications are made to soil before the crop is planted. They are typically made before or during seedbed preparation and before the crop is sown. This treatment method is not usually relevant in natural ecosystems or noncrop production systems where the soil is rarely bare.

Preemergence applications are made to the soil after the crop or desirable plant is sown but before emergence of the crop or weeds. This method has some utility in restoration programs where disturbance of resident vegetation is extreme and reestablishment of a tolerant native species is the goal.

Postemergence treatments are applied to both crop and weeds after they have germinated and emerged from the soil or, depending on the characteristics of

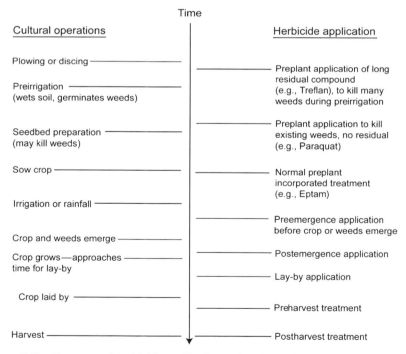

Figure 7.19 Sequence of herbicide application, cultural practices, and stages of crop and weed growth. (From McHenry and Norris 1972, *Study Guide for Agricultural Pest Control Advisors on Weed Control.* Publication 4050. Copyright 1972. University of California Press, with permission.)

the herbicide, may be applied just to weeds or invasive plants. Usually this term implies that an application of the herbicide will be made early in the development of the plants. Where herbicides are used, postemergence herbicide applications are the most common timing of chemical treatment for invasive plant suppression in forest plantations, rangeland restoration, and in some instances invasive weed control in natural ecosystems. However, other postemergence applications at later stages of development are also possible, especially in agriculture. They include the following:

- *Lay-by*—a herbicide application made to row crops that is the last equipment operation in the field until harvest.
- *Preharvest*—an application of herbicide made prior to harvest usually to desiccate the crop foliage and remove weeds that might interfere with harvesting operations.
- *Postharvest*—a herbicide application that is made to control weeds after harvest but which is not strictly part of the weed control program for the next crop.

Because of the importance of herbicides to modern weed control in agroeco-systems and their increasing use for invasive plant control and restoration in natural resource production areas or natural ecosystems, various topics pertaining to herbicide use will be discussed in the next chapter. These topics include basic aspects of herbicide chemistry, classification, selectivity, application, toxicology, safety, and regulation.

SUMMARY

Vegetation management is the fostering of beneficial vegetation and the suppression of undesirable plants. Weed control is only one component of vegetation management. Most decisions about weed/invasive plant management are based on weed responsiveness to tools, opportunities to improve productivity, and profitability. These three elements must be considered iteratively for sound decisions about whether to prevent, eradicate, or control weeds. Weed control methods remove, suppress, or destroy vegetation, which also can modify or disrupt the habitat of other organisms, pests and nonpests. Thus, it is important for weeds, invasive plants, and weed control tools to be considered as a component of management programs that involve other beneficial and nonbeneficial organisms.

Developing and assessing vegetation management priorities, especially in ecosystems other than agriculture, can be difficult. In these cases, assessments of risk can be informed by inventories, surveys, and monitoring of areas subject to invasive plants. Risk assessments are usually conducted at the species or plant community levels and the impacts of action, as well as inaction, can be determined in this way. A framework to combine research with management of invasive plants is most likely to produce successful management outcomes in natural ecosystems.

Physical, cultural, biological, and chemical methods are used to control weeds and invasive plants. Weed control practices influence plant communities by direct reductions of plant density and by alteration of species composition. Reductions in plant density by weed control procedures are usually obvious; however, shifts in weed species composition from weed control are more subtle. All tools used to control weeds alter the species composition of crop–weed stands to some extent. Herbicide resistance is an extreme example of a shift in weed species composition or change in genotype frequency in a species following prolonged use of the same weed control chemical.

Many tools are available for the physical disruption, suppression, or elimination of vegetation. These tools or techniques are fire and flame; manual pulling, hoeing, and cutting; and various mechanical implements. Tillage is a principal means of seedbed preparation and weed control in agriculture and site preparation in natural resource production systems. Many tools have been developed to accomplish tillage effectively. However, some problems exist with conventional tillage systems, and other alternative systems that reduce or eliminate tillage have been devised. Chaining, dredging, flooding, and artificial mulches are other physical methods of weed control. These tools are used effectively in certain circumstances and locations to control weeds.

Cultural practices used to control weeds are prevention, crop rotation, competition, living mulches (cover crops) and smother crops, and harvesting. In general, any practices that favor growth of desirable plant species will disfavor weed abundance, unless a crop is grown sequentially for a number of years and weed control or cultural practices do not vary. Quarantines and weed laws also represent preventive methods of weed control.

Biological control utilizes natural enemies to suppress weed species. There are defined protocols that must be met before new natural enemies can be introduced into the United States for biological weed control. Numerous successful natural enemy introductions have effectively controlled certain weed species, but some problems have occurred, such as feeding on desirable plants by introduced natural enemies. Solutions to this problem are possible, including careful matching of weed/invasive plant populations with natural enemies using molecular techniques. Both allelopathy and mycoherbicides may become important biological tools for weed suppression in the future.

Herbicides are phytotoxic chemicals that are used to suppress weeds. Herbicides are the most used form of pesticide in the United States. Although not without problems, herbicides are a popular form of weed control for farmers and other land managers because of their ease of use and effectiveness.

8

HERBICIDES

Herbicides are chemicals used to suppress or kill unwanted vegetation and are only one of the many types of pesticides, which include insecticides, fungicides, rodenticides, and others. Herbicides are used to reduce weeds in cropland, forest plantations, rangeland, and many other situations such as roadsides and rights-of-way where weed growth is sometimes a problem. They are also increasingly used to assist in management and restoration of areas previously invaded by invasive plants. Herbicides have become a major technological tool and are responsible, at least in part, for significant increases in crop production during the last quarter to half century. The U.S. Department of Agriculture (USDA) estimates that herbicides represent over 80% of all pesticides used in the United States (Short and Colborn 1999). Because of the importance of these chemicals to modern agriculture and natural resource management, this chapter describes how they are developed, regulated, and classified as well as their uses, characteristics, and environmental fate.

HERBICIDES AS COMMERCIAL PRODUCTS

Few herbicides, if any, are initially synthesized solely for their *phytotoxicity* or plant-killing properties. Rather, most manufacturers prepare and "screen" numerous chemical structures for a variety of purposes, including potential herbicidal activity. It is likely that a single chemical manufacturer will synthesize and test

Ecology of Weeds and Invasive Plants. By Steven R. Radosevich, Jodie S. Holt, and Claudio M. Ghersa
Copyright © 2007 John Wiley & Sons, Inc.

thousands of potential herbicides in a single year. It is during this primary synthesis and screening that a chemical is identified as a potential herbicide.

Herbicide development, following discovery, is a process of systematic chemical modification and examination for biological activity. It is an empirical procedure, based on both experimentation and experience, where chemists systematically add various substituent groups to the parent compound. Each of these "new" chemicals is also tested to find the material with greatest biological potential for plant susceptibility. Further laboratory, greenhouse, and field experiments are then conducted with the most promising materials to determine plant selectivity, soil persistence, and other physical and biological characteristics that influence the fate of the chemical in the environment.

The primary agency responsible for registration and regulation of herbicide development in the United States is the Environmental Protection Agency (EPA), which enforces federal laws requiring pesticides to be effective and safe. Herbicides must kill unwanted vegetation but not injure crops or other desirable plants. They must not enter the food chain or cause adverse effects to the environment. The necessary data to meet federal requirements enforced by the EPA require experiments on toxicology, biology, chemistry, and biochemical degradation of the chemical. In addition, the effects of the chemical on air and water quality, soil microorganisms, wildlife, and fish must be determined by the pesticide manufacturer.

Laws for Herbicide Registration and Use in United States

There are two laws in the United States that provide the authority to regulate pesticide development and use. These laws are the Federal Insecticide, Fungicide and Rodenticide Act (FIFRA) and some portions of the Food, Drug, and Cosmetic Act (FDCA). The FIFRA enforces the concept that any benefits from pesticide use must be in balance with concerns about public health and environmental impacts. It provides for registration and cancellation of pesticides, creates a classification system for pesticides based on toxicity, and allows states to regulate pesticides in a manner consistent with federal regulations. The FDCA requires the establishment of tolerances for pesticides in food, feed, fiber, and water.

Pesticides, including herbicides, cannot be distributed or sold in the United States unless they are registered with the EPA. Pesticides are classified by the EPA as being for either *general* or *restricted* use. The criteria for classification as a restricted use pesticide are (1) danger or impairment of public health, (2) hazard to farm workers, (3) hazard to domestic animals and crops, or (4) damage to subsequent crops by persistent residues in the soil.

In addition, hazard to surface and ground water supplies is an important criterion for herbicide regulation and restriction. Many herbicides are not as toxic as some other types of pesticides, such as insecticides. However, certain herbicides are very toxic and must be used with extreme caution. Furthermore, laboratory tests with animals indicate that some herbicides may be toxic following chronic exposure for several months or years.

Pesticide uses and environmental concerns often vary among U.S. states. For this reason, the regulatory aspects of herbicide use are influenced strongly by state laws. Every state has pesticide worker safety and restricted materials regulations that specify safe worker practices for individuals who handle or apply pesticides. These regulations are implemented to reduce the risk of pesticide exposure to people. It is the employers' responsibility to provide a safe working environment for their employees and to see that they are following safe practices as defined by the law. In most states, it is the state's Department of Food and Agriculture or a similar agency that has the responsibility for pesticide regulation and worker safety.

Information on Herbicide Label. The label printed on an herbicide container is considered to be a legal document that specifies how the material should be used to ensure its safety and effectiveness. All labels must show clearly the following information:

- Product trade name
- Name of the registrant (usually the manufacturer of the product)
- Net weight or measure of the contents
- EPA registration number
- Registration number of the formulation plant or factory
- Ingredients statement containing the name and percentage of active ingredient of the product
- Percentage of inert ingredients
- Use classification, that is, general or restricted
- Warning or precautionary statement

Toxicity Categories of Herbicides. Warning and precautionary statements on the pesticide label are concerned with human toxicity and environmental, physical, and chemical hazards associated with each material. A *toxicity category* is assigned to every pesticide based on levels of hazard indicators. Each toxicity category and its indicator are shown in Table 8.1. The signal word *danger* is required for a pesticide meeting any criterion for toxic category I. Toxicity category II materials require the signal word *warning*, while pesticides in categories III and IV use the word *caution*. Both federal and state laws require that pesticides be used in accordance with the instructions printed on the label.

Voluntary and Legislative Restrictions on Herbicide Use

The EPA has the primary responsibility for gathering use and risk data from chemical manufacturers for registration and classification of pesticides, including herbicides, according to U.S. toxicity categories. However, there are other national and international organizations that adopt a more cautious view of

TABLE 8.1 Signal Words Used in Labeling

Signal Word	Oral Toxicity		Dermal Toxicity			Inhalation LC$_{50}$
	LD$_{50}$	Lethal Dose for 150-lb Person	Toxicity LD$_{50}$	Eye Effects	Skin Effects	
Danger	Up to and including 50 mg/kg	A taste to a teaspoonful	Up to and including 200 mg/kg	Corrosive: corneal not reversible	Corrosive	Up to 2000 μg/L
Warning	50–500 mg/kg	A teaspoonful to 1 oz	200–2000 mg/kg	Corneal opacity: reversible within 7 days; irritation present for 7 additional days	Severe irritation at 72 h	2000–20,000 μg/L
Caution	>500 mg/kg	>1 oz	>2000 mg/kg	No corneal opacity: no irritation; or reversible within 7 days	Mild to moderate	To 20,000 μg/L

Note: LD$_{50}$ values are stated in milligrams of pesticide per kilogram of body weight; 1 mg/kg = 1 part per million (ppm). LC$_{50}$ values are stated in micrograms of the compound per liter of air.

Source: Whitford (2002), *The Complete Book of Pesticide Management: Science, Regulation, Stewardship, and Communication.* John Wiley and Sons, NY. Copyright 2002 with permission of John Wiley & Sons Inc.

herbicide (pesticide) use. For example, the World Health Organization (WHO) maintains lists of prohibited pesticides that are *persistent or toxic* or *whose derivatives remain biologically active and accumulate in the food chain beyond their intended use.* All organizations of organic food growers also maintain lists of chemicals that farmers may use if they wish to market their products as organic. Others, such as some land management certification organizations, have adopted policies that limit the pesticides that can be used by their membership based on the potential risk of the chemical to the environment, humans, and other organisms. The basis for the restriction is usually information from published sources on the chemical and biological properties of the pesticide, which are outlined below. These organizations generally provide a mechanism for derogation (exemption) if it can be demonstrated that the use of the material is less hazardous than suggested by the properties of the chemical alone. Other organizations, such as the U.S. Forest Service (USFS), are legislatively mandated to perform an environmental risk assessment for all pesticides used on land under their jurisdiction, according to the National Environmental Protection Act (NEPA).

The difference between the approach taken by voluntary organizations and that of the agrichemical industry and the EPA in regulating pesticide use is both philosophical and technical. In some cases, as with organic food organizations, all or nearly all artificially manufactured pesticides are prohibited, whereas inorganic or naturally occurring chemicals may be used. In other cases, restrictions are based on the potential risks of use determined from published information, regardless of whether the material is organic, inorganic, artificial, or naturally constructed. What is usually not considered in cases of voluntary restriction of herbicide use is the potential for reducing exposure to the chemical, or *hazard.* The EPA considers hazard explicitly in its registration procedure, while hazard is only the basis for derogation for other organizations.

Properties of Herbicides that Affect Human, Animal, and Environmental Safety

Commercial herbicide products often contain several different ingredients, but testing for immediate (*acute*) and long-term (*chronic*) toxicity is usually only performed on the *active ingredient* of the product, that is, the component of the product believed to actually affect the target organism. The other *inert ingredients* are discussed later in this chapter. The criteria for assessing the possible effects of herbicides (and pesticides) on human, animal, and environmental safety are *acute* and *chronic toxicity*, including *carcinogenicity, endocrine disruption, mutagenicity and reproductive disorders, biological magnification*, and *persistence in the environment* (Briggs 1992, Whitford 2002). Although only brief discussion of properties that affect herbicide safety is possible here, this subject has been reviewed in depth by Briggs (1992), Kamrin (1997), Short and Colborn (1999), Whitford (2002), and D'Mello (2003).

Toxicity. Toxicological testing evaluates whether exposure to an herbicide (pesticide) will produce acute effects (e.g., eye or skin irritation, neurotoxicity) or chronic effects (e.g., impaired liver function, reproductive abnormalities, cancer). Toxicological evaluations are conducted with experimental animals exposed to various levels of the pesticide for various lengths of time from hours to years (Whitford 2002). The duration, magnitude, and frequency of the doses determine the severity of the effect to the test animals, which is then extrapolated to humans and other organisms. Toxicologists follow the basic premise that all chemicals are toxic at some dose; a dose–response curve establishes the gradation of effects from increasing doses of a chemical. These dose responses generally follow a bell-shaped curve (Figure 8.1*a*) that ranges from low to high doses and establishes the occurrence of highly susceptible to resistant individuals.

Acute Toxicity. For a measured response such as death the percentage of test animals that die increases proportionately as the dose increases. The LD_{50} is a common measure used to define acute toxicity, defined as the lethal dose for 50% of the animals tested (Briggs 1992, Whitford 2002). Some common terms used in pesticide toxicology are *reference dose* (RfD) or *margin of exposure* (MOE), *no observed adverse effect level* (NOAEL), and the *lowest observed adverse effect level* (LOAEL). According to Whitford, a dose–response curve has three distinct regions (Figure 8.1*b*), no detectable response, increased linear response, and maximum (plateau) response. With the possible exception of some types of cancer, most toxicological responses occur at or above a threshold dose level. The threshold level of responses at the beginning of the linear region of the curve shown in Figure 8.1*b* describes the NOAEL and LOAEL. The point of the response curve (plateau) where an increase in dose no longer produces an increase in response is the *maximum observed adverse affect level* (MOAEL). The RfD or MOE is determined by comparing the anticipated level of human exposure to the lowest NOAEL. Thus,

$$MOE = RfD = \frac{NOAEL}{\text{estimated human exposure}} \tag{8.1}$$

An RfD is typically extrapolated to humans from animal studies using a safety factor of 100 after finding the NOAEL. Because of inadequacies of using only acute testing, RfD, and LD_{50} values, much more comprehensive toxicological testing of pesticides is now performed (Briggs 1992, Whitford 2002).

Subchronic and Chronic Toxicity. Chronic studies of toxicity measure the effects of daily exposure to a pesticide over a one- to two-year time period. Subchronic effects are measured from repeated exposure over weeks or months. Chronic toxicity research is typically performed on rats and dogs and examines the cumulative effects of a pesticide on body organs, such as lungs, kidneys, liver, and so on. (Whitford 2002). The RfD and LD_{50} values are calculated in the same manner as for acute toxicity.

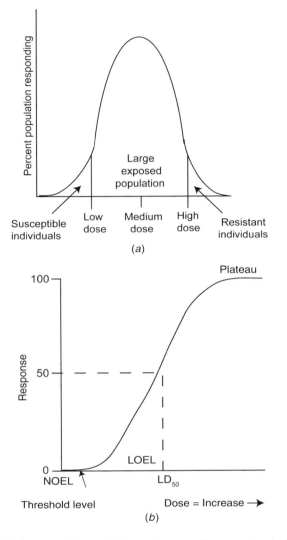

Figure 8.1 (*a*) Living organisms exhibit various reactions to chemicals. (*b*) Standard dose–response curve. (Modified from Whitford 2002, *The Complete Book of Pesticide Management: Science, Regulation, Stewardship, and Communication.* John Wiley and Sons, NY.)

Carcinogenicity, Mutagenicity, Reproductive Disorders, and Endocrine Disruption. Carcinogenicity studies specifically examine the potential of a pesticide to cause cancer in test animals. Preference is given to the routes of exposure that are most likely in humans. Abnormal growths or tissues are looked for and examined for malignancy. No minimum amount of a carcinogen has been found below which it has no effect (Briggs 1992). Adverse effects on *reproduction* include tendency to abort, reduced offspring weight, and birth defects. Diseases and behavioral and learning disorders in offspring are also reproductive dysfunctions. Toxicological studies on reproduction address pesticide influence on the fetus and interference with normal reproductive processes. Two types of studies are common: developmental/teratological and reproduction/fertility (Whitford 2002). To determine if a pesticide is a *mutagen*, in vivo and in vitro genotoxicity tests are performed that examine the potential of the chemical to damage either chromosomes or genes. These tests are also valuable as screens for possible carcinogens since most mutagens are also carcinogens. Pesticides will be classified as highly toxic and be highly restricted in use by the EPA (Table 8.1) if they are found to be carcinogenic or mutagenic or to cause reproductive disorders.

The endocrine system consists of glands (e.g., pituitary, pineal, thyroid, adrenal, ovaries, and testes) that secrete hormones into the blood. These hormones signal other cells to turn on or off processes of metabolism, development, stress, and reproduction (Whitford 2002). Some pesticides are believed to be endocrine disruptors that mimic hormones to interrupt normal functional processes, although this subject is still under debate. The EPA has established a four-tiered approach to examine and assess the hazard of pesticides in mammals, wildlife, and fish, since hormones in these organisms are the best understood (Ecobichon 1996, Whitford 2002).

Biological Magnification. *Biomagnification* is the buildup of herbicide (pesticide) in food chains (Briggs 1992). In this process very small quantities of chemical in soil, air, or water may be taken up by plants and small animals that are eaten by larger animals, and so on, to the top of the food chain. Amounts accumulate at each stage, so the small amounts present in each prey species add together until the top animal receives the grand total, which can be several orders of magnitude higher than in the surrounding environment (Briggs 1992). Since humans are one of the organisms at the top of the food chain, biological magnification is a major reason to avoid persistent herbicides, or those that resist transformation to less damaging substances.

Two terms are used to examine pesticide propensity to bioaccumulate in the environment. These are *biological concentration factor* (BCF) and K_{ow}. According to Kamrin (1997), BCF is the concentration of a substance in a living organism in relation to the concentration of that substance in the surrounding environment.

$$BCF = \frac{concentration\ in\ organism}{environmental\ concentration} \tag{8.2}$$

Also according to Kamrin (1997), K_{ow} is the octanol–water partition coefficient, or a measure of how well a chemical is distributed at equilibrium between octanol (oillike substances) and water. Since it is a measure of lipophilicity, it is used to predict the concentration of a chemical in the fat or lipid fraction of organisms. When the logarithm of the K_{ow} is 3 or more for a chemical, the octanol–water differential gradient is 1000 : 1. At this level, a chemical is very likely to concentrate up the food chain (Shaw and Chadwick 1998). In general, BCF is considered to be more sensitive to biological accumulation than K_{ow}, but the value of 1000 is used to assess the potential of herbicides to biologically magnify in both cases.

Persistence. Persistence is the length of time an herbicide will remain in the environment, whether it stays where it is put or, as we will see later, whether it moves through air, water, soil, or living organisms. Persistence time usually applies only to the active ingredient of the chemical and is measured on a determination of the *half-life* of a given dose of the herbicide (Briggs 1992, Kamrin 1997, Whitford 2002). The amount of time required to degrade 50% of the applied material is its half-life. A standard definition explained by Kamrin establishes the following half-life criteria for persistence:

Nonpersistent or weakly persistent	Less than 30 days
Moderately persistent	30–100 days
Strongly persistent	Greater than 100 days

However, environmental and soil conditions always affect the length of time an herbicide actually remains in the environment. Persistence in soil depends on soil characteristics, temperature, adsorption, and pH. As discussed later, some herbicides may persist in soil but because of adsorption or binding be effectively unavailable to exert toxic effects in the ecosystem. On the other hand, such herbicides, although "inactive" to the target organism, after a short period of time could still be available and affect nontarget organisms or biomagnify. Persistence is an important question for pesticide regulators and land mangers that see the benefits of herbicide use but do not care to impact the environment needlessly. Whitford (2002) indicates that pesticide persistence is influenced by both the pesticide half-life and the sorptive capacity of the soil (see herbicide fate, below). He suggests a nine-point set of criteria (high, medium, or low sorption vs short, intermediate, or long half-life) to determine pesticide persistence.

Voluntary Selection Criteria for Herbicide Use

The application of pesticides may have real or perceived effects on a greater number of people than just the user of the products or owner of the land on which they are used. (This topic is discussed in greater depth in Chapter 9). However, it is unclear whether regulations or policies to assure pesticide safety to humans, animals, and the environment are ever considered collectively. One example where the broader impacts of pesticide use are considered is by the Forest Stewardship Council (FSC), a voluntary international organization comprised of

TABLE 8.2   Criteria, Indicators, and Thresholds for Herbicide Use According to Forest Stewardship Council

Criterion[a]	Indicator	Threshold for Inclusion on FSC List of Highly Hazardous Pesticides
Quantitative or Semiquantitative		
Acute toxicity to mammals	WHO toxicity class (active ingredients)	If acute oral LD_{50} for rats ≤ 200 mg/kg b.w. (body weight)
	Acute toxicity (oral LD_{50} for rats)	WHO toxicity class 1a, 1b
	(Acute) reference dose (RfD)	—
Acute toxicity to aquatic organisms	Aquatic toxicity (LC_{50})	If $LC_{50} < 50$ μg/L
Chronic toxicity to mammals	Reference dose	If RfD < 0.01 mg/kg day
Persistence in soil or water	Half-life in soil or water (DT_{50})	If $DT_{50} \geq 100$ days, strongly persistent
Biomagnification, bioaccumulation	Octanol–water partition coefficient (KOW) or bioconcentration factor (BCF) or bioaccumulation factor (BAF)	If KOW > 1000, i.e., $\log(KOW) > 3$
Carcinogenicity	IARC carcinogen; EPA carcinogen	If listed in any category below: (a) International Agency for Research on Cancer (IARC) within group 1, agent (mixture) is carcinogenic to humans, or group 2A, agent (mixture) is probably carcinogenic to humans (IARC 2004) (b) EPA defined as chemical that is within group A, human carcinogen (EPA 1986) (c) EPA defined as chemical that can reasonably be expected to be carcinogenic to humans (chemicals categorized by EPA into group B2, see below)
Endocrine-disrupting chemical (EDC)	EDC listed by EPA and NTP (National Toxicology Program)	If classified as EDC by NTP or EPA
Mutagenicity to mammals	Not specified any further	If mutagenic to any species of mammals

(Continued)

TABLE 8.2 *Continued*

Criterion[a]	Indicator	Threshold for Inclusion on FSC List of Highly Hazardous Pesticides
	Qualitative	
Specific chemical class	Chlorinated hydrocarbon (definition from Radosevich et al. 2002)	If chemical meets definition from Radosevich et al. 2002.
	Compounds which contain only carbon, hydrogen, and one or more halogen, AND/OR	Note: the 2002 policy includes the statement that "not all organo-chlorines exceed the stated thresholds for toxicity, persist-ence or bioaccumulation, and they are not included in this list of prohibited pesticides, but they should be avoided."
	Organic molecules with hydrogen and carbon atoms in linear or ring carbon structure, containing carbon-bonded chlorine, which may also contain oxygen and/or sulfur but not phosphorus or nitrogen.	However, the current list of highly hazardous pesticides does not include organochlorines unless they are excluded on the basis of other indicators.
Heavy metals	Lead (Pb), cadmium (Cd), arsenic (As), and mercury (Hg)	If pesticide contains any heavy metal as listed
Dioxins (residues or emissions)	Equivalents of 2,3,7,8-TCDD	If contaminated with any dioxins at a level of 10 part per trillion (corresponding to 10 ng/kg) or greater of tetrachlorodibenzo-p-dioxin (TCDD) equivalent or if it produces such an amount of dioxins when burned
International legislation	Banned by international agreement	If banned by international agreement

[a]All references are cited in FSC (2006).
Source: Forest Stewardship Council (2006) Bonn, Germany, with permission.

forest products producers, suppliers, and environmental organizations. This organization adopted a policy to reduce the use of pesticides in forests certified by them. The FSC uses WHO and EPA toxicity classifications in addition to all of the criteria listed above to determine what pesticides (including herbicides) are satisfactory to use, according to their criteria. The criteria, indicators, and thresholds for pesticide use according to this organization are given in Table 8.2.

CHEMICAL PROPERTIES OF HERBICIDES THAT AFFECT USE

Most herbicides are organic chemicals, primarily made of carbon and hydrogen atoms. The carbon atoms of organic molecules bind together to form "chains." The simplest organic compound is methane, which is composed of a single carbon atom that is bonded to four hydrogen atoms (CH_4). If a hydrogen atom in methane is replaced by another carbon, ethane (C_2H_6), a two-carbon chain, is formed. Long chains of interconnected carbon atoms can be made in this way. The chains may be straight, branched, or cyclic.

Organic compounds composed of only carbon and hydrogen, such as those described above, are *hydrocarbons*. Hydrocarbons are *saturated* when all available bonds are occupied by an atom of carbon or hydrogen. *Unsaturation* occurs when two carbon atoms share more than one bond. There may be only a few or many double and triple bonds in an organic molecule. Acetylene (C_2H_2) and benzene (C_6H_6) are examples of unsaturated hydrocarbons. Benzene is a common constituent of many herbicides. Organic chemicals that are arranged in an unsaturated ring configuration (e.g., benzene) are also called *aromatic* hydrocarbons.

Only a few elements other than carbon and hydrogen are found in organic compounds, including herbicides. These elements are oxygen, nitrogen, sulfur, phosphorus, and the halogens (chlorine, fluorine, iodine, and bromine). Organic chemicals having other atoms than carbon as part of their ring structure are called *heterocyclic* hydrocarbons. Many herbicides are heterocyclic compounds and, in addition, most herbicides contain at least one halogen atom as part of their molecular structure. Alcohols (R–OH), organic acids (R–COOH), and esters (R–O–R) are forms of organic compounds that dramatically influence chemical and physical properties. These structures tend to be highly reactive and influence such properties as water solubility, electrical charge, and potential to vaporize.

Chemical Structure

Each herbicide has inherent chemical properties that influence its ability to kill plants. The biologically active portion of a commercially manufactured herbicide is called the *active ingredient.* This is the fundamental molecular composition and configuration of the herbicide. In addition to biological activity, chemical and physical properties of the herbicide can determine the method of application and use. For example, the active ingredient of 2,4-D is the acid form of that herbicide. However, the herbicide is rarely sold or applied in that form because it does not penetrate leaves or kill plants as well as other forms of the chemical.

An active ingredient can be altered slightly by chemical processes, such as esterification, which may improve biological activity, alter the method of application, or influence the herbicide's fate in the environment. Herbicides that are derived from alcohols, phenols, and organic acids are often more soluble in water than those that are not. In contrast, ester forms of herbicides are relatively more soluble in oil or organic solvents and have a tendency to produce vapors. The loss of herbicide as vapor is called *volatility*, which is related to the vapor pressure of the chemical.

Organic salts of herbicides may also be formed during the manufacturing process. The acid form and two organic salts of 2,4,-D are shown below. These forms of 2,4,-D are rather unlikely to volatilize but are soluble in water. The ester forms of 2,4-D are soluble in organic solvents and are more likely to vaporize than the organic salts. However, the size of the ester linkage to the parent 2,4-D acid molecule also influences the degree of volatility of the chemical. For example, the isobutyl ester is highly volatile, whereas the butoxyethyl ester of 2,4-D (below) has much lower volatility characteristics.

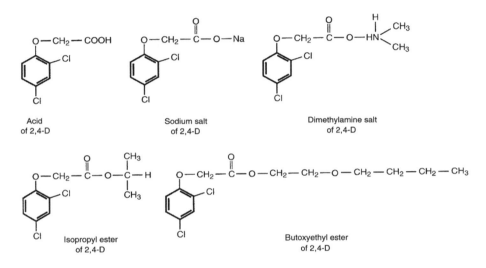

As just demonstrated for 2,4-D, it is possible that rather small changes in chemical structure can significantly alter the properties, uses, and effectiveness of herbicides. The chemical and physical properties of some common organic herbicides are presented in Table 8.3.

Water Solubility and Polarity

If a chemical is soluble in water, a solution forms when the two substances are mixed. The solvent action of water is based on the ability of water molecules to form hydrogen bonds and dipole–dipole interactions with other molecules and ions. Many chemicals, like alcohols, organic acids, phosphates, nitrates, chlorates, ammonium compounds, and sugar, are held in solution with water by hydrogen bonding. Water also is electrically asymmetrical or *polar*, since the centers of positive and negative charges are located at different molecular points (Figure 8.2). Other types of chemicals that are also polar readily dissolve in water due to such dipole–dipole interactions.

As a general rule, polar substances dissolve in other polar substances. Ionizable salts, like table salt (NaCl), dissolve in water in this way. Herbicides that are produced as salts are quite water soluble and are usually formulated to be applied in

TABLE 8.3   Summary of Information About Some Herbicides

Common Name	Leaching Class[a]	Water Solubility[b]	Volatility[c]	Site of Uptake	Soil Persistence
Ureas					
Diuron	2	42	None	Roots and foliage	6 months
Tebuthiuron	2	2,300	None	Roots	12–15 months
Triazines					
Atrazine	3	34	Low	Roots, some foliage	6 months
Simazine	3	3.5	None	Roots (little if any foliage)	6 months
Prometon	3	750	Low	Roots and foliage	6 months to many years
Prometryn	2	48	Low	Roots and foliage	2 months
Metribuzin	3–4	1,220	Low	Roots and foliage	2 months
Uracils-pyrimidines					
Bromacil	4	815	None	Roots	6 months
Terbacil	3	710	None	Roots	6 months
Acylanilides					
Propanil	2	50,000	None	Foliage	1–3 days
Pyridazinones					
Pyrazon/ chloridazon	2	400	Low	Roots	1–2 months
Bis-carbamates					
Phenmedipham	1	10	None	Foliage	1 month
Desmedipham	1	7	None	Foliage	1 month
Dinitroanilines or toluidines					
Benefin	1	1	Low	Roots	>6 months
Ethalfluralin	1	0.3	Moderate	Roots	2–3 months
Oryzalin	2	2	Low	Roots	2 months
Pendimethalin	1	0.5	Moderate	Primarily roots	~4 months incorporated
Triflualin	1	1	High	Roots	3 months

[a]Leachability divided into five classes: class 1, immobile → class 5, very mobile.
[b]Values are in ppm for unformulated molecules.
[c]Only none (insignificant), low, moderate, and high are used.
Source: Adapted from Zimdahl (1999), *Fundamentals of Weed Science*, Academic Press, San Diego, CA.

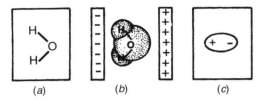

Figure 8.2 Dipolar structure of water molecules. (*a*) Hydrogen atoms are positively charged (though not ionized in the ordinary sense) and oxygen is negatively charged. (*b*) Orientation of water molecules in electrical field. (*c*) Simple diagram of a polar molecule, such as water, showing positive and negative portions. (From Slabaugh and Parsons 1966, *General Chemistry.* Copyright 1966 with permission of John Wiley & Sons Inc.)

water. In contrast, nonpolar substances are practically insoluble in water. For example, oil is a nonpolar solvent and does not mix well with water because of differences in polarity. Thus, herbicides that are soluble in nonpolar, oil like solvents are not very soluble in water. The solubilities of various herbicides in water and other solvents are listed in Table 8.3.

Water is the primary substance used to disperse, that is, spray, herbicides. Therefore, the water solubility of an herbicide determines, to some extent, the type of product that is formulated and how it is applied. Water solubility is also important because it influences herbicide movement in the soil profile, which is discussed later.

Volatility

The tendency of chemicals to volatilize is determined by their vapor pressure, which is measured and expressed in milligrams of mercury (Hg). Herbicides with low vapor pressures (e.g., 10^{-5} mg Hg) are relatively nonvolatile, while those with high vapor pressures (e.g., 10^{-3} mg Hg) volatilize readily (Table 8.3). Both chemical form and formulation influence the ability of herbicides to volatilize.

Formulations

The active ingredient of many herbicides is unsuitable to use as a commercial product. Therefore, it must be refined by the manufacturer prior to sale and use. The final product, or *formulation*, contains the active ingredient of the herbicide and "inert" ingredients, such as solvents, emulsifiers, diluents, and so on, that enhance the marketability or biological activity of the chemical. Herbicides usually are formulated for ease of transportation and application in water or as dry material for granule applications.

Some herbicides are formulated as a number of different products, all containing the same active ingredient. These products are developed to enhance the particular chemical or physical properties of the herbicide, improve weed control or herbicide selectivity, reduce animal toxicity, or provide an economic advantage to the manufacturer. Large differences in effectiveness, rate of application, hazard, or cost often exist among such herbicide formulations.

Formulations used as liquid sprays include *water-soluble powders* (SP) and *liquids* (SL), *emulsifiable concentrates* (EC), *wettable powders* (WP), *water-dispersible liquids* or *flowable* materials (WDL or F), and *granules* (G). Some herbicide active ingredients are formulated dry as granules or pellets for direct application without dilution in water. These formulations often have low concentrations of active ingredient and are less hazardous than other formulations for this reason.

Carriers and Adjuvants for Herbicide Applications. Herbicides are always applied as a mixture with some other material, like water, oil, or a dry carrier such as certain types of clays, vermiculite, plant residues, starch polymers, and some types of dry fertilizers. The *carrier* is used to dilute and disperse the herbicide over the field.

An *adjuvant i*s a material that is mixed with a spray solution or suspension to improve the performance, handling, or application of herbicides. Adjuvants are designed to be inert chemicals and are classified according to their use, rather than chemical or physical properties. For many herbicide products, the adjuvants are formulated with the active ingredient at the time of manufacture. At other times, it is desirable to add a specific material, such as a surfactant, to improve or enhance the performance of the formulated product. Caution should be used when selecting additional adjuvants for use with herbicide products since a loss of selectivity or change in toxicity activity can result. Terms often used to describe adjuvants include *activator*, *additive*, *dispersing agent*, *emulsifier*, *spreader*, *sticker*, *surfactant*, *thickener*, and *wetting agent*.

HERBICIDE CLASSIFICATION

There are approximately 140 herbicide active ingredients (Vencill 2002, Mallory-Smith and Retzinger 2003). Most of these basic forms of herbicides are further refined and formulated, creating several hundred commercial products. Because of the number of herbicides available, it is necessary to distinguish among them somehow. Herbicides are classified most often according to similarities in (1) *chemical structure*, (2) *use*, and (3) *effects on plants*. Herbicides are also classified according to toxicity or hazard level, as discussed earlier.

Classification Based on Chemical Structure

Classification systems based on chemical structure categorize herbicides by *chemical similarity*. This is the classification system used in the Weed Science Society of America (WSSA) *Herbicide Handbook* (Vencill 2002), which provides a description of the various herbicides that are used in the United States.

However, new herbicide development and registration are a continual process, so even this survey is likely to be both incomplete and outdated. The primary use, formulations, water solubility, and acute oral toxicity for each herbicide are provided in the *Herbicide Handbook* (Vencill 2002). In addition, the principal manner in which the herbicides of each chemical group suppress plants is described.

Every herbicide is named in three ways. Since herbicides are chemicals, each active ingredient has a *chemical name* to describe its chemical structure. The chemical constituents that make up the herbicide active ingredient can be determined and similarities to other chemicals can be found in this way. Herbicides are also commercial products. Therefore, each herbicide has a *trade name* given by its manufacturer that distinguishes it from other products and assists in its sale. However, some herbicides may be manufactured by several companies and each gives its product a different trade name. To avoid confusion, herbicides also are provided a *common name* by the WSSA. This common name refers to all herbicide products that have the same active ingredient. The herbicides in the WSSA *Herbicide Handbook* are organized according to both chemical structure and common name.

Classification Based on Use

When herbicides are classified based on how they are used, they are first characterized as being either selective or nonselective (Figure 8.3). *Selective* herbicides are chemicals that suppress or kill certain weeds without significantly injuring an associated crop or other desirable plants. Usually some weed species are also not injured by selective herbicides. *Nonselective* herbicides, in contrast, result in suppression of treated vegetation with no plant survival intended. Many herbicides occur in both categories, however, because differential phytotoxicity among plants (selectivity) is not an absolute characteristic of the chemical. Rather, selectivity depends upon the rate (dose) of herbicide applied, method of application, and many other plant and environmental factors. Herbicide selectivity is an important principle of modern weed control and is discussed later in this chapter. Herbicides that are always nonselective are soil-applied fumigants and certain chemicals used for aquatic weed control. A partial list of herbicides based on how they are used is provided in Table 8.4.

Soil-Applied Herbicides. Soil-applied herbicides (Figure 8.3 and Table 8.4) are applied before planting (*preplant*), before crop or weed emergence (*preemergence*), or after the plants emerge (*postemergence*) (Figure 7.19). Soil-applied herbicides must be moved into the soil profile by water or mechanical incorporation to be effective since they are usually taken up by plant roots, underground structures, or seed. The phytotoxicity of soil-applied herbicides depends on inherent plant tolerance to the chemical, location of the herbicide in the soil, and depth of plant roots. Some soil-applied herbicides are applied as bands, either

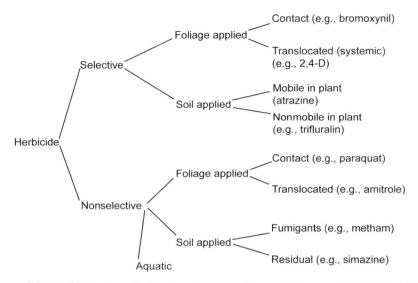

Figure 8.3 Herbicide classification based on use. (From McHenry and Norris 1972, *Study Guide for Agricultural Pest Control Advisors on Weed Control*, Publication 4050. Copyright 1972, University of California Press, with permission.)

over or between crop rows, to enhance selectivity and decrease costs of application.

Foliage-Applied Herbicides. Foliage-applied herbicides (Figure 8.3 and Table 8.4) injure plants when the chemical is applied to the leaves or stems. Some herbicides injure only the portion of the plant actually touched by the chemical or spray solution and are called *contact* herbicides. Herbicides in this category are usually applied to foliage and movement in plants is limited. Paraquat is an example of a foliage-applied contact herbicide. In some cases, herbicides may be directed away from desirable vegetation like crops or applied in shields to minimize foliage exposure to these chemicals.

Some soil-applied and many foliage-applied herbicides move in treated plants along with the products of photosynthesis and other materials in the phloem during translocation. Thus, *translocated* (Figure 8.3 and Table 8.4) or *systemic* herbicides move in the plant after application. Herbicides of this use type often effectively suppress root, rhizome, or shoot growth, which is usually at some distance from the point of application in a treated plant.

Soil Residual Herbicides. Soil residual herbicides are chemicals applied to the soil that are selective at some rates and conditions but at higher rates suppress plant growth for several months to years (Table 8.3). These herbicides were once called soil sterilants, but this nomenclature is discouraged now since the herbicides do not "sterilize" the soil but kill plant seedlings for a prolonged period of

TABLE 8.4 Partial List of Herbicides Based on Use

Common Name	Trade Name	Common Name	Trade Name
		Aquatic Herbicides	
Acrolein	Magnacide	Endothall	Hydrothol, Aquathol
Copper chelate	Komeen, Cutrine	Fluridone	Sonar
Copper sulfate	Several	Glyphosate	Aquamaster
Diquat	Diquat, Reward		
		Foliage Applied Contact Herbicides	
Ametryne	Evik	Endothall	Several
Bentazon	Basagran	Ethofumesate[a]	Nortron, Prograss
Bromoxynill	Buctrill	Glufosinate	Finale, Liberty
Diclofop	Hoelon	Oxyfluorfen[a]	Goal
Difenzoquat	Avenge	Paraquat	Paraquat, Gramoxone
Diquat	Diquat, Reward		
		Foliage Applied Translocated Herbicides	
Chlorsulfuron[a]	Glean, Telar	Imazapyr[a]	Pursuit
Clopyralid	Transline	MCPA[b]	Several
2,4-D[b]	Several	MSMA[b]	Several
2,4-DB[b]	Butoxone	Phenmedipham	Betamix
Dicamba	Banvel	Picloram	Tordon, Amdon
Fluazifop	Fusilade	Propanil	Stam
Glyphosate	Roundup, others	Sethoxydim	Poast
Halosulfuron	Sledgehammer	Triclopyr	Garlon, Turflon
		Soil Applied Herbicides	
Alachlor	Lasso	Naptalam	Alanap
Atrazine	Aatrex, others	Norflurazon	Solicam, Zorial
Bensulide	Betasan, Prefar	Oryzalin	Surflan
Bromacil	Hyvar	Oxadiazon	Ronstar
Butylate	Sutan	Pebulate	Tillam
Cycloate	Ro-Neet	Pendimethalin	Prowl
DCPA[b]	Dacthal	Prometon	Pramitol
Dichlobenil	Casoron	Prometryn	Caparol
Diuron	Karmex, others	Propachlor	Ramrod
EPTC[b]	Eptam	Pyrazon	Pyramin
Fluometuron	Cotoran	Simazine	Princep, others
Hexazinone	Velpar	Sodium chlorate	Several
Linuron	Lorox	Tebuthiuron	Spike
Metolachlor	Dual Magnum	Terbacil	Sinbar

(Continued)

TABLE 8.4 *Continued*

Common Name	Trade Name	Common Name	Trade Name
Metribuzin	Lexone, Sencor	Thiobencarb	Bolero
Molinate	Ordram	Triallate	Far-go
Napropamide	Devrinol	Trifluralin	Treflan
		Soil Applied Fumigants	
Dazomet	Basimid	Methyl bromide	Brom-o-gas, others
Metham sodium	Vapam		

[a]Also has pre-emergence activity.

[b]2,4-D, 2,4-dichlorophenoxy acetic acid; 2,4-DB, 4-2,4-dichlorophenoxy butanoic acid; MCPA, 4-chloro-2-methylphenoxy acetic acid; MSMA, monosodium salt of methylarsonic acid; DCPA, dimethyl 2,3,5,6-tetrachloro-1,4-benzenedicarboxylate; EPTC, S-ethyl dipropyl carbamothioate.

Source: Adapted from Gowgani et al. (1989) and Vencill (2002). Other trade names are possible; no endorsement is intended.

time. Most of the herbicides in this category translocate to some degree in germinating seedlings.

Soil Fumigants. Soil fumigants (Figure 8.3) are gasses applied to the soil that kill all vegetative plant growth. These herbicides are usually applied prior to crop planting and weed emergence. Small germinating weeds are most susceptible to the treatment, while some dormant seed tolerate the chemicals. The length of time soil remains weed free following soil fumigation depends upon the chemical used, amount applied, soil type, soil water status, and extent of weed seed dissemination from adjacent areas. Soil fumigation is usually expensive and used only for high-value crops. Also, due to the relatively high toxicity level of these chemicals, many are being reviewed for possible cancellation of registration.

Aquatic Herbicides. Most herbicides used for aquatic weed control are nonselective (Figure 8.3 and Table 8.4). These materials are applied either directly to the water or to the soil in canals, ponds, and lakes.

Classification Based on Biological Effect in Plants

The way herbicides kill or suppress plants is another method of classification. Herbicides are grouped as hormone-type growth regulators, cell division inhibitors, photosynthesis inhibitors, pigment synthesis inhibitors, lipid synthesis (cell membrane) inhibitors, or inhibitors of cell metabolism including amino acid biosynthesis. This method of classification requires that the cause of plant injury for specific herbicides be known. Although knowledge about specific biochemical changes that result from herbicide use is sometimes incomplete, enough information is usually available about specific herbicides to place them into such broad categories of cellular dysfunction and symptomology. The subject of herbicide mode and mechanism of action is described by Devine et al. (1993), Anderson (1996), Roe et al. (1997), and Böger et al. (2002). The reader is referred to these excellent texts for more discussion on this subject.

HERBICIDE SYMPTOMS AND SELECTIVITY

Since herbicides alter the ability of plants to grow, various structural features of plants change following exposure to these chemicals. These visible changes in plant morphology are the *symptoms* of herbicide effects. Selectivity depends on the degree of plant tolerance to the herbicide.

Symptoms

The symptoms of herbicide exposure include abnormal cell development and division, epinasty (twisting), chlorosis and necrosis, albinism, altered geotropic and phototropic responses, and reduced formation of cuticle and waxes.

Abnormal Tissues and Twisted Plants. *Epinasty* is the bending or twisting of stems and leaves. This symptom is most characteristic of herbicides such as 2,4-D, dicamba, triclopyr, and picloram that interfere with hormonal regulation in plants. Increased tillering of shoots or callus formation on roots is sometimes a response to low rates of these herbicides. Other formative symptoms occur during development and include thickened coleoptiles or leaves, multiple shoot formation at internodes, reduced internode length, and abnormal seedling development. Herbicides of the chlorinated aliphatic acid, chloroacetamide, dinitroaniline, nitrile, and thiocarbamate groups cause such formative symptoms in treated plants. Epinasty and formative symptoms are caused by abnormal cell division, cell enlargement, and tissue differentiation as a result of herbicide exposure.

Disruption of Cell Division. The process of cell multiplication (mitosis) is inhibited by many herbicides, including chemicals belonging to the carbamate, thiocarbamate, and dinitroaniline groups. The symptoms of these herbicides range from suppressed root or shoot development in whole plants to aberrant and multinucleate cells. Although the general symptoms of mitotic disruption are similar for a large number of herbicides, detailed microscopic examinations have shown important differences in the mechanism of action among specific chemicals. The primary site of toxic action of mitotic inhibitors is located in meristematic regions of plants, such as forming root tips or buds.

Chlorosis, Necrosis, and Albinism. *Chlorosis* is a common symptom of herbicides that inhibit photosynthesis. Chlorosis is the bleached yellow appearance of leaves following the degradation of chlorophyll in treated plants. Some herbicides cause chlorosis along the leaf veins, while chlorosis between veins is the symptom of other herbicides. *Necrosis*, or tissue death, is the most advanced or extreme case of chlorosis. It often takes several days to weeks for chlorosis and finally necrosis to develop following treatment. Some herbicides that produce these symptoms in plants are members of the urea, uracil, and triazine groups.

Other herbicides such as glyphosate, DSMA, MSMA, and some phenoxy-type herbicides can cause chlorosis. This symptom is caused by a lack of chlorophyll, which makes other leaf pigments more obvious following herbicide application. *Albinism* results when new foliage is devoid of chlorophyll. Amitrole is an herbicide that causes this symptom of herbicide exposure. A few herbicides kill plant foliage so rapidly that only necrosis results from herbicide treatment. A general dysfunction of cell membranes is responsible for such a rapid and dramatic symptom. Contact-type herbicides, such as paraquat, diquat, bromoxynil, and endothall, are responsible for this type of herbicide injury.

Altered Geotropic and Phototropic Responses. *Geotropism* is the ability of plants to orient and grow downward in response to gravity. The ability of plants to grow toward light is *phototropism*. Naptalam has been reported to alter these responses.

Reduced Leaf Waxes. Cuticular and epicuticular waxes are complex structures that cover the foliage of plants. The primary role of these structures is to restrict water loss from the plant. Herbicides in the thiocarbamate and aliphatic acid groups inhibit epicuticular wax formation, in addition to causing other morphological symptoms already mentioned.

Selectivity

Some plants are inherently tolerant to certain herbicides while others have evolved resistance after repeated exposure to a chemical. Tolerant and resistant plants usually degrade or metabolize the chemical to nonphytotoxic substances. In some cases of resistance, such as to the triazine herbicides, the herbicide does not affect the site of toxic action in treated plants. Although tolerance and resistance are common, herbicide selectivity among plants is often conditional; that is, it depends on the rate and timing of herbicide application, placement of the herbicide relative to the location and stature of the crop or other desirable plants and weeds, and numerous other plant and environmental factors that influence herbicide performance. Some of the factors that influence herbicide selectivity are as follows:

- Physiological or biochemical tolerance to the herbicide
- Herbicide rate (dose)
- Time of application
- Stage of weed and crop or other plant development

- Weather patterns
- Variation in microenvironment or microtopography
- Variation in resource level
- Soil type and pH

Many of the principles and practices of how herbicides used to attain selective chemical weed control are discussed in weed science texts (Akobundu 1987, Anderson 1996, Ross and Lembi 1999, Gressel 2002, Monaco et al. 2002). These principles involve the role of plant morphology and physiology, chemical properties, and environmental factors in the differential susceptibility of plants to herbicides.

Plant Factors of Herbicide Selectivity. Plant factors that influence the way weeds and crops or other desirable plants respond to herbicides are *genetic inheritance*, *age*, *growth rate*, *morphology*, and *physiological* and *biochemical processes*. The most effective use of herbicides results from considering these factors when selecting an herbicide or application method.

Genetic Inheritance. Plant species within a genus usually respond to herbicides in a similar manner, while responses to herbicides by plants in different genera often vary. The reason is that plants with similar taxonomic traits often have similar genetic and enzymatic components. Thus, crops and weeds that belong to the same genera are usually susceptible to the same herbicides. For example, herbicides that do not injure tomato also fail to control nightshade (*Solanum* spp.) weed species because the crop and weeds are members of the same taxonomic family (Solanaceae) and have similar biochemistry. This rule of thumb is not absolute, however, because varieties of many crops are known to respond differentially to herbicides.

HERBICIDE-RESISTANT CROPS. The development of crops that are resistant to herbicides is a relatively new way to improve weed control in agriculture (Schulz et al. 1990, Duke 1996, Powles and Shaner 2001, Gressel 2002). Herbicide-resistant crops can be created by standard methods of plant breeding, but the use of genetic engineering techniques is more usual. Glyphosate and glufosinate are herbicides most used for the manufacture of herbicide-resistant crops. For example, soybean, corn, cotton, sugarbeet, and canola are available as glyphosate-resistant cultivars and some are now widely planted in the United States. There are several potential benefits of herbicide-resistant crops:

- Increased margin of safety (selectivity) to the crop
- Avoiding crop injury due to herbicide carryover in treated soils
- Increased options for weed control when the number of herbicides is limited, such as in minor crops

- Control of particularly problematic weeds
- Increased window time for herbicide application
- Reduction of environmental damage by using newer, less toxic herbicides
- Possibly lower costs

A number of potentially detrimental effects can result from the use of herbicide-resistant crops (HrCs) (Dyer et al. 1993, Powles and Shaner 2001, Ellstrand 2003, Martinez-Ghersa et al. 2003). This subject is considered more completely in Chapter 9. In addition, the safety of transgenically modified foods is currently under debate in many countries (McHughen 2000, Gasson 2003, Pusztai et al. 2003), which brings the precautionary principle (Chapter 2) to bear on marketing these products.

Plant Age and Growth Rate. Weed seedlings or young plants are usually killed more easily than large or mature vegetation. In addition, some preemergence herbicides that suppress seed germination are often not effective when used to control larger, better established plants. Plants that are growing rapidly generally are more susceptible to herbicides than are plants growing slowly.

Morphology. The morphology or growth habit of plants can determine the degree of sensitivity to some herbicides. Morphological differences in root structure, location of growing points, and leaf properties between crops or other desirable plants and weeds can determine the selectivity pattern of some herbicides. Annual weeds in a perennial crop, meadow, or pasture usually can be controlled by herbicides because of their different root distribution and structure compared to those of perennial plants. For example, perennial crops such as alfalfa can recover from moderate herbicide injury to foliage whereas annual weeds, because of their small size and shallow root system, will be killed by the same herbicide application.

The meristematic regions of most grasses, such as cereal crops and grass weeds, are located at the base of the plant or even below the soil surface. The growing points are protected from herbicide exposure by the foliage or soil that surrounds them. Thus, herbicide that contacts only foliage may injure some leaves but will not typically impair the ability of the plant to grow. In contrast, most dicot plants have their meristems exposed at shoot tips and leaf axils. For this reason, these plants are more susceptible than grasses to foliage-applied herbicides, especially contact herbicides.

Leaf properties of some plants can impart selectivity to certain herbicides, while other plants are effectively controlled. Spray droplets do not adhere well to the surfaces of narrow, upright, waxy leaves that characterize many monocot plants like cereals, onion, and most grasses. Thus, spray droplets do not adequately cover such leaves following herbicide application and the effect of the herbicide is reduced. In contrast, dicot plants have relatively wide leaves that are usually horizontal to the main stem. Leaves of dicot plants, therefore, intercept more spray than leaves of grasses and spray droplets spread more evenly over dicot foliage. Herbicide effectiveness is best when spray interception and coverage are greatest.

Physiological and Biochemical Processes. The physiology of a plant influences the ability of an herbicide to enter it following application. This process is called *absorption.* The extent of herbicide movement in a plant (*translocation*) after it has been absorbed is also a physiological process. Both absorption and translocation are important processes governing herbicide activity and vary markedly among plant species. Generally, plant species that absorb and translocate herbicides readily are most easily killed by them.

Biochemical and biophysical processes are also important plant factors determining herbicide selectivity. A process called *adsorption* can be responsible for differential herbicide susceptibility among plant species. During this process an herbicide is bound so tightly by cellular constituents (usually cell walls) that it cannot be translocated readily and thus is inactivated (Figure 8.4*a*). Membrane stability is another biochemical/biophysical process that results in herbicide selectivity among plants. In this case, the cell membranes of tolerant plants can

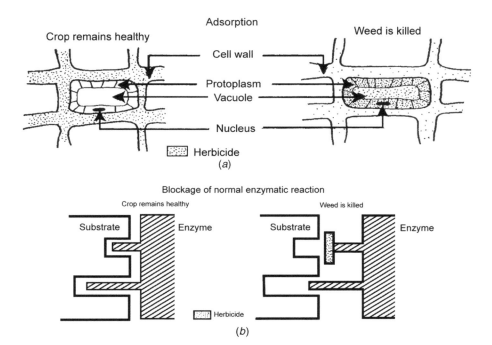

Figure 8.4 (*a*) Selectivity based on physical binding (adsorption in plants). On the left, herbicide is adsorbed by cell walls and is prevented from reaching the cytoplasm of a treated plant. On the right, herbicide is not adsorbed by the cell wall and reaches the cytoplasm. (*b*) Selectivity based on enzyme inactivation. On the left, herbicide does not interfere with enzyme reaction and metabolism. On the right, herbicide alters the structure and attachment of the enzyme and upsets metabolic processes. (From Ashton and Harvey 1971, *Selective Chemical Weed Control*, University of California Circular 558. Copyright 1971, University of California Press, with permission.)

withstand the disruptive action of the herbicide. The ability of carrot to withstand the toxicity of certain oils is an example of this form of herbicide selectivity.

Enzyme inactivation, herbicide activation, and herbicide inactivation are biochemical processes that can occur in plants in response to herbicide treatment. Since all plants cannot perform these processes equally well, they also form the basis for herbicide selectivity. *Enzyme inactivation* occurs when an herbicide reduces the activity of a particular enzyme in a plant (Figure 8.4*b*). Thus, some plants are killed but others are not. *Herbicide activation* results when a nontoxic chemical is transformed in a plant to an herbicide. An example of this process is the transformation of 2,4-DB into 2,4-D in susceptible plants but not tolerant ones. *Herbicide inactivation* occurs when an herbicide is degraded in a treated plant to nontoxic materials. There are many examples of this form of differential sensitivity of plants to herbicides.

Chemical Factors of Herbicide Selectivity. As discussed earlier, the *structure* and *formulation* of the herbicides themselves can influence the tolerance of plants to them. Only small changes in molecular configuration of an herbicide are needed to modify its chemical properties and also its effects on plants. Differences between two herbicides, benefin and trifluralin, offer an example of this type of herbicide selectivity. The only difference between the two herbicides is that a CH_2 group is moved from one side of the molecule to the other. However, benefin will control many weeds without harming lettuce while trifluralin kills lettuce even at low rates.

The *formulation* of an herbicide is also an important consideration for herbicide selectivity. As already discussed, herbicides are formulated in a number of different ways to improve transportation, storage, application, or marketing. Herbicide formulation also may enhance herbicide selectivity by increasing toxicity in susceptible plants or decreasing activity in tolerant ones. The uses of granule formulations, emulsifiers, or surfactants are examples of herbicide formulations used to improve selectivity.

An herbicide may also be directed away from susceptible plants, such as a crop, which imparts a type of selectivity. The uses of *shields* and *directed sprays* are examples of this positional type of herbicide selectivity.

Environmental Factors of Herbicide Selectivity. Factors of the environment that influence herbicide selectivity are *soil type, rainfall and irrigation patterns*, and *temperature*. Soil type and the amount of precipitation or irrigation determine the location of herbicides in the soil profile. In general, herbicides tend to move more readily in sandy soil than in clay and in wet soil than in dry. Temperature and soil moisture also determine the rate of herbicide degradation in the soil and the rate of plant growth. Warmer temperatures and wetter soil conditions promote more rapid microbial and chemical degradation of herbicides than do cooler or drier conditions. Warm temperatures and moist soils also promote more rapid plant growth and thus more rapid onset of herbicidal injury.

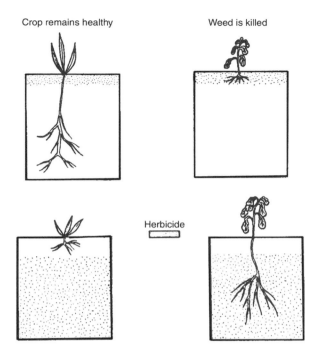

Crop remains healthy   Weed is killed

Herbicide

Figure 8.5   Selectivity based on herbicide placement in soil. Top, deep-rooted crop (left) is not affected by herbicide that remains near the soil surface. Shallow-rooted weed is killed by herbicide that stays near the surface. Bottom, shallow-rooted crop (left) remains alive if the herbicide moves below its rooting zone. Deep-rooted plants are killed when herbicide is leached into the deeper zones of the soil. (From Ashton and Harvey 1971, *Selective Chemical Weed Control*, University of California Circular 558. Copyright 1971, University of California Press, with permission.)

Some soil-applied herbicides that are not biochemically selective may be functionally selective by their placement in the soil profile (Figure 8.5). This type of selectivity requires differential rooting habits between desirable plants and weeds and an understanding of the factors that influence vertical herbicide movement (leaching) in the soil. The placement of herbicides in the soil relative to the roots of desirable plants (e.g., crops) and weeds is an important principle of herbicide selectivity. The factors that influence herbicide movement in soil are considered later.

HERBICIDE APPLICATION

Herbicides must be applied accurately and uniformly to an area of land or foliage to be effective because too much of the chemical can damage desirable plants while too little will not provide acceptable weed control. It is also necessary for the herbicide to arrive at the targeted area and not be displaced by drift, volatility, leaching, or runoff. Damage to susceptible plants or chemical residues in food, feed, water, or soil may result if herbicide displacement occurs. Improved accuracy can be achieved by proper calibration and operation of herbicide application equipment.

Both ground and aerial applications of herbicides are used in agriculture, forestry, and range management. Frequently, ground applications are made by tractor-drawn sprayers but hand applications of herbicides are also common in some locations and weed control situations. Aircraft are also used to apply herbicides, but special precautions and equipment are necessary when herbicides are applied by aircraft.

Proper Rate (Dose)

Herbicides are used within a specified range of doses or rates. The rate of herbicide usually is expressed as the amount of chemical per unit of ground area to be covered. Common units of herbicide rates are pounds per acre or kilograms per hectare. In addition, the rate of herbicide applied is usually expressed in terms of chemical active ingredient as well as the amount of commercial product. The reason to calibrate herbicide application equipment is to assure that the chemical needed for optimal weed control is spread uniformly over the specified area. The procedure for calibration is similar for both ground and aircraft applications [see Ross and Lembi (1999) and Zimdahl (1999) for specific examples]. However, some special calibration techniques are necessary for aircraft because of the speed and extent of area covered by this equipment.

Proper Distribution

The uniformity obtained from an herbicide application depends on several factors, including topography of the land, type and quality of equipment, skill of the operator, and certain weather conditions, especially wind and temperature. Unfortunately, it is often impossible to determine the degree of uniformity until after an herbicide has been applied. Strips of injured desirable plants or uninjured weeds in a treated field indicate poor uniformity of application. If poor application is suspected, both the equipment and its operation should be examined to determine where improvements can be made.

Application Equipment

Herbicides are usually applied using some form of sprayer that is specially adapted for aircraft (fixed-wing or helicopter), ground, or manual applications. Common features of an herbicide sprayer are *tank*, *agitation device*, *pump*, *pressure regulator*, *hoses*, and *nozzles*. These basic components and arrangement are described in most weed science texts. There is also equipment developed for special applications of herbicides that allows more efficient herbicide application,

Figure 8.6 Aerial application of herbicide by helicopter. (Photograph by R. G. Wagner, Oregon State University.)

minimal chemical loss, enhancement of weed control, or improved herbicide selectivity. These are described in Radosevich et al. (1997).

Herbicides represent approximately 25% of all aircraft pesticide applications made in the United States. The major advantages of aerial applications over those made using ground vehicles or manual operations are the ability of aircraft to cover large areas rapidly. Most agricultural applications of herbicides by air are made with fixed-wing aircraft. Airplanes with a load capacity of about 1.5–2.0 tons are the most common aircraft used. Helicopters are used less extensively for agricultural applications than fixed-wing aircraft. However, helicopter applications are prevalent for many forestry and rangeland uses where terrain often limits the access of airplanes (Figure 8.6).

FATE OF HERBICIDES IN ENVIRONMENT

The fate of herbicides in the environment has become a serious concern for farmers, foresters, other land managers, scientists, and the general public. Figure 8.7 demonstrates the diversity and interrelationships of environmental processes that lead to herbicide movement, detoxification, degradation, or persistence. In order for herbicides to be effective, they must persist long enough to kill weeds. However, if persistence is too long, herbicide injury to nontarget plants and other organisms, undesirable residues in food or feed crops, or contamination of various components of air, soil, or water may result. The length of time herbicides persist in a field following application is determined by the following three factors: (1) *displacement* or movement to other environmental compartments, (2) *adsorption* to soil, and (3) *decomposition* or degradation.

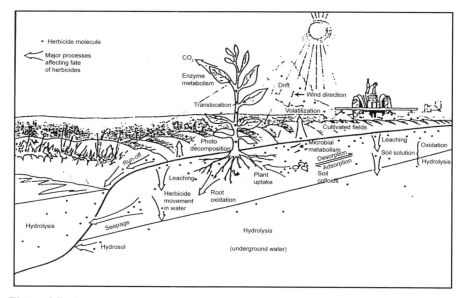

Figure 8.7   Interrelations of processes that lead to displacement, adsorption, or decomposition of herbicides in environment. (From Akobundu 1987, *Weed Science in the Tropics: Principles and Practices.* Copyright 1987 with permission of John Wiley & Sons Inc.)

Knowledge about environmental processes is necessary in order to understand the patterns of persistence and loss of herbicides. This same information can also be used to avoid displacement of herbicides from their intended site of application or to improve their decomposition. Minimal environmental contamination results when the processes that regulate herbicide fate are understood because application methods or management techniques can usually be devised that reduce herbicide impacts to most environmental compartments (Figure 8.7).

Herbicide Displacement in Environment

Displacement of an herbicide from its intended site of application or through the soil profile was once considered to be a loss of the chemical from the environment. However, this assumption is not necessarily true if a proportion of the herbicide applied simply moved to another environmental compartment (air, soil, surface water, groundwater) from the treated field. Chemical displacement can be in any direction, vertical or lateral, in the soil, across the soil, into plants, or into the atmosphere (Figure 8.7). Movement into the atmosphere is determined by herbicide vapor gradients and air circulation patterns. Movement on and in the soil is determined primarily by the flow of water, characteristics of the soil colloids, and water solubility of the herbicide. Water solubility of the chemical is also a factor that results in herbicide displacement to surface water or groundwater.

As discussed previously, herbicide movement is, at times, desirable and several techniques were described that use movement to improve herbicide selectivity. At other times, herbicide movement results in the chemical being where it is not intended. Unnecessary or unexpected environmental contamination resulting from herbicide use can be reduced if the processes of herbicide displacement, adsorption, and decomposition are incorporated into application techniques.

Herbicide Movement in Air

Drift. Drift is the physical displacement of an herbicide as particles or droplets from the intended target during application. Since most herbicides must move through the air in order to reach target vegetation or the soil surface, the opportunity for drift is always present. Drift is of concern because it may result in herbicide injury to plants not intended for control, herbicide residues on adjacent crops or wild plants gathered as food, or contamination of land or surface water.

Methods to Reduce Drift. Herbicides are most often applied as a mixture with water and dispensed as a spray of droplets. Most of the droplets fall rapidly to the targeted vegetation or soil, but some do not. Figure 8.8 shows the pattern of spray coverage for a typical aircraft application. The greatest potential for drift exists when droplets are small, usually less than 100 μm in size. Fine droplets can remain in the air for a long time and travel for an indefinite distance. A major way to reduce drift of herbicides is to apply them with equipment systems or nozzle types that produce relatively large droplets (over 100 μm).

During ground application, less herbicide is lost by drift when nozzles are close to the soil surface or vegetation being sprayed. When herbicides are applied inside shields, the potential for drift is substantially reduced. Wind and temperature inversions also influence the potential for herbicides to drift during and after application. For this reason, herbicides are normally applied when the wind speed

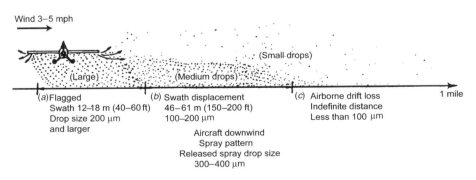

Figure 8.8 Effect of drop size and wind condition on aircraft swath displacement. (Modified from Akesson and Walton 1989, California Weed Conference. Thompson Publications, Fresno, CA.)

is less than 5 mph and during the morning when temperature inversions are least likely. In addition, spraying downwind results in less drift than spraying upwind or against a cross-wind. Some herbicide formulations are less likely to drift than others. For example, granule and pellet formulations will not drift as far as droplets. Some herbicide applicators mix drift-retardant chemicals in the spray solution/suspension to create large droplets that fall to the target area more rapidly. Such spray additives can reduce coverage of the herbicide on the foliage of treated plants, however, and sometimes diminish weed control.

USE OF BUFFER ZONES. Injury to susceptible crops or sensitive areas can often be avoided by precautionary measures before and during herbicide applications. For example, sensitive areas, such as streambeds and riparian areas in forest plantations and range environments, are protected by wide buffer zones that are adjacent to the area being treated. Buffer zones vary in size according to the herbicide being applied, method of application, and local ordinances and regulations. Aerial applications of herbicides generally require wider buffer zones than other methods. Similarly, if susceptible crops, desirable plants, or homes are adjacent to an area to be treated with an herbicide, it is wise to leave an untreated strip of land between the two areas, which reduces the unintended consequences of drift.

Volatility. Volatilization is the change of a solid or liquid into a gas. All chemicals, including herbicides, can volatilize (vaporize) depending upon the vapor pressure of the chemical and temperature. As pointed out earlier, some herbicides volatilize readily, while others volatilize very little. Because volatility is an important source of herbicide displacement from treated soil and vegetation, herbicide labels and precautions should be followed closely to reduce air contamination. Usually, volatile soil-applied herbicides are mechanically mixed with dry soil to minimize losses to the environment as vapors.

Incorporation of certain herbicides, such as the thiocarbamates, into the soil enhances adsorption and thus diminishes displacement via volatility. This practice also reduces the potential for vertical and lateral herbicide displacement in the soil with water. Incorporation of both volatile and nonvolatile herbicides into the soil often improves weed control as well. There are several methods to mechanically mix or incorporate herbicides into the soil.

Herbicides in Soil. Herbicides are often applied directly to the soil or mixed with it for weed control. Other herbicides eventually settle on the soil during or after herbicide application. Herbicide runoff from treated foliage and the decomposition of herbicide-injured plants are other sources of herbicide entry into the soil system. Since most herbicides either are applied to the soil for weed control or eventually arrive there, herbicide interactions within the soil system are very important aspects of herbicide persistence and fate in the environment (Anderson 1996, Zimdahl 1999).

Characteristics of Soil. Soil is the substance on the surface of the earth in which plants grow. It contains the basic mineral parent material from the weathering of rock. Soil also contains water, gases, organic matter, and numerous types of living organisms. Any parcel of land, farm, forest, and so on, is probably composed of several different *soil types* that vary according to physical structure, texture, profile, organic matter content, and fertility. All soils have four basic components, the solid, liquid, gaseous, and biological phases. Soil is also a dynamic system and, therefore, subject to the entry and loss of substances like herbicides, which can associate and interact with all four phases. See Radosevich et al. (1997) for greater discussion of soil properties and phase interaction with herbicides.

Herbicide Adsorption to Soil. Adsorption is the adhesion of chemicals to the surfaces of solids. In soil, adsorption of herbicides is a colloidal process that involves the negatively charged particles of clay and organic matter (Figure 8.9). *Desorption* is the release of herbicide molecules into the soil solution. Through adsorption and desorption an equilibrium forms that regulates the amount of herbicide on soil colloids and in the soil solution (Figure 8.9). The amount of herbicide adsorbed to the soil depends upon the amount of clay, organic matter, and moisture present in each soil type and the ionization properties of each herbicide. A convenient way to measure an herbicide's tendency to adsorb to soil particles is the adsorption coefficient K_{oc}, which is calculated according to the formula (Kamrin 1997)

$$K_{oc} = \frac{\text{concentration adsorbed/concentration dissolved}}{\% \text{ organic carbon in soil}} \tag{8.3}$$

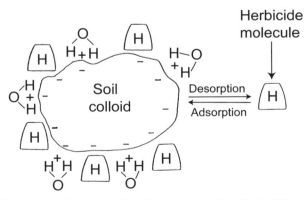

Figure 8.9 Schematic adsorption of molecules on soil colloids. Water molecules can compete with herbicide molecules for adsorption sites on colloids. (From Adkins 1995, *Weed Science Lecture Guide*, with permission of S. Adkins, University of Queensland, Brisbane, Australia.)

Adsorption of herbicides to soil colloids is the most important process affecting herbicide availability to plants and persistence in the soil, and virtually all herbicides are adsorbed to some extent. Adsorption also influences the amount and rate of microbial degradation of herbicides in soil. When adsorbed, herbicides are not available for plant uptake, lateral or vertical movement (leaching), or degradative processes. Herbicides first must be desorbed from soil particles in order for other soil processes, such as those just mentioned, to happen.

INFLUENCE OF SOIL COLLOIDS ON ADSORPTION. Most herbicides are adsorbed by either clay or humus particles. However, humus contains many more adsorptive sites than clay. Usually, only a small percentage of humus, 1% or less, is enough to affect the adsorptive properties of a soil. Because of the sensitivity of herbicide adsorption to the level of soil organic matter, both plant injury and herbicide persistence can be affected significantly by this factor. Herbicide recommendations and manufacturers' labels usually caution users concerning the activity and persistence of herbicides when applied to certain soils. Often the basis for such precautions is the amount of clay and organic matter present in the soil. Sometimes activated carbon is added to soil to reduce the phytotoxicity of an herbicide to a susceptible crop or other plants. Activated carbon decreases herbicide availability by providing a large amount of sites for adsorption.

INFLUENCE OF WATER ON ADSORPTION. Water effectively interferes with adsorption by displacing herbicide ions or molecules from soil colloids. A thin film of water can surround soil colloids in wet soils, making herbicide adsorption difficult. Herbicide that is not adsorbed will either remain in the soil solution, where leaching, degradation, or absorption by plants will occur, or volatilize into the air.

The equilibrium between the herbicide adsorbed to soil colloids and that in the soil solution (Figure 8.9) is also influenced by the water solubility and vapor pressure of the chemical. Herbicides with poor water solubility often adsorb more readily to soil than highly soluble chemicals. Usually within a class of herbicides, adsorption is inversely proportional to water solubility. This means that within an herbicide class herbicides with low water solubility usually adsorb to soil best. However, water solubility is not a good indicator of adsorptive ability if the chemistry among herbicides being compared varies widely.

Volatile chemicals, such as the thiocarbamate herbicides, are easily displaced from soil colloids by soil moisture. These herbicides are prone to loss from the soil as vapors unless special precautions are taken to enhance adsorption, such as incorporation into dry soil. Herbicide persistence in the soil is increased in this case because the volatile herbicides are not displaced to the atmosphere. For the same reason, weed control is also enhanced by incorporating thiocarbamate herbicides into dry soil. As soil is wetted by precipitation or irrigation, the herbicides are desorbed and made available for plant uptake.

INFLUENCE OF CHEMICAL CHARGE ON ADSORPTION. Most herbicides act in the soil as either weak acids or bases. The degree that various herbicides ionize depends on the soil pH and the ionization constants of each chemical. Since the colloidal component of soil is negatively charged, herbicides that form a positive ion when dissolved in water (cations) are adsorbed readily to soil. Herbicides that form cationic molecules (e.g., paraquat) are bound so tightly to soil that they are unavailable for plant uptake. Under acid conditions (soil pH of 5 or 6), many herbicides undergo *protonation*, which makes neutral and anionic chemicals relatively more attractive to the negatively charged soil colloids. Under neutral or alkaline conditions (pH 7 and above), the opposite reaction occurs. Generally, herbicides will be repelled by colloids in alkaline soils, making them relatively more available for plant uptake or movement with soil water.

Over the pH range of most soils, many herbicides behave as though neutral in charge. Adsorption of herbicides with neutral charge is determined by chemical properties such as water solubility, vapor pressure, or molecular size and shape.

Herbicide Movement with Water

Lateral Displacement. Lateral, or horizontal, movement of herbicides from soil can occur with surface *runoff* water (Figure 8.7). However, many agricultural fields are flat or have only small inclines, which do not favor lateral displacement of soil-applied chemicals. This is not the case with many forest or rangeland sites, but because of herbicide adsorption to soil colloids, lateral movement on slopes is still difficult. Some lateral displacement of herbicides adsorbed to soil particles has been demonstrated when precipitation patterns are intense and movement of soil particles results. Lateral movement of herbicides with surface water most often occurs when the chemical is applied over a stream bed. Therefore, care must be taken during herbicide applications to assure that herbicides are never applied near active or dry streams. Stream beds and riparian zones are protected from herbicide exposure by regulations that specify the size of untreated buffer strips adjacent to them.

Vertical Displacement. The vertical displacement with water of substances in the soil is called *leaching* (Figure 8.7). Actually, herbicide leaching can be in any direction in the soil profile, depending on where the chemical is placed and the direction of water flow. In fields that are irrigated by ditches or rills, lateral and sometimes even upward leaching of herbicides through soil may result in loss of herbicide selectivity and poor weed control. Upward water movement concentrates the herbicide in the crop row, at the soil surface, or along the ditch sides.

Downward leaching of herbicides in the soil profile is most usual because of the percolating action of water from rain or irrigation through the soil profile. The most important factor that influences herbicide leaching is adsorption (Figures 8.7 and 8.9). Herbicide that is adsorbed is not present in the soil solution and, therefore, cannot leach until desorption occurs. Leaching is also dependent on the water solubility of the herbicide. This observation is especially true within a chemical category of herbicides. Generally, herbicides that are relatively insoluble in water leach poorly and remain near the soil surface. For many such herbicides, the chemical tends to concentrate in the soil profile within a few centimeters of the soil surface. This vertical band of herbicide-treated soil is caused by the adsorption of dissolved herbicide in the soil solution as it moves through the soil profile.

The depth and amount of herbicide leached vertically is influenced not only by the adsorptive capacity of the soil but also by the amount and long-term duration of percolating water from irrigation and rainfall. The accumulation of herbicides deep in the soil profile or in groundwater also depends on the tendency of herbicides to degrade in the soil environment. Deep leaching was sometimes used in the past to "remove" herbicide residues from the rooting zone of susceptible crops or as a means of increasing crop tolerance to herbicides. This practice is now discouraged.

Herbicides in Groundwater. Residues of herbicides in groundwater supplies have been found in some regions where the chemicals are used repeatedly and extensively. The residues arise because even a minute amount of herbicide in the soil solution can leach throughout the soil profile, regardless of whether adequate time has been allowed between herbicide applications for soil adsorption and degradative processes to work sufficiently. Decreasing the reliance of farmers or other land managers on a single herbicide, greater use of crop rotations and other cultural methods of weed control, and use of different herbicides or weed control techniques probably will reduce but not eliminate the incidence of herbicide residues in groundwater. More discussion on this important topic is presented in EPA (1993) and Radosevich et al. (1997).

Herbicide Decomposition in Environment

The final factor that accounts for herbicide fate is *decomposition*. This is the process of destruction of the original herbicide molecule and usually loss of herbicidal activity. After degradation, parts of the original herbicide structure remain as different molecules. These breakdown products ultimately may be decomposed further to simple organic molecules, but more complex breakdown products may be incorporated into organic residues. Often, degradative processes occur in the soil. However, they are not restricted to soil and may occur in water, air, plants, microbes, and animals.

Persistence curves exist for most herbicides. These curves usually document actual destruction of a particular herbicide in soil, water, or air through time and are used to calculate its half-life (Figure 8.10). However, the half-life of an herbicide is not absolute because it depends on the soil type, temperature, and concentration of the herbicide applied. Thus, herbicide decomposition, like many other aspects of herbicide science, is relative to a number of other factors. Most organic herbicides are degradable so they do not accumulate appreciably in the soil, having half-lives that range from only a few days to several months (Figure 8.10), unless repeated applications are made. However, small concentrations or residues of herbicides can persist in soil or water for a long time and build up when repeated applications are made (Figure 8.11).

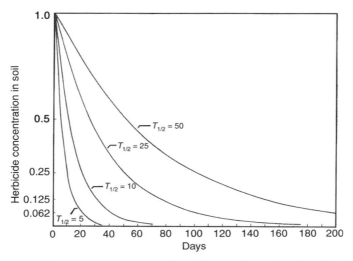

Figure 8.10 Persistence curves of herbicides with half-lives of 5, 10, 25, and 50 days. Actual half-life curves for 2,4-D and dicamba in Oklahoma soils approximate 5- and 25-day curves, respectively. (From Altom and Stritzke 1973, *Weed Sci.* 21:556–560. Copyright 1973. Weed Science Society of America. Reprinted by permission of Alliance Communications Group, a division of Allen Press Inc.)

Photochemical Decomposition. Some herbicides undergo photochemical reactions when exposed to sunlight that result in degradation of the chemical molecule. However, *photolysis* is a unique feature of each herbicide. Some herbicides are photochemically decomposed with relative ease, while others do not degrade well in sunlight. Photochemical degradation is an important process of herbicide loss from surfaces, such as soil and leaves.

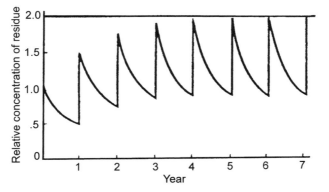

Figure 8.11 Residue pattern for single annual herbicide application and half-life of one year. (Modified from American Chemical Society, 1976.)

Chemical Decomposition. Herbicides can decompose in soil by purely chemical means. However, it is often difficult to distinguish between chemical and microbial decomposition. Chemical decomposition of herbicides may result from the following types of reactions: *oxidation–reduction, hydrolysis,* and the *formation of water-insoluble salts and chemical complexes.* These reactions are unique for each herbicide. Herbicide decomposition has been the subject of reviews by Kearney and Kaufman (1988), Devine et al. (1993), and Roe et al. (1997).

Microbial Decomposition. Many types of organisms can degrade herbicides. However, microorganisms are the primary agents responsible for the degradation of herbicides in the soil system. Organic herbicides are subject to a wide array of soil microorganisms that utilize these compounds as a source of carbon, nutrients, and energy. In the process of using herbicides as a food source, microorganisms also destroy the herbicidal properties of the chemical and often break the material down into less complex substances. A portion of the herbicide molecule may be incorporated into the microorganisms themselves, which in turn is decomposed as the microbes die and are recycled in the soil.

Herbicide degradation by microorganisms is due to specific enzymes that are often secreted by the microorganism to break down complex organic products. The following degradative reactions occur as herbicides are decomposed by microbes. These reactions result in alterations of the parent herbicide molecule (Anderson 1996):

Dehalogenation	Removal of chlorine, bromine, or other halogen atoms
Dealkylation	Removal of organic side chains
Hydrolysis	Removal of amides or esters
Beta oxidation	Cleavage of carbon units in twos or units of two
Ring hydroxylation	Addition of hydroxyl ($-OH$) groups to the aromatic ring
Ring cleavage	Breaking structure of aromatic ring
Reduction	Addition of hydrogen to NO_2 groups under anaerobic conditions

Numerous factors influence microbiological herbicide degradation in the soil. These factors include soil moisture and soil conditions such as temperature, aeration, pH, and organic matter content. Generally, if any of these factors of the soil is reduced, the rate of herbicide decomposition also diminishes. Other factors that influence decomposition are the dose and structure of the herbicide, adsorption to soil colloids, and composition and density of the microbe population.

Effects of Herbicides on Soil Microflora and Fauna. The microflora and fauna in soil are large and diverse. It is, therefore, not surprising that some herbicides adversely affect certain species of microbes and soil-borne animals. Generally, it is believed that negative effects of herbicides on microbial populations are reversible, meaning that the population level or composition of species is decreased for awhile but later improves. Beneficial organisms known to be affected negatively by specific herbicides include nitrogen-fixing bacteria and some mycorrhizal fungi. In addition, some plant pathogens have increased as a result of herbicide use.

Reduction of Herbicides in Agriculture and Natural Resource Production Systems

Agricultural fields, forest tree plantations, and rangeland grazing systems, like all other ecological systems, are regulated by positive- and negative-feedback responses among the organisms that comprise them and their environment (Figure 7.6). Negative-feedback responses naturally regulate population sizes within ecosystems, and when they are diminished exponential growth of other populations results (Figure 2.7 and Equations 2.3 and 2.4). This exponential population growth causes an invasive process called an infestation, infection, or epidemic, depending on the organism and the point of view of the land manager. Understanding the elements of natural population regulation is a keystone of integrated pest management (IPM), as we will see in Chapter 9, and is crucial for reducing herbicide use in agriculture and natural resource production systems.

As described in Chapter 9, IPM programs consist of some basic elements:

1. Acquisition of knowledge about the biology and population dynamics of the target pest organism
2. Monitoring of target pest population levels
3. Determining acceptable injury and action thresholds of the disease, insect, animal, or weed or invasive plant
4. Employment of an acceptable population control method for the organism

The goal of herbicide reduction in any production system is to maintain an ecologically healthy, viable system while continuing to maintain or restore productivity. Ways to accomplish this goal are considered in Chapter 9 for agricultural, forest and forest plantation, and rangeland production systems.

SUMMARY

Herbicides are chemicals that kill or suppress the growth of plants. Most herbicides are discovered by the systematic synthesis and testing of chemicals for biological activity. The EPA is the primary organization responsible for the regulation of herbicides in the United States. The FIFRA and the FDCA are two major laws regulating herbicide use in this country. However, there are also other organizations that voluntarily or legislatively restrict the use of herbicides by their members. Properties of herbicides that affect human, animal, and environmental safety are acute and chronic toxicity, including carcinogenicity, endocrine disruption, mutagenicity and reproductive disorders, biological magnification, and persistence in the environment.

Important chemical properties of herbicides include chemical form, water solubility, and vapor pressure. Most herbicides are modified or formulated before sale and use as commercial products. The formulation of an herbicide contains the active ingredient of the chemical and various inert ingredients, such as solvents and emulsifiers. Herbicides may be applied dry, as granules or pellets, but are usually applied as a liquid spray. Herbicides are most effectively used when they are applied accurately and uniformly. The use of properly designed, calibrated, and operated equipment assures both accuracy and uniformity of application.

Herbicides are classified according to similarities in chemical structure, use, and biological processes inhibited. Plants often exhibit characteristic symptoms of cell disruption, hormone imbalance, leaf chlorosis, necrosis, or albinism following herbicide application. These symptoms reflect the inhibition of a vital plant process in exposed plants. Most herbicides are selective, meaning that some plants are injured by them while others are not. However, selectivity is relative and depends on the rate of herbicide applied and numerous other plant, chemical, and environmental factors. The plant factors that influence herbicide selectivity are genetic inheritance, age, growth rate, morphology, physiology, and certain biochemical/biophysical processes. Development of herbicide-resistant crops is a recent innovation for weed control in agriculture which has benefits but has also raised concerns. Herbicide structure or configuration, formulation, and placement also influence differential herbicide sensitivity among treated plants. Environmental factors that affect the movement of herbicides in the soil also affect selectivity patterns among desirable plants and weed species.

The fate of herbicides in the environment is an issue that should be understood and addressed when herbicides are used. The soil acts as an important buffer governing the persistence of most herbicides in the environment. Since herbicides are either applied directly to the soil or eventually arrive there by runoff or in plant residues, herbicide persistence is influenced by the soil processes of adsorption, movement, and decomposition. Herbicides are adsorbed most readily by dry soil. Adsorption is also affected by the chemical charge and water solubility of each herbicide. The amount of organic matter and clay in soil also influences herbicide adsorption dramatically. Herbicides that are adsorbed to soil colloids are not available for absorption by plants, movement in the environment, or degradation.

Movement of herbicides in the environment results from drift, volatility, and lateral (runoff) and vertical (leaching) displacement in soil by water. Herbicide displacement from chemical drift can be reduced by using appropriate equipment or spray additives that produce large spray droplets. Buffer zones around sensitive areas are also an effective way to minimize the impacts of spray drift. Herbicide leaching in soil may result in herbicide contamination to groundwater supplies if appropriate care is not taken. Herbicide decomposition can occur by photochemical, chemical, or microbiological means. Degradation by microbes is probably the most important mechanism for herbicide decomposition in soil. Some herbicides also may adversely affect beneficial microflora and fauna or enhance temporarily the incidence of plant diseases.

Methods and approaches are now being explored to reduce herbicides use in agriculture, forest, and rangeland systems. The reasons for this change in research and management emphasis to more ecologically based management in these systems is largely due to concerns about the unintended impacts of herbicides on human health, environmental quality, and direct and indirect effects on nontarget plants, animals, and microbes.

9

SYSTEMS APPROACHES FOR WEED AND INVASIVE PLANT MANAGEMENT

Weeds and invasive plants are first and foremost part of complex biological and socioeconomic systems that include agricultural, forestry, and range management systems as well as natural ecosystems. While their impacts can be severe, weeds are only one component of these systems and as such they interact with, influence, and are influenced by many other components of the entire system (Figure 7.6). Thus, the way humans manage their lands influences what plants grow there and for how long. Weeds are also a consequence of human perception (Chapter 1), so the way people think about weeds influences their notions of economic and ecological harm that may be caused by these plants. Aldo Leopold, the great American naturalist and author, for example, takes exception in his essay titled "What Is a Weed" with those who list many of the native plants of Iowa as weeds (Leopold 1943). Leopold points out that most of the plants listed in the *Iowa Bulletin* to which he objects have substantial value for wildlife, soil cover, fertility, and beauty. He grants that some plants do "enormous harm to cropland" but also points out that most weeds arise from overgrazing, soil exhaustion, or the "needless disturbance" of more advanced successional stages of vegetation.

The simplest approach to deal with weeds has been to control them directly in order to reduce their abundance. However, this tactic can be inadequate, especially when all biological, social, and economic factors are not considered (Chapter 7). Forcella and Harvey (1983), for example, note that the incidence of weeds and invasive plants has increased over the past 50 years in spite of all the effort to control them, a process that continues today. In order to understand how weeds fit into a particular ecosystem, it is necessary to determine how the processes and factors within the system interact to favor or disfavor undesired vegetation and then to organize the processes into a management framework (Figure 7.10) (Parker et al. 1999).

The remainder of this chapter integrates further the ecological, management, and social principles described in Chapter 2 and throughout this book with current vegetation management principles in production (agriculture, forestry, and rangeland) and natural systems. It is clear that the novel systems described below are not identical, some often missing elements incorporated by the others. Ideally, however, scientists and managers from each discipline will learn from each other as the common goal of better, more effective, and ecologically aware vegetation management options are explored.

CYCLES OF LAND USE, EXPANSION, AND INTENSIFICATION FOR PRODUCTION

Land uses for agriculture, forestry, and rangeland grazing have been coevolving with human societies for millennia, resulting in reoccurring cycles of land expansion and production intensification (Figure 9.1). Historic requirements for human food, fiber, and shelter (Perlin 1989) were met most simply by expanding to new areas available for land use, which often meant displacement of indigenous cultures (Crosby 1994). In the case of agriculture, however, a different kind of expansion also occurred over the last century since technological intensification also allowed cropping in areas that were previously unsuitable for food production, such as wetlands, deserts, and forests (Merchant 1980, McKibben 1999, 1995). Similar cycles of expansion and intensification have occurred in forestry and rangeland production, but with different phases than in agriculture.

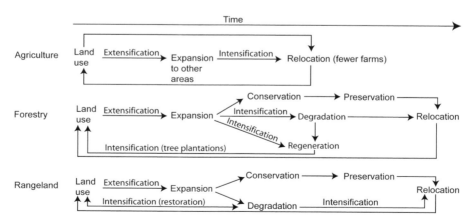

Figure 9.1 Processes of land use, land expansion, and technological intensification in agricultural, forest, and range production systems.

When each production system nears the end of its grand phase of expansion and there is no more land to use, social and economic pressures are usually met by intensifying management activities (Figure 9.1). For example, in forestry, limitation in timber resources usually results first in relocation, then in attempts to conserve and preserve forests as land becomes more limited. These approaches are followed by technologies to regenerate tree stands (Perlin 1989). Range managers, after decades to centuries of domestic grazing on native grasslands, now resort to restoration of lands occupied by invasive plants (Figure 9.1). Thus, as the human population grows and cultural pressures increase, especially over the last century, scientists and land managers continue to try to meet societal needs for food, fiber, and shelter with ever more technology. A consequence of increased technologies used solely for production, however, is the simplification of environmental and biological diversity that assures the ecological balance and natural services that human society also needs. The necessity of biological and environmental feedback mechanisms and ecosystem services, such as an abundance of clean air and water, is only now being recognized by some segments of modern human society.

The cycles of land expansion and technological intensification are not free of environmental and climatic adversity, however, which generally decreases yield and sometimes the availability of other ecological services. Outstanding among these adversities are weeds, undesirable insects and other herbivores, and pathogens. The impacts of invasive plants in natural ecosystems are now widely recognized (Rejmánek 2000, Sakai et al. 2001, Pyke and Knick 2003). The predominant tools to suppress unwanted organisms, especially in agriculture, for the last six to seven decades have been pesticides (Briggs 1992, Liebman 2001, Whitford 2002). However, closer examination revealed unforeseen problems with the intensification of pesticide use and differing philosophical views questioned the expansion–intensification model (Carson 1962, Briggs 1992, vanden Bosch 1992) and its associated dogma of "production at any cost." It was into this framework that *integrated pest management (IPM)* was conceived in the early 1970s (Table 2.2), which has at its core knowing how crop and other organisms interact with each other and a commonly shared environment in a production system. According to vanden Bosch (1992), a founder of the IPM concept, "Integrated control is simply rational pest control: the fitting together of information, decision-making criteria, methods, and materials with naturally occurring pest mortality into effective and redeeming pest-management systems."

Evolution of Modern Integrated Pest Management

Even during the colonial era (Crosby 1994) international transport moved organisms around the globe, which frequently caused human disease epidemics or introduced exotic pests into new environments. Even then, there was awareness that at least some pest and disease problems originated by the disruption of biological checks and balances of natural systems, and deliberate introductions of natural enemies were initiated to control pest outbreaks (Doutt 1964, Simmonds et al. 1976). The earliest concept of IPM, therefore, consisted of using insecticides that were consistent with the biological control of insects (Stern et al. 1959, Smith and vanden Bosch 1967). This concept was later expanded to encompass all pest control tactics.

Since using pesticides in a compatible way with biological controls implies knowledge of how organisms interact with each other and their environment (Figure 7.6), scientists and policymakers argued that IPM should be framed around ecological concepts. Based on this view, the U.S. National Academy of Science (NAS) (National Research Council 1996) recommended that ecologically based pest management replace the IPM concept. In 1998, the U.S. Department of Agriculture (USDA) defined IPM with a suite of management activities—prevention, avoidance, monitoring, and suppression of pests. Norris et al. (2003) indicate that both ideas (NAS and USDA) fail to incorporate either control strategies or pest management disciplines, which were included in the original IPM concept. Norris et al. further extends the IPM concept to describe it as a decision support system that aids in the selection of pest control tactics. Ideally, multiple tactics are coordinated into a comprehensive management strategy. Decision support should be based on cost-benefit analyses that consider the interests of and impacts on producers, society, and the environment. Interestingly, Norris et al. (2003) do not refer to using ecological principles for pest management, which is a component of the original notion of IPM. However, the ecological perspective may be implied in the above definition since ecological principles can be used even when pest management is rationally based on economics and social values.

Evolution of Weed Science

In many respects, the evolution of weed science is the history of modern weed control, which has progressed from hand pulling and primitive hand tools to animal and fossil fuel powered implements to chemical herbicides. For example, in the *Boke of Husbandry* Fitzherbert (1523) enumerates a number of weeds that "doe moche harme," which Salisbury (1961) infers were some of the most common weeds in England nearly 500 years ago. Salisbury also provides several enlightening examples of the shifting patterns of the past and present weed flora in Great Britain that resulted from the modernization (intensification) of agriculture and the introduction by the Romans of what is now considered to be the native weed flora of Great Britain. Zimdahl (1999) notes that the development of weed science as an academic discipline is a relatively recent event that began during the mid-twentieth century, was dominated by the U.S. land grant system of higher education, and is highly chemically oriented. During this same time period, however, other scientists—botanists and plant scientists—also studied weeds as unique and interesting organisms. Although not known as weed scientists, these people devoted much of their careers to the study of weed biology, life cycles, and evolution. Through such pioneering efforts a new era of understanding and activity in weed science began to emerge that focuses on management based on biological and ecological principles. Today, this approach has taken on even greater urgency due to concerns about the impacts of weeds and invasive plants in complex natural ecosystems where options for chemical and other technological methods of weed control are often limited.

Even though weed scientists began exploring the concept of IPM soon after its inception, practices in which chemical and mechanical weed control tools and tactics are used in conjunction with other approaches for vegetation management are still difficult to find, especially in agriculture and forestry. However, integration of weed control tools with other biological and environmental control agents and methods is more common in rangeland production systems (Rew and Pokorny 2006, Krueger-Mangold et al. 2006) and for the management of invasive species in general (Sakai et al. 2001). Nonetheless, efforts are now being made in each production system to incorporate weed control tools into approaches that maintain or enhance productivity, ecological feedbacks, balances, and ecosystem services, which is a concept first introduced by Levins in 1986 (Table 2.2).

APPROACHES FOR PEST AND WEED MANAGEMENT

Levins (1986) proposed three models of pest management that continue today, which he labeled industrial, IPM, and ecological agriculture (Table 2.2). In the *industrial* model, the expansion and intensification of agriculture incorporated more land into production and compensated for yield reductions caused by soil erosion, low soil fertility, or increased weeds, insects, rodents, and diseases. However, studies soon showed that while such technological innovations often increased yield, costs of production also increased (Pimentel et al. 1992) even while demand for food and fiber commodities was well below supply (Vasey 1992), especially in developed nations. The new production systems also disrupted social and economic structures in poorer nations that lacked transportation networks to distribute the increased production among their human populations (Lewontin 1982, Alstrom 1990). Nevertheless, Liebman (2001) suggests several reasons why modern farmers still rely heavily on herbicides as a technological advancement in modern agriculture:

- Apparent low risk and ease of chemical weed management
- Aggressive marketing of chemical solutions to weed management problems
- Externalization of environmental and human health costs of agrichemical technologies
- Agricultural policies that foster input-intensive agricultural practices

Liebman and his associates (2001) go on to describe the benefits and procedures of an approach they call *ecological weed management*, which is discussed later in this chapter. It is interesting that as agriculture is now turning away from Levins's industrial model as the only way to efficiently and economically produce food, forestry has embraced it (Hayes et al. 2005). These observations will also be discussed later.

Levins (1986), writing over 20 years ago, also indicates that IPM was significant in the evolution of modern agriculture because it posed an alternative to the pesticide-saturating strategy of controlling pests only using chemicals. The *IPM model* was seen as a step along a path from the high-level intervention of industrial agriculture toward an ecologically rational production system, which Levins called *ecological agriculture* (Table 2.2). He envisioned IPM as an alternative, intermediate approach leading away from the confrontation of cropping practices with the many living things present in agricultural fields and toward softened strategies of conditional steady states and coexistence.

Integrated Weed Management

As noted in Chapter 7, development of herbicides began shortly after World War II and chemical weed control became a frequent agricultural practice (Chapter 8) throughout the twentieth century. Zimdahl (1999) notes that weed science was first influenced by the ideas of IPM during the early 1970s as opportunities for biological weed control became evident (Coombs et al. 2004). However, most weed studies at that time still focused on the action of chemical weed control products and simple crop–weed competition experiments to determine yield losses from weeds (Barberi 2002). Although progress has been made toward development, assessment, and implementation of integrated pest and disease management systems, weed science still lags behind other pest management systems in this respect (Mortensen et al. 2000), and after almost five decades weed control is still not clearly connected to biological principles (Messersmith and Adkins 1995, Buhler et al. 2000, Norris and Kogan 2000, Liebman et al. 2001, Hatcher and Melander 2003, Krueger-Mangold et al. 2006).

Levels of Integrated Weed Management. Cardina et al. (1999) identified three levels of research within integrated weed management (IWM):

Level 1—individual components of crop–weed systems
Level 2—research on multiple management tools
Level 3—cropping systems design

Frequently, studies in IWM search for ways to reduce herbicide doses or applications (level 1) using a variety of cropping alternatives (level 2), such as varying plant densities or plant spacing or altering soil fertilization to increase yields and reduce external inputs and costs (Liebman and Davis 2000, Bostrom and Fogelfors 2002). Such studies address questions about tool effectiveness and the individual components of management, such as weed thresholds, critical periods of control, or interrow cultivation (Swanton et al. 2002). While admirable, studies conducted at levels 1 and 2 are usually insufficient to understand how weed control impacts an entire cropping system.

TABLE 9.1   Nonchemical Methods of Managing Weeds and Ecological Principles upon Which Each is Based

Ecological Principle	Weed Control Practice
Reduce inputs to and increase outputs from soil seed bank	Prevention, soil solarization, weed control before seed set
Allow crop earlier space (resource) capture	Early cultivation, using crop transplants, choice of planting date
Reduce weed growth and thus space capture	Cultivation, mowing, mulching
Maximize crop growth and adaptability	Choice of crop variety, early planting
Minimize intraspecific competition of crop, maximize crop space capture	Choice of seeding rate, choice of row spacing
Maximize competitive effects of crop on weed	Planting smother or cover crops
Modify environment to render weeds less well adapted	Rotation of crops, rotation of control methods
Maximize efficiency of resource utilization by crops	Intercropping

Source: Holt 1996 (unpublished), in Radosevich et al. (1997). *Weed Ecology: Implications for Management*, 2nd ed. Copyright 1997 with permission of John Wiley and Sons, Inc.

For example, crop rotation is considered to be an important strategy for IWM since it alters the environment and thus selection on weeds (Liebman 2001, Buhler 2002). However, studies about the effects of rotation on weeds continue to be contradictory (Liebman and Gallandt 1997, Liebman and Ohno 1998, Mortensen et al. 2000, Liebman 2001). Recently Swanton et al. (2002, p. 504) stated that "current research has apparently reached the threshold of Level 3, where the focus is on reducing seed banks through cropping system design." The importance of seed banks to weed persistence, discussed extensively in Chapter 5, was also recognized by system-level studies conducted by Cardina et al. (1999) and Buhler et al. (2000).

Ecological Principles to Design Weed Management Systems. Holt (unpublished) considers several nonchemical methods to manage agricultural weeds and lists the ecological principles on which they are based (Table 9.1). In addition, Radosevich et al. (1997), based on Altieri and Liebman (1988), provide the following description of ecological principles that should be included when designing IWM systems. The chapter in this text where each principle is discussed is indicated in parentheses. These principles hold for managed forests, forest plantations, and rangeland production systems as well as agriculture and clearly have applications for natural ecosystems as well.

- Crop monocultures seldom use all of the environmental resources available for plant growth. The resulting ecological niches (Chapter 2) are, therefore, susceptible to invasion by weeds and should be protected (Chapters 3, 6, and 7).

- Weed populations are either active, as photosynthesizing plants, or dormant, usually as seed. Thus, the seed bank as well as the above-ground vegetation should be considered when determining weed abundance (Chapters 2 and 5).
- The reproductive capacity and seed survival of weeds/invasive plants determine the composition and species abundance of the succeeding weed populations (Chapters 2 and 5). In intensive production systems, weed/invasive plant populations are a direct reflection of the crop (desirable plants) and how it is managed (Chapter 7). In developing countries, where mixed cropping is traditionally practiced in contrast to more intensive forms of crop production, the more complex cropping patterns disfavor weeds (Chapter 6). Additionally, in less intensive cropping systems than those presently practiced in many developed nations, weed management has less relevance than all cropping practices together in determining the weed (early successional) community composition (Chapters 2 and 9).
- The cropping pattern can be a powerful agent to reduce weed densities by preemption of environmental resources to crops or other desirable plants (Chapter 6). Weed control tactics, while usually reducing the abundance of weeds, can also shift the composition, density, and spatial distribution of weeds in a field (Chapter 7). These shifts in spatial or temporal dynamics of weed populations and communities ultimately may increase weed abundance and, therefore, crop competition or species diversity over the long term (Chapters 5 and 6). Although shifts in species composition often seem inevitable, the practice of growing crops with divergent life cycles and using rotational sequences of crops and weed control tactics are effective strategies to prevent any one weed species from becoming dominant (Chapter 9). This practice reduces selection pressure on weed populations (Chapter 4).
- Single measures of weed density are usually not adequate to determine weed impacts (Chapters 2, 3, and 7). Most crops (desirable plants) have weed thresholds, expressed as density, amount of biomass, or period of time before significant loss in yield results (Chapter 6). However, these thresholds vary among cropping systems, desirable species, weeds/invasive plants, and environmental constraints. The thresholds may also need to be adjusted when criteria other than crop loss are used to determine economic, ecological, or social effects of weeds/invasive plants (Chapters 2, 7, and 9).
- Suppressed crop or desirable plant growth cannot always be explained by crop–weed competition (Chapters 5 and 6). At times, weeds may simply indicate deteriorated soil or resource base (Chapters 3 and 9). Allelopathy may also be a mechanism through which weeds affect growth of desirable plants and vice versa (Chapter 6). In addition, crops and weeds may coexist without crop yields being reduced economically and beneficial effects are possible from some crop–weed associations (Chapters 2 and 6).
- Weed populations and communities are regulated by a combination of factors (Chapter 3), for example, environmental stress, interference, and weed control activities, and regulation varies with stage of the weeds' life cycle

(Chapter 5). Thus, control of weeds can be achieved by direct or indirect means (Chapter 7). During direct control the undesired plant is physically or chemically removed, while indirect controls rely on biological functions for regulation of the crop/desirable plant, weeds, and associates (Chapters 7, 8, and 9). Both strategies have been successful, although indirect regulation is most appropriate for complex cropping systems and long-term weed suppression (Chapters 2, 5, and 7).

Future Directions in Integrated Weed Management. As seen in Chapter 4, no production system is static since evolutionary processes are always at work among the species in the biotic communities where they exist. In addition, social changes over time alter peoples' needs and goals for the land. These evolutionary changes in the flora of agro- and natural ecosystems must be recognized by agricultural and natural resource scientists in order to maintain production and limit undesired consequences of weeds and the tools used to control them. Gentler technologies, as suggested by Levins (1986) in Table 2.2, however, are often site specific. Understanding the interactions among biotic and abiotic components of ecosystems is always hard and even more difficult to translate into reliable management strategies. Even though ecological approaches have been criticized for these reasons (Peters 1991), there is little doubt that IWM must continue to focus on the complexity and integration of patterns resulting from human activities in agro- and natural ecosystems (Buhler et al. 2000, Norris et al. 2003).

Opportunities for IWM exist in four interrelated and scale-dependent areas: (1) *population dynamics* of weeds and invasive plants, (2) *community-level interactions*, particularly food webs, (3) spatial and temporal *patterns of plant succession* in agricultural and natural ecosystems, and (4) understanding *evolutionary patterns* promoted by intensive management practices.

Population Dynamics of Weeds and Invasive Plants. The ecological processes and patterns that determine the function and stability of plant populations during invasions are known to be scale dependent (Table 2.3) and range from individual species to landscapes (Naylor et al. 2005, Rinella and Luschei (2007)). Additionally, disturbance is believed to be fundamental for successful introduction and colonization/naturalization phases of the plant invasion process. (See Chapters 2 and 3 for a discussion of the process of plant invasions.) Hobbs and Humphries (1995) suggest a framework to set priorities for management of invasive plants based on land value and the degree of disturbance (risk of invasion), which could also include arable land (Figure 2.12). The four sections of Figure 2.12 demonstrate clear management priorities; however, actual management is usually much more difficult (Chapter 3) than suggested by Hobbs and Humphries (1995). Most tools and practices for weed suppression in agricultural and natural resource production systems are also important components of any management strategy for invasive plants (Radosevich et al. 1997, Ross and Lembi 1999, Sheley and Petroff 1999, Muyt 2001). But tools, whether physical, mechanical, chemical, or biological, are not benign (Chapters 7 and 8). The tools used to eradicate, control, or contain weeds or invasive plants can have important impacts on agro- and natural ecosystems, besides their effects on the undesirable plants of concern. In fact, there may be a balance or threshold of the benefits from control/containment of weeds or invasive plants in an ecosystem and the potential adverse effects of the disturbance to the ecosystem associated with tool use (Chapter 7).

Recent research on invasive plants (Neubert and Caswell 2000, Caswell and Takada 2004) indicates that biological parameters, especially plant demographics and factors of long-distance dispersal, determine the extent and rate of weed population expansion across landscapes. However, descriptions of biological and environmental characteristics of most weeds and invasive plants are often lacking or only general in nature (e.g., PLANTS database, http//plants.usda.gov). Thus, it has been difficult to determine which species are most likely to be invasive or to develop general predictive models for weed or invasive plant expansion (Mack 1996, Rejmánek and Richardson 1996, Enserink 1999, Neubert and Caswell 2000, Caswell and Takada 2004). The process of weed invasion/infestation may be best studied mechanistically by considering that *establishment, competition,* and *dispersion* are three closely linked processes that assure persistence of weed species in human production systems (Parker 2000, Dekker 2003). In this way, demographic parameters and population interactions of both wild and domesticated species become relevant and useful to predict risks and performance of weeds/invasive plants (Lambrecht-McDowell and Radosevich 2005).

Community-Level Interactions, Particularly Trophic Levels and Food Webs. Soil, plants, consumers, carnivores, and decomposers are all part of natural and agricultural ecosystems (Figure 7.6). Many of the processes regulating the behavior of individuals, populations, and communities in these systems are influenced by soil factors that first affect plants (producers), which in turn propagate effects upward to consumers in the food chain (Figure 9.2). Other processes are influenced from the top down by carnivores or consumers that, in turn, affect plants and soil. In agricultural systems, except for birds and mammals, most individual consumers are small (e.g., bacteria, fungi, arthropods, nematodes), compared to crops and weeds on which they feed. A single plant can host complete populations, even communities, of these organisms. For this reason, even small changes in plant populations or community composition can produce great impacts upward in the food web (Root 1973, May 2001, Marshall et al. 2003). Conversely, large changes in the consumer populations at the top of the web have relatively little impact on the plant community, although biotic interactions among individual consumers can be variable because of relative size differences among predators and their prey.

Because of general size differences between consumers (predators) and producers (prey), biological control by natural enemies for insect pests is more likely to be successful than for plants. Nevertheless, plants often have components with large size differences, such as seed and spores that are the smallest individuals in plant populations. Impacts by herbivores and predators are largest during those life-cycle stages (Crawley 1992, Nurse et al. 2003, Coombs et al. 2004). Seed

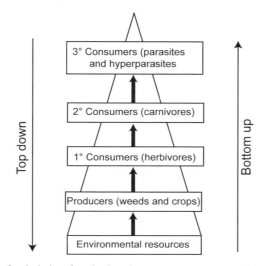

Figure 9.2 Basic food chain of agricultural ecosystem (crop, small vertebrates and birds, arthropods, nematodes, fungi, and bacteria). Thick arrows indicate energy–matter flow and narrow arrows indicate direction of controls of food chain (bottom up or top down). (Modified from Polis and Winemiller 1996, *Food Webs: Integration of Pattern and Dynamics.* Chapman and Hall, NY.)

predation may be most important for regulating seed bank densities of plants with large seed, in zero-tillage systems (Ghersa et al. 2000, Davis et al. 2003, Harrison et al. 2003), or in particular agricultural landscapes where field margins are managed for the habitat of seed predators (Marshall and Moonen 2002).

The structure of insect communities is highly sensitive to management practices (Messersmith and Adkins 1995, de la Fuente et al. 2003), yet there is still little information on how food web structure affects such herbivores or competition among weeds and desirable plants. Biological control of weeds typically has often focused only on the pest plant and the control organism, paying relatively little attention to environmental conditions (both biotic and abiotic) (Hasan and Ayres 1990, Coombs et al. 2004). Development of conditions that are suppressive to weeds could include manipulations that enhance the detrimental effects of indigenous pathogens or insects on weed propagules (Kennedy and Kremer 1996, Kremer 1998) or reduce the competitive ability of weeds relative to the crop (Schroeder et al. 2004). Shifts from pesticides to using herbivore-repellant crops could increase generalist plant consumers in agricultural production systems and selectively increase consumption of weeds. This management practice has been successful with some traditionally used crop species (Altieri and Liebman 1986, Liebman and Davis 2000) that are toxic or deterrent to plant-feeding insects. These crops impact weeds because insect herbivores use them as their main food source, rather than the crop species.

The development of certain genetically modified organisms (GMOs) creates an opportunity to manage naturally occurring diseases and insect herbivores for weed suppression. Johnsongrass (*Sorghum halepense*), for example, is an important perennial weed that harbors many plant diseases and insect herbivores as well as several species of detrimental bacteria, fungi, mites, nematodes, and viruses (McWhorter 1989). Thus, many control programs against this plant focus on minimizing disease transmittal through generalist insect vectors and reduction of its competitive effect on crops (King and Hagood 2003). The development of GMO crops resistant to the viruses and insects could seemingly concentrate the impact of shared pests to only weeds. There are many efforts to use biological control agents against weeds by "flooding ecosystems" with specific arthropods, fungi, or bacteria (Spencer 2000, Coombs et al. 2004); however, there are few examples that address how existing generalist consumers are affected by the introduction of newly developed crops.

Spatial and Temporal Patterns of Plant Succession. Succession is a process in which the biotic community changes through time (Chapter 2). Although this process is well known, few studies focus on patterns of succession in agricultural lands (Swanton et al. 1993, Ghersa and León 1999, Svejcar 2003), except in shifting and cyclic agricultural systems, mixed cropping systems, and grazing systems in rangeland. In addition, some descriptions exist of early succession including exotic species in forestry as a result of harvesting practices (Halpern and Spies 1995). Rosenzweig (1995) indicates that the structure of plant communities is strongly influenced by time (duration) and area. A short time frame for a community could be due to recent or frequent disturbance, for example, while area occupied is strongly influenced by disturbance resulting in fragmentation. Thus, species richness increases as these factors increase, while richness is lower when an area occupied by a plant community decreases in time or is fragmented. However, as succession progresses over an extended time period (tens of decades), species richness usually declines again (Grime 1989, Kimmins 1997, Perry 1997).

EFFECT OF FRAGMENTATION AND LONGEVITY ON WEED SUCCESSION. Cropping systems, despite their often recurrent tillage activities, follow similar successional patterns as natural ones, that is, with increasing species numbers occurring over time and area (Perry 1997, Ghersa and León 1999). For example, a patch of arable land inside natural grasslands or forests generally has few weeds, but as crops are grown there over time the number of weeds increases (Ghersa and León 1999, Suarez et al. 2001, Hyvonen et al. 2003). Theoretically, crop rotation should also reduce weed community richness since rotation produces a fragmentation effect on the plant community in both time and space.

Weed competition is related to both species richness and dominance (Chapter 6) since species with superior competitive traits should dominate under most forms of management. Paradoxically, crop rotations carried out for long periods of time contribute to a diversity of habitats in which many plants species, including weeds, find refuge. Therefore, rotations result in a land use mosaic that enhances weed dominance in particular fields more than others. This paradox was first demonstrated by May (1974, 2001) and was empirically shown by Ghersa and León (1999). In these studies, extended periods in which monocultures were grown over large areas increased the presence of weeds while overall species diversity on the landscape declined. No studies have been conducted to evaluate the relative importance of the area covered by a particular cropping system or rotation pattern to determine if more or less dominant weed communities are generated.

SHIFTS IN WEED SPECIES COMPOSITION. Shifts in weed communities are frequent events in modern agriculture (Chapters 7 and 8). Buhler et al. (2000) and Manley et al. (2000) emphasize that it is the ability of weed communities to shift in response to control practices that requires integrated and diverse approaches to weed management. Despite the large volume of empirical and theoretical information on ecological succession, only a few researchers have studied change in weed communities using succession as a framework (Swanton et al. 1993, Sheley and associates 1996, 2005, 2006).

Succession not only encompasses variation in species number but also includes qualitative changes in a plant community (Table 2.5 and Figures 1.5 and 2.10). During early succession species have relatively high growth rates and short life spans and allocate a large proportion of their resources to reproduction (ruderals). During midstages, allocation is shared among structures that allow individual plants to maintain competitive ability and to reproduce (competitors). Finally, at late stages, species have low growth rates and longer life spans and are mainly adapted to overcome stress (stress tolerant) (Grime 1989). Impoverishment of soil fertility in agriculture, due to chemical and physical changes caused by cropping, drives the qualitative changes in weed flora by reducing crop species dominance usually achieved through the availability of soil resources (Martinez-Ghersa et al. 2000b, Suarez et al. 2001). As fertility declines, weed communities gain diversity, that is, species diversity increases by incorporating stress-tolerant plants while dominance of the crop is reduced. With high fertility, however, dominance by the crop or a few weed species increases at the expense of overall species richness (Figure 1.5), which is shrunken by competitive exclusion (Chapter 2). Changes in soil-tilling practices, especially if the stratification of the topsoil is affected (e.g., no-till conditions), also have profound effects on successional patterns of weeds (Swanton et al. 1993, de la Fuente et al. 2003). Under rangeland conditions where native plants are often stress tolerant, reduction in soil fertility usually reduces the incidence of ruderal and competitive invasive plants while favoring native flora (Sheley and associates 1996, 2005, 2006).

FIELD PATTERNS AND WEED/INVASIVE PLANT OCCURRENCE. Specific patterns that result from spatial and temporal differences in land use are important factors controlling the occurrence and rate of change of weeds in fields and landscapes (Forman 1995, Le Coeur et al. 2002). Persistence of weeds in agricultural land is strongly dependent on dispersal mechanisms (Maxwell and Ghersa 1992, Cousens and Mortimer 1995, Neubert and Caswell 2000, Caswell and Takada 2004); thus, the location of invaded patches (propagule sources) and the perimeter-to-area ratio of noninvaded areas (sinks) are important factors for plant community stability. Networks of roads, railways, and waterways cover agricultural and natural landscapes. Fences also create orthogonal shapes, frequently forming linear-shaped fields that may be kilometers long but only a few meters wide (Figure 9.3). Many of these linear fields are dominated by perennial species of herbs and woody plants, despite the fact that most cropped or grazed landscapes are abundantly covered with annual plants. This observation is frequently attributed to birds perching on fences and power lines, which deposit propagules of the perennial species, and to the relatively less disturbed condition of these unmanipulated habitats. It has been demonstrated empirically and by modeling that other factors than birds, however, can be more important in causing the successional patterns in long fields (Ghersa and León 1999, Le Coeur et al. 2002).

Plants that have short life cycles and allocate resources to the production of propagules with effective dispersal characteristics (Table 2.5) lose individuals and biomass when established in narrow linear fields because most seed and litter fall outside of the field (Figure 9.3). However, species with a long life cycle and little reproductive allocation generally remain in the field once established because individuals retain most of the biomass in their body structures. Over the long term, perennials stay and ruderals become locally extinct from such long areas, whereas the opposite is true for fields of square or circular design.

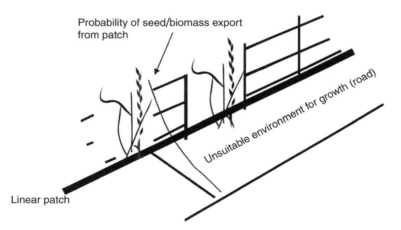

Figure 9.3 Edge effects in linear patches. Seed, biomass, or other propagules are lost to unsuitable environments (roads or frequently disturbed croplands). Dark line represents linear patch along a fence line that is <1 m wide and extends hundreds of meters/ kilometers. Short-lived species with high dispersal ability export most of their seed to unsuitable environments; long-lived species, once established, occupy the linear patch for long periods and continue changing succession pattern.

On the other hand, disturbance along roads and other human corridors is believed to be a major factor in the establishment of invasive plants in rangeland and some forest systems. It is important to understand how such "human corridors" affect the propagule source–sink relationships that influence plant invasions. Linear shaped, large fields and small fields of any shape are most efficient at exporting propagules from their boundaries (Cousens and Mortimer 1995, Rew et al. 2006).

Evolutionary Patterns Promoted by Intensive Management Practices. The process of adaptation that occurs in agricultural weeds and invasive plants is best understood by considering the importance of three fundamental plant population attributes: (i) *genetic variation*, (ii) *breeding system*, and (iii) *selective forces* imposed on the weeds by intensive (agricultural) practices. Each of these attributes occurs at the population level, but their effects are expressed at the community level as well. They are discussed in depth in Chapter 4.

GENETIC VARIATION. The response of plants to natural or human selection depends on the presence of heritable variation at the population level. Evolution implies the selection of phenotypes that are best adapted to the selecting agent. Because of selection, genotypic and phenotypic variations are reduced in successive populations, which may affect their fitness, that is, the ability for future generations to invade new agricultural or natural ecosystems. Whether a weed species is plastic with a high level of genotypic variation or has evolved toward more specialized phenotypes is an important question. The former implies that the species can invade and adapt to many different environments, while the latter suggests that invasion might be genotype and site specific with species maintaining several to many genotypes to avoid extinction under variable environmental conditions.

BREEDING SYSTEM. Plant evolutionists speculate about the role of the breeding system on the ability of plants to invade new areas and become weeds. A well-discussed adaptive advantage of weeds is the linking of uniparental reproduction (self-fertilization or agamospermy) with occasional genetic recombination (Baker 1965, 1974, Bartlett et al. 2002). According to Baker, many weed species utilize breeding systems adapted for inbreeding or vegetative reproduction to produce stable duplicates of successful genotypes. These strategies are usually coupled with occasional outcrossing for recombination to occupy new niches in new or changed environments. It is believed that cheatgrass (*Bromus tectorum*) follows such a strategy when invading sagebrush (*Artemisia*)–dominated grassland in the western United States (Figure 9.4). Cheatgrass populations increase in size dramatically following removal of native perennial grasses by overgrazing or fire (Young and Evans, 1976, D'Antonio and Vitousek 1992), and most of the vegetation in the sagebrush–cheatgrass community dies after being so seriously disturbed. The surviving cheatgrass seed give raise to plants at sites of high resource availability. There, the phenology of the species changes, which increases the probability for cross-pollination. Because each invading plant is essentially an inbred line, heterosis enters the weed population during the second year after

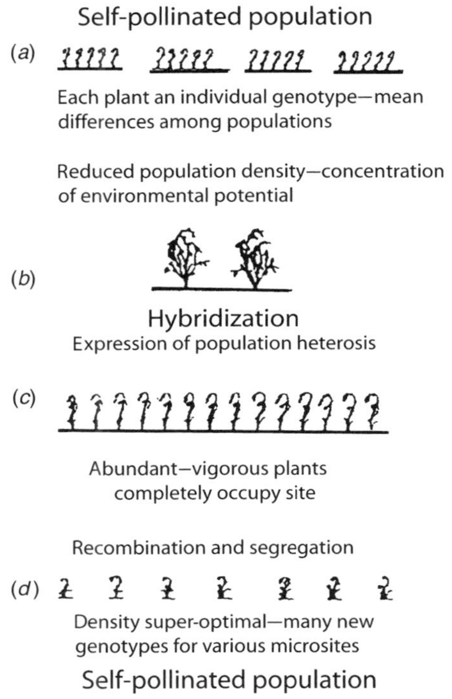

Self-pollinated population

(a) *ʔʔʔʔʔ*   *ʔʔʔʔʔ*   *ʔʔʔʔʔ*   *ʔʔʔʔʔ*

Each plant an individual genotype—mean
differences among populations

Reduced population density—concentration
of environmental potential

(b)

Hybridization
Expression of population heterosis

(c) *ʔʔʔʔʔʔʔʔʔʔʔʔʔʔ*

Abundant—vigorous plants
completely occupy site

Recombination and segregation

(d) *ʔ   ʔ   ʔ   ʔ   ʔ   ʔ   ʔ*

Density super-optimal—many new
genotypes for various microsites

Self-pollinated population

Figure 9.4 Model for hybridization in largely self-pollinated populations of *Bromus tectorum*. Environmental concentration can be caused by fallow operations for weed control or by wildfires. (From Young and Evans 1976, *Weed Sci.* 24:186–190. Copyright 1976. Weed Science Society of America. Reprinted by permission of Alliance Communications Group, a division of Allen Press Inc.)

disturbance. Following this hybrid generation, recombination occurs and wide expression of genetic variation allows the species to occupy a great number of sites throughout the disturbed habitat. Successful genotypes resume self-pollination, duplicating and increasing their frequency (Figure 9.4). Although there is no general rule (Chapter 4), examples in the literature suggest that greater numbers of weed species possess these or similar breeding system attributes than can be found in the general flora of a region.

AGRICULTURAL PRACTICES AS SELECTIVE FORCES. There are many examples that point out the effects of cropping systems on associated weed communities (Radosevich and Holt 1984, Radosevich et al. 1997, Martinez-Ghersa et al. 2000b, Mohler 2001). Some general observations are presented here to account for this repeated pattern.

The collapse of a native plant community following the introduction of agriculture is usually followed by a floristically simpler crop–weed community. Therefore, agricultural intensification into new areas tends to produce sudden decreases in the number of species as compared to the original grassland, forest, or woodland communities. In addition, vestiges of native species typically remain and some recruitment continues to occur in the disturbed area. Both native and exotic species coevolve with crops according to the extent of disturbance in the production system and the degree of variation in the species present. Farming once involved rotating areas of high human disturbance with areas of low disturbance, such as in fallow or grazed land (Radosevich et al. 1997), resulting in time periods over which succession or evolution could occur. However, the rate of change in farming practices within a given geographic area may exceed the rate at which weed species can adapt genetically to new habitats (Rejmánek 1989) (but see the discussion on herbicide-resistant weeds in Chapter 8). Thus, coevolution of crops and associated weeds or native species may be limited to periods when no dramatic technological changes occur in a production system or region.

Present agricultural technology has high transformative power in that it alters deficiencies in soil, water, and nutrients and disturbs enormous areas of land by removing natural or spontaneous vegetation. However, once a production system is adopted, no other significant land use changes usually occur for extended periods of time. The resulting large, nearly homogeneous habitats become suitable for an increasing number of exotic invading and reoccurring native species, which in turn are probably regulated by short- and long-term modifications in technology and the environment. When an environment is stabilized by intensive agricultural practices, selection pressures change the genetic structure of weed/invasive plant populations. Selection tends to favor traits that result in vacant niches in the production system being filled. Only then can coevolution operate as a structure- and function-determining force in the intensively managed environment.

Changes in seed dormancy, morphology, phenology, herbicide resistance, and so on, are all well-documented evolutionary consequences of intensive management practices observed in weed/invasive plant populations (Figure 9.5). Weeds result when selective forces are imposed on plants living in habitats where humans manipulate a significant proportion of the environmental variation. Weed problems usually begin when human control over these species is lost and they spread into other managed or natural areas (Chapters 1, 5, and 7). This view brings the coevolution of weeds and crops into a new framework that considers past changes in the weed flora and allows forecasting of future weed communities due to changes in crop production technology. Such a framework was used by Dekker (2003) in studies of the foxtail (*Setaria*) species group (Chapter 4). Using his study of *Setaria* spp., Dekker concluded the following about how to manage weeds that exist now under the selective forces of intensive agriculture:

1. The history of *Setaria* indicates that invasion and local adaptation in the North American Midwest is a continuous, ongoing process. Within half a century giant foxtail invaded American maize fields and became widespread, showing that *Setaria* spp. are highly adapted to changes in management practices.

2. Selection, adaptation, and consequential evolution of *Setaria* spp. will likely continue because weed control is the fundamental selective force shaping and determining the conditions of existence for this weed group.

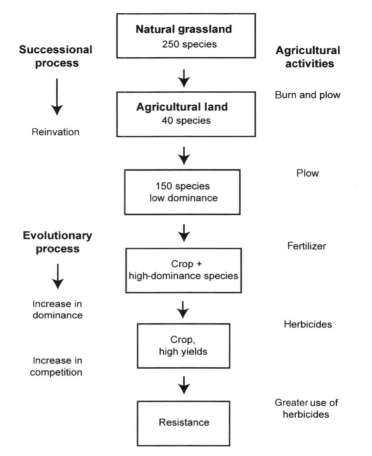

Figure 9.5 Successional and evolutionary changes through agricultural history. (From Martínez-Ghersa et al. 2000b. *Plant. Sp. Biol.* 15:127–137. Copyright 2000. Blackwell Publishing Ltd, reproduced with permission.)

3. The last and perhaps most important finding of this study is related to the traits and adaptations that surviving *Setaria* plants acquire during selection. Dekker believes that *Setaria* spp. will continue to be problems that farmers confront. Therefore, weed management is the management of weed selection pressure (Holt 2002).

NOVEL ECOSYSTEMS

Novel ecosystems result when species occur in combinations and relative abundances that have not occurred previously in a given biome (Hobbs et al. 2006). These novel ecosystems are composed of new combinations of plants and animals and arise through the deliberate or inadvertent introduction of species from other regions. These new ecosystems have the potential for changing past ecosystem functions and the services and uses provided to humans. Hobbs et al. (2006) provide several examples of newly formed ecosystems, which include exotic pine invasions into fynbos shrublands of South Africa, rain-shadow tussock grasslands

TABLE 9.2   Examples of Novel Ecosystems

Ecosystem Type	Description	Reference
Puerto Rico's "new" forests	Regenerating forests on degraded lands, composed largely of nonnative species and exhibiting multiple successional pathways	Aide et al. 1996, Zimmerman et al. 2000, Lugo 2004
Brazil's tropical savannas (the Cerrado)	Savannas transformed extensively by increased fire and introduction of grass species such as *Melinis minutiflora*	Hoffmann and Jackson 2000, Klink and Moreira 2002
Mediterranean pine woodlands	Woodlands with altered dynamics due to changing climatic conditions coupled with altitudinal range shifts in herbivores	Peñuelas et al. 2002, Hodar et al. 2003
Rivers in the western United States	Rivers altered by regulation, altered flows and invasive species	Ward and Stanford 1979, Scott and Lesch 1996, Postel et al. 1998, Kowalewski et al. 2000
Tropical agroforestry systems	Diverse combinations of native and nonnative perennial plants used locally to derive ecosystem goods and services	Ewel et al. 1991, Ewel 1999
Kelp forests	Removal of keystone species (sea otter) results in shift to novel ecosystem state	Simenstad et al. 1987, Estes and Duggins 1995
Near-shore ocean floors invaded by *Caulerpa*	Invasion by the alga *Caulerpa* in the Mediterranean and elsewhere leads to a novel ecosystem and monospecific dominance	Davis et al. 1997, Meinesz 1999
San Francisco Bay	An estuary now dominated almost entirely by nonnative species, with entirely novel species combinations	Carlton 1989, Cohen and Carlton 1998

Note: The breadth of ecosystem types involved and the range of causal factors leading to the novel system are indicated with relevant literature sources. The list is intended not to be comprehensive but merely to indicate the pervasiveness of novel ecosystems.
Source: Hobbs et al. (2006). All references are cited in original source. *Global Ecol. Biogeogr.* 15:1–17. Copyright 2006. Blackwell Publishing Ltd, reproduced with permission.

of New Zealand, secondary salinization in southern Australia that leads to replacement of native vegetation with impoverished exotic vegetation, and others described in Table 9.2. These authors indicate that novel ecosystems arise from the degradation and plant invasion of "wild" or natural/seminatural systems or from the abandonment of intensively managed production systems (also see Chapter 3 for more discussion). Perhaps the evolution of most crop–weed production systems (Chapters 3 and 5) also represent early examples of novel ecosystem development.

Hobbs et al. (2006) believe that either biotic or abiotic thresholds are crossed when such novel ecosystems form, making the restoration or reversal of them difficult, if not impossible. They suggest that this final point argues for the following:

- Conserving less impacted places now so they do not change into some new and possibly less desirable form
- Not wasting precious resources (time, energy, money, etc.) on what may be a hopeless quest to "fix" those systems for which there is little chance of recovery
- Exploring whatever useful aspects these new or novel ecosystems might have, since some will probably have more useful kinds of functions than others

It is clear that humans all over the planet are assisting in the development of new ecosystems. Such ecosystems are not emerging de novo. Instead they are emerging from within preexisting systems that are naturally dynamic over both long and short time scales. It is imperative that humans learn how to manage these ecosystems and utilize them for benefit to society (Hobbs et al. 2006).

NOVEL WEED/INVASIVE PLANT MANAGEMENT SYSTEMS

The interrelationships among biotic and abiotic components of ecosystems (Chapter 3) are hard to comprehend and even more difficult to translate into effective management strategies (Figure 7.10). At this point in time, there are few, if any, weed management systems that fully integrate production with ecological principles. However, some new management systems are being tried in agriculture, managed forests, and rangeland that rely on better understanding and use of ecological principles. It is apparent that each of these developing weed/invasive plant management systems must adapt to the local realities of environment, biology, economics, and production practices. It also is quite likely that soil disturbance and herbicides will remain as important tools of most management and restoration efforts for perhaps the next decade or more. The reliance on these tools should diminish as better understanding of biology and environment is incorporated into novel weed management systems.

Agriculture

Much of the science and management in agriculture over the last half of the twentieth century has been to improve crop productivity and efficiency. Liebman et al. (2001) point out that much of this effort on arable lands has been directed at weed control, if not outright weed eradication, with herbicides. They indicate that herbicide use is now the conventional norm for growing food throughout the midwestern United States (Chapters 1 and 8). Direct weed control with herbicides has reduced weed competition, increased crop production, and improved farm labor efficiency, but it has also caused substantial costs in environmental pollution, threats to human health, and a growing dependence of farmers on purchased inputs and artificial subsidies (Liebman 2001). Liebman et al. (2001) introduce the concept of a novel weed management system in agriculture that is more reliant on ecological processes and less reliant on herbicides, which they call *ecological weed management*. The ecological processes they discuss in their text with applications for weed control are competition, allelopathy, herbivory, and disease, insect, weed seed and seedling responses to soil disturbance, and plant succession.

Liebman et al. (2001) believe that ecological weed management, like conventional herbicide use, will not eliminate weeds from agricultural fields or totally eliminate all herbicide uses. However, they suggest that it has the potential to effectively reduce weed density, limit weed competitive ability, and prevent undesirable shifts in agricultural weed community composition. For example, crop rotations of corn and soybean with oat/forage legumes (Figure 9.6) create greater stress and mortality to weed species, thus reducing density and weed seed in the soil (Covarelli and Tei 1988, Blackshaw 1994, Kelner et al. 1996, Liebman and Ohno 1998). The concept relies on the following:

1. Acquisition and use of biological (ecological) information
2. Multiple tactical options for weed suppression
3. Improved farmer decision making
4. Careful implementation of general ecological principles to site-specific conditions

Farmers using the ecological weed management approach clearly assume a larger role and responsibility for assuring weed management success and crop productivity than when simply using herbicide-driven weed management systems. Liebman (2001) also notes that ecological weed management is still in its infancy with many important research questions remaining to be answered. He indicates that significant changes in educational modes and government policies are necessary for ecological weed management to be implemented on a broad scale in agriculture.

Managed Forests and Forest Plantations

As in conventional farming, many foresters believe that it is untenable to regenerate forests without herbicides, because dramatic changes in tree and shrub

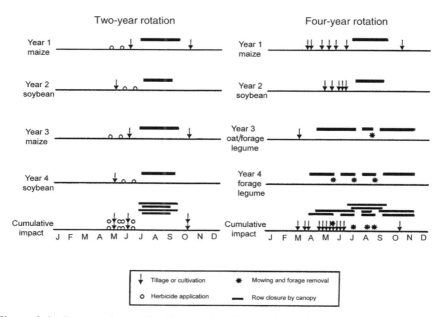

Figure 9.6 Stress and mortality factors affecting weeds in simple two-year rotation and more diverse four-year rotation. Both rotations are suitable for the midwestern United States. The two-year rotation uses conventional weed management practices, whereas the four-year rotation uses practices commonly found on low-external-input farms. The forage legume (e.g., alfalfa or red clover) used in the four-year rotation is planted with oat, which is harvested for grain in July. As indicated by symbols shown on lines labeled "cumulative impact," the timing of stress and mortality factors is more varied and the duration of canopy cover is greater in the more diverse rotation. (Adapted from Kelner et al. 1996 and Derksen 1997, in Liebman et al. 2001, *Ecological Management of Agricultural Weeds.* Copyright 2001. Reprinted with permission of Cambridge University Press, NY.)

composition often result from harvest practices such as clearcutting. However, maintaining forests in a mixed-species, uneven-aged condition is one of the best protections against weed and invasive plant infestations (Kimmins 1997, Perry 1997), and harvesting practices that change species dominance usually set forest structure and composition back to an early stage of succession (Figures 2.4 and 2.6). Herbicides are then used to reduce the rotation time between clearcuts by decreasing the densities or growth rates of undesirable tree, shrub, or herbaceous species and improving the survival and growth rates of artificially or naturally regenerated marketable trees (Stewart et al. 1984, Walstad and Kuch 1987). Thus, a clearcut-herbicide treadmill can be created in which the harvesting practice creates a weed problem that dictates continued routine herbicide use across an ownership. A key to avoiding and eliminating such overuse of herbicide is to employ a management plan that does not include harvesting approaches that create a routine vegetation management problem. Dregson and Stevens (1997) list

29 biological/environmental, managerial, or philosophical guidelines for ecologically responsible forest restoration.

Kimmins (1997) introduces the concept of *ecological rotation* (Figure 3.4) for harvest and subsequent regeneration of forest trees; this is the period of time required for a site to return to the predisturbance ecological condition or some other desired seral stage. Kimmins points out that whenever a forest or forest plantation is harvested, the site reverts to earlier stages of the successional sequence. This can be economically desirable and even ecologically appropriate if the disturbance is not excessive, that is, damaging to future forest productivity. Figure 9.7 compares the successional sequences of a single high degree of logging disturbance with a moderate level of disturbance by logging, the latter being repeated at a frequency greater than the ecological rotation. In terms of the renewability of the resources (trees), the single severe disturbance in Figure 9.7 immediately impairs the system, making the resource less renewable. The moderate disturbance with a short rotation initially sustains the resources by changing species composition, but as successively earlier seral stages are created, this treatment creates the same effect as the severe disturbance (Kimmins 1997). Low disturbance levels by periodic thinning or selective harvesting can maintain forest site dominance with desirable tree species and reduce the competitive influence of understory plants, requiring little, if any, herbicide use.

Harrington (1992) and associates also describe how forest vegetation management can be accomplished without herbicides. Topics addressed in the workshop proceedings include microenvironmental changes, fire, mechanical disturbance, animal grazing, mulches, and hand cutting or pulling.

When Limited Herbicide Use Is Acceptable. Unfortunately, forest landowners are sometimes faced with areas of degraded land that resulted from the

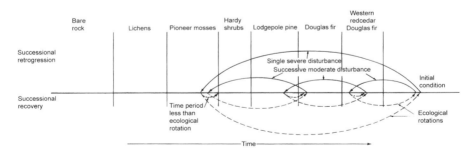

Figure 9.7 Concept of ecological rotation expressed in terms of successional retrogression and recovery. The figure depicts the successional consequences of either a single excessive disturbance or a series of successive, moderate harvesting disturbances in conjunction with rotations equal to or less than the ecological rotation. Successional retrogression and post-disturbance recovery are shown in hypothetical xerarch succession at low elevation in southern coastal British Columbia. The same events are expressed graphically in Figure 3.4. (Modified from Kimmins 1997, *Forest Ecology: A Foundation for Sustainable Management.* 2nd Ed. Prentice Hall, NJ.)

management practices of past owners or a major catastrophic event like a wildfire. Changes in soil structure and fertility as well as species composition usually occur from such events. In these cases and if an abiotic threshold has not been crossed (Kimmins 1997, Hobbs et al. 2006), methods that do not use herbicides may be ineffective for regeneration of forests or restoring productivity and ecological balance among species in a timely manner. Forest managers might need to temporarily use herbicides for such restorative purposes. If herbicides are found to be necessary, specific sites for application should be identified and the application should be restricted to specific plants or stands. The criteria for such site selection would be a quantifiable risk assessment (as discussed in Chapter 7) of impacts to desirable tree species and potential impacts to nontarget plants, animals, and their habitats. Adequate buffers to minimize herbicide drift, runoff, or leaching to riparian areas, waterways, or areas of human habitation also should be made (Chapter 8). Indigenous and recreational food sources also exist in forests. Herbicide applications should not be made to such food sources, and if they are, established residue tolerances should not be exceeded.

Rangeland

Exotic species dominate and continue to spread throughout millions of hectares of rangeland in the western United States (Sheley and Petroff 1999). These grasslands were typically dominated by perennial bunchgrass species and shrubs, primarily sagebrush, but cultivation, grazing, and altered fire regimes have resulted in vast areas now dominated by exotic grass and forb species (Parks et al. 2005a). Many of the native bunchgrass communities are so altered that they may have crossed an ecological threshold into a different stable vegetation state (Bunting et al. 2002, Johnson and Swanson 2005, Hobbs et al. 2006). A concept of successional management has been proposed as a framework for developing ecologically based vegetation management strategies on rangeland (Sheley et al. 1996, 2005, Krueger-Mangold et al. 2006).

Pickett et al. (1987) provided a successional hierarchy to which ecological processes, modifying factors, and causes of succession were added by Krueger-Mangold et al. (2006; Table 9.3). Sheley et al. (1996) and Krueger-Mangold et al. (2006) argue that rangeland plant communities dominated by exotic plants could be restored by strategically addressing the three causes of succession in Table 9.3. This hierarchy also provides a basis for weed ecologists and land manager to develop, test, and implement integrated management strategies for rangeland restoration.

Successional management has been tested for invasive plants over the past several years (Krueger-Mangold et al. 2006), and restoring rangeland systems was found to be most effective when the causes of succession were addressed sequentially (R. Sheley, J. S. Jacobs, and T. J. Svejcar, unpublished). The potential efficacy of biological control was evaluated by examining how natural enemies affect dispersal, stress, and interference of target and nontarget plants (Sheley and Rinella 2001). In addition, augmentative restoration of rangelands inhabited by

TABLE 9.3 Causes of Succession, Contributing Processes and Components, and Modifying Factors in Expanded Successional Management Framework

Causes of Succession	Processes and Components	Modifying Factors
Site availability	Disturbance: *tolerance (Connell and Slatyer 1977), fluctuating resource availability (Davis et al. 2000)*	Size, severity, time intervals, patchiness, predisturbance history, **shallow tillage, grazing with multiple types of livestock**
Species availability	Dispersal: *inhibition (Connell and Slatyer 1977), initial floristic composition (Egler 1954)*	Dispersal mechanisms and landscape features, **dispersal vectors, seedbed preparation, seeding in phases**
	Propagule pool: *inhibition, initial floristic composition*	Land use, disturbance interval, species life history, **assessment of propagule pool, seed coating**
Species performance	Resource supply: *facilitation (Connell and Slayer 1977), resource ratio hypothesis (Tilman 1977, 1982, 1984, 1988)*	Soil, topography, climate, site history, microbes, litter retention, **soil resource assessment, soil impoverishment, R^***
	Ecophysiology: *vital attributes (Noble and Slatyer 1980)*	Germination requirements, assimilation rates, growth rates, genetic differentiation, **comparison between native and introduced environments, seed priming**
	Life history: *tolerance K- and r-strategists (MacArthur 1962)*	Allocation, reproduction timing and degree, **sensitivity analysis**
	Stress: *tolerance, C-S-R (Grime 1979), community assembly theory (Diamond 1975)*	Climate, site history, prior occupants, herbivory, natural enemies, **identifying abiotic and biotic filters, seeding species-rich mixtures**
	Interference: *inhibition*	Competition, herbivory, allelopathy, resource availability, predators, other level interactions, **cover crops, assisted succession**

Note: Successional models and relevant citations are listed in italics under processes. Boldfaced modifying factors are additional modifying factors proposed in Krueger-Mangold et al. 2006. *All references are cited in original source.*

Source: Krueger-Mangold et al. (2006). *Weed Sci.* 54:597–605. Copyright 2006. Weed Science Society of America. Reprinted by permission of Alliance Communications Group, a division of Allen Press Inc.

374 SYSTEMS APPROACHES FOR WEED AND INVASIVE PLANT MANAGEMENT

invasive plants was implemented by assessing the state of each ecological process (Table 9.3) associated with succession on an experimental site. If a process was damaged, it was amended to allow restoration of native grasses and forbs (Bard et al. 2004). Although this management framework is still in development and modification, it offers land managers practical methods for modifying ecological processes that direct plant composition away from invasive plants toward desired plant assemblages.

VALUE SYSTEMS IN AGRICULTURAL AND NATURAL ECOSYSTEM MANAGEMENT

Value systems are the way people think about activities (what we do and how we do it) and technologies (what we do it with). They are actually the underlying principles through which we judge our own and other people's actions. Although weed control and invasive plant management are arguably only small parts of land management decisions, it is important to examine where they are placed within this larger context. Ferre (1988) and List (2000) believe that there are four fundamental value systems that influence decisions about the management of agricultural and natural resources (Castle 1990):

- *Material Well-being.* Any activity or technology should be of benefit or utility to society. Implicit in this value system is the belief that society will be better off with than without the new tool, tactic, approach, or activity.
- *Sanctity of Nature.* An activity should not proceed if risk or damage to the environment is likely to result (environmentalism). Nonintervention is a key ingredient of this value system, and although some human involvement is recognized and accepted, it should be minimally disruptive to avoid adverse consequences.
- *Individual Rights.* Individual liberty and property entitlement are the primary concerns of this value system. The marketplace is often posed as the best way for society to accept or reject activities, through demand for the products produced. Government interventions are often seen as a problem, not a solution.
- *Justice as Fairness.* All people have equal use of the earth's resources and benefits from them should be distributed equitably. The issue is not bounded by time or geography. Thus, impacts of activities and technologies on other parts of the world or future generations are of concern.

Each of these value systems raises different questions about our efforts to manage land, raise crops, or manipulate vegetation. However, the value system most accepted by occidental cultures seems to be the first one in the above list, material well-being or utilitarianism. Explicit in that value system is that people will be better off with than without a particular tool or activity. Nonetheless, people usually do not view such benefits identically, so an approach has been devised to assess the relative benefits and costs in such situations. This procedure is called *cost–benefit* or, more recently, *risk–benefit* analysis (see Chapter 2 for more discussion). Risk–benefit analysis is used during both registration and cancellation proceedings for herbicides (see below), in which the potential benefits of a product for weed control are weighed against its potential risks to human health and harm to the environment.

Role of Human Institutions in Weed Management

Value systems play an important role in maintaining existing or developing new ways of thinking about agriculture, other natural resource production systems, and natural ecosystems, because innovations in technology originate with a set of ethical values that tend to justify existing tools or the need for new approaches (Ferre 1988). Thus, a technology can easily become rooted in the supposition that its underlying values are either universal or take priority over other values in society. A common assumption, for example, in weed control is that pest (weed) control increases food abundance (material well-being) with negligible harmful effects to either the environment (sanctity of nature) or people (justice as fairness). The assumption also fails to ask who benefits (individual rights and justice as fairness), except in a limited way. Studies have shown, however, that many activities meant to control pests have caused at least as many problems through harmed wildlife, contaminated soil, watershed erosion, and pest resistance (Briggs 1992, D'Mello 2003) as have been solved by limiting losses due to pests (Pimentel 1986, Pimentel et al. 1992).

Natural ecosystems may represent a special case because if they are managed at all they are not managed for a marketable product. Unlike agriculture and some other natural resource production systems, the value of natural ecosystems to society lies in the ecosystem services they provide, which include maintenance of biodiversity, wildlife habitats, pure water and air, and recreational opportunities and aesthetic benefits (discussed in Chapter 1). Thus, human values pertaining to natural ecosystems can be quite different in both quantity and quality from those occurring in production systems. It is not surprising, therefore, that disagreements arise about the appropriateness of applying the tools and tactics for weed control developed in production systems to control of invasive plants in natural systems. Below are three examples of how social values influence weed/invasive plant management. Each of these examples has been discussed more thoroughly in Radosevich et al. (1997).

The 2,4,5-T Controversy. In the mid-1960s the first significant social debate over the use and public safety of a tool used exclusively for vegetation management began. The controversy centered on 2,4,5-T (2,4,5-trichlorophenoxy acetic acid), then the predominant herbicide used for shrub and tree suppression in young forest tree plantations, and was fueled, perhaps, by the fact that 2,4,5-T was a major ingredient in Agent Orange—a defoliant used by the U.S. Army during the Vietnam war.

This debate and subsequent litigation lasted for two decades until early 1985, when the U.S. Environmental Protection Agency (EPA) canceled the chemical's registration and, therefore, use of the herbicide in the United States. This effectively ended one of the most controversial and turbulent times for forestry in the United States and Canada (McGee and Levy 1988), except perhaps the social debate about clearcut logging that is still underway in those and other countries.

The 2,4,5-T dispute also marked a turning point in the history of weed science because never before had the discipline's assumptions about the societal good and safety of its tools met with such intense social activism, litigation, and even violence as occurred over 2,4,5-T (Walstad and Dost 1986). There is now little doubt that science, technology, and social values played important roles in the activities and decisions surrounding the use and eventual removal of 2,4,5-T. Nor is there much doubt that this episode in the history of weed science left its mark on the discipline, especially on those involved directly in the controversy. It was also one of the "battles" that led the way toward much of the public concern, mistrust, and opposition to pesticides. Finally, as pointed out by McGee and Levy (1988), this issue, although often stated in scientific terms, was trans-scientific. In other words, issues of this type are simply beyond the ability of science to answer and must be addressed in the policy and political arenas as part of the ever-ongoing discussion about what constitutes "the public good."

Atrazine and Water Quality. In 1993, the EPA completed an economic benefits–environmental risks assessment to determine the consequences of a national ban on atrazine and the entire group of triazine herbicides. Atrazine and the other triazine herbicides were chosen for this analysis because they were then the most widely used herbicides in the corn-growing region of the United States, and water quality and public safety benchmarks were being exceeded by weed control practices in the region.

The EPA used an ecological–economic modeling approach, called megamodeling, to accomplish its analysis. The tool developed consisted of four interactive components (evaluation criteria, agricultural decision, fate and transport, and policy specification) and eight interfaced models. The response functions of each model summarized the relationships among chemical concentrations within an environment and a set of variables such as weather, soil conditions, tillage, and chemical properties that define the transport and fate of the chemical (EPA 1993). The final analysis summarized economic benefits and environmental risk indicators to compare policy scenarios (EPA 1993).

In deciding whether or not the risks of atrazine use outweighed its social benefits, the EPA needed to determine the magnitude of the benefits and whether or not curtailment would lead to a net environmental gain or loss under the different policy options (EPA 1993). There is little doubt that substantial benefit arose from the use of atrazine and triazines and that hardship would occur if they were banned. For example, corn acreage and corn yields were both predicted to decline by about 3% owing to a shift in weed control practices to other herbicides and increased tillage. The costs of weed control in corn were also expected to increase by $6 to $8 per acre. Even though soybean acreage and soybean yields were expected to increase, the national economic welfare was expected to decline by $365 million under an atrazine ban and $526 million under a triazine ban. Under both bans, crop producers in the midwestern United States would bear the largest share of the economic burden.

Agency scientists concluded that a high level of surface loading of soils with other triazine herbicides would increase under an atrazine-only ban, resulting in chemical concentrations exceeding water quality benchmarks. This problem would be reduced under a total triazine ban. They also noted that for groundwater all concentrations would be below the long-term benchmarks for both projections. Exposure of aquatic vegetation to herbicides was always high, often having 20-fold greater values than EPA standards. These values were not projected to change because they were most indicative of corn and soybean culture in the midwestern United States rather than specific herbicide uses. Soil erosion was projected to increase under the bans, due to greater use of conventional tillage instead of reduced-till or no-till systems of corn production.

In 1995, the EPA called for a two-year special public review of atrazine and the triazine herbicides. Over that time period, testimonies were given and further analyses performed. The results of the analysis and review resulted in no action by the EPA, nor did the decision result in litigation by those who perceived the process to be unlawful or unfair.

Herbicide-Resistant Crops. A decade and a half ago, advances in molecular genetics, biochemistry, and cell physiology opened the possibility of creating crops resistant to herbicide (HRCs). For example, herbicide-resistant cultivars of glyphosate or glufosinate are now available for nearly every major crop in the United States. Although the potential benefits of HRCs through improved production have been described (Chapter 8), there is also concern about their impact on agroecosystem production and stability, rural and urban economies and cultures, and human health (Dekker and Comstock 1992, Martinez-Ghersa et al. 2003). Martinez-Ghersa et al. (2003) summarize recent findings about HRC use below:

- Increased crop yields rarely result from HRC use (Altieri and Rosset 1999, Benbrook 1999, Moser et al. 2000).
- Weed control efficiency often stays the same or decreases (Owen 1997, Medlin and Shaw 2000).
- Declines in farmer income sometimes occur (Owen 1997).
- The use of weed control alternatives is unlikely (Owen 1997, Pringnitz 2001, U.S. Economic Research Service 2002).
- There are concerns about loss of marketability and human safety of HRCs (Dekker and Comstock 1992, Gasson 2003, Pusztai et al. 2003).
- The likelihood of resistant weeds evolving increases with greater HRC use (Owen 1997, Powles et al. 1998, Dill et al. 2000, Hall et al. 2000, Lee and Ngim 2000, Van Gressel 2001, Pringnitz 2001).

- Transfer of herbicide-resistant genes into weeds and other plants could occur (Raybould and Gray 1993, Jørgensen et al. 1996, Mikkelsen et al. 1996).
- Further uniformity and reduction in species diversity in agriculture might occur (Altieri 2000, Dale et al. 2002, Martinez-Ghersa et al. 2003).
- Genetically modified foods might produce allergens (Kaeppler 2000).
- New and restrictive cultivars of HRCs might result in genetic erosion (Tripp 1996).
- There are possible detrimental effects on nontarget organisms from repeated herbicide applications (Losey et al. 1999).

There is also the question of who benefits from HRC technology—herbicide and seed production companies, pesticide distributors, or farmers (Dekker and Comstock 1992, Radosevich et al. 1992, Martinez-Ghersa et al. 2003). In addition, the safety of transgenically modified foods is currently under debate in many countries (McHughen 2000, Gasson 2003, Pusztai et al. 2003), which brings the precautionary principle to bear on the marketing of these products.

Martinez-Ghersa et al. (2003) indicate that the concerns raised over a decade ago about HRC use are still relevant today. Even though adoption of HRCs has risen dramatically since their commercial introduction, there is still no evidence of overall reductions in production costs or enhanced yields, although it may be easier for farmers to use HRCs than not. Furthermore, current knowledge about the potential biological risks from adoption of HRCs is still insufficient to address most concerns, and the widespread use of HRCs has supplanted detailed field testing to gain fuller understanding of biological impacts before release. Aside from some potential marketing advantages to companies with specific herbicides, both potential benefits and potential risks from HRCs are uncertain at best. Until such uncertainties are eased, the above discussion raises legitimate biological and social concerns about the consequences of widespread use of HRCs. The precautionary principle (Ferre 1988, Strauss et al. 2000; Chapter 2) is one way to guide such discussion. It forces examination before harm can be done (Strauss et al. 2000, Martinez-Ghersa et al. 2003) and in this way protects good science from becoming bad technology.

Consequences of Human Values on Weed and Invasive Plant Management

Weed scientists, like many other scientists and land managers in applied natural resource disciplines, often engage in social debates with other members of society about values and perceptions of how to produce food, fiber, or shelter, conserve biodiversity, or maintain ecosystem services. The focus of these debates is often the tools and tactics (means) used to grow crops, produce wood products, or manage grazing lands. Although the debate is well formed in all areas of natural resource management, it is especially well developed within the agricultural community (Figure 2.3), perhaps because of agriculture's almost exclusive emphasis, until recently, on marketable yield and production efficiency as societal goals (ends). Periodically, it is important to reexamine such goals to determine if they have changed, if current technologies continue to meet societal expectations, or if unforeseen consequences now impact their use. In the following pages, the role of weed management is examined within the context of evolving agricultural, forestry, and rangeland production systems.

Simplification, Deterioration, and Loss of Biological Regulation in Agriculture

Agricultural systems in some parts of the world are well developed and backed up with substantial modern technological paraphernalia, while elsewhere modern agricultural systems are just beginning. In its early stages, agriculture is tightly coupled with the productivity of ecosystems and humans compete for food with other guilds of organisms (insects, mammals, etc.). This competition for subsistence creates negative feedbacks (see the discussion of cybernetics in Chapter 2) that limit human social and agricultural development. In these systems, the acquisition of food is the primary focus of activity and the amount of cultivated area is determined by the level of energy that can be transformed into human and animal labor to grow and harvest crops (Merchant 1980). In modern agriculture, farming is a more open system in which materials and energy (e.g., fossil fuels) pass from an external supplier to an external buyer and money and energy put into the system often exceed money and energy coming out (Lewontin 1982). In addition to the development of capitalism, the technological and social changes resulting from industrialization substantially alter relationships between humans and other components of the agroecosystem. Feedbacks to the social system come from political, economic, and academic institutions rather than from daily farming (biotic) interactions with the environment. As a result, coevolution between the ecosystem and land management is less predominant (Ghersa et al. 1994c).

Modern agriculture is no longer limited strictly by characteristics of the biome. In fact, the use of information, expressed as power and technology, allows farms to flourish in deserts, former rainforests, and drained wetlands and even on the reclaimed ocean floor. Technology, moreover, has allowed farmers to correct for deficiencies in soil nutrients, disturb enormous areas of land, and modify topography to create a gigantic, nearly homogeneous habitat suitable for a relatively small, but productive, group of crop species (Chapter 1). This use of technology stabilizes agriculture over the short term and increases crop harvests per unit area (Hall et al. 1992, Way 1977), while also creating a simplified agroecosystem of early seral communities (Chapter 2). Thus, a farming system maintained under continuous disturbance will deteriorate unless it is subsidized with external inputs such as synthetic fertilizers, manure, or irrigation. Deterioration can result from accelerated erosion (Pimentel et al. 1976), pollution by excess fertilization and chemical residues (Tivy 1990), or reduction in the biotic activity of the soil (Woodmansee 1984). Ecosystem deterioration can also result from either losses or addition of resources, such as soil fertility, to a production system.

Weeds and Invasive Plants as Symptoms of Ecosystem Dysfunction

The continued deterioration of ecosystems is considered by Hobbs et al. (2006) to be responsible for the occurrence of novel ecosystems throughout the world. Hobbs et al. indicate that species additions resulting from increased human activities are the main symptom of such deterioration, a contention supported by MacDougall and Turkington (2005).

Weed Occurrence on Deteriorating Soil Base. In intensively farmed or managed areas, equilibrium is possible between nutrient gains and losses (Tivy 1990). Because of intensive soil disturbance and seasonal interruption of the system, however, these areas probably do not retain all of their natural feedback loops and, therefore, are more likely to lose nutrients and organic matter than gain them. Many management practices have, in fact, coevolved with such a deteriorating pattern of the soil resource base. For example, the deterioration of soil in the rolling pampas of Argentina (Radosevich et al. 1997) and presence of old fields in the sagebrush-dominated grasslands of North America (Endress et al. 2007 in press) suggest relationships between past human use patterns and the occurrence of weeds and invasive plants.

Surveys in the United States indicate that the use of herbicides has increased twice as much as any other agricultural input since 1950 (Aspelin and Grube 1999, NRC 1989, USDA 1999). Such sustained demand for herbicides could be due to farmers attempting to compensate for the deterioration of their soil resource (Gunsolus and Buhler 1999, Liebman and Mohler 2001). In many cases, crop yields are more sensitive to weed removal than to the addition of fertilizer (Appleby et al. 1976, Hall et al. 1992, Tollenaar et al. 1994, McKenzie 1996, Liebman and Mohler 2001). The elimination of weeds under such circumstances would probably appear more reasonable to agronomists than trying to replace unknown nutrient losses, especially if the limiting resource cannot be identified readily.

Other Examples of Ecosystem Deterioration

Departure from Natural Disturbance. Exotic plants can also invade an ecosystem when regulatory mechanisms (Figures 1.8 and 2.9) are modified or depart radically from the natural disturbance regime (D'Antonio et al. 1999). This kind of ecosystem deterioration was demonstrated in Chapters 1 and 3 by examining the influence of changes in fire and grazing regimes on ecosystem invasibility (D'Antonio and Vitousek 1992).

Reductions in Forest Canopy by Fire and Logging. Large reductions in forest canopy that increase the abundance of native and exotic shrubs and herbaceous plants can also result in ecosystem deterioration (Kimmins 1997). For example, exotic species in the Pacific coastal Douglas fir region of the United States are associated with high-light environments and disturbance (DeFerrari and Naiman

1994, Heckman 1999, Perendes and Jones 2000). Therefore, opportunities for exotic species establishment are created by disturbances that open the forest canopy, such as fire, forest thinning and harvest, and road building and maintenance. Exotic species that establish after fire or other canopy disturbance must originate from the soil seed bank, be transported to the site by logging or fire suppression equipment, or be dispersed from populations located along nearby roads, in riparian corridors, or in intentionally opened habitats. Gradually, intense competition from shading by residual native species and regenerating conifers can eliminate some plants, including exotic species, from understory plant communities (Halpern and Spies 1995, Oliver and Larson 1996), especially when recently harvested sites are densely planted with Douglas fir (Schoonmaker and McKee 1988) or other native conifers.

Environmental and Biotic Thresholds. Broad-scale occurrence of invasive plants may also suggest that an environmental or biological threshold from past or present human disturbances has been crossed (Perry 1997, Kimmins 1997, Alpert et al. 2003) from which return to a relatively native state will be difficult. In Chapter 3, the various factors that influence invasibility of habitats and plant communities by exotic plants were discussed. Alpert et al. (2003) believe that plant invasions occur when both environmental stresses and disturbances depart from natural levels (Figure 3.3). Briske et al. (2003) indicate that such threshold-level departures will most likely be difficult to overcome without considerable expenditure in manager time, energy, and money (Hobbs and Humphries 1995, Hobbs et al. 2006).

Socioeconomic Influences on Weed and Invasive Plant Management

As seen in Figure 2.2, human social systems are arranged according to function and scale. Levels of human and institutional activity in social systems are determined by differences in process rates (e.g., adoption of new technology, cultural invasion, education), just as in ecological systems. Modern weed management approaches, tools, and tactics in every production system designed to "fix" cropland infested with weeds or to restore natural ecosystems invaded by exotic plants are now just as likely to result from political or social factors as from ecological ones. The problem is twofold (Figure 9.8) because either (1) policy formation is too slow to respond adequately to eminent ecological threats or (2) the ecological system cannot respond or responds only temporarily to policies such as quarantines of exotic plants. Either scenario can result in apparent failure and thus abandonment of procedures to manage ecosystems. It is important for land managers to respond to positive feedbacks, as evidenced by increases in undesirable plants that are caused by social and political agendas, although individual land managers may have little to say about such directives.

Future Challenges for Scientists, Farmers, and Land Managers

History indicates that agricultural and other natural resource production systems will continue to try to provide for the human needs of food, fiber, and shelter

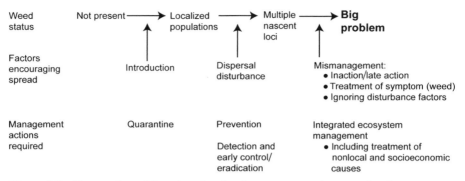

Figure 9.8 Phases of weed invasion showing factors encouraging spread and management actions required at each stage. (From Hobbs and Humphries 1995, *Conserv. Biol.* 9:761–770. Copyright 2000. Blackwell Publishing Ltd, reproduced with permission.)

through the expansion–intensification model of production. But as human population increases, people will also continue to search for ways to generate sustainable products, buffer resource consumption, and protect the environment and the services that ecosystems give them. These are the challenge of modern farmers and resource managers, because they will need to combine classical notions of production with the ecological concepts of biodiversity, stability, and functionality of ecosystems. This too is the challenge for weed/invasive plant management. While progress is being made, finding ways to manage weeds based on an understanding of how ecosystems work is still far from being achieved.

Experiments should refocus to examine the complexity of existing and novel ecosystems, addressing how production practices affect natural controls and their checks and balances. Involuntary long-term, large-scale "experiments" are now being carried out as the expansion–intensification model proceeds into most parts of the world. New areas are being converted into cropland and production forests, new species are being intentionally or accidentally introduced into natural ecosystems, and new technologies like no-tillage and pesticide-resistant, genetically modified organisms are being introduced into older agricultural areas without a real understanding of their consequences. We hope through this book that researchers, farmers, and land mangers who embrace the concepts of IPM and ecologically based management will concentrate their efforts on these "involuntary, landscape-scale experiments" using the principles of ecology and empirical study to derive productive, environmentally sound strategies of production and management.

SUMMARY

Weeds and invasive plants are components of most production systems, generally reducing productivity of desirable plants but also affecting biodiversity and perhaps the functioning of ecosystems. Weeds are also a consequence of how people perceive them. Agriculture, forestry, and range management go through phases of expansion and intensification as they coevolve with human needs for the land. Modern pest control in agriculture arose from the need to diminish biological hazards to crops, which generally decreased yields. Thus, both the concepts and practices of IPM, weed science, and later IWM evolved as the need to intensify agricultural practices increased. Early in its evolution, IPM was viewed as an intermediate step toward ecological agriculture as an approach to contain pests. Several future directions for development of IWM have been identified that include (1) population dynamics of weeds and invasive plants, (2) community-level interactions, particularly food webs, (3) spatial and temporal patterns of plant succession in agricultural and natural ecosystems, and (4) evolutionary patterns promoted by intensive management practices. Ecological principles for the design of weed management systems in agriculture, forestry, and range management have also been identified.

Novel ecosystems are new combinations of species that have not existed previously in a region or biome. Their occurrence is believed to be enhanced by the intentional or accidental activities of people. Novel weed management systems in agriculture, managed forests, and rangeland that attempt to substitute understanding of ecological principles for herbicides have been identified and need further development.

Value systems are the way people think about human activities and technologies. There are four fundamental value systems that affect agricultural and natural resource management (1) material well-being, (2) sanctity of nature, (3) individual rights, and (4) justice as fairness. Each of these value systems raises different questions about human efforts to raise crops and manage the land, even to control weeds. The controversy over 2,4,5-T use, influence of atrazine on water quality, and release of HRCs are three examples of how society interacts and impacts weed management. It is possible that many modern weed and invasive plant management practices have coevolved to maintain productivity on a deteriorating soil resource base or depart significantly from natural disturbance or stress regimes of natural ecosystems. In the future, land managers and weed scientist will be called upon to develop new methods of weed management that consider environmental and societal concerns as well as those of production and efficiency.

REFERENCES

Abbott, R. J., L. C. Smith, R. I. Milne, R. M. Crawford, K. Wolff, and J. Balfour. 2000. Molecular analysis of plant migration and refugia in the Arctic. *Science* 289: 1343–1346.

Abrahamson, W. G. 1980. Demography and vegetative reproduction. Pp. 89–106 in O. T. Solbrig (Ed.), *Demography and Evolution in Plant Populations.* University of California Press, Berkeley.

Ackakaya, H. R., M. A. Burgman, and L. R. Ginzburg. 1999. *Applied Population Ecology: Principles and Computer Exercises Using RAMAS Ecolab,* 2nd ed. Sinauer Associates, Sunderland, MA.

Adkins, S. 1995. *Weed Science Lecture Guide.* University of Queensland, Brisbane, Australia.

Akesson, N. B. and S. V. Walton, Jr. 1989. Aircraft. Pp. 155–162 in E. A. Kurtz (Ed.), *Principles of Weed Control in California,* 2nd ed. Thompson Publications, Fresno, CA.

Akobundu, I. O. 1987. *Weed Science in the Tropics: Principles and Practices.* Wiley, New York.

Aldo Leopold Wilderness Research Institute (ALWRI). 2003. *Research Needs for Managing Non-Native Species in Wilderness Areas.* ALWRI, Missoula, MT.

Aldrich, R. J. 1984. *Weed-Crop Ecology: Principles in Weed Management.* Breton, North Scituate, MA.

Alien Plant Working Group. 2002. *Weeds Gone Wild: Alien Plant Invaders of Natural Areas.* Plant Conservation Alliance. Available: http://www.nps.gov/plants/alien/bkgd.htm.

Ecology of Weeds and Invasive Plants. By Steven R. Radosevich, Jodie S. Holt, and Claudio M. Ghersa
Copyright © 2007 John Wiley & Sons, Inc.

Allard, R. W. 1996. Genetic basis of the evolution of adaptedness in plants. *Euphytica* 92:1–11.

Allen, E. B., W. W. Covington, and D. A. Falk. 1997. Developing the conceptual basis for restoration ecology. *Restor. Ecol.* 5:275–276.

Allen, M. F. 1991. *The Ecology of Mycorrhizae.* Cambridge University Press, New York.

Allen, M. F., W. Swenson, J. I. Querejeta, L. M. Egerton-Warburton, and K. K. Treseder. 2003. Ecology of mycorrhizae: A conceptual framework for complex interactions among plants and fungi. *Annu. Rev. Phytopathol.* 41:271–303.

Allessio Leck, M., V. T. Parker, and R. L. Simpson (Eds.). 1989. *Ecology of Soil Seed Banks.* Academic, San Diego, CA.

Alonso-Blanco, C., L. Bentsink, C. J. Hanhart, H. B. E. Vries, and M. Koornneef. 2003. Analysis of natural allelic variation at seed dormancy loci of *Arabidopsis thaliana. Genetics* 164:711–729.

Alpert, P., E. Bone, and C. Holzapfel. 2000. Invasiveness, invasibility and the role of environmental stress in the spread of non-native plants. *Perspect. Plant Ecol. Evol. Syst.* 3:52–66.

Alpert, P., C. Holzapfel, and C. Slominski. 2003. Differences in performance between genotypes with different degrees of resource sharing in *Fragaria chiloensis. J. Ecol.* 91:27–35.

Alstrom, S. 1990. *Fundamentals of Weed Management in Hot Climate Peasant Agriculture.* Crop Production Science, Uppsala, Sweden.

Altieri, M. A. 1999. The ecological role of biodiversity in agroecosystems. *Agric. Ecosyst. Environ.* 74:19–31.

Altieri, M. A. 2000. The ecological impacts of transgenic crops on agroecosystem health. *Ecosyst. Health* 6:13–23.

Altieri, M. A. and M. Liebman. 1986. Insect, weed and plant disease management in multiple cropping systems. Pp. 183–218 in C. A. Francis (Ed.), *Multiple Cropping Systems.* Macmillan, New York.

Altieri, M. A. and M. Liebman (Eds.). 1988. *Weed Management in Agroecosystems: Ecological Approaches.* CRC Press, Boca Raton, FL.

Altieri, M. A. and P. Rosset. 1999. Ten reasons why biotechnology will not ensure food security, protect the environment and reduce poverty in the developing world. Mindful Organization. Available: http://www.Mindful.org/GE/Ten-Reasons-Why-Not.

Altieri, M. A., A. van Schoonhoven, and J. A. Doll. 1977. The ecological role of weeds in insect pest management systems: A review illustrated by bean (*Phaseolus vulgaris*) cropping systems. *PANS (Pest Artic. News Summ.)* 23:195.

Altom, J. D. and J. F. Stritzke. 1973. Degradation of dicamba, picloram, and four phenoxy herbicides in soils. *Weed Sci.* 21:556–560.

Anderson, J. A., M. E. Sorrells, and S. D. Tanksley. 1993. RFLP analysis of genomic regions associated with resistance to pre-harvest sprouting in wheat by RFLPs. *Crop Sci.* 33:453–459.

Anderson, R. N. 1968. *Germination and Establishment of Weeds for Experimental Purposes.* Humphrey, Geneva, NY.

Anderson, W. P. 1996. *Weed Science: Principles and Applications,* 3rd ed. West, St. Paul, MN.

Andow, D. A. 1983. Effects of agricultural diversity on insect populations. Chapter 5 in W. Lockeretz (Ed.), *Environmentally Sound Agriculture*. Praeger, New York.

Andow, D. A. 1988. Management of weeds for insect manipulation in agroecosystems. Pp. 265–301 in M. A. Altieri and M. Liebman (Eds.), *Weed Management in Agroecosystems: Ecological Approaches*. CRC Press, Boca Raton, FL.

Andow, D. A. and C. Zwahlen. 2006. Assessing environmental risks of transgenic plants. *Ecol. Lett.* 9:196–214.

Appleby, A. P., P. D. Olson, and D. R. Colbert. 1976. Winter wheat yield reduction from interference by Italian ryegrass. *Agron. J.* 68:463–466.

Ashton, F. M. and W. A. Harvey. 1971. *Selective Chemical Weed Control*. Circular 558. University of California, Berkeley.

Ashton F. M. and T. J. Monaco. 1991. *Weed Science: Principles and Practices*. 3rd ed. John Wiley and Sons, New York, NY.

Ashton, P. A. and R. J. Abbott. 1992. Multiple origins and genetic diversity in the newly arisen allopolyploid species, *Senecio cambrensis* Rosser (Compositae). *Heredity* 68:25–32.

Aspelin, A. L. and A. H. Grube. 1999. *Pesticide Industry Sales and Usage: 1996 and 1990 Market Estimates*. Office of Prevention, Pesticides and Toxic Substances, Publication No. 7330R-99-001. U.S. Environmental Protection Agency, Washington, DC.

Atkinson, I. A. E. and E. K. Cameron. 1993. Human influence on the terrestrial biota and biotic communities of New Zealand. *Trends Ecol. Evol.* 8:447–451.

Augustine, D. J. and S. J. McNaughton. 1998. Ungulate effects on the functional species composition of plant communities: Herbivore selectivity and plant tolerance. *J. Wildl. Manage.* 62:1165–1183.

Auld, B. A. 1984. Weed distribution. *Proc. 7th Austral. Weeds Conf.* 1:173–176.

Auld, B. A. and R. G. Coote. 1990. Invade: Towards the simulation of plant spread. *Agric. Ecosyst. Environ.* 30:121–128.

Auld, B. A., J. Hosking, and R. E. McFadyen. 1983. Analysis of the spread of tiger pear and parthenium weed in Australia. *Aust. Weeds* 2:56–60.

Auld, T. D., D. A. Keith, and R. A. Bradstock. 2000. Patterns in longevity of soil seed banks in fire-prone communities of south-eastern Australia. *Aust. J. Bot.* 48: 539–548.

Auld, B. A., K. M. Menz, and N. M. Monaghem. 1978. Dynamics of weed spread: Implications for policies of public control. *Prot. Ecol.* 1:41–148.

Auld, B. A., K. M. Menz, and C. A. Tisdell. 1987. *Weed Control Economics*. Academic, London.

Auld, B. A. and C. A. Tisdell. 1988. Influence of spatial distribution of weeds on crop yield loss. *Plant Prot. Q.* 3:31.

Bailey, L. H. and E. Z. Bailey. 1941. *Hortus the Second*. Macmillan, New York.

Bais, H. P., R. Vepachedu, S. Gilroy, R. M. Callaway, and J. M. Vivanco. 2003. Allelopathy and exotic plant invasion: From molecules and genes to species interactions. *Science* 301:1377–1380.

Bais, H. P., T. S. Walker, F. R. Stermitz, R. A. Hufbauer, and J. M. Vivanco. 2002. Enantiomeric-dependent phytotoxic and antimicrobial activity of ($\pm$)-catechin. A rhizosecreted racemic mixture from spotted knapweed. *Plant Physiol.* 128:1173–1179.

Baker, H. G. 1965. Characteristics and modes of origin of weeds. Pp. 147–168 in H. G. Baker and G. L. Stebbins (Eds.), *The Genetics of Colonizing Species.* Academic, New York.

Baker, H. G. 1974. The evolution of weeds. *Annu. Rev. Ecol. Syst.* 5:1–24.

Baker, H. G. 1995. Aspects of the genecology of weeds. Pp. 189–224 in A. R. Kruckeberg, R. B. Walker, and A. E. Leviton (Eds.), *Genecology and Ecogeographic Races.* Pacific Division of the American Association for the Advancement of Science, San Francisco, CA.

Baker, H. G. and G. L. Stebbins (Eds.). 1965. *The Genetics of Colonizing Species.* Academic, New York.

Balandier, P., C. Collet, J. H. Miller, P. E. Reynolds, and S. M. Zedaker. 2006. Designing forest vegetation management strategies based on the mechanisms and dynamics of crop tree competition by neighboring vegetation. *Forestry* 79:3–27.

Ballaré, C. L., A. L. Scopel, C. M. Ghersa, and R. A. Sánchez. 1987. The demography of *Datura ferox* L. in soybean crops. *Weed Res.* 27:91–102.

Ballaré, C. L., A. L. Scopel, C. M. Ghersa, and R. A. Sánchez. 1988. The fate of *Datura ferox* seeds in the soil as affected by cultivation, depth of burial and degree of maturity. *Ann. Appl. Biol.* 112:337–345.

Ballaré, C. L., A. L. Scopel, and R. A. Sánchez. 1990. Far-red radiation reflected from adjacent leaves: An early signal of competition in plant canopies. *Science* 247: 329–332.

Barberi, P. 2002. Weed management in organic agriculture: Are we addressing the right issues? *Weed Res.* 42:177–193.

Barbour, M. G., J. H. Burk, W. D. Pitts, F. S. Gilliam, and M. W. Schwartz. 1999. *Terrestrial Plant Ecology*, 3rd ed. Benjamin Cummings, Menlo Park, CA.

Bard, E. B., R. L. Sheley, J. S. Jacobsen, and J. J. Borkowski. 2004. Using ecological theory to guide the implementation of augmentative restoration. *Weed Technol.* 18:1246–1249.

Barnes, J. P., A. R. Putnam, and B. A. Burke. 1986. Allelopathic activity of rye (*Secale cereale* L). Pp. 271–286 in A. R. Putnam and C. S. Tang (Eds.), *The Science of Allelopathy.* Wiley, New York.

Barrett, S. C. H. 1983. Crop mimicry in weeds. *Econ. Bot.* 37:255–282.

Barrett, S. C. H. 1988. Genetics and evolution of agricultural weeds. Pp. 57–75 in M. A. Altieri and M. Liebman (Eds.), *Weed Management in Agroecosystems: Ecological Approaches.* CRC Press, Boca Raton, FL.

Barrett, S. C. H. and B. C. Husband. 1990. Genetics of plant migration and colonization. Pp. 254–277 in A. H. D. Brown, M. T. Clegg, A. L. Kahler, and B. S. Weir, (Eds.), *Plant Population Genetics, Breeding and Genetic Resources.* Sinauer Associates, Sunderland, MA.

Barrett, S. C. H. and B. J. Richardson 1986. Genetic Attributes of invading species. Pp. 21–33 in R. Groves and J. J. Burdon (Eds.), *Ecology of Biological Invasions, An Australian Perspective.* Australian Academy of Sciences, Canberra.

Barrett, S. C. H. and D. E. Seaman. 1980. The weed flora of California rice fields. *Aquat. Bot.* 9:351–376.

Barrett, S. C. H. and J. S. Shore. 1989. Isozyme variation in colonizing plants. Pp. 106–126 in D. Soltis and P. Soltis (Eds.), *Isozymes in Plant Biology.* Dioscorides Press, Portland, OR.

Barthell, J. F., J. M. Randall, R. W. Thorp, and A. M. Wenner. 2001. Promotion of seed set in yellow star-thistle by honey bees: Evidence of an invasive mutualism. *Ecol. Appl.* 11:1870–1883.

Bartlett, E., S. J. Novak, and R. N. Mack. 2002. Genetic variation in *Bromus tectorum* (Poaceae): Differentiation in the eastern United States. *Am. J. Bot.* 89:602–612.

Baruch, Z., A. B. Hernández, and M. G. Montilla. 1989. Dinámica del crecimiento, fenología y repatición de biomassa en gramíneas nativas e introducidas de una sabana Neotropical. *Ecotropicos* 2:1–13.

Baskin, C. C. and J. M. Baskin. 1998. *Seeds: Ecology, Biogeography and Evolution of Dormancy and Germination.* Academic, New York.

Baskin, J. M. and C. C. Baskin. 1989. Physiology of dormancy and germination in relation to seed bank ecology. Pp. 53–66 in M. Allessio Leck, V. T. Parker, and R. L. Simpson (Eds.), *Ecology of Soil Seed Banks.* Academic, San Diego, CA.

Bastiaans, L., M. J. Kropff, J. Goudriaan, and H. H. van Laar. 2000. Design of weed management systems with a reduced reliance on herbicides poses new challenges and prerequisites for modeling crop-weed interactions. *Field Crops Res.* 67:161–179.

Bayer, R. J. 1990. Investigations into the evolutionary history of the *Antennaria rosea* (Asteraceae-Inuleae) polyploidy complex. *Plant Syst. Evol.* 169:97–110.

Bazzaz, F. A. 1979. The physiological ecology of plant succession. *Annu. Rev. Ecol. Syst.* 10:351–371.

Bazzaz, F. A. 1990. Plant-plant interactions in successional environments. Pp. 240–265 in J. B. Grace and D. Tilman (Eds.), *Perspectives on Plant Competition.* Academic, San Diego, CA.

Beal, W. J. 1911. The vitality of seeds buried in the soil. *Proc. Sec. Promot. Agric. Sci.* 31:21–23.

Becker, E. M., P. de la Bastide, R. L. Hahn, S. F. Shamoun, and W. E. Hintz. 2000. Molecular markers for monitoring mycoherbicides. Pp. 301 in N. R. Spenser (Ed.), *X International Symposium on Biological Control of Weeds,* Bozeman, MT. Advanced Litho Printing, Great Falls, MT.

Beckwith, E. G. 1854. Report of explorations for a route for the Pacific railroad of the line of the forty-first parallel of north latitude. In Vol. II, House Rep. Ex. Doc. No. 91, 33rd Congress, 2nd session.

Begon, M. and M. Mortimer. 1986. *Population Ecology: A Unified Study of Animals and Plants.* Blackwell Scientific, London.

Begon, M., M. Mortimer, and D. J. Thompson. 1996. *Population Ecology: A Unified Study of Animals and Plants,* 3rd ed. Blackwell Scientific, London.

Bekker, R. M., J. P. Bakker, U. Grandin, R. Kalamees, P. Milberg, P. Poschlod, K. Thompson, and J. H. Willems. 1998. Seed size, shape and vertical distribution in the soil: Indicators of seed longevity. *Funct. Ecol.* 12:834–842.

Belsky, A. J. and J. L. Gelbard. 2000. *Livestock Grazing and Weed Invasions in the Arid West.* Oregon Natural Desert Association, Bend, OR.

Benbrook, C. 1999. Evidence of the magnitude and consequences of the roundup ready soybean yield drag from university-based varietal trials in 1998. AgbioTech InfoNet Technical Paper No. 1. Available: http://www.biotech-info.net/RR_yield_drag_98.pdf.

Benech Arnold, R. L. and R. A. Sánchez. 1994. Modeling weed seed germination. Pp. 545–566 in J. Kigel and G. Galili (Eds.), *Seed Development and Germination.* Marcel Dekker, New York.

Benech Arnold, R. L., R. A. Sánchez, F. Forcella, B. Kruk, and C. M. Ghersa, 2000. Environmental control of dormancy in weed seed banks in soil. *Field Crops Res.* 67:105–122.

Bennett, A. C., A. J. Price, M. C. Sturgill, G. S. Buol, and G. G. Wilkerson. 2003. HADSSTM, Pocket HERBTM, and WebHADSSTM: Decision aids for field crops. *Weed Technol.* 17:412–420.

Bertness, M. D. and R. Callaway. 1994. Positive interactions in communities. *Trends Ecol. Evol.* 9:191–193.

Bever, J. D. 2003. Soil community feedback and the coexistence of competitors: Conceptual framework and empirical tests. *New Phytol.* 157:465–473.

Bhatt, G. M., F. W. Ellison, and D. J. Mares, 1993. Inheritance studies in dormancy in three wheat crosses. Pp. 274–278 in J. E. Kruger and D. E. Laberge (Eds.), *Third International Symposium on Pre-Harvest Sprouting in Cereals.* Westview, Boulder, CO.

Bhowmik, P. C. and K. Inderjit, 2003. Challenges and opportunities in implementing allelopathy for natural weed management. *Crop Prot.* 22:661–671.

Billings, W. D. 1990. *Bromus tectorum*, a biotic cause of ecosystem impoverishment in the Great Basin. Pp. 301–322 in G. M. Woodwell (Ed.), *The Earth in Transition: Patterns and Processes of Biotic Impoverishment.* Cambridge University Press, New York.

Blackman, V. H. 1919. The compound interest law and plant growth. *Ann. Bot. (London)* 33:353.

Blackshaw, R. E. 1994. Rotation affects downy brome (*Bromus tectorum*) in winter wheat (*Triticum aestivum*). *Weed Technol.* 8:728–732.

Blair, A. C., B. D. Hanson, G. R. Brunk, R. A. Marrs, P. Westra, S. J. Nissen, and R. A. Hufbauer. 2005. New techniques and findings in the study of a candidate allelochemical implicated in invasion success. *Ecol. Lett.* 8:1039–1047.

Bleasdale, J. K. A. 1967. Systematic designs for spacing experiments. *Exp. Agric.* 3: 73–85.

Blossey, B. 1999. Before, during and after: The need for long-term monitoring in invasive plant management. *Biol. Invas.* 1:301–311.

Blossey, B. and R. Nötzold. 1995. Evolution of increased competitive ability in invasive nonindigenous plants: A hypothesis. *Ecology* 83:887–889.

Böger, P., K. Wakabayashi, and D. Hirai (Eds.). 2002. *Herbicide Classes in Development: Mode of Action, Targets, Genetic Engineering, Chemistry.* Springer-Verlag, Berlin.

Booth, B. D., S. D. Murphy, and C. J. Swanton. 2003. *Weed Ecology in Natural and Agricultural Systems.* CAB International, Wallingford, United Kingdom.

Booth, B. D. and C. J. Swanton. 2002. Assembly theory applied to weed communities. *Weed Sci.* 50:2–13.

Bostrom, U. and H. Fogelfors. 2002. Long-term effects of herbicide application strategies on weeds and yield in spring-sown cereals. *Weed Sci.* 50:196–203.

Boudry, P., M. Mörchen, P. Saumitou-Laprade, P. Vernet, and H. Van Dijk. 1993. The origin and evolution of weed beets: Consequences for the breeding and release of herbiciresistant transgenic sugar beets. *Theor. Appl. Genet.* 87:471–478.

Bouquet F., B. Legeard, and F. Peureux. 2002. CLPS-B—A constraint solver for B. Pp. 188–204 in J. P. Katoen and P. Stevens (Eds.), *Tools and Algorithms for the Construction and Analysis of Systems.* LNCS 2280. Springer-Verlag, New York.

Brenchley, W. E. 1920. *Weeds of Farm Land.* Longmans, Green, London.

Brenchley, W. E. and K. Warington. 1930. The weed seed population of arable soil. I. Numerical estimation of viable seeds and observations on their natural dormancy. *J. Ecol.* 18:235–272.

Brenchley, W. E. and K. Warington. 1933. The weed seed population of arable soil. II. Influence of crop, soil, and methods of cultivation upon the relative abundance of viable seeds. *J. Ecol.* 21:103–127.

Brenchley, W. E. and K. Warington. 1945. The influence of periodic fallowing on the prevalence of viable weed seeds in arable soil. *Ann. Appl. Biol.* 32:285–296.

Breton, P. and H. Tremblay. 1990. *Le Secteur Forestier Suedois.* Rapport de mission d'etude. Forest Canada, Region du Quebec.

Briese, D. T. 2000. Classical biological control. Pp. 161–192 in B. Sindel (Ed.), *Australian Weed Management Systems.* R. G. and F. J. Richardson, Melbourne, Australia.

Briggs, S. A. 1992. *Basic Guide to Pesticides: Their Characteristics and Hazards.* Hemisphere, Washington, DC.

Briske, D. D., S. D. Fuhlendorf, and F. E. Smeins. 2003. Vegetation dynamics on rangelands: A critique of the current paradigms. *J. Appl. Ecol.* 40:601–614.

Brooker, R., Z. Kikvidze, F. I. Pugnaire, R. M. Callaway, P. Choler, C. J. Lortie, and R. Michalet. 2005. The importance of importance. *Oikos* 109:63–70.

Brown, A. H. D. 1990. Genetic characterization of plant mating systems. Pp. 145–162 in A. H. D. Brown, M. T. Clegg, A. L. Kahler, and B. S. Weir (Eds.), *Plant Population Genetics, Breeding, and Genetic Resources.* Sinauer, Sunderland, MA.

Brown, A. H. D. and J. J. Burdon. 1983. Multilocus diversity in an outbreeding weed, *Echium plantagineum* L. *Aust. J. Biol. Sci.* 36:503–509.

Brown, A. H. D. and D. R. Marshall. 1981. Evolutionary changes accompanying colonization in plants. Pp. 351–363 in G. G. E. Scudder and J. L. Reveal (Eds.), *Evolution Today.* Proceedings of the 2nd International Congress of Systematics and Evolutionary Biology. Hunt Institute for Botanical Documentation, Carnegie–Mellon University Press, Pittsburgh, PA.

Brown, J. K. and J. Kapler-Smith (Eds.). 2000. *Wildland Fire in Ecosystems: Effect of Fire on Flora.* Gen. Tech. Rep. RMRST-GTR-42-Vol.2. USDA Forest Service, Rocky Mountain Research Station, Ogden, UT.

Buckley, Y. M., H. L. Hinz, D. Matthies, and M. Rees. 2001. Interactions between density-dependent processes, population dynamics and control of an invasive plant species, *Tripleurospermum perforatum* (scentless chamomile). *Ecol. Lett.* 4:551–558.

Buhler, D. D. 2002. 50th anniversary invited article: Challenges and opportunities for integrated weed management. *Weed Sci.* 50:273–280.

Buhler, D. D., M. Liebman, and J. J. Obrycki. 2000. Theoretical and practical challenges to an IPM approach to weed management. *Weed Sci.* 48:274–280.

Bunting, S. C., J. K. Kingery, M. A. Hemstrom, M. A. Schroeder, R. A. Gravenmier, and W. J. Hann. 2002. *Altered Rangeland Ecosystems in the Interior Columbia Basin.* PNW-GTR-553. USDA Forest Service, Pacific Northwest Research Station, Portland, OR.

Burdon, J. J. and A. H. D. Brown. 1986. Population genetics of *Echium plantagineum* L.: Target weed for biological control. *Aust. J. Biol. Sci.* 39:369–378.

Burger, J. C., S. Lee, and N. C. Ellstrand. 2006. Origin and genetic structure of feral rye in the western United States. *Mol. Evol.* 15:2527–2539.

Burkholder, P. R. 1952. Cooperation and conflict among primitive organisms. *Am. Sci.* 40:601–631.

Bussan, A. J. and W. E. Dyer. 1999. Herbicides and rangeland. Pp. 116–132 in R. L. Sheley and J. K. Petroff (Eds.), *Biology and Management of Noxious Rangeland Weeds.* Oregon State University Press, Corvallis, OR.

Byers, J. E., S. Reichard, J. M. Randall, I. M. Parker, C. S. Smith, W. M. Lonsdale, I. A. E. Atkinson, M. Williamson, E. Choresky, and D. Hayes. 2002. Directing research to reduce the impacts of nonindigenous species. *Conserv. Biol.* 6:630–640.

Callaway, R. M. and E. T. Aschehoug. 2000. Invasive plants versus their new and old neighbors: A mechanism for exotic invasion. *Science* 290:521–523.

Callaway, R. M. and W. M. Ridenour. 2004. Novel weapons: Invasive success and the evolution of increased competitive ability. *Front. Ecol. Environ.* 2:436–443.

Callihan, R. H., T. S. Prather, and F. E. Northam. 1993. Longevity of yellow starthistle (*Centaurea solstitialis*) achenes in soil. *Weed Technol.* 7:33–35.

Campbell, B. D. and J. P. Grime. 1992. An experimental test of plant strategy theory. *Ecology* 73:15–29.

Cannas, S. A., D. E. Marco, and S. A. Paez. 2003. Modelling biological invasions: Species traits, species interactions, and habitat heterogeneity. *Math. Biosci.* 183:93–110.

Canner, S. R. and L. J. Wiles. 2002. Weed reproduction model parameters may be estimated from crop yield loss data. *Weed Sci.* 50:763–772.

Cardina, J., T. M. Webster, C. P. Herms, and E. E. Regnier. 1999. Development of weed IPM: Levels of integration for weed management. Pp. 239–267 in D. D. Buhler (Ed.), *Expanding the Context of Weed Management.* Hayworth, New York.

Carneggie, D. M., B. J. Schrumpf, and D. M. Mouat. 1983. Rangeland applications. Pp. 2325–2364 in R. N. Colwell (Ed.), *Manual of Remote Sensing.* 2nd ed. American Society of Photogrammetry and Remote Sensing, Falls Church, VA.

Carpinelli, M. F. 2005. The absolute-log method of quantifying relative competitive ability and niche differentiation. *Weed Technol.* 19:972–978.

Carson, R. 1962. *Silent Spring.* Houghton Mifflin, New York.

Casagrandi, R. and M. Gatto. 1999. A mesoscale approach to extinction risk in fragmented habitats. *Nature* 400:560–562.

Castle, E. 1990. Toward a philosophy of natural resource management: A case for pluralism. Unpublished special report. College of Forestry, Oregon State University, Corvallis, OR.

Castro, J., R. Zamora, J. A. Hodar, and J. M. Gomez. 2002. Use of shrubs as nurse plants: A new technique for reforestation in Mediterranean mountains. *Restor. Ecol.* 10:297–305.

Caswell, H. 2001. *Matrix Population Models*, 2nd ed. Sinauer Associates, Sunderland, MA.

Caswell, H., R. Lensink, and M. G. Neubert. 2003. Demography and dispersal: Life table response experiments for invasion speed. *Ecology* 84:1968–1978.

Caswell, H. and T. Takada. 2004. Elasticity analysis of density-dependent matrix population models: The invasion exponent and its substitutes. *Theor. Popul. Biol.* 65:401–411.

Cavers, P. B. and D. L. Benoit. 1989. Seed banks in arable lands. Pp. 309–328 in M. Allessio Leck, V. T. Parker, and R. L. Simpson (Eds.), *Ecology of Soil Seed Banks.* Academic, San Diego, CA.

Cavers, P. B., R. H. Groves, and P. E. Kaye. 1995. Seed population dynamics of *Onopordum* over one year in southern New South Wales. *J. Appl. Ecol.* 32:425–433.

Chancellor, R. J. and N. C. B. Peters. 1970. Seed production by *Avena fatua* populations in various crops. Pp. 7–11 in *Proceedings 10th British Weed Control Conference.* British Crop Protection Council, Nottingham, United Kingdom.

Chancellor, R. J. and N. C. B. Peters. 1972. Germination periodicity, plant survival and seed production in populations of *Avena fatua* L. growing in spring barley. Pp. 218–225 in *Proceedings 11th British Weed Control Conference.* British Crop Protection Council, Brighton, United Kingdom.

Chaneton, E. J., S. B. Perelman, M. Omacini, and R. J. C. León. 2002. Grazing, environmental heterogeneity, and alien plant invasions in temperate Pampa grasslands. *Biol. Invas.* 4:7–24.

Chapin, F. S. III, O. E. Sala, I. C. Burke, J. P. Grime, D. U. Hooper, W. K. Lauenroth, A. Lombard, H. A. Mooney, A. R. Mosier, S. Naeem, S. W. Pacala, J. Roy, W. L. Steffen, and D. Tilman. 1998. Ecosystem consequences of changing biodiversity. *Bioscience* 48:45–52.

Charudattan, R. and C. J. DeLoach, Jr. 1988. Management of pathogens and insects for weed control in agroecosystems. Pp. 245–264 in M. A. Altieri and M. Liebman (Eds.), *Weed Management in Agroecosystems: Ecological Approaches.* CRC Press, Boca Raton, FL.

Checkland, P. 1981. *Systems Thinking, Systems Practice.* Wiley, New York.

Cheng, H. H. 1992. A conceptual framework for assessing allelochemicals in the soil environment. Pp. 21–58 in S. J. H. Rizvy and V. Rizvy (Eds.), *Allelopathy: Basic and Applied Aspects.* Chapman and Hall, London.

Chepil, W. S. 1946. Germination of weed seeds. I. Longevity, periodicity of germination and vitality of seeds in cultivated soil. *Sci. Agric.* 26:307–346.

Chiariello, N. R., H. A. Mooney, and K. Williams. 1991. Growth, carbon allocation and cost of plant tissues. Pp. 327–365 in R. W. Pearcy, J. R. Ehleringer, H. A. Mooney, and P. W. Rundel (Eds.), *Plant Physiological Ecology. Field Methods and Instrumentation.* Chapman and Hall, New York.

Chippendale, J. F. 1991. Potential returns to research on rubber vine (*Cryptostegia grandiflora*). M.S. Thesis. University of Queensland, Brisbane, Australia.

Chittapur, B. M., C. S. Hunshal, and H. Shenoy. 2001. Allelopathy in parasitic weed management: Role of catch and trap crops. *Allelopathy J.* 8:147–159.

Clements, F. E. 1916. *Plant Successions: An Analysis of the Development of Vegetation.* Carnegie Institution of Washington Publ. 242. Carnegie Institution of Washington, Washington, DC.

Clements, F. E. 1936. Nature and structure of the climax. *J. Ecol.* 24:252–284.

Coble, H. D. and D. A. Mortensen. 1991. The threshold concept and its application to weed science. *Weed Technol.* 6:191–195.

Cole, E. C. and M. Newton. 1987. Fifth-year responses of Douglas fir to crowding and non-coniferous competition. *Can. J. For. Res.* 17:181–186.

Connell, J. H. 1990a. Apparent versus "real" competition in plants. Pp. 9–26 in J. B. Grace and D. Tilman (Eds.), *Perspectives on Plant Competition*. Academic, San Diego, CA.

Connell, J. H. and R. O. Slatyer. 1977. Mechanisms of succession in natural communities and their role in community stability and organization. *Am. Nat.* 111:1119–1144.

Cook, R. 1980. The biology of seeds in the soil. Pp. 107–129 in O. T. Solbrig (Ed.), *Demography and Evolution in Plant Populations*. University of California Press, Berkeley.

Coombs, E. M., J. K. Clark, G. L. Piper, and A. F. Cofrancesco, Jr. 2004. *Biological Control of Invasive Plants in the United States*. Oregon State University Press, Corvallis, OR.

Corbin, J. D. and C. M. D'Antonio. 2004. Competition between native perennial and exotic annual grasses: Implications for an historical invasion. *Ecology* 85:1273–1283.

Cornelissen, J. H., C. P. Castro-Diaz, and A. L. Carnelli. 1998. Variation in relative growth rate among woody species. Pp. 363–392 in H. Lambers, H. Poorter, and M. M. I. van Vuuren (Eds.), *Inherent Variation in Plant Growth*. Backhuys, Leiden, Netherlands.

Cousens, R. 1985. A simple model relating yield loss to weed density. *Ann. Appl. Biol.* 107:239–252.

Cousens, R. 1991. Aspects of the design and interpretation of competition (interference) experiments. *Weed Technol.* 5:664–673.

Cousens, R. and M. Mortimer. 1995. *Dynamics of Weed Populations*. Cambridge University Press, New York.

Cousens, R., N. C. B. Peters, and C. J. Marshall. 1984. Models of yield loss-weed density relationships. Pp. 367–374 in *Proceedings 7th International Symposium on Weed Biology, Ecology and Systematics*. Columa, Paris.

Covarelli, G. and F. Tei. 1988. Effet de la rotation culturale sur la flore adventice du mais. Pp. 477–484 in *VIIeme Colloque International sur la Biologie, l'Ecologie et la Systematique des Mauvaises Herbes*, Vol. 2. Comité Français de Lutte contre les Mauvaises Herbes, Paris.

Crafts, A. S. 1975. *Modern Weed Control*. University of California Press, Berkeley.

Crawley, M. J. 1983. *Herbivory: The Dynamics of Plant-Animal Interactions*. University of California Press, Berkeley.

Crawley, M. J. 1987. What makes a community invasible? Pp. 429–454 in A. J. Gray, M. J. Crawley, and P. J. Edwards (Eds.), *Colonization, Succession and Stability*. Blackwell Scientific, Oxford.

Crawley, M. J. 1989. *Biological Invasions: A Global Perspective*. Wiley, New York.

Crawley, M. J. 1992. Seed predators and plant population dynamics. Pp. 157–191 in M. Fenner (Ed.), *Seeds: The Ecology and Regeneration of Plant Communities*. CAB International, Wallingford, United Kingdom.

Cromar, H. E., S. E. Murphy, and C. J. Swanton. 1999. Influence of tillage and crop residue on post-dispersal predation of weed seeds. *Weed Sci.* 47:184–194.

Cronk, Q. C. B. and J. L. Fuller. 1995. *Plant Invaders, A 'People and Plants' Conservation Manual*. Chapman and Hall, London.

Cronquist, A. 1988. *The Evolution and Classification of Flowering Plants*, 2nd ed. New York Botanical Garden, New York.

Crooks, J. and M. E. Soule. 1999. Lag times in population explosions of invasive species: Causes and implications. Pp. 103–125 in O. T. Sandlund, P. J. Schei, and A. Viken (Eds.), *Invasive Species and Biodiversity Management*. Kluwer Academic, Dordrecht, Netherlands.

Crosby, A. W. 1994. *Ecological Imperialism: The Biological Expansion of Europe, 900–1900*. Cambridge University Press, New York.

Cruden, R. W. 1977. Pollen-ovule ratios: A conservative index of breeding systems in flowering plants. *Evolution* 31:22–46.

Cruttwell McFadyen, R. E. 2000. Successess in biological control of weeds. Pp. 3–14 in N. R. Spenser (Ed.), *X International Symposium on Biological Control of Weeds*. Bozeman, MT. Advanced Litho Printing, Great Falls, MT.

Cussans, G. W. 1987. Weed management in cropping systems. Pp. 337–347 in D. Lemerle and A. R. Leys (Eds.), *Proceedings 8th Australlian Weeds Conference*, Weed Society of New South Wales, Sydney.

Daehler, C. C. 1998. Variation in self-fertility and the reproductive advantage of self-fertility for an invading plant (*Spartina alterniflora*). *Evol. Ecol.* 12:553–568.

Daehler, C. C. and D. A. Carino. 2000. Predicting invasive plants: Prospects for a general screening system based on current regional models. *Biol. Invas.* 2:93–102.

Daehler, C. C., J. S. Denslow, S. Arnari, and H. C. Huang. 2004. A risk assessment system for screening out invasive pest plants from Hawaii and other Pacific islands. *Conserv. Biol.* 18:360–368.

Dale, P. J., B. Clarke, and E. M. G. Fontes. 2002. Potential for the environmental impact of transgenic crops. *Nature Biotechnol.* 409:567–574.

D'Antonio, C. M., T. L. Dudley, and R. N. Mack. 1999. Disturbance and biological invasion: Direct effects and feedbacks. Pp. 413–452 in L. R. Walker (Ed.), *Ecosystems of Disturbed Ground*. Elsevier, Amsterdam, Netherlands.

D'Antonio, C. M., N. E. Jackson, D. D. Horvitz, and R. Hedberg. 2004. Invasive plants in wildland ecosystems: Merging the study of invasion processes with management needs. *Front. Ecol. Environ.* 2:513–521.

D'Antonio, C. M., D. C. Odion, and C. M. Tyler. 1993. Invasion of maritime chaparral by the introduced succulent *Carpobrotus edulis*: The roles of fire and herbivory. *Oecologia* 95:14–21.

D'Antonio, C. M. and P. M. Vitousek. 1992. Biological invasions by exotic grasses, the grass/fire cycle, and global change. *Annu. Rev. Ecol. Syst.* 23:63–87.

Darmency, H., C. Ouin, and J. Pernes. 1987. Breeding foxtail millet (*Setaria italica*) for quantitative traits after interspecific hybridization and polyploidization. *Genome* 29:453–456.

Darwin, C. 1859. *The Origin of Species by Means of Natural Selection*. Murray, London.

Davis, A. S., P. M. Dixon, and M. Liebman. 2003. Cropping system effect on giant foxtail (*Setaria faberi*) demography: II. Retrospective perturbation analysis. *Weed Sci.* 51:930–939.

Davis, A. S., A. I. Kathleen, S. G. Hallett, and K. A. Renner. 2006. Weed seed mortality in soils with contrasting agricultural management histories. *Weed Sci.* 54:291–297.

Davis, M. 2005. Invasion biology 1958–2004: The pursuit of science and conservation. Pp. 1–27 in M. W. McMahon and T. Fukami (Eds.), *Conceptual Ecology and Invasion Biology: Reciprocal Approaches to Nature*. Kluwer, London.

Davis, M. A., J. P. Grime, and K. Thompson. 2000. Fluctuating resources in plant communities: A general theory of invasibility. *J. Ecol.* 88:528–534.

Davis, M. A., K. Thompson, and J. P. Grime. 2005. Invasibility: The local mechanism driving community assembly and species diversity. *Ecography* 28:696–704.

Davy, A. J., R. Scott, and C. V. Cordazzo. 2006. Biological flora of the British Isles: *Cakile maritima* Scop. *J. Ecol.* 94:695–711.

Dayan, F. E., J. G. Romagni, and S. O. Duke. 2000. Investigating the mode of action of natural phytotoxins. *J. Chem. Ecol.* 26:2079–2094.

Dean, S. J., P. M. Holmes, and P. W. Weisset. 1986. Seed biology of invasive alien plants in South Africa and southwest Africa/Namibia. Pp. 157–170 in I. A. W. MacDonald, F. J. Kruger, and A. A. Ferrar (Eds.), *The Ecology and Management of Biological Invasions in Southern Africa*. Oxford University Press, Cape Town, South Africa.

DeAngelis, D. L. and J. C. Waterhouse. 1987. Equilibrium and nonequilibrium concepts in ecological models. *Ecol. Monogr.* 57:1–21.

De Blois, S., G. Domon, and A. Bouchard. 2001. Environmental, historical, and contextual determinants of vegetation cover: A landscape perspective. *Landscape Ecol.* 16:421–436.

De Clerck-Floate, R. 1997. Cattle as dispersers of hound's-tongue on rangeland in southeastern British Columbia. *J. Range Manage.* 50:239–243.

Deen, W., R. Cousens, J. Warringa, L. Bastiaans, P. Carberry, K. Rebel, S. Riha, C. Murphy, L. Benjamin, C. Cloughley, J. Cussans, F. Forcella, T. Hunt, P. Jamieson, J. Lindquist, and E. Wang. 2003. An evaluation of four crop:weed competition models using a common data set. *Weed Res.* 43:116–129.

De Ferrari, C. M. and R. J. Naiman. 1994. A multi-scale assessment of the occurrence of nonindigenous plants on the Olympic Peninsula, Washington. *J. Veget. Sci.* 5:247–258.

Dekker, J. 2003. Evolutionary biology of the foxtail (*Setaria*) species-group. Pp. 65–114 in K. Inderjit (Ed.), *Weed Biology and Management*. Kluwer Academic, Dordrecht, Netherlands.

Dekker, J. and G. Comstock. 1992. Ethical and environmental considerations in the release of herbicide resistant crops. *Agric. Hum. Values* 9:1–13.

De la Fuente, E. B., S. A. Suarez, and C. M. Ghersa. 2003. Weed and insect communities in wheat crops with different management practices. *Agron. J.* 95:1542–1549.

De la Fuente, E. B., S. A. Suarez, and C. M. Ghersa. 2006. Soybean-weed community composition and richness between 1995 and 2003 in the Rolling Pampas, Argentina. *Agric. Ecosyst. Environ.* 115:229–236.

Derksen, D. A. 1997. Weeds. Pp. 24–28 in D. Domitruk and B. Crabtree (Eds.), *Zero-Tillage: Advancing the Art*. Manitoba-North Dakota Zero Tillage Farmers Association, Brandon, Manitoba, Canada.

De Wet, J. M. J. and J. R. Harlan. 1975. Weeds and domesticates: Evolution in the man-made habitat. *Econ. Bot.* 29:99–107.

De Wit, C. T. 1960. On competition. *Versl. Landbouwk. Onderz. Ned.* Vol. 66, Pp. 1–82.

Devine, M. D., S. O. Duke, and C. Fedtke. 1993. *Physiology of Herbicide Action*. Prentice-Hall, Inc., Englewood Cliffs, NJ.

Didham, R. K., J. M. Tylianakis, M. A. Hutchison, R. M. Ewers, and N. J. Gemmell. 2005. Are invasive species the drivers of ecological change? *Trends Ecol. Evol.* 20:470–474.

Dill, G., S. Baerson, L. Casagrande, Y. Feng, R. Brinker, T. Reynolds, N. Taylor, D. Rodriguez, and Y. Teng. 2000. Characterization of glyphosate resistant *Eleusine indica* biotypes from Malaysia. In *Proceedings 3rd International Weed Science Congress*. Foz do Iguassu, Brazil International Weed Science Society, Oxford, MS.

Dilworth, M. J. and A. R. Glenn (Eds.). 1991. *Biology and Biochemistry of Nitrogen Fixation.* Elsevier Science, New York.

DiTomaso, J. M. 2000. Invasive weeds in rangeland: Species, impacts and management. *Weed Sci.* 48:255–265.

DiTomaso, J. M., M. L. Brooks, E. B. Allen, R. Minnich, P. M. Rice, and G. B. Kyser. 2006. Control of invasive weeds with prescribed burning. *Weed Technol.* 20:535–548.

DiTomaso, J. M. and D. W. Johnson (Eds.). 2006. *The Use of Fire as a Tool for Controlling Invasive Plants.* Cal-IPC Publication 2006-01. California Invasive Plant Council, Berkeley, CA.

D'Mello, J. P. F. 2003. *Food Safety: Contaminants and Toxins.* CAB International, Wallingford, United Kingdom.

Donohue, K., E. Hammond, P. D. Messiqua, M. S. Heschel, and J. Schmitt. 2000. Density dependence and population differentiation of genetic architecture in *Impatiens capensis* in natural environments. *Evolution* 54:1969–1981.

Doutt, R. L. 1964. The historical development of biological control. Pp. 21–44 in P. DeBach (Ed.), *Biological Control of Insect Pests and Weeds*. Reinhold, New York.

Downey, R. K., A. J. Klassen, and G. R. Stringam. 1980. Rapeseed and mustard. Pp. 495–509 in W. R. Fehr and H. H. Hadley (Eds.), *Hybridization of Crop Plants*. American Society of Agronomy and Crop Science Society of America, Madison, WI.

Drake, J. A., C. R. Zimmerman, T. Purucker, and C. Rojo. 1999. On the nature of the assembly trajectory. Pp. 233–250 in E. Weir and P. A. Keddy (Eds.), *Ecological Assembly Rules: Perspectives, Advances, Retreats*. Cambridge University Press, New York.

Dregson, A. R. and V. Stevens. 1997. Ecologically responsible reforestation and ecoforestry. Pp. 69–76 in A. R. Dregson and D. M. Taylor (Eds.), *Ecoforestry: The Art and Science of Sustainable Forest Use*. New Society Publishers, Gabriola Island, Canada.

Duke, S. O. 1986. Microbially produced phytotoxins as herbicides—A perspective. Pp. 287–304 in A. R. Putnam and C. S. Tang (Eds.), *The Science of Allelopathy*. Wiley, New York.

Duke, S. O. (Ed.). 1996. *Herbicide-Resistant Crops. Agricultural, Environmental, Economic, Regulatory, and Technical Aspects*, Lewis, Boca Raton, FL.

Duke, S. O. and H. K. Abbas. 1995. Natural products with potential use as herbicides. Pp. 348–362 in K. Inderjit, M. M. Dakshini and F. A. Einhellig (Eds.), *Allelopathy: Organisms, Processes, and Applications*. American Chemical Society Symposium Series No. 582. American Chemical Society, Washington, DC.

Duke, S. O., A. M. Rimando, S. R. Baerson, B. E. Scheffler, E. Ota, and R. G. Belz. 2002. Strategies for the use of natural products for weed management. *J. Pest. Sci.* 27:298–306.

Duke, S. O., J. G. Romagni, and F. E. Dayan. 2000. Natural products as sources for new mechanisms of herbicide action. *Crop Prot.* 19:583–589.

Dukes, J. S. and H. A. Mooney. 1999. Does global change increase the success of biological invaders? *Trends Ecol. Evol.* 14:135–139.

Duvel, J. W. T. 1903. Seeds buried in soil. *Science* 17:872–873.

Dwire, K. A., C. G. Parks, M. L. McInnis, and B. J. Naylor. 2006. Seed production and dispersal of sulfur cinquefoil, an invasive non-native species in northeast Oregon. *Range Ecol. Manage.* 59:63–72.

Dyer, W. E., F. D. Hess, J. S. Holt, and S. O. Duke. 1993. Potential benefits and risks of herbicide-resistant crops produced by biotechnology. *Hort. Rev.* 15:365–408.

Ebert, T. A. 1999. *Plant and Animal Populations: Methods in Demography.* Academic, San Diego, CA.

Ecobichon, D. J. 1996. Toxic effects of pesticides. Pp. 774–784 in C. D. Klaassen (Ed.), *Casarett and Doulls Toxicology.* McGraw-Hill, New York.

Eddington, G. E. and W. W. Robbins. 1920. Irrigation water as a factor in the dissemination of weed seeds. *Colo. Exp. Sta. Bull.* 253:25.

Edwards, G. W. and J. W. Freebairn. 1982. The social benefits from an increase in part of an industry. *Rev. Mark. Agric. Econ.* 50:193–210.

Ehrlich, I. and G. Becker. 1972. Market insurance, self insurance, and self protection. *J. Pol. Econ.* 80:623–648.

Ellis, R. G. 1994. Alien Study Group. *BSBI News* 65:44–45.

Ellstrand, N. C. 2003. *Dangerous Liaisons? When Cultivated Plants Mate with Their Wild Relatives.* The John Hopkins University Press, Baltimore, MD.

Ellstrand, N. C. and M. L. Roose. 1987. Patterns of genotypic diversity in clonal plant species. *Am. J. Bot.* 74:123–131.

Ellstrand, N. C. and K. A. Schierenbeck. 2000. Hybridization as a stimulus for invasiveness in plants? *Proc. Natl. Acad. Sci. USA* 97:7043–7050.

Ellstrand, N. C., R. Whitkus, and L. H. Rieseberg. 1996. Distribution of spontaneous plant hybrids. *Proc. Natl. Acad. Sci. USA* 93:5090–5093.

Elton, C. S. 1958. *The Ecology of Invasions by Animals and Plants.* Methuen, London.

Emerson, R. W. 1878. *Fortune of the Republic.* Houghton and Osgood, Boston, MA.

Enache, A. J. and R. D. Ilnicki. 1990. Weed control by subterranean clover (*Trifolium subterraneum*) used as a living mulch. *Weed Technol.* 4:534–538.

Endress, B. A., B. J. Naylor, C. G. Parks, and S. R. Radosevich. In press. Landscape factors influencing the abundance and dominance of the invasive plant *Potentilla recta.* *Rangeland Ecol. Manage.*

Enserink, M. 1999. Biological invaders sweep in. *Science* 285:1834–1836.

Esau, K. 1965. *Plant Anatomy,* 2nd ed. Wiley, New York.

Everitt, J. H., D. E. Escobar, and M. R. Davis. 2001. Reflectance and image characteristics of selected noxious rangeland species. *J. Range Manage.* 54:106–120.

Ewel, J. J., D. O'Dowd, J. Bergelson, and C. C. Curtis. 1999. Deliberate introduction of species: Research needs. *Bioscience* 49:619–630.

Fargione, J., C. S. Brown, and D. Tilman. 2003. Community assembly and invasion: An experimental test of neutral versus niche processes. *Proc. Natl. Acad. Sci. USA* 100:8916–8920.

Fargione, J. and D. Tilman. 2006. Plant species traits and capacity for resource reduction predict yield and abundance under competition in nitrogen-limited grassland. *Funct. Ecol.* 20:533–540.

Fenner, M. 1985. *Seed Ecology*. Chapman and Hall, London.

Fenner, M. 1994. Ecology of seed banks. Pp. 507–528 in J. Kigel and G. Galili (Eds.), *Seed Development and Germination*. Marcel Dekker, New York.

Fennimore, S. A., W. E. Nyquist, G. E. Shaner, R. W. Doerge, and M. E. Foley. 1999. A genetic model and molecular markers for wild oat (*Avena fatua* L.) seed dormancy. *Theor. Appl. Genet.* 99:711–718.

Ferre, F. 1988. *Philosophy of Technology*. Prentice-Hall, Englewood Cliffs, NJ.

Finnoff, D., J. F. Shogren, B. Leung, and D. Lodge. 2005. Risk and nonindigenous species management. *Rev. Agric. Econ.* 27:475–482.

Firbank, L. G. and A. R. Watkinson. 1985. On the analysis of competition within two-species mixtures of plants. *Ecology* 22:503–517.

Firbank, L. G. and A. R. Watkinson. 1987. On the analysis of competition at the level of the individual plant. *Oecologia* 71:308–317.

Firbank, L. G. and A. R. Watkinson. 1990. On the effect of competition: From mono-cultures to mixtures. Pp. 165–192 in J. B. Grace and D. Tilman (Eds.), *Perspectives on Plant Competition*. Academic, San Diego, CA.

Fischer, N. H. 1986. The function of mono and sesquiterpenes as plant germination and growth regulators. Pp. 203–218 in A. R. Putnam and C. S. Tang (Eds.), *The Science of Allelopathy*. Wiley, New York.

Fischer, R. A. and R. E. Miles. 1973. The role of spatial pattern in the competition between crop plants and weeds: A theoretical analysis. *Math. Biosci.* 18:335–350.

Fisher, R. A. 1937. The wave of advance of advantageous genes. *Ann. Eugen.* 7:353–369.

Fitzherbert, A. 1523. *Boke of Husbandry*. In E. S. Salisbury. 1961. *Weeds and Aliens*. Collins, London, UK.

Flinn, K. M. and M. Vellend. 2005. Recovery of forest plant communities in post-agricul-tural landscapes. *Front. Ecol. Environ.* 3:243–250.

Flores, J. and E. Jurado. 2003. Are nurse-protege interactions more common among plants from arid environments? *J. Veget. Sci.* 14:911–916.

Foley, M. E. 2001. Seed dormancy: An update on terminology, physiological genetics, and quantitative trait loci regulating germinability. *Weed Sci.* 49:305–317.

Food and Agriculture Organization. 2001. *Global Forest Resource Assessment*. Food and Agriculture Organization of the United Nations (FAO) Forestry Paper No. 140.

Forcella, F. and S. J. Harvey. 1983. Relative abundance in an alien weed flora. *Oecologia* 59:292–295.

Forest Stewardship Council. 2006. *FSC Pesticide Policy: Proposed Revisions*. FSC-Dis-1-006. Forest Stewardship Council, Bonn, Germany.

Forman, R. T. T. 1995. *Land Mosaics: The Ecology of Landscapes and Regions*. Cam-bridge University Press, New York.

Forman, R. T. T. and M. Gordon. 1981. *Landscape Ecology*. Wiley, New York.

Fosberg, M. A., J. G. Goldhammer, D. Rind, and C. Price. 1990. Global change: Effect of forest ecosystems and wildfire severity. Pp. 483–486 in J. G. Goldammer (Ed.), *Fire in the Tropical Biota: Ecosystem Processes and Global Challenges*. Springer-Verlag, Berlin.

Foster, D. R., G. Motzkin, and B. Slater. 1998. Land-use history as long-term broad-scale disturbance: Regional forest dynamics in central New England. *Ecosystems* 1:96–119.

Fowler, N. 1981. Competition and coexistence in a North Carolina grassland. II. The effect of the experimental removal of species. *J. Ecol.* 69:843–845.

Franklin, A. B., B. R. Noon, and T. L. George. 2002. What is habitat fragmentation? *Stud. Avian Biol.* 25:20–29.

Freckleton, R. P. and A. R. Watkinson. 2001. Asymmetric competition between plant species. *Funct. Ecol.* 15:615–623.

Froud-Williams, R. J., R. J. Chancellor, and D. S. H. Drennan. 1983. Influence of cultivation regime upon buried weed seeds in arable cropping systems. *J. Appl. Ecol.* 20:199–208.

Fukai, S. and B. R. Trenbath. 1993. Processes determining intercrop productivity and yields of component crops. *Field Crops Res.* 34:247–271.

Gajíc, D., S. Malencic, M. Vrbaški, and S. Vrbaški. 1986. Fragm. Herb. Jugoslavica 63:121 in E. L. Rice, 1986, Allelopathic growth stimulation. Pp. 23–42 in A. R. Putnam and C. S. Tang (Eds.), *The Science of Allelopathy.* Wiley, New York.

Gashwiler, J. S. 1967. Conifer seed survival in a western Oregon clearcut. *Ecology* 48:431–433.

Gasson, M. J. 2003. The safety evaluation of genetically modified foods. Pp. 329–346 in J. P. F. D'Mello (Ed.), *Food Safety: Contaminants and Toxins.* CAB Intenational, Wallingford, United Kingdom.

Gaudet, C. L. and P. A. Keddy. 1988. A comparative approach to predicting competitive ability from plant traits. *Nature* 334:242–243.

Gause, G. F. 1934. *The Struggle for Existence.* Williams and Wilkins, Baltimore, MD.

Gerowitt, B., E. Bertke, S.-K. Hespelt, and C. Tute. 2003. Towards multifunctional agriculture—weeds as ecological goods. *Weed Res.* 43:227–235.

Ghersa, C. M. 2006. La sucesión ecológica en los agroecosistemas pampeanos. Pp. 195–214 in M. Oesterheld, M. R. Aguiar, C. M. Ghersa, and J. M. Paruelo (Eds.), *Modelos y Significado Ecológico.* Heterogeneidad de la vegetación de los agroecosistemas, Facultad de Agronomía, Buenos Aires, Argentina.

Ghersa, C. M., E. de la Fuente, S. Suarez, and R. J. C. León. 2002. Woody species invasion in the Rolling Pampa grasslands, Argentina. *Agric. Ecosyst. Environ.* 88:271–278.

Ghersa, C. M. and J. S. Holt. 1995. Using phenology prediction in weed management. *Weed Res.* 35:461–470.

Ghersa, C. M. and R. J. C. León. 1999. Successional changes in the agroecosystems of the Rolling Pampas. Pp. 487–502 in L. R. Walker (Ed.), *Ecosystems of Disturbed Ground.* Elsevier, Amsterdam, Netherlands.

Ghersa, C. M., M. A. Martinez-Ghersa, T. G. Brewer, and M. L. Roush. 1994a. Selection pressures for diclofop-methyl resistance and germination time of Italian ryegrass. *Agron. J.* 86:823–828.

Ghersa, C. M., M. A. Martinez-Ghersa, J. J. Casal, M. Kaufmann, M. L. Roush, and V. A. Deregibus. 1994b. Effect of light treatments on winter wheat and Italian ryegrass (*Lolium multiflorum*) competition. *Weed Technol.* 8:37–45.

Ghersa, C. M., M. A. Martinez-Ghersa, E. H. Satorre, M. L. Van Esso, and G. Chichotky. 1993. Seed dispersal, distribution and recruitment of seedlings of *Sorghum halepense* (L) Pers. *Weed Res.* 33:79–88.

Ghersa, C. M., S. B. Perelman, S. Burkart, and R. J. C. León. In press. Floristic and structural changes related to opportunistic soil tilling and pasture planting in grassland communities of the Flooding Pampa. *Biodivers. Conserv.*

Ghersa, C. M. and M. L. Roush. 1993. Searching for solutions to weed problems. Do we study competition or dispersion? *BioScience* 43:104–109.

Ghersa, C. M., M. L Roush, S. R. Radosevich, and S. M. Cordray. 1994c. Coevolution of agroecosystems and weed management. *BioScience* 44:85–94.

Ghersa, C. M., E. H. Satorre, R. L. Benech Arnold, and M. A. Martínez-Ghersa. 2000. Advances in weed management strategies. *Field Crops Res.* 67:95–105.

Gilbert, B. and M. J. Lechowicz. 2005. Invasibility and abiotic gradients: The positive correlation between native and exotic plant diversity. *Ecology* 87:1848–1855.

Gill, D. E., L. Chao, S. L. Perkins, and J. B. Wolf. 1995. Genetic mosaicism in plants and clonal animals. *Annu. Rev. Syst. Ecol.* 26:423–444.

Glass, E. H. 1975. *Integrated Pest Management: Rationale, Potential, Needs and Implementation.* Entomol. Soc. Am. Spec. Pub. pp. 75–102. Entomological Society of America, College park, MD.

Gleason, H. A. 1926. The individualistic concept of plant association. *Bull. Torrey Bot. Club* 53:783–794.

Glenn-Lewin, D. C. and E. van der Maarel. 1992. Patterns and processes of vegetation dynamics. Pp. 11–59 in D. C. Glenn-Lewin, R. L. Peet, and T. T. Veblen (Eds.), *Plant Succession—Theory and Prediction.* Chapman and Hall, New York.

Goldberg, D. E. 1990. Components of resource competition in plant communities. Pp. 27–49 in J. B. Grace and D. Tilman (Eds.), *Perspectives on Plant Competition.* Academic, San Diego, CA.

Goldberg, D. E. and A. M. Barton. 1992. Patterns and consequences of interspecific competition in natural communities: A review of field experiments with plants. *Am. Nat.* 139:771–801.

Goldberg, D. E., T. Rajaniemi, J. Gurevitch, and A. Stewart-Oatmen. 1999. Empirical approaches to quantifying interaction intensity: Competition and facilitation along productivity gradients. *Ecology* 80:1118–1131.

Goldberg, D. E. and P. A. Werner. 1983. Equivalence of competitors in plant communities: A null hypothesis and a field experimental approach. *Am. J. Bot.* 70:1098–1104.

Goldberg, P. A. 1987. Neighbourhood competition in an old-field community. *Ecology* 8:1211–1223.

Goldberg, P. A. and K. Landa. 1991. Competitive effect and response: Hierarchies and correlated traits in early stages of competition. *J. Ecol.* 79:1013–1030.

Gomez-Aparicio, L., R. Zamora, J. M. Gomez, J. A. Hodar, J. Castro, and E. Baraza. 2004. Applying plant facilitation to forest restoration: A meta-analysis of the use of shrubs as nurse plants. *Ecol. Appl.* 14:1128–1138.

Goodwin, B. J., A. J. McAllister, and L. Fahrig. 1999. Predicting invasiveness of plant species based on biological information. *Conserv. Biol.* 13:422–426.

Gordon, D. R. 1998. Effects of invasive, non-indigenous plant species on ecosystem processes: Lessons from Florida. *Ecol. Appl.* 8:975–989.

Goslee, S. C., D. P. C. Peters, and K. G. Beck. 2006. Spatial prediction of invasion success across heterogeneous landscapes using an individual-based model. *Biol. Invas.* 8:193–200.

Gotelli, N. J. 2001. *A Primer of Ecology*, 3rd ed. Sinauer Associates, Sunderland, MA.

Gouyon P. H., R. Lumaret, G. Valdeyron, and P. Vernet. 1983. Reproductive strategies and disturbance by man. Pp. 214–225 in H. A. Mooney and M. Gordon (Eds.), *Disturbance and Ecosystems: Components of Response.* Springer, Berlin.

Gowgani, G., F. A. Holmes, and F. O. Colbert. 1989. Herbicides. Pp.173–197 in *Principles of Weed Control in California.* 2nd ed. California Weed Conference. Thompson Publications, Fresno, CA.

Grace, J. B. 1990. On the relationship between plant traits and competitive ability. Pp. 51–65 in J. B. Grace and D. Tilman (Eds.), *Perspectives on Plant Competition.* Academic, San Diego, CA.

Grace, J. B. 1991. A clarification of the debate between Grime and Tilman. *Funct. Ecol.* 5:583–587.

Grace, J. B. 1995. On the measurement of plant competition intensity. *Ecology* 76: 305–308.

Grace, J. B. and D. Tilman. 1990. *Perspectives on Plant Competition.* Academic, San Diego, CA.

Grandin, U. 2001. Short-term and long-term variation in seed bank/vegetation relations along an environmental and successional gradient. *Ecography* 24:731–741.

Grant, M. C. 1993. The trembling giant. *Discover* (Oct):83–88.

Grant, M. C., J. B. Mitton, and Y. B. Linhart. 1992. Even larger organisms. *Nature* 360:216.

Grant-Down, R. T. and H. G. Dickinson. 2005. Epigenetics and its implications for plant biology. 1. The epigenetic network in plants. *Ann. Bot.* 96:1143–1164.

Gratkowski, H. and P. Lauterbach. 1974. Releasing Douglas-firs from varnishleaf ceanothus. *J. For.* 72:150–152.

Gray, A. 1879. The predominance and pertinacity of weeds. *Am. J. Sci.* 118:161–167.

Gray, A. J., R. N. Mack, J. L. Harper, M. B. Usher, K. Joysey, and H. Kornberg. 1986. Do invading species have definable genetic characteristics? *Philos. Trans. R. Soc. Lond.* 314:655–674.

Gressel, J. 2002. *Molecular Biology of Weed Control.* Taylor and Francis, London.

Grime, J. P. 1974. Vegetation classification by reference to strategies. *Nature* 250:26–31.

Grime, J. P. 1977. Evidence for the existence of three primary strategies in plants and its relevance to ecological and evolutionary theory. *Am. Nat.* 11:1169–1194.

Grime, J. P. 1979. *Plant Strategies and Vegetation Processes.* Wiley, New York.

Grime, J. P. 1989. Seed banks in ecological perspective. Pp. xv–xxii in M. Allessio Leck, V. T. Parker, and R. L. Simpson (Eds.), *Ecology of Soil Seed Banks.* Academic, San Diego.

Grime, J. P. and S. H. Hillier. 1981. Predictions based upon the laboratory characteristics of seeds. P. 6 in *Annual Report* 1981. Unit of Comparative Plant Ecology (NERC), University of Sheffield, Sheffield, England.

Grime, J. P. and R. Hunt. 1975. Relative growth rate: Its range and adaptive significance in a local flora. *J. Ecol.* 63:393–422.

Grodzinski, A. M. 1992. Allelopathic effects of cruciferous plants in crop rotation. Pp. 77–85 in S. J. H. Rizvi and V. Rizvi (Eds.), *Allelopathy: Basic and Applied Aspects.* Chapman and Hall, London.

Grotkopp, E., M. Rejmánek, and T. L. Rost. 2002. Toward a causal explanation of plant invasiveness: Seedling growth and life-history strategies of 29 pine (*Pinus*) species. *Am. Nat.* 159:397–419.

Groves, R. H. 1986. Invasion of Mediterranean ecosystems by weeds. Pp. 129–145 in B. Dell, S. J. M. Hopkins, and B. B. Lamont (Eds.), *Resilience in Mediterranean-Type Ecosystems.* Junk, Dordrecht, Netherlands.

Gu, X. Y., F. S. Kianian, and M. E. Foley. 2004. Multiple loci and epistasis control genetic variation for seed dormancy in weedy rice (*Oriza sativa*). *Genetics* 166: 1503–1516.

Guenzi, W. D. and T. McCalla. 1962. Inhibitions of germination and seedling development by crop residues. *Proc. Soil Sci. Soc. Am.* 26:456–458.

Gunsolus, J. L. and D. D. Buhler. 1999. A risk management perspective on integrated weed management. *J. Crop Prot.* 2:167–187.

Gurevitch, J., L. L. Morrow, A. Wallace, and J. S. Walsh. 1992. A meta-analysis of competition in field experiments. *Am. Nat.* 140:539–572.

Gurevitch, J., S. M. Scheiner, and G. A. Fox. 2002. *The Ecology of Plants.* Sinauer Associates, Sunderland, MA.

Haas, H. and J. C. Streibig. 1982. Changing patterns of weed distribution as a result of herbicide use and other agronomic factors. Pp. 57–80 in H. M. LeBaron and J. Gressel (Eds.), *Herbicide Resistance in Plants.* Wiley, New York.

Haase, P. 2001. Can isotropy *vs.* anisotropy in the spatial association of plant species reveal physical *vs.* biotic facilitation? *J. Veget. Sci.* 12:127–136.

Hakam, N. and J. P. Simon. 2000. Molecular forms and thermal and kinetic properties of purified glutathione reductase from two populations of barnyardgrass (*Echinochloa crus-galli* (L.) Beauv.: Poaceae) from contrasting climatic regions in North America. *Can J. Bot.* 78:969–980.

Hakansson, S. 1988. Behovet av ogr sbek mpning-bed mningsgrunder och prognosm jligheter. Pp. 21–31 in *Ogr s̲- och v xtskyddskonfer-enserna. Gemensam och v xtskyddsdel.* Uppsala, Sweden.

Hall, A. J., C. M. Rebella, C. M. Ghersa, and P. M. Culot. 1992. Field crop systems of the Pampas. Pp. 413–450 in C. J. Pearson (Ed.), *Ecosystems of the World. Field Crop Ecosystems.* Elsevier, New York.

Hall, L., K. Topinka, J. Huffman, L. Davis, and A. Good. 2000. Pollen flow between herbicideresistant *Brassica napus* is the cause of multiple-resistant *B. napus* volunteers. *Weed Sci.* 48:688–694.

Hall, L. S., P. R. Krausman, and M. L. Morrison. 1997. The habitat concept and a plea for standard terminology. *Wildl. Soc. Bull.* 25:173–182.

Halpern, C. B. and T. A. Spies. 1995. Plant species diversity in natural and managed forests of the Pacific Northwest. *Ecol. Appl.* 5:913–934.

Hamrick, J. L. and M. J. Godt. 1990. Allozyme diversity in plant species. Pp. 43–63 in A. H. D. Brown, M. T. Clegg, A. L. Kahler, and B. S. Weir (Eds.), *Plant Population Genetics, Breeding, and Genetic Resources.* Sinauer, Sunderland, MA.

Hamrick, J. L., M. J. Godt, and S. L. Sherman-Broyles. 1992. Factors influencing levels of genetic diversity in woody plant species. *New For.* 6:95–124.

Hamrick, J. L. and L. R. Holden. 1979. Influence of microhabitat heterogeneity on gene frequency distribution and gametic phase disequilibrium in *Avena barbata*. *Evolution* 33:521–533.

Hanfling, B. and J. Kollmann. 2002. An evolutionary perspective of biological invasions. *Trends Ecol. Evol.* 17:545–546.

Hanski, I. 1999. *Metapopulation Ecology*. Oxford University Press, New York.

Hanski, I. and M. E. Gilpin (Eds.). 1997. *Metapopulation Biology: Ecology, Genetics, and Evolution*. Academic, San Diego, CA.

Harper, J. L. 1956. The evolution of weeds in relation to the resistance to herbicides. Pp. 179–188 in *Proceedings 3rd British Weed Control Conference*. British Crop Protection Council, London, United Kingdom.

Harper, J. L. 1964. The individual in the population. *J. Ecol.* 52(Suppl.):149–158.

Harper, J. L. 1977. *The Population Biology of Plants*. Academic, London.

Harper, J. L. 1981. The concept of population in modular organisms. Pp. 53–77 in *Theoretical Ecology: Principles and Applications*, 2nd ed. Blackwell Scientific, Oxford.

Harper, J. L. 1984. *Modules, Branches and the Capture of Resources*. Yale University Press, New Haven, CT.

Harper, J. L. and A. D. Bell. 1979. The population dynamics of growth forms in organisms with modular construction. Pp. 29–52 in R. M. Anderson, B. D. Turner, and R. L. Taylor (Eds.), *Population Dynamics*. Blackwell Scientific, Oxford.

Harper, J. L. and J. White. 1971. The dynamics of plant populations. Pp. 41–62 in P. J. den Boer and G. R. Gradwell (Eds.), *Dynamics of Populations. Proceedings Advanced Study Institute on Dynamics of Numbers in Populations. Oosterbeek*, The Netherlands, Pudoc, Wageningen.

Harper, R. M. 1944. Preliminary report on the weeds of Alabama. *Bull. Geol. Surv. Ala.* 53:275.

Harrington, J. F. 1972. Seed storage and longevity. Pp. 145–245 in T. T. Kozlowski (Ed.), *Seed Biology*. Academic Press, New York.

Harrington, T. 1992. *Forest Vegetation Management without Herbicides*. Workshop Proceedings. Oregon State University, Corvallis, OR.

Harrison, S. 1999. Native and alien species diversity at the local and regional scales in a grazed California grassland. *Oecologia* 121:99–106.

Harrison, S. K., E. E. Regnier, and J. T. Schmoll. 2003. Post dispersal predation of giant ragweed (*Ambrosia trifida*) seed in no-tillage corn. *Weed Sci.* 51:955–964.

Harrod, J. R. 2001. The effects of invasive and noxious plants on land management in eastern Oregon and Washington. *Northwest Sci.* 75:85–90.

Hasan, S. and P. G. Ayres. 1990. The control of weeds through fungi: Principles and prospects. *New Phytol.* 115:201–222.

Hashem, A., S. R. Radosevich, and M. L. Roush. 1996. Effect of proximity factors on competition between winter wheat (*Triticum aestivum*) and Italian ryegrass (*Lolium multiflorum*). *Weed Sci.* 46:181–190.

Hatcher, P. E. and B. Melander. 2003. Combining physical, cultural and biological methods: Prospects for integrated non-chemical weed management strategies. *Weed Res.* 43:303–322.

Hawkes, C. V., J. Belnap, C. D'Antonio, and M. K. Firestone. 2006. Arbuscular mycorrhizal assemblages in native plant roots change in the presence of invasive exotic grasses. *Plant Soil* 281:369–380.

Hayes, J. P., S. H. Schoenholtz, M. J. Hartley, G. Murphy, R. F. Powers, D. Berg, and S. R. Radosevich. 2005. Environmental consequences of intensively managed forest plantations in the Pacific Northwest. *J. For.* 103:83–87.

He, J. S. and F. A. Bazzaz. 2003. Density-dependent responses of reproductive allocation to elevated atmospheric CO_2 in *Phytolacca americana*. *New Phytol.* 157:229–239.

Heap, I. 2006. *International Survey of Herbicide Resistant Weeds*. Available: http://www.weedscience.com, September 20, 2006.

Heckman, C. W. 1999. The encroachment of non-indigenous herbaceous plants into the Olympic National Forest. *Northwest Sci.* 73:264–276.

Hegde, S. G., J. D. Nason, J. M. Clegg, and N. C. Ellstrand. 2006. The evolution of California's wild radish has resulted in the extinction of its progenitors. *Evolution* 60:1187–1197.

Heger, T. and L. Trepl. 2003. Predicting biological invasions. *Biol. Invas.* 5:313–321.

Heiser, C. B. and T. W. Whitaker. 1948. Chromosome number, polyploidy, and growth habit in California weeds. *Am. J. Bot.* 35:179–186.

Henderson, L. 1991. Alien invasive *Salix* spp. (willows) in the grassland biome of South Africa. *S. Afr. For. J.* 157:91–95.

Hengeveld, R. 1989. *Dynamics of Biological Invasions*. Chapman and Hall, New York.

Hiebert, R. D. 1997. Prioritizing invasive plants and planning for management. Pp. 195–210 in J. O. Luken and J. W. Thieret (Eds.), *Assessment and Management of Plant Invasions*. Springer-Verlag, New York.

Hierro, J. L. and R. M. Callaway. 2003. Allelopathy and exotic plant invasion. *Plant Soil* 256:29–39.

Higgins, S. I., S. T. Pickett, and W. J. Bond. 2000. Predicting extinction risks for plants: Environmental stochasticity can save declining populations. *Trends Ecol. Evol.* 15: 516–520.

Higgins, S. I., D. M. Richardson, and R. M. Cowling. 1996. Modeling invasive plant spread: The role of plant-environment interactions and model structure. *Ecology* 77:2043–2054.

Hitchcock, A. S. and G. L. Clothier. 1898. *Kans. Exp. Sta. Bot. Bull.* 80:113–169.

Hobbs, N. T. 1996. Modification of ecosystems by ungulates. *J. Wildl. Manage.* 60:695–713.

Hobbs, R. J. 1991. Disturbance as a precursor to weed invasion to native vegetation. *Plant Prot. Q.* 6:99–104.

Hobbs, R. J. and L. Atkins. 1988. The effect of disturbance and nutrient addition on native and introduced annuals in the western Australian wheat belt. *Aust. J. Ecol.* 13:171–179.

Hobbs, R. J. and L. F. Huenneke. 1992. Disturbance, diversity, and invasion: Implications for conservation. *Conserv. Biol.* 6:324–337.

Hobbs, R. J. and S. E. Humphries. 1995. An integrated approach to the ecology and management of plant invasions. *Conserv. Biol.* 9:761–770.

Hobbs, R. J. and H. A. Mooney. 1991. Effects of rainfall variability and gopher disturbance on serpentine annual grassland dynamics. *Ecology* 72:478–491.

Hobbs, R. J., A. Salvatore, J. Aronson, J. S. Baron, P. Bridgewater, V. A. Cramer, P. R. Epstein, J. J. Ewel, C. A. Klink, A. E. Lugo, D. Norton, D. Ojima, D. M. Richardson, E. S. Sanderson, F. Valladares, M. Villa, R. Zamora, and M. Zobel. 2006. Novel ecosystems: Theoretical and management aspects of the new ecological world order. *Global Ecol. Biogeog.* 15:1–7.

Hodkinson, D. J. and K. Thompson. 1997. Plant dispersal: The role of man. *J. Appl. Ecol.* 34:1484–1496.

Holling, C. S. 1986. The resilience of terrestrial ecosystems: Local surprise and global change. Pp. 292–317 in W. C. Clark and R. E. Munn (Eds.), *Sustainable Development of the Biosphere.* Cambridge University Press, New York.

Holm, L., J. Doll, E. Holm, J. Pancho, and J. Herberger. 1997. *World Weeds. Natural Histories and Distribution.* Wiley, New York.

Holm, L. G., D. L Plucknett, J. V. Pancho, and J. P. Herberger. 1977. *The World's Worst Weeds: Distribution and Biology.* University Press of Hawaii, Honolulu.

Holmes, E. E., M. A. Lewis, J. E. Banks, and R. R. Veir. 1994. Partial differential equations in ecology: Spatial interactions and population dynamics. *Ecology* 75:17–29.

Holmes, M. G. and H. Smith. 1975. The function of phytochrome in plants growing in the natural environment. *Nature* 254:512–514.

Holmes, P. M. 2002. Depth distribution and composition of seed banks in alien-invaded and uninvaded fynbos vegetation. *Aust. Ecol.* 27:110–120.

Holmgren, M., M. Scheffer, and M. A. Huston. 1997. The interplay of facilitation and competition in plant communities. *Ecology* 78:1966–1975.

Holt, J. S. 1991. Applications of physiological ecology to weed science. *Weed Sci.* 39:521–528.

Holt, J. S. 1992. History of identification of herbicide-resistant weeds. *Weed Technol.* 6:615–620.

Holt, J. S. and H. M. LeBaron. 1990. Significance and distribution of herbicide resistance. *Weed Technol.* 4:141–149.

Holt, J. S. and D. R. Orcutt. 1991. Functional relationships of growth and competitiveness in perennial weeds and cotton (*Gossypium hirsutum*). *Weed Sci.* 39:575–584.

Holt, J. S., S. B. Powles, and J. A. M. Holtum. 1993. Mechanisms and agronomic aspects of herbicide resistance. *Annu. Rev. Plant Physiol. Plant Mol. Biol.* 44:203–229.

Holzner, W., I. Hayashi, and J. Glauninger. 1982. Reproductive strategy of annual agrestals. Pp. 111–121 in W. Holzner and M. Numata (Eds.), *Biology and Ecology of Weeds.* Junk, The Hague.

Holzner, W. and R. Immonen. 1982. Europe: An overview. Pp. 203–226 in W. Holzner and M. Numata (Eds.), *Biology and Ecology of Weeds.* Junk, The Hague.

Holzner, W. and M. Numata (Eds.). 1982. *Biology and Ecology of Weeds.* Junk, The Hague.

Honnay, O., K. Verheyen, and M. Hermy. 2002. Permeability of ancient forest edges for weedy plant species invasion. *For. Ecol. Manage.* 161:109–122.

Horn, H. S. 1971. *The Adaptive Geometry of Trees.* Princeton University Press, Princeton, NJ.

Horwith, B. 1985. A role for intercropping in modern agriculture. *BioScience* 35:286–291.

Howard, T. G. and D. E. Goldberg. 2001. Competitive response hierarchies for germination, growth, and survival and their influence on abundance. *Ecology* 82:979–990.

Huenneke, L. F., S. P. Hamburg, R. Koide, H. A. Mooney, and P. M. Vitousek. 1990. Effect of soil resources on plant invasion and community structure in California serpentine grassland. *Ecology* 71:478–491.

Hughes, G. and L. V. Madden. 2003. Evaluating predictive models with application in regulatory policy for invasive weeds. *Agric. Syst.* 76:755–774.

Hughes, J. and A. J. Richards. 1988. The genetic structure of populations of sexual and asexual *Taraxacum* dandelions. *Heredity* 60:161–172.

Humphries, S. E., R. H. Groves, and D. S. Mitchell. 1991. Plant invasions of Australian ecosystems. A status review and management directions. Pp. 1–127 in D. W. Walton et al. (Eds.), *Plant Invasions: The Incidence of Environmental Weeds in Australia* (Kowari). Australian National Parks and Wildlife Service, Canberra.

Hunt, R. 1978. *Plant Growth Analysis.* Edward Arnold, London.

Hunt, R. 1982. *Plant Growth Curves. The Functional Approach to Plant Growth Analysis.* University Park Press, Baltimore, MD.

Huston, M. A. 1994. *Biological Diversity: The Coexistence of Species on Changing Landscapes.* Cambridge University Press, New York.

Huston, M. and T. Smith. 1987. Plant succession: Life history and competition. *Am. Nat.* 130:168–198.

Hutchinson, C. M. and M. E. McGiffen. 2000. Cowpea cover crop mulch for weed control in desert pepper production. *Hort. Sci.* 35:196–198.

Hyde, E. O. C. 1954. The function of the hilum in some Papilionaceae in relation to the ripening of the seed and the permeability of the testa. *Ann. Bot. N.S.* 18: 241–256.

Hyvonen, T., E. Ketoja, and J. Salonen. 2003. Changes in the abundance of weeds in spring cereal fields in Finland. *Weed Res.* 43:348–356.

Inderjit, K. 1999. *Allelopathy: Principles, Procedures, Processes, and Promises for Biological Control.* Advances in Agronomy 67. Academic, New York.

Inderjit, K. 2005. Soil microorganisms: An important determinant of allelopathic activity. *Plant Soil* 274:227–236.

Inderjit, K. and R. Callaway. 2003. Experimental designs for the study of allelopathy. *Plant Soil* 256:1–11.

Inderjit, K., M. M. Dakshini, and F. A. Einhellig (Eds.). 1995. *Allelopathy: Organisms, Processes, and Applications.* American Chemical Society Symposium Series No. 582. American Chemical Society, Washington, DC.

Inderjit, K., M. M. Dakshini, and C. L. Foy (Eds.). 1999. *Principles and Practices in Plant Ecology: Allelochemical Interaction.* CRC Press, Boca Raton, FL.

Inderjit, K. and E. T. Nilsen. 2003. Bioassays and field studies for allelopathy in terrestrial plants: Processes and problems. *Crit. Rev. Pl. Sci.* 22:221–238.

Inderjit, K. and J. Weiner. 2001. Plant allelochemical interference or soil chemical ecology? *Perspect. Plant Ecol. Evol. Syst.* 4:3–12.

Izaguirre, M. M., C. A. Mazza, M. Biondini, I. T. Aldwin, and C. L. Ballaré. 2006. Remote sensing of future competitors: Impacts on plant defenses. *Proc. Natl. Acad. Sci. USA* 103:7170–7174.

Jackson, J. B. C., L. W. Buss, and R. E. Cook (Eds.). 1985. *Population Biology and Evolution of Clonal Organisms.* Yale University Press, New Haven, CT.

Jackson, S. T. 1997. Documenting natural and human-caused plant invasions using paleoe-cological methods. Pp. 37–55 in J. O. Luken and J. W. Thieret (Eds.), *Assessment and Management of Plant Invasions*. Springer-Verlag, New York.

Jana, S., M. K. Upadhyaya, and S. N. Acharya. 1988. Genetic basis of dormancy and differential response to sodium azide in *Avena fatua* seeds. *Can. J. Bot.* 66:635–641.

Janzen, D. H. 1977. What are dandelions and aphids? *Am. Nat.* 111:586–589.

Janzen, D. H. 1984. Dispersal of small seeds by big herbivores: Foliage is the fruit. *Am. Nat.* 123:338–353.

Jasieniuk, M. 2004. *Invasive Plant Management: Email Survey of Research Needs for Eco-logically Sound Management of Invasive Plants*. Center for Invasive Plant Management. Available: http://www.weedcenter.org/management/mariearticle.html.

Johansson, M. E., C. Nilsson, and E. Nilsson, 1996. Do rivers function as corridors for plant dispersal? *J. Veget. Sci.* 7:593–598.

Johnson, C. G. and D. K. I. Swanson. 2005. *Bunchgrass Plant Communities of the Blue and Ochoco Mountains: A Guide for Managers*. USDA Forest Service, Pacific North-west Research Station, Portland, OR. PNW-GTR-641.

Johnson D. E. 1999. Surveying, mapping, and monitoring noxious weeds on rangelands. Pp. 19–35 in R. L. Sheley and J. K. Petroff (Eds.), *Biology and Management of Noxious Rangeland Weeds*. Oregon State University Press, Corvallis, OR.

Johnson, L. P. V. 1935. The inheritance of delayed germination in hybrids of *Avena fatua* and *A. sativa*. *Can. J. Res.* 13:367–387.

Jolliffe, P. A. 1997. Are mixed populations of plant species more productive than pure stands? *Oikos* 80:595–602.

Jolliffe, P. A. 2000. The replacement series. *J. Ecol.* 88:371–385.

Jolliffe, P. A., A. N. Minjas, and V. E. Runeckles. 1984. A reinterpretation of yield relationships in replacement series experiments. *J. Appl. Ecol.* 21:227–243.

Jørgensen, R. B. and B. Andersen. 1994. Spontaneous hybridization between oilseed rape (*Brassica napus*) and weedy *B. campestris* (Brassicaceae): A risk of growing genetically modified oilseed rape. *Am. J. Bot.* 81:1620–1626.

Jørgensen, R. B., B. Andersen, L. Landbo, and T. R. Mikkelsen. 1996. Spontaneous hybridiz-ation between oilseed rape (*Brassica napus*) and weedy relatives. *Acta Hort.* 407:193–200.

Jørgensen, S. E. 2006. Towards a thermodynamics of biological systems. *Int. J. Eco-dynamics* 1:9–27.

Julien, M. H. and M. W. Griffiths. 1998. *Biological Control of Weeds: A World Catalogue of Agents and their Target Weeds*, 4th ed. CAB International, Wallingford, United Kingdom.

Kaeppler, H. F. 2000. Food safety assessment of genetically modified crops. *Agron. J.* 92:793–797.

Kalkhan, M. A. and T. J. Stohlgren. 2000. Using multi-scale sampling and spatial cross-correlation to investigate patterns of plant species richness. *Environ. Monit. Assess.* 64:591–605.

Kamrin, M. A. 1997. *Pesticide Profiles*, Toxicity, Environmental Impacts, and Fate. CRC Press, Boca Raton, FL.

Kareiva, P. 1996. Developing a prediction ecology for nonindigenous species and ecologi-cal invasions. *Ecology* 77:1651–1652.

Karssen, C. M. 1970a. The light promoted germination of *Chenopodium album* L. III. The effect of the photoperiod during growth and development of the plants on the dormancy of the produced seeds. *Acta Bot. Neerl.* 19:81–94.

Karssen, C. M. 1970b. The light promoted germination of *Chenopodium album* L. VI. Pfr requirements during different stages of the germination process. *Acta Bot. Neerl.* 19:297–312.

Kartesz, J. and A. Farstad. 1999. Exotic vascular plant species: Where do they occur? Pp. 51–54 in T. H. Ricketts, E. Dinerstein, D. M. Olson, and C. J. Loucks (Eds.), *Terrestrial Ecoregions of North America: A Conservation Assessment.* Island, Washington, DC.

Kearney, P. C. and D. D. Kaufman (Eds.). 1988. *Herbicides: Chemistry, Degradation and Mode of Action*, 2nd ed. Vols. 1 and 2. Marcel Dekker, New York.

Keating, B. A. and P. S. Carberry. 2003. An overview of APSIM, a model designed for farming systems simulation. *Eur. J. Agron.* 18:267–288.

Keddy, P., K. Nielsen, E. Weiher, and R. Lawson. 2002. Relative competitive performance of 63 species of terrestrial herbaceous plants. *J. Veget. Sci.* 13:5–16.

Keddy, P. A. 1992. A pragmatic approach to functional ecology. *Funct. Ecol.* 6:621–626.

Keddy, P. A. 2001. *Competition*, 2nd ed. Kluwer Academic, Dordrecht, Netherlands.

Keddy, P. A., L. Twolan-Strutt, and I. C. Wisheu. 1994. Competitive effect and response rankings in 20 wetland plants—Are they consistent across 3 environments? *J. Ecol.* 82:635–643.

Keller, L. F. and D. M. Waller. 2002. Inbreeding effects in wild populations. *Trends Ecol. Evol.* 17:230–241.

Kelner, D., L. Juras, and D. Derksen. 1996. Integrated weed management: Making it work on your farm. Saskatchewan Agriculture and Food Extension Bulletin, Manitoba Agriculture and Agriculture and Agri-Food Canada. Available: http://www.agr. gov.sk.ca/docs/production/weedmgt.asp.

Kennedy, P. B. 1903. Summer ranges of eastern Nevada sheep. *Nevada Agric. Exp. Sta. Bull.* 55. Pp. 1–53.

Kennedy, A. C. and R. J. Kremer. 1996. Microorganisms in weed control strategies. *J. Prod. Agric.* 9:480–484.

Kidd, F. and E. West. 1919. Physiological pre-determination: The influence of the physiological condition of the seed upon the subsequent growth and upon the yield. IV. Review of the literature. Chapter III. *Ann. Appl. Biol.* 5:220–251.

Kie, J. G., R. T. Bowyer, and K. M. Stewart. 2003. Ungulates in western forests: Habitat requirements, population dynamics, and ecosystem processes. Pp. 296–340 in C. J. Zabel and R. G. Anthony (Eds.), *Mammal Community Dynamics in Coniferous Forests of Western North America: Management and Conservation.* Cambridge University Press, New York.

Kimmins, J. P. 1997. *Forest Ecology: A Foundation for Sustainable Management.* 2nd Ed. Prentice-Hall, Upper Saddle River, NJ.

King, J. J. 1966. *Weeds of the World: Biology and Control.* Interscience, New York.

King, S. A. and R. T. Buckney. 2001. Exotic plants in the soil-stored seed bank of urban brushland. *Aust. J. Bot.* 49:717–720.

King, S. R. and E. S. Hagood. 2003. The effect of johnsongrass (*Sorghum halepense*) control method on the incidence and severity of virus diseases in glyphosate-tolerant corn (*Zea mays*). *Weed Technol.* 17:503–508.

Klemmedson, J. O. and J. G. Smith. 1964. Cheatgrass (*Bromus tectorum* L). *Bot. Rev.* 30:226–291.

Knapp, E. E., M. A. Goedde, and K. J. Rice. 2001. Pollen-limited reproduction in blue oak: Implications for wind pollination in fragmented populations. *Oecologia* 128:48–55.

Knops, J. M. H., D. Tilman, N. M. Haddad, S. Naeem, C. E. Mitchell, J. Haarstad, M. E. Ritchie, K. M. Howe, P. B. Reich, E. Siemann, and J. Groth. 1999. Effects of plant species richness on invasion dynamics, disease outbreaks, insect abundances and diversity. *Ecol. Lett.* 2:286–293.

Knowe, S. A., S. R. Radosevich, and R. G. Shula. 2005. Prediction and projection model for young Douglas-fir plantations and associated vegetation in the Coast Ranges and Cascade Mountains of the Pacific Northwest. *West. J. Appl. For.* 20:11–93.

Kogan, M. (Ed.). 1986. *Ecological Theory and Integrated Pest Management Practice.* Wiley, New York.

Kolar, C. S. and D. M. Lodge. 2001. Progress in invasion biology: Predicting invaders. *Trends Ecol. Evol.* 16:199–204.

Kornas, J. 1990. Plant invasions in Central Europe: Historical and ecological perspectives. Pp. 19–36 in F. di Castri, A. J. Hansen, and M. Debussche (Eds.), *Biological Invasions in Europe and the Mediterranean Basin.* Kluwer Academic, Dordrecht, Netherlands.

Kot, M., M. A. Lewis, and P. van den Driessche. 1996. Dispersal data and spread of invading organisms. *Ecology* 77:2027–2042.

Kotanen, P. M. 1996. Revegetation following soil disturbance in a California meadow: The role of propagule supply. *Oecologia* 108:652–662.

Kowarik, I. 1995. Time lags in biological invasions with regard to the success and failure of alien species. Pp. 15–38 in P. Pyšek, K. Prach, M. Rejmánek, and M. Wade (Eds.), *Plant Invasions: General Aspects and Special Problems.* SPB Academic, Amsterdam, Netherlands.

Kremer, R. J. 1998. Microbial interactions with weed seeds and seedlings and its potential for weed management. Pp. 161–179 in J. L. Hatfield, D. D. Buhler, and B. A. Stewart (Eds.), *Integrated Weed and Soil Management.* Ann Arbor Press, Chelsea, MI.

Kremer, R. J. and J. Li. 2003. Developing weed-suppressive soils through improved soil quality management. *Soil Till. Res.* 72:193–202.

Krishna, K. R. 2005. *Mycorrhizas: A Molecular Analysis.* Science, Enfield, NH.

Kriticos, D. J., J. R. Brown, G. Maywald, I. D. Radford, D. M. Nicholas, R. W. Sutherst, and S. W. Adkins. 2003. SPANDX: A process-based population dynamics model to explore management and climate change impacts on an invasive alien plant, *Acacia nilotica. Ecol. Model.* 163:187–208.

Kropff, M. J. and L. A. P. Lotz. 1992. Optimization of weed management systems: The role of ecological models of interplant competition. *Weed Technol.* 6:462–470.

Kropff, M. J. and H. H. van Laar (Eds.). 1993. *Modelling Crop-Weed Interactions.* CAB International, Wallingford, United Kingdom.

Krueger-Mangold, J. M., R. L. Sheley, and T. J. Svejcar. 2006. Toward ecologically-based invasive plant management on rangeland. *Weed Sci.* 54:597–605.

Kuijt, J. 1969. *The Biology of Parasitic Flowering Plants.* University of California Press, Berkeley, CA.

Kulmatiski, A., K. H. Beard, and J. M. Stark. 2004. Finding endemic soil-based controls for weed growth. *Weed Technol.* 18(Suppl. S.):1353–1358.

Lambrecht-McDowell, S. C. and S. R. Radosevich. 2005. The population demographics and trade-offs to reproduction for an invasive and noninvasive species of *Rubus. Biol. Invas.* 7:281–295.

Landis, D. A., F. D. Menalled, A. C. Costamagna, and T. K. Wilkinson. 2005. Manipulating plant resources to enhance beneficial arthropods in agricultural landscapes. *Weed Sci.* 53:902–908.

La Peyre, M. K. G., J. B. Grace, E. Hahn, and I. A. Mendelssohn. 2001. The importance of competition in regulating plant species abundance along a salinity gradient. *Ecology* 82:62–69.

Larson, B. M. H. 2005. The war of roses: Demilitarizing invasion biology. *Front. Ecol. Environ.* 2:495–500.

Lass, L. W, D. C. Thill, B. Shaffi, and T. S. Prather. 2002. Detecting spotted knapweed (*Centaurea maculosa*) with hyperspectral remote sensing technology. *Weed Technol.* 16:426–432.

Latore, J., P. Gould, and A. M. Mortimer. 1998. Spatial dynamics and critical patch size of annual plant populations. *J. Theor. Biol.* 190:277–285.

Leake, J., D. Johnson, D. Donnelly, G. Muckle, L. Boddy, and D. Read. 2004. Networks of power and influence: The role of mycorrhizal mycelium in controlling plant communities and agroecosystem functioning. *Can. J. Bot.* 82:1016–1045.

Leather, G. R. and F. A. Einhellig. 1986. Bioassays in the study of allelopathy. Pp. 133–146 in A. R. Putnam, and C. S. Tang (Eds.), *The Science of Allelopathy.* John Wiley and Sons, New York, NY.

LeBaron, H. M. and J. Gressel (Eds.). 1982. *Herbicide Resistance in Plants.* Wiley, New York.

Le Coeur, D., J. Baudry, F. Burel, and C. Thenail. 2002. Why and how we should study field boundary biodiversity in an agrarian landscape context. *Agric. Ecosyst. Environ.* 89:23–40.

Lee, C. E. 2002. Evolutionary genetics of invasive species. *Trends Ecol. Evol.* 17:386–391.

Lee, L. J. and J. Ngim. 2000. A first report of glyphosate-resistant goosegrass (*Elusine indica* L. Gaertn) in Malaysia. *Pest Manag. Sci.* 56:336–339.

Lejeune, K. D. and T. R. Seastedt. 2001. *Centaurea* species: The forb that won the west. *Conserv Biol.* 15:1568–1574.

Leopold, A. 1943. What is a weed? The weed flora of Iowa. Bulletin no. 4. Pp. 306–309 in S. L Hader and J. B. Callicott (Eds.), 1991. *The River of the Mother of God and Other Essays by Aldo Leopold.* University of Wisconsin Press, Madison, WI.

Leps, J. and V. Hadincova. 1992. How reliable are our vegetation analyses? *J. Veget. Sci.* 3:119–124.

Lesica, P. and B. Martin. 2003. Effects of prescribed fire and season of burn on recruitment of the invasive plant, *Potentilla recta*, in a semiarid grassland. *Restor. Ecol.* 11:516–523.

Leslie, P. H. 1945. On the use of matrices in certain population mathematics. *Biometrika* 33:183–212.

Leung, B., D. M. Lodge, D. Finnoff, J. F. Shogren, M. A. Lewis, and G. Lamberti. 2002. An ounce of prevention or a pound of cure: Bioeconomic risk analysis of invasive species. *Proc. R. Soc. Lond. Series B-Biol. Sci.* 269: 2407–2413.

Levin, D. A. and H. W. Kerster. 1974. Gene flow in seed plants. *Evol. Biol.* 7:139–220.

Levine, J. M. 2000. Species diversity and biological invasions: Relating local process to community pattern. *Science* 288:852–854.

Levine, J. M. and C. M. D'Antonio. 1999. Elton revisited: A review of evidence linking diversity and invasibility. *Oikos* 87:15–26.

Levine, J. M., M. Vila, C. M. D'Antonio, J. S. Dukes, K. Grigulis, and S. Lavorel. 2003. Mechanisms underlying the impact of exotic plant invasions in terrestrial ecosystems. *Proc. R. Soc. Lond. Series B Biol. Sci.* 260:1059–1062.

Levins, R. 1986. Perspectives in integrated pest management. Pp. 1–18 in M. Kogan (Ed.), *Ecological Theory and Integrated Pest Management Practice.* Wiley, New York.

Levins, R. 1970. Extinction. Pp. 75–107 in M. Gerstenhaber (Ed.), *Some Mathematical Questions in Biology: Lecture Notes on Mathematics in the Life Sciences.* American Mathematical Society, Providence, RI.

Levins, R. and J. H. Vandermeer. 1994. The agroecosystem embedded in a complex ecological community. Pp. 341–362 in C. R. Carroll, J. H. Vandermeer, and P. M. Rosset (Eds.), *Agroecology.* McGraw-Hill, New York.

Lewis, J. 1973. Longevity of crop and weed seed: Survival after 20 years in the soil. *Weed Res.* 13:179–191.

Lewis, M. A., M. G. Neubert, H. Caswell, J. S. Clark, and K. Shea. 2005. *A Guide to Calculating Discrete-Time Invasion Rates from Data.* Demography and Dispersal Working Group, University of California, Santa Barbara.

Lewontin, R. 1982. Agricultural research and the penetration of capital. *Sci. People* 1: 12–17.

Liebman, M. 1986. Ecological suppression of weeds in intercropping systems: experiments with barley, pea, and mustard. Ph.D. Dissertation. University of California, Berkeley.

Liebman, M. 1988. Ecological suppression of weeds in intercropping systems: A review. Pp. 197–212 in M. A. Altieri and M. Liebman (Eds.), *Weed Management in Agroecosystems: Ecological Approaches.* CRC Press, Boca Raton, FL.

Liebman, M. 2001. Managing weeds with insects and pathogens. Pp. 375–408 in M. Liebman, C. M. Mohler, and C. P. Staver (Eds.), *Ecological Management of Agricultural Weeds.* Cambridge University Press, New York.

Liebman, M. and A. S. Davis. 2000. Integration of soil, crop and weed management in low-external-input farming systems. *Weed Res.* 40:27–47.

Liebman, M. and E. R. Gallandt. 1997. Many little hammers: Ecological management of crop-weed interactions. Pp. 291–343 in L. E. Jackson (Ed.), *Ecology in Agriculture.* Academic, San Diego, CA.

Liebman, M. and C. M. Mohler. 2001. Weeds and the soil environment. Pp. 210–268 in M. Liebman, C. M. Mohler, and C. P. Staver (Eds.), *Ecological Management of Agricultural Weeds.* Cambridge University Press, New York.

Liebman, M., C. M. Mohler, and C. P. Staver. 2001. *Ecological Management of Agricultural Weeds.* Cambridge University Press, New York.

Liebman, M. and T. Ohno. 1998. Crop rotation and legume residue effects on weed emergence and growth: Implications for weed management. Pp. 181–221 in J. L. Hatfield, D. D. Buhler, and B. A. Stewart (Eds.), *Integrated Weed and Soil Management.* Ann Harbor, Chelsea, MI.

Liebman, M. and C. P Staver. 2001. Crop diversification for weed management. Pp. 322–374 in M. Liebman, C. M. Mohler, and C. P. Staver (Eds.), *Ecological Management of Agricultural Weeds*. Cambridge University Press, New York.

Linde, M., S. Diel, and B. Neuffer. 2001. Flowering ecotypes of *Capsella bursa-pastoris* (L.) Medik (Brassicaceae) analyzed by a cosegregation of phenotypic characters (QTL) and molecular markers. *Ann. Bot.* 87:91–99.

Linhart Y. B. and M. C. Grant. 1996. Evolutionary significance of local genetic differentiation in plants. *Annu. Rev. Ecol. Syst.* 27:237–277.

List, P. C. (Ed.). 2000. *Environmental Ethics and Forestry*. Temple University Press, Philadelphia, PA.

Lonsdale, W. M. 1999. Global patterns of plant invasions and the concept of invasibility. *Ecology* 80:1522–1536.

Loope, L. L. and D. Mueller-Dombois. 1989. Characteristics of invaded islands with special reference to Hawaii. Pp. 257–280 in J. A. Drake, F. di Castri, R. H. Groves, F. J. Kruger, H. A. Mooney, M. Rejmánek, and M. H. Williams (Eds.), *Biological Invasions: A Global Perspective*. Wiley, New York.

Losey, E. J., L. S. Rayor, and M. E. Carter. 1999. Transgenic pollen harms monarch larvae. *Nature* 399:214.

Lotka, A. J. 1925. *Elements of Physical Biology*. Williams and Wilkins, Baltimore, MD.

Louda, S. M. 1989. Predation in the dynamics of seed regeneration. Pp. 25–51 in M. Allessio Leck, V. T. Parker, and R. L. Simpson (Eds.), *Ecology of Soil Seed Banks*. Academic Press, San Diego, CA.

Louda, S. M. 2000. *Rhinocyllus conicus*—Insights to improve predictability and minimize risk of biological control of weeds. Pp. 187–194 in N. R. Spenser (Ed.), *X International Symposium on Biological Control of Weeds*, Bozeman, MT. Advanced Litho Printing, Great Falls, MT.

Louda, S. M., KH. Keeler, and R. D. Holt. 1990a. Herbivore influences on plant performance and competitive interactions. Pp. 414–444 in J. B. Grace and D. Tilman (Eds.), *Perspectives on Plant Competition*. Academic, San Diego, CA.

Louda, S. M., R. W. Pemberton, and M. T. Johnson. 2003. Nontarget effects—The Achilles' Heel of biological control? Retrospective analysis to reduce risk associated with biocontrol introductions. *Annu. Rev. Entomol.* 48:365–396.

Loyd, D. G. 1965. Evolution of self compatibility and racial differentiation in *Leavenworthia* (Cruciferae). *Contrib. Gray Herb. Harvard Univ.* 195:3–134.

Lueschen, W. E. and R. N. Anderson. 1980. Longevity of velvetleaf seeds in soil under agricultural practices. *Weed Sci.* 28:341–346.

Luken, J. O. and J. W. Thieret. 1997. *Assessment and Management of Plant Invasions*. Springer, New York.

MacArthur, R. H. 1962. Generalized theorems of natural selection. *Proc. Natl. Acad. Sci. USA* 48:1893–1897.

MacDougall, A. S. and R. Turkington. 2005. Are invasive species the drivers or passengers of change in degraded ecosystems? *Ecology* 86:42–55.

Mack, R. N. 1981. Invasion of *Bromus tectorum* L. into western North America: An ecological chronicle. *Agro-Ecosystems* 7:145–165.

Mack, R. N. 1984. Invaders at home on the range. *Nat. Hist.* 93:40–47.

Mack, R. N. 1986. Alien plant invasion into the intermountain west: A case history. Pp. 191–213 in H. A. Mooney and J. A. Drake (Eds.), *Ecology of Biological Invasions of North America and Hawaii.* Ecological Studies 58. Springer-Verlag, New York.

Mack, R. N. 1991. The commercial seed trade: An early disperser of weeds in the United States. *Econ. Bot.* 455:257–273.

Mack, R. N. 1995. Understanding the process of weed invasions: The influence of environmental stochasticity. Pp. 65–74 in C. Stirton (Ed.), *Weeds in a Changing World.* British Crop Protection Council Symposium Proceedings No. 64, Brighton, United Kingdom.

Mack, R. N. 1996. Predicting the identity and fate of plant invaders: Emergent and emerging approaches. *Biol. Conserv.* 78:107–121.

Mack, R. N. 2005. Assessing biotic invasions in time and space: The second imperative. Pp. 179–208 in H. A. Mooney, R. N. Mack, J. A. McNeely, L. E. Neville, P. J. Schei, and J. K. Waage (Eds.), *Invasive Alien Species. A New Synthesis.* Island, Washington, DC.

Mack, R. N., D. Simberloff, W. M. Lonsdale, H. Evans, M. Clout, and F. A. Bazzaz. 2000. Biotic invasions: Causes, epidemiology, global consequences and control. *Ecol. Appl.* 10:179–200.

MacLeod, D. G. 1962. Some anatomical and physiological observations on two species of *Cuscuta. Trans. Bot. Soc. Edin.* 39:302–315.

Maddonni, G. A., M. E. Iglesias Pérez, J. Cárcova, and M. A. Ghersa, 1999. Maize flowering dynamic in soils with contrasting agricultural history. *Maydica* 44:141–147.

Magurran, A. E. 1988. *Ecological Diversity and Its Measurements.* Princeton University Press, Princeton, NJ.

Mallory-Smith, C. A. and E. J. Retzinger. 2003. Revised classification of herbicides by site of action for weed resistance management strategies. *Weed Technol.* 17:605–619.

Manley, B. S., H. P. Wilson and T. E. Hines. 2000. Weed management and crop rotations influence populations of several broadleaf weeds. *Weed Sci.* 49:106–122.

Maron, J. L., V. Montserrat, R. Bommarco, S. Elmendorf, and P. Beardsley. 2004. Rapid evolution of an invasive plant. *Ecol. Monogr.* 74:261–280.

Marshall, D. R. and S. K. Jain. 1967. Cohabitation and relative abundance of two species of wild oats. *Ecology* 48:656–659.

Marshall, E. J. P., V. K. Brown, P. J. W. Lutman, G. R. Squire, and L. K. Ward. 2003. The role of weeds in supporting biological diversity with in crop fields. *Weed Res.* 43:77–89.

Marshall, E. J. P. and A. C. Moonen. 2002. Field margins in northern Europe: Their functions and interactions with agriculture. *Agric. Ecosyst. Environ.* 89:5–21.

Martinez-Ghersa, M. A. and C. M. Ghersa. 2006. The relationship of propagule pressure to invasion potential in plants. *Euphytica* 148:87–96.

Martinez-Ghersa, M. A., C. M. Ghersa, R. L. Benech Arnold, and R. A. Sanchez. 2000a. Adaptive traits regulating dormancy and germination of invasive species. *Plant Sp. Biol.* 15:127–137.

Martinez-Ghersa, M. A., C. M. Ghersa, and E. H. Satorre. 2000b. Coevolution of agricultural systems and their weed companions: Implications for research. *Field Crops Res.* 67:181–190.

Martinez-Ghersa, M. A., C. A. Worster, and S. R. Radosevich. 2003. Concerns a weed scientist might have about herbicide-tolerant crops: A revisitation. *Weed Technol.* 17:202–210.

Marushia, R. G. and J. S. Holt. 2006. The effects of habitat on dispersal patterns of an invasive thistle, *Cynara cardunculus. Biol. Invas.* 8:577–593.

Mashhadi, H. R. and S. R. Radosevich. 2003. Invasive plants. Pp. 1–28 in K. Inderjit (Ed.), *Weed Biology and Management.* Kluwer Academic, Dordrecht, Netherlands.

Mason, J. M. 1932. Weed survey of the prairie provinces. *Dom. Can. Nat. Res. Council Rep.* 26:34.

Mater, J. 1992. A paradigm shift for marketing the forest industry—From public relations to research. 46th Annual meeting, June, 1992, Forest Products Research Society, Charleston, South Carolina.

Maxwell, B. D. 2005. Monitoring of non-indigenous plant species. Chapter 5 in *Invasive Plant Management: CIPM Online Textbook.* Available: http://www.weedcenter.org/textbook/5_maxwell_monitoring.html.

Maxwell, B. D. and C. M. Ghersa. 1992. The influence of weed seed dispersion versus the effect of competition on crop yield. *Weed Technol.* 6:196–204.

Maxwell, B. D., L. J. Rew, and R. Aspinall. 2003. Exotic plant survey and monitoring: Methods to discover distribution and low frequency occurrence. In T. Phillippi and R. Doren (Eds.), *Proceedings for Detecting Invasive Exotic Plants, Workshop and Conference.* Florida State University, Miami, FL.

Maxwell, B. D., M. V. Wilson, and S. R. Radosevich. 1988. Population modeling approach for evaluating leafy spurge (*Euphorbia esula*) development and control. *Weed Technol.* 2:132–138.

May, R. M. 1974. *Stability and Complexity in Model Ecosystems.* Princeton University Press, Princeton, NJ.

May, R. M. 2001. *Stability and Complexity in Model Ecosystems.* 2nd ed. Princeton University Press, Princeton, NJ.

Mayr, E. 1982. Speciation and macroevolution. *Evolution.* 36:1119-1132.

Mazia, C. N., E. J. Chaneton, C. M. Ghersa, and R. J. C. León. 2001. Limits to tree species invasion in grassland and forest plant communities. *Oecologia* 128:594–602.

McDowell, S. C. L. and D. P. Turner. 2002. Reproductive effort in invasive and noninvasive *Rubus. Oecologia* 133:102–111.

McGee, A. B. and E. Levy. 1988. Herbicide use in forestry: Communication and information gaps. *J. Environ. Manage.* 26:111–126.

McHenry, W. B. and R. F. Norris. 1972. *Study Guide for Agricultural Pest Control Advisers on Weed Control.* Publication 4050. University of California, Berkeley.

McHughen, A. 2000. *Pandora's Picnic Basket. The Potential and Hazards of Genetically Modified Foods.* Oxford University Press, Oxford.

McKenzie, F. R. 1996. Influence of applied nitrogen on weed invasion of *Lolium perenne* pastures in a subtropical environment. *Aust. J. Exp. Agric.* 36:657–660.

McKibben, B. 1999. *The End of Nature.* Random House, New York.

McNeely, J. A. 1999. The great reshuffling: How alien species help feed the global economy. Pp. 11–31 in O. T. Sandlund, P. J. Schei, and Å. Viken (Eds.), *Invasive Species and Biodiversity Management.* Based on a selection of papers presented at the Norway/UN Conference on Alien Species, Trondheim, Norway. Population and Community Biology Series, Vol. 24, Kluwer Academic, Dordrecht, Netherlands.

McPherson, G. R. and S. DeStefano. 2003. *Applied Ecology and Natural Resource Management*. Cambridge University Press, New York.

McWhorter, C. G. 1989. History, biology and control of johnsongrass. *Rev. Weed Sci.* 4:87–115.

Medlin, C. R. and D. R. Shaw. 2000. Economic comparison of broadcast and site-specific herbicide applications in non-transgenic and glyphosate-tolerant *Glycine max. Weed Sci.* 48:653–661.

Menalled, F. D., P. C. Marino, K. A. Renner, and D. A. Landis. 2000. Post-dispersal weed seed predation in Michigan crop fields as a function of agricultural landscape structure. *Agric. Ecosyst. Environ.* 77:193–202.

Mendoza, M. R., P. E. Gundel, M. A. Martinez-Ghersa, and C. M. Ghersa. 2005. Dormición en semillas de *Lolium multiflorum* asociada a la exposición al herbicida diclofop-metil y al nivel de resistencia. Actas XXX Jornadas Argentinas de Botánica, Rosario, Argentina.

Menge, B. A. and J. P. Sutherland. 1987. Community regulation: Variation in disturbance, competition, and predation in relation to environmental stress and recruitment. *Am. Natur.* 130:730–757.

Merchant, E. 1980. *The Death of Nature*. Harper and Row, New York.

Messersmith, C. G. and S. W. Adkins. 1995. Integrating weed-feeding insects and herbicides for weed control: *Weed Technol.* 9:199–208.

Michelena, R. O., C. B. Irurtia, F. A. Vavruska, R. Mon, and A. Pittaluga. 1989. Degradación de suelos en el Norte de la Región Pampeana. INTA Pub. Tecn. 6.0. Proyecto de Agricultura Conservacionista. Centros regionales de Buenos Aires Norte, Córdoba, Entre Ríos y Santa Fé.

Michener, W. K. 1997. Quantitatively evaluating restoration experiments: Research design, statistical analysis, and data management considerations. *Restor. Ecol.* 5:324–337.

Mikkelsen, T. R., B. Andersen, and R. B. Jørgensen. 1996. The risk of transgene spread. *Nature* 380:31.

Milberg, P. 1990. What is the maximum longevity of seeds? *Svensk Botanisk Tidskrift* 84:323–352.

Miller, J. A. 1991. Biosciences and ecological integrity. *BioScience* 41:206–210.

Miller, T. E. and P. A. Werner. 1987. Competitive effects and responses in plants. *Ecology* 68:1201–1210.

Milne, R. A. and R. J. Abbott. 2000. Origin and evolution of invasive naturalized material of *Rhododendron ponticum* L. in the British islands. *Mol. Ecol.* 9:541–556.

Milton, S. J. 2003. Emerging ecosystems: A washing-stone for ecologists, economists, and sociologists? *S. Afr. J. Sci.* 99:404–406.

Mitchell, R. J., R. H. Marrs, and M. H. D. Auld, 1998. A comparative study of the seed-banks of heathland and successional habitats in Dorset, Southern England. *J. Ecol.* 86:588–596.

Mitton, J. B. and M. C. Grant. 1996. Genetic Variation and the natural history of quaking aspen. *Bioscience* 46:25–31.

Mohler, C. L. 1996. Ecological basis for the cultural control of annual weeds. *J. Prod. Agric.* 9:468–474.

Mohler, C. L. 2001. Mechanical management of weeds. Pp. 139–209 in M. Liebman, C. M. Mohler, and C. P. Staver (Eds.), *Ecological Management of Agricultural Weeds*. Cambridge University Press, New York.

Molisch, H. 1937. *Der Einfluss einer Pflanze auf die andere-Allelopathie.* Fischer, Jena.

Molvar, E. M., R. T. Bowyer, and V. Van Ballenberghe. 1993. Moose herbivory, browse quality, and nutrient cycling in an Alaskan treeline community. *Oecologia* 94:472–479.

Monaco, T. J., S. C. Weller, and F. M. Ashton. 2002. *Weed Science Principles and Practices*, 4th ed. Wiley, New York.

Moody, K. 1988. Developing appropriate weed management strategies for small scale farmers. Pp. 319–330 in M. A. Altieri and M. Liebman (Eds.), *Weed Management in Agroecosystems: Ecological Approaches.* CRC Press, Boca Raton, FL.

Moody, M. E. and R. N. Mack. 1988. Controlling the spread of plant invasions: The importance of nascent foci. *J. Appl. Ecol.* 25:1009–1021.

Moran, G. F. and D. R. Marshall. 1978. Allozyme uniformity within and variation between races of the colonizing species *Xanthium strumarium* L. (Noogoora Burr.). *Aust. J. Biol. Sci.* 31:283–291.

Mortensen, D. A., L. Bastiaans, and M. Sattin. 2000. The role of ecology in the development of weed management systems: An outlook. *Weed Res.* 40:49–62.

Mortensen, D. A., G. A. Johnson, and L. J. Young. 1993. Weed distribution in agricultural fields. Pp. 113–124 in P. Robert and R. H. Rust (Eds.), *Site Specific Crop Management.* American Society of Agronomy Crop Science Society of America-Soil Science Society of America, Madison, WI.

Moser, H. S., W. B. McCloskey, J. C. Silvertooth, P. Dugger, and D. Richter. 2000. Performance of transgenic cotton varieties in Arizona. Pp. 497–499 in D. Richter and P. Dugger (Eds.) *Proceedings of the Beltwide Cotton Conference*, Vol. 1. National Cotton Council of America, San Antonio, TX.

Moss, S. R. 2002. Herbicide resistant weeds. Pp. 225–252 in R. E. L. Naylor (Ed.), *Weed Management Handbook*, 9th ed. Blackwell, London.

Mueller-Dombois, D. and H. Ellenberg. 1974. *Aims and Methods of Vegetation Ecology.* Wiley, New York.

Musselman, L. J. (Ed.). 1987. *Parasitic Weeds in Agriculture.* CRC Press, Boca Raton, FL.

Muyt, A. 2001. *Bush Invaders of Southeast Australia.* R. G. and F. J. Richardson, Meredith, Victoria, Australia.

Naeem, S., J. M. H. Knops, D. Tilman, K. M. Howe, T. Kennedy, and S. Gale. 2000. Plant diversity increases resistance to invasion in the absence of covarying extrinsic factors. *Oikos* 91:97–108.

Nathan, R. and C. Muller-Landau. 2000. Spatial patterns of seed dispersal, their determinants and consequences for recruitment. *Trends Ecol. Evol.* 15:278–285.

National Research Council. 1989. *Alternative Agriculture.* National Academy Press, Washington, DC.

National Research Council. 1996. *Ecologically Based Pest Management: New Solution for a New Century.* National Academy Press, Washington, DC.

National Research Council. 2000. *The Future Role of Pesticides in US Agriculture.* National Academy Press, Washington, DC.

NAWMA. 2002. North American invasive plant mapping standards. Available: http://www.nawma.org.

Naylor, B. J., B. A. Endress, and C. G. Parks. 2005. Multi-scale detection of sulfur cinquefoil using aerial photography. *Rangeland Ecol. Manag.* 58:447–451.

Neill, R. L. and E. L. Rice. 1971. Possible role of *Ambrosia psilostachya* on pattern and succession in old-fields. *Am. Midland Nat.* 86:344.

Nelder, J. A. 1962. New kinds of systematic designs for spacing studies. *Biometrics* 18:283–307.

Neubert, M. G. and H. Caswell. 2000. Demography and dispersal: Calculation and sensitivity analysis of invasion speed for structured populations. *Ecology* 81:1613–1628.

Neuffer, B. and H. Hurka. 1999. Colonization history and introduction dynamics of *Capsella bursa-pastoris* (Brassicaceae) in North America: Isozymes and quantitative traits. *Mol. Ecol.* 8:1667–1681.

Ngouajio, M., M. E. McGiffen, and C. M. Hutchinson. 2003. Effect of cover crop management system on weed populations in lettuce. *Crop Prot.* 22:57–64.

Nicollier, G. F., D. F. Pope, and A. C. Thompson. 1985. Pp. 207–218 in A. C. Thompson (Ed.), *The Chemistry of Allelopathy*. American Chemical Society, Washington, DC.

Nielson, E. T. and C. H. Muller. 1980. A comparison of the relative naturalization ability of two *Schinus* species in Southern California. *Bull. Torrey Bot. Club* 107:51–56.

Norris, R. F. 1982. Interactions between weeds and other pests in the agroecosystem. Pp. 343–406 in J. L. Hatfield and I. J. Thomason (Eds.), *Biometeorology in Integrated Pest Management*. Academic, New York.

Norris, R. F. 1992. Case history for weed competition/population ecology: Barnyardgrass (*Echinochloa crus-galli*) in sugarbeets (*Beta vulgaris*). *Weed Technol.* 6:220–222.

Norris, R. F., E. P. Caswell-Chen, and M. Kogan. 2003. *Concepts in Integrated Pest Management*. Pearson Education, Upper Saddle River, NJ.

Norris, R. F. and M. Kogan. 2000. Interactions between weeds, arthropod pests, and their natural enemies in managed ecosystems. *Weed Sci.* 48:94–158.

Novak, S. J. and R. N. Mack. 1993. Genetic variation in *Bromus tectorum* (Poaceae): Comparison between native and introduced populations. *Heredity* 71:167–176.

Novak, S. J., R. N. Mack, and D. E. Soltis. 1991. Genetic variation in *Bromus tectorum* (Poaceae): Population differentiation in its North American range. *Am. J. Bot.* 78:1150–1161.

Novak, S. J., R. N. Mack, and P. S. Soltis. 1993. Genetic variation in *Bromus tectorum* (Poaceae): Introduction dynamics in North America. *Can. J. Bot.* 71:1441–1448.

Nurse, R. E., B. D. Booth, and C. J. Swanton. 2003. Predispersal seed predation of *Amaranthus retroflexus* and *Chenopodium album* growing in soybean fields. *Weed Res.* 43:260–268.

O'Connor, T. G. 1997. Micro-site influence on seed longevity and seedling emergence of bunchgrass (*Themeda triandra*) in a semi-arid savanna. *Afr. J. Range Forage Sci.* 14:7–11.

Odum, E. P. 1971. *Fundamentals of Ecology*, 3rd ed. Saunders, Philadelphia, PA.

Odum, S. 1965. Germination of ancient seeds: Floristical observations and experiments with archaeologically dated soil samples. *Dan. Bot. Ark.* 24:2.

Odum, S. 1974. Seeds in ruderal soils, their longevity and contribution to the flora of disturbed ground in Denmark. Pp. 1131–1141 in *Proceedings 12th British Weed Control Conference*. British Crop Protection Council, London, United Kingdom.

Ogden, J. 1970. Plant population structure and productivity. *Proc. N.Z. Ecol. Soc.* 17:1–9.

Olckers, T., H. G. Zimmermann, and J. H. Hoffmann. 1998. Integrating biological control into the management of alien invasive weeds in South Africa. *Pesticide Outlook* 9: 9–16.

Oliver, C. D. and B. C. Larson.1996. *Forest Stand Dynamics*. Pacific Southwest Research Station, Redding, CA. Wiley, New York.

Oliver, W. W. and R. F. Powers. 1978. Growth models for ponderosa pine. I. Yield of unthinned plantations in northern California. For. Ser. Res. Paper PSW-133, USDA Forest Service.

Olson, B. E., R. T. Wallander, and R. W. Kott. 1997. Recovery of leafy spurge seed from sheep. *J. Range Manag.* 50:10–15.

O'Neill, R. V., D. L de Angelis, J. B. Waide, and T. F. Allen. 1986. *A Hierarchical Concept of Ecosystems.* Princeton University Press, Princeton, NJ.

Oregon Department of Agriculture, Plant Division, Noxious Weed Control Program. 2000. *Economic Analysis of Containment Programs, Damages, and Production Losses from Noxious Weeds in Oregon.* Oregon Department of Agriculture, Corvallis, OR.

Organic Farming Research Organization. 2002. *Final Results of the Third Biennial National Organic Farmers' Survey*, March 22. Available: http//www.ofrf.org/publications/survey/1997.html.

Orians, G. H. 1975. Diversity, stability, and maturity in natural ecosystems. Pp. 139–150 in W. H. van Dobben and R. H. Lowe-McConnell (Eds.), *Unifying Concepts in Ecology.* W. Junk, The Hague.

Ortega, Y. K. and D. E. Pearson. 2005. Weak *vs.* strong invaders of natural plant communities: Assessing invasibility and impact. *Ecol. Appl.* 15:651–661.

Owen, M. D. K. 1997. North American development of herbicide-tolerant crops. *Proc. Br. Crop Protection Conf. Brighton* 3:955–963.

Owen, S. J. 1997. *Ecological Weeds on Conservation Land in New Zealand: A Database.* Department of Conservation, Wellington, NZ.

Palmblad, I. G. 1968. Competition in experimental populations of weeds with emphasis on the regulation of population size. *Ecology* 49:26–34.

Panetsos, C. A. and H. G. Baker. 1968. The origin of variation in "wild" *Raphanus sativus* (Cruciferae) in California. *Genetica* 38:243–274.

Panetta, F. D. and N. D. Mitchell. 1991. Homoclime analysis and the prediction of weediness. *Weed Res.* 31:273–284.

Parker, I. M. 2000. Invasion dynamics of *Cytisus scoparius*: A matrix model approach. *Ecol. Appl.* 10:726–743.

Parker, I. M., D. Simberloff, W. M. Lonsdale, K. Goodell, M. Wonham, P. M. Kareiva, M. H. Williamson, B. Von Holle, P. B. Moyle, J. E. Byers, and L. Goldwasser.1999. Impact: Toward a framework for understanding the ecological effects of invaders. *Biol. Invas.* 1:3–19.

Parker, K. C. and J. L. Hamrick. 1992. Genetic diversity and clonal structure in a columnar cactus *Lophocereus schottii. Am. J. Bot.* 79:86–96.

Parks, C. G., S. R. Radosevich, B. A. Endress, B. J. Naylor, D. Anzinger, L. J. Rew, B. D. Maxwell, and K. A. Dwire. 2005a. Natural and land-use history of the northwest mountain regions (USA) in relation to patterns of plant invasions. *Perspect. Plant Ecol. Syst.* 7:137–158.

Parks, C. G., M. L. Wisdom, and J. G. Kie. 2005b. Ungulates as contributors of non-native plant invasions: A review. Appendix D in *Environmental Impact Statement for Region 6*. USDA Forest Service Invasive Plant Program, May 2005, LaGrand, OR.

Parsons, W. T. and E. G. Cuthbertson. 1992. *Noxious Weeds of Australia*. Melbourne, Australia.

Patrick, Z. A. 1971. Phytotoxic substances associated with the decomposition in soil of plant residues. *Phytopathology* 53:152–161.

Pattison, R. R., G. Goldstein, and A. Ares. 1998. Growth, biomass allocation and photosynthesis of invasive and native Hawaiian rainforest species. *Oecologia* 117: 449–459.

Pauly, P. J. 1987. *Controlling Life: Jacque Loeb and the Engineering Ideal in Biology*. University of California Press, Berkeley.

Pavlychenko, T. K. 1937a. Quantitative study of the entire root system of weed and crop plants under field conditions. *Ecology* 18:62–79.

Pavlychenko, T. K. 1937b. The soil washing method in quantitative root study. *Can. J. Res.* 15:33–57.

Pavlychenko, T. K. 1940. Investigations relating to weed control in Western Canada P. 9 in R. O. Whyte (Ed.), *The Control of Weeds*. Impleria Bureau of Pastures and Forage Crops, Aberystwyth, Wales.

Pemberton, R. W. 2000. Predictable risk to native plants in weed biological control. *Oecologia* 125:489–494.

Penfound, W. T. and T. T. Earle. 1948. The biology of the water hyacinth. *Ecol. Monogr.* 18:447–472.

Perelman, S. B., S. E. Burkart, and R. J. C. León. 2003. The role of a native tussock-grass (*Paspalum quadrifarium*) in structuring plant communities in the Flooding Pampa grasslands, Argentina. *Biodivers. Conserv.* 12:225–238.

Perendes, L. A. and J. A. Jones. 2000. Role of light availability and dispersal in exotic plant invasion along roads and streams in the H. J. Andrews Experimental Forest, Oregon. *Conserv. Biol.* 14:64–75.

Perlin, J. 1989. *A Forest Journey: The Role of Wood in the Development of Civilization*. Harvard University Press, Cambridge, MA.

Perrin, B., K. Wiele, and L. Iles. 1993. Public attitudes towards herbicides and their implications for the public involvement strategy for the vegetation management alternatives program. VMAP Technical Report 93-04. Queen's Printer, Toronto, Canada.

Perry, D. A. 1997. *Forest Ecosystems*. The John Hopkins University Press, Baltimore, MD.

Peters, R. H. 1991. *A Critique for Ecology*. Cambridge University Press, New York.

Peterson, A. T. 2003. Predicting the geography of species' invasions via ecological niche modeling. *Q. Rev. Biol.* 78:419–433.

Peterson, A. T. and D. A. Vieglais. 2001. Predicting species invasions using ecological niche modeling: New approaches from bioinformatics attack a pressing problem. *BioScience* 51:363–371.

Pheloung, P. C. 2001. Weed risk assessment for plant introductions into Australia. Pp. 83–92 in R. H. Groves, F. D. Panetta, and J. G. Virue (Eds.), *Weed Risk Assessment*. CSIRO Publishing, Collingweed Victoria, Australia.

Pheloung, P. C., P. A. Williams, and S. R. Halloy. 1999. A weed risk assessment model for use as a biosecurity tool evaluating plant introductions. *J. Environ. Manag.* 57: 239–251.

Pianka, E. R. 1970. On *r*- and *K*-selection. *Am. Nat.* 104:592–597.

Pianka, E. R. 2000. *Evolutionary Ecology*, 6th ed. Benjamin Cummings, San Francisco, CA.

Pickett, S. T. A., S. L. Collins, and J. J. Armesto. 1987. Models, mechanisms and pathways of succession. *Bot. Rev.* 53:335–371.

Pickett, S. T. A., V. T. Parker, and P. L. Fiedler. 1992. The new paradigm in ecology: Implications for conservation biology above the species level. Pp. 65–88 in P. L. Fiedler and S. J. Jain (Eds.), *Conservation Biology: The Theory and Practice of Nature Conservation, Preservation and Management*. Chapman and Hall, New York.

Pickett, S. T. A. and P. S. White (Eds.). 1985. *Ecology of Natural Disturbance and Patch Dynamics*. Academic, San Diego, CA.

Piggin, C. M. and A. W. Sheppard. 1995. Pp. 87–110 in R. H. Groves, R. C. H. Shepherd, and R. G. Richardson (Eds.), *The Biology of Australian Weeds*. Vol. 1. R. G. and F. J. Richardson Publications, Melbourne.

Pimentel, D. 1986. Agroecology and economics. Pp. 299–319 in M. Kogan (Ed.), *Ecological Theory and Integrated Pest Management Practice*. Wiley, New York.

Pimentel, D., H. Acquay, M. Biltonen, P. Rice, M. Silva, J. Nelson, V. Lipner, S. Giordano, A. Horowitz, and M. D'Amore. 1992. Environmental and economic impacts of pesticide use. Pp. 277–278 in S. A. Briggs (Ed.), *Basic Guide to Pesticides: Their Characteristics and Hazards*. Hemisphere, Washington, DC.

Pimentel, D., L. Lach, R. Zunig, and D. Morrison. 1999. Environmental and economic costs of nonindigenous species in the United States. *BioScience* 50:53–65.

Pimentel, D. E., C. Terhune, R. Dyson-Hudson, S. Rochereau, R. Samis, E. Smith, D. Denman, D. Reifschnieder, and M. Shepard. 1976. Land degradation: Effects on food and energy resources. *Science* 194:149–155.

Pimentel, D., C. Wilson, C. McCullum, R. Huang, P. Dwen, J. Flack, Q. Tran, T. Saltman, and B. Cliff. 1997. Economic and environmental benefits of biodiversity. *BioScience* 47:747–757.

Pimm, S. L. 1991. *The Balance of Nature?* University of Chicago Press, Chicago, IL.

Pitafi, B. A. and J. A. Roumasset. 2005. The resource economics of invasive species. *Proceedings Northeastern Agricultural Resource Economics Association workshop on Invasive Species*. Annapolis, MD. Available: http://www.arec.umd.edu/llynch/personal%20files/NAREA-IS-work.htm.

Polis, G. A. and K. O. Winemiller. 1996. *Food Webs. Integration of Pattern and Dynamics*. Chapman and Hall, New York.

Poorter, H. and C. Remkes. 1990. Leaf area ratio and net assimilation rate of 24 wild species differing in relative growth rate. *Oecologia* 83:553–559.

Popay, A. I. and G. W. Ivens. 1982. East Africa. Pp. 345–372 in W. Holzner and M. Numata (Eds.), *Biology and Ecology of Weeds*. W. Junk, The Hague.

Popay, A. I. and E. H. Roberts. 1970a. Ecology of *Capsella bursa-pastoris* and *Senecio vulgaris* in relation to germination behavior. *J. Ecol.* 58:123–139.

Popay, A. I. and E. H. Roberts. 1970b. Factors involved in the dormancy and germination of *Capsella bursa-pastoris* and *Senecio vulgaris. J. Ecol.* 58:103–122.

Post, W. M., C. C. Travis, and D. L. DeAngelis. 1985. Mutualism, limited competition, and positive feedback. Pp. 305–325 in D. H. Boucher (Ed.), *The Biology of Mutualism: Ecology and Evolution.* Oxford University Press, New York.

Postgate, J. 1998. *Nitrogen Fixation*, 3rd ed. Cambridge University Press, New York.

Power, M. E., D. Tilman, J. A. Estes, B. A. Menge, W. J. Bond, L. S. Mills, G. Daily, J. C. Castilla, J. Lubchenco, and R. T. Paine. 1996. Challenges in the quest for keystones. *BioScience* 46:609–620.

Powles, S. B. and J. A. M. Holtum. 1994. *Herbicide Resistance in Plants. Biology and Biochemistry.* CRC Press, Boca Raton, FL.

Powles, S. B., D. F. Lorraine-Colwill, J. J. Delow, and C. Preston. 1998. Evolved resistance to Italian ryegrass. *Weed Sci.* 46:604–607.

Powles, S. B. and D. L. Shaner. (Eds.). 2001. *Herbicide Resistance and World Grains.* CRC Press, Boca Raton, FL.

Prasada Rao, K. E., J. M. J. de Wet, D. E. Brink, and M. H. Mengesha. 1987. Intraspecific variation and systematics of cultivated *Setaria italica*, foxtail millet (Poaceae). *Econ. Bot.* 41:108–116.

Prather, T. 2006. How can you incorporate risk assessment into invasive plant management? In M. McFadzen (Coordinator), *Understanding and Assessing Plant Invasions: A Framework for Prioritizing Management Strategies.* Online workshop, Center for Invasive Plant Management, Bozeman, MT. Available: http://www.weedcenter.org/education/syllabus_07.htm.

Price, S. C., K. M. Shumaker, A. L Kahler, R. W. Allard, and J. E. Hill. 1984. Estimates of population differentiation obtained from enzyme polymorphisms and quantitative characters. *J. Hered.* 75:141–142.

Prieur-Richard, A-H., S. Lavorel, K. Grigulis, and A. Dos Santos. 2000. Plant community diversity and invasibility by exotics: Invasion of Mediterranean old fields by *Conyza bonariensis* and *Conyza canadensis. Ecol. Lett.* 3:412–422.

Pringnitz, B. A. 2001. *Issues in Weed Management for 2002.* Extension Publication PM 1898. Iowa State University Extension Service, Ames, IA.

Prinzing, A., W. Durka, S. Klotz, and R. Bradl. 2002. Which species become alien? *Evol. Ecol. Res.* 4:385–405.

Probert, R. J. 1992. The role of temperature in germination ecophysiology. Pp. 285–325 in M. Fenner (Ed.), *Seeds: The Ecology of Regeneration in Plant Communities.* CAB International, Wallingford, United Kingdom.

Purves, D. W. and J. Dushoff. 2005. Directed seed dispersal and metapopulation response to habitat loss and disturbance: Application to *Eichhornia paniculata. J. Ecol.* 93: 658–669.

Pusztai, A., S. Bardocz, and S. W. B. Ewen. 2003. Genetically modified foods: Potential human health effects. Pp. 347–373 in J. P. F. D'Mello (Ed.), *Food Safety: Contaminants and Toxins.* CAB International, Wallingford, United Kingdom.

Putman, R. 1994. *Community Ecology.* Chapman and Hall, London.

Putnam, A. R. and C. S. Tang. (Eds.) 1986. *The Science of Allelopathy.* Wiley, New York.

Putnam, A. R. and L. A. Weston. 1986. Adverse impacts of allelopathy in agricultural systems. Pp. 43–56 in A. R. Putnam and C. S. Tang (Eds.), *The Science of Allelopathy.* Wiley, New York.

Putwain, P. D. and J. L. Harper. 1970. Studies in the dynamics of plant populations. III. The influence of associated species on populations of *Rumex acetosella* L. in grasslands. *J. Ecol.* 58:251–264.

Putwain, P. D. and A. M. Mortimer. 1995. Evolution and spread of herbicide resistant weeds. The role of selection pressure. P. 115 in *Proceedings International Symposium Weed and Crop Resistance to Herbicides.* Cordoba, Spain, University of Cordoba.

Pyke, D. A. and S. T. Knick. 2003. Plant invaders, global change and landscape restoration. Pp. 278–288 in N. Allsopp. A. R. Palmer, S. J. Milton, K. P. Kirkman, G. I. H. Kerley, and D. R. Brown (Eds.), *Proceedings of the VII International Rangelands Congress.* International Rangeland Congress, Durban, South Africa.

Pyšek, P. 1997. Clonality and plant invasions: Can a trait make a difference? Pp. 405–427 in H. de Kroon and J. van Groenendael (Eds.), *The Ecology and Evolution of Clonal Plants.* Backhuys, Leiden.

Pyšek, P., K. Prach, and P. Smilauer. 1995. Relating invasion success to plant traits: An analysis of the Czech alien flora. Pp. 39–60 in P. Pyšek, K. Prach, M. Rejmánek, and M. Wade (Eds.), *Plant Invasions—General Aspects and Special Problems.* SPB Academic, Amsterdam, Netherlands.

Pyšek, P., D. M. Richardson, and M. Williamson. 2004. Predicting and explaining plant invasions through analysis of source area floras: Some critical considerations. *Divers. Distrib.* 10:179–187.

Quigley, T. M. and S. J. Arbelbide. 1997. An assessment of ecosystem components in the interior Columbia Basin and portions of the Klamath and Great Basins. General Technical Report PNW-GTR-405. USDA Forest Service, Pacific Northwest Research Station, Portland, OR.

Quinn, L. D. 2006. Ecological Correlates of Invasion by *Arundo donax* Ph.D. Dissertation, University of California, Riverside.

Radosevich, S. R. 1987. Methods to study interactions among crops and weeds. *Weed Technol.* 1:190–198.

Radosevich, S. R., B. A. Endress, and C. G. Parks. 2005. Defining a regional approach for invasive plant research and management. Pp. 141–166 in K. Inderjit (Ed.), *Ecological and Agricultural Aspects of Invasive Plants.* Birkhäuser-Verlag, Basel, Netherlands.

Radosevich, S. R. and C. M. Ghersa. 1992. Weeds, crops, and herbicides: A modern day "neckriddle." *Weed Technol.* 6:788–795.

Radosevich, S. R., C. M. Ghersa, and G. Comstock. 1992. Concerns a weed scientist might have about herbicide-tolerant crops. *Weed Technol.* 6:635–639.

Radosevich, S. R., D. E. Hibbs, and C. M. Ghersa. 2006. Effect of species mixtures on growth and stand development of Douglas-fir and red alder. *Can. J. For. Res.* 36:768–782.

Radosevich, S. R. and J. S. Holt. 1984. *Weed Ecology: Implications for Vegetation Management.* Wiley, New York.

Radosevich, S. R., J. S. Holt, and C. M. Ghersa. 1997. *Weed Ecology: Implications for Management,* 2nd ed. Wiley, New York.

Radosevich, S. R. and M. L Roush. 1990. The role of competition in agriculture. Pp. 341–363 in J. B. Grace and D. Tilman (Eds.), *Perspectives on Plant Competition.* Academic, San Diego, CA.

Radosevich, S. R. and R. Shula. 1994. Implementation of weed control in IPM. Pp. 58–70 in *Pesticide Risk Reduction and Strategic Planning Forum.* Pest Management Alternatives Office, Val-Morin, Quebec, Canada.

Radosevich, S. R., N. L. Smith, and F. Kegel. 1975. Johnsongrass control in field corn. West. Soc. Weed Sci. Res. Prog. Rep. Pp. 62–64. Western Society of Weed Science, Logan, U.T.

Radosevich, S. R., M. M. Stubbs, and C. M. Ghersa. 2003. Plant invasions—process and patterns. *Weed Sci.* 51:254–259.

Ramsey E. W., G. A. Nelson, S. K. Sapkota, E. B. Seeger, and K. D. Martella, 2002. Mapping Chinese tallow with color-infrared photography. *Photogramm. Eng.* 68: 251–255.

Randall, J. M. 1996. Weed control for the preservation of biological diversity. *Weed Technol.* 10:370–383.

Randall, J. M. 1997. Defining weeds of natural areas. Pp. 18–25 in J. O. Luken and J. W. Thieret (Eds.), *Assessment and Management of Plant Invasions.* Springer-Verlag, New York.

Rapoport, E. H. 1991. Tropical versus temperate weeds: A glance into the present and future. Pp. 441–451 in P. S. Ramakrishnan (Ed.), *Ecology of Biological Invasion in the Tropics.* International Scientific Publications, New Delhi, India.

Raybould, A. F. and A. J. Gray. 1993. Genetically modified crops and hybridization with wild relatives: A UK perspective. *J. Appl. Ecol.* 30:199–219.

Reichard, S. E. 1997. Prevention of invasive plant introductions on national and local levels. Pp. 215–240 in J. O. Luken and J. W. Thieret (Eds.), *Assessment and Management of Plant Invasions.* Springer, New York.

Reichard, S. E. and C. W. Hamilton. 1997. Predicting invasions of woody plants introduced into North America. *Conserv. Biol.* 11:193–203.

Reineke, L. H. 1933. Perfecting a stand-density index for even-aged forests. *J. Agric. Res.* 46:627–638.

Rejmánek, M. 1989. Invasibility of plant communities. Pp. 369–388 in J. A. Drake, H. A. Mooney, F. di Castri, R. Groves, F. J. Kruger, M. Rejmánek, and M. Williamson. (Eds.), *Biological Invasions: A Global Perspective.* Wiley, New York.

Rejmánek, M. 1999. Invasive plant species and invasible ecosystems. Pp. 79–109 in O. T. Sandlund, P. J. Schei, and A. Vilken (Eds.), *Invasive Species and Biodiversity Management.* Kluwer Academic, Dordrecht, The Netherlands.

Rejmánek, M. 2000. Invasive plants: Approaches and predictions. *Aust. Ecol.* 25:497–506.

Rejmánek, M. and M. J. Pitcairn. 2002. When is eradication of exotic pest plants a realistic goal? Pp. 249–253 in C. R. Veitch and M. N. Clout (Eds.), *Turning the Tide: The Eradication of Invasive Species.* IUCN/SSC Invasive Species Specialist Group. IUCN, Gland, Switzerland, and Cambridge.

Rejmánek, M. and J. Randall. 1994. Invasive alien plants in California: 1993 summary and comparison with other areas in North America. *Madroño* 41:161–177.

Rejmánek, M. and D. M. Richardson. 1996. What attributes make some plant species more invasive? *Ecology* 77:1655–1661.

Rejmánek, M., G. R. Robinson, and E. Rejmankova. 1989. Weed-crop competition: Experimental designs and models for data analysis. *Weed Sci.* 37:276–284.

Rew, L. J., B. D. Maxwell, and R. Aspinall. 2005. Predicting the occurrence of nonindigenous species using environmental and remotely sensed data. *Weed Sci.* 53:236–241.

Rew, L. J., B. D. Maxwell, F. L. Dougher, and R. Aspinall. 2006. Searching for a needle in a haystack: Evaluating survey methods for non-indigenous plant species. *Plant Invas.* 8:523–539.

Rew, L. J. and M. L. Pokorny (Eds.). 2006. *Inventory and Survey Methods for Nonindigenous Plant Species.* Montana State University Extension Service, Bozeman, MT.

Reynolds, H. L. 1999. Plant interactions: Competition. Pp. 648–676 in F. I. Pugnaire and F. Vallandares (Eds.), *Handbook of Functional Plant Ecology.* Marcel Dekker, New York.

Rice, E. L. 1984. *Allelopathy*, 2nd ed. Academic, Orlando, FL.

Rice, E. L. 1995. *Biological Control of Weeds and Plant Diseases: Advances in Applied Allelopathy.* University of Oklahoma Press, Norman, OK.

Rice, P. M. 2004. *Fire as a Tool for Controlling Nonnative Invasive Plants: A Review of Current Literature.* Montana State University, Center for Invasive Plant Management, Bozeman, MT.

Richards, C. M. 2000. Inbreeding depression and genetic rescue in a plant metapopulation. *Am. Nat.* 155:383–394.

Richardson, D. M. and W. J. Bond. 1991. Determinants of plant distribution: Evidence from pine invasions. *Am. Nat.* 137:639–668.

Richardson, D. M., P. Pyšek, M. Rejmánek, M. G. Barbour, F. D. Panetta, and C. J. West. 2000. Naturalization and invasion of alien plants: Concepts and definitions. *Divers. Distrib.* 6:93–107.

Richardson, D. M., P. A. Williams, and R. J. Hobbs. 1994. Pine invasions in the southern hemisphere: Determinants of spread and invadability. *J. Biogeogr.* 21:511–527.

Rinella, M. J. and E. C. Luschei. In Press. The within-site and regional impacts of leafy spurge (*Euphorbia esula*): Hierarchical Bayesian methods estimate invasive weed impacts at pertinent spatial scales. Biol. Invas.

Ritchie, M. A. and H. Olff. 1999. Spatial scaling laws yield a synthetic theory of biodiversity. *Nature* 400:557–560.

Rizvi, S. J. H. and V. Rizvi (Eds.) 1992. *Allelopathy: Basic and Applied Aspects.* Chapman and Hall, London.

Robbins, W. W., M. K. Bellue, and W. S. Ball. 1951. *Weeds of California.* California State Department of Agriculture, Documents and Publications, Sacramento, CA.

Robbins, W. W., A. S. Crafts, and R. N. Raynor. 1942. *Weed Control.* McGraw-Hill, New York.

Roberts, E. H. 1972a. Cytological, genetical and metabolic changes associated with loss of viability. Pp. 253–306 in E. H. Roberts (Ed.), *Viability of Seeds.* Syracuse University Press, Syracuse, NY.

Roberts, E. H. 1972b. Storage environment and the control of viability. Pp. 14–58 in E. H. Roberts (Ed.), *Viability of Seeds.* Syracuse University Press, Syracuse, NY.

Roberts, H. A. and P. M. Feast. 1972. Emergence and longevity of seeds of annual weeds in cultivated and undisturbed soil. *J. Appl. Ecol.* 10:133–143.

Roberts, H. A. and P. M. Feast. 1973. Changes in the number of viable weed seeds in soil under different regimes. *Weed Res.* 13:298–303.

Roberts, H. A. and M. E. Ricketts. 1979. Quantitative relationship between the weed flora after cultivation and the seed population in the soil. *Weed Res.* 19:269–275.

Robertson, J. H. 1954. Half-century changes of northern Nevada ranges. *J. Range Manag.* 7:117–121.

Rodgers, N. K., G. A. Buchanan, and W. E. Johnson. 1976. Influence of row spacing on weed competition with cotton. *Weed Sci.* 24:410–413.

Roe, R. M., J. D. Burton, and R. J. Kuhr (Eds.). 1997. *Herbicide Activity: Toxicology, Biochemistry and Molecular Biology*. IOS Press, Amsterdam, The Netherlands.

Root, R. B. 1973. Organization of a plant-arthropod association in simple and diverse habitats: The fauna of collards (*Brassica oleracea*). *Ecol. Monogr.* 43:95–124.

Rosenthal, S. S., D. M. Maddox, and K. Brunetti. 1989. Biological control methods. Pp. 77–99 in California Weed Conference, *Principles of Weed Control in California*, 2nd ed. Thomson, Fresno, CA.

Rosenzweig, M. L. 1995. *Species Diversity in Space and Time*. Cambridge University Press, New York.

Ross, M. A. and C. A. Lembi. 1985. *Applied Weed Science*. Burgess, Minneapolis, MN.

Ross, M. A. and C. A. Lembi. 1999. *Applied Weed Science*, 2nd ed. Burgess, Minneapolis, MN.

Roush, M. L. 1988. Models of a four-species annual weed community: Growth, competition, and community dynamics. Ph.D. Dissertation. Oregon State University, Corvallis, OR.

Roush, M. L. and S. R. Radosevich. 1985. Relationships between growth and competitiveness of four annual weeds. *J. Appl. Ecol.* 22:895–905.

Roush, M. L., S. R. Radosevich, R. G. Wagner, B. O. Maxwell, and T. D. Petersen. 1989. A comparison of methods for measuring effects of density and proportion in plant competition experiments. *Weed Sci.* 37:268–275.

Roy, S., J. P. Simon, and F. J. Lapointe. 2000. Determination of the origin of the cold-adapted populations of barnyard grass (*Echinochloa crus-galli*) in eastern North America: A total-evidence approach using RAPD DNA and DNA sequences. *Can. J. Bot.* 78:1505–1513.

Ruiz, R. A., D. C. Vacek, P. E. Parker, W. E. Wendel, U. Schaffner, R. Sobhian, and D. C. Sands. 2000. Using randomly amplified polymorphic DNA chain reaction (RAPD-PCR) to match natural enemies to their host plants. Pp. 289–294 in N. R. Spenser (Ed.), *X International Symposium on Biological Control of Weeds*, Bozeman, MT. Advanced Litho Printing, Great Falls, MT.

Sagar, G. R. and A. M. Mortimer. 1976. An approach to the study of the population dynamics of plants with special reference to weeds. *Ann. Appl. Biol.* 1:1–47.

Sakai, A. K., F. W. Allendorf, J. S. Holt, D. M. Lodge, J. Molofsky, K. A. With, S. Baughman, R. J. Cabin, J. E. Cohen, N. C. Ellstrand, D. E. McCauley, P. O'Neill, I. M. Parker, J. N. Thompson, and S. G. Weller. 2001. The population biology of invasive species. *Annu. Rev. Ecol. Syst.* 32:305–332.

Salisbury, E. J. 1942a. *The Reproductive Capacity of Plants*. G. Bell, London.

Salisbury, E. J. 1961. *Weeds and Aliens*. Collins, London.

Sammul, M., K. Kull, L. Oksanen, and P. Veromann. 2000. Competition intensity and its importance: Results of field experiments with *Anthoxanthum odoratum*. *Oecologia* 125:18–25.

Sarukhan, J. and M. Gadgil. 1974. Studies on plant demography: *Ranunculus repens* L., *R. bulbosus* L. and *R. acris* L. III. A mathematical model incorporating multiple modes of reproduction. *J. Ecol.* 62:921–936.

Sauer, J. and G. Struik. 1964. A possible ecological relation between soil disturbance, light flash, and seed germination. *Ecology* 45:884–886.

Sauer, J. D. 1988. *Plant Migrations: The Dynamics of Geographic Patterning in Seed Plant Species.* University of California Press, Berkeley.

Sax, D. F. 2001. Latitudinal gradients and geographic ranges of exotic species: Implications for biogeography. *J. Biogeogr.* 28:139–150.

Sax, D. F. 2002. Native and naturalized plant diversity are positively correlated in scrub communities of California and Chile. *Divers. Distrib.* 8:193–210.

Schlapfer, F., B. Schmid, and I. Deidl. 1999. Expert estimates about effects of biodiversity on ecosystem processes and services. *Oikos* 84:346–352.

Schoonmaker, P. and A. McKee. 1988. Species composition and diversity during secondary succession of coniferous forests in the western Cascade Mountains of Oregon. *For. Sci.* 34:960–979.

Schroeder, J., S. H. Thomas, and L. W. Murray. 2004. Root-knot nematodes affect annual and perennial weed interactions with chile pepper. *Weed Sci.* 52:28–46.

Schroeder, J., S. H. Thomas, and L. W. Murray. 2005. Impacts of crop pests on weeds and weed-crop interactions. *Weed Sci.* 53:918–922.

Schulz, A., F. Wengenmayer, and H. M. Goodman. 1990. Genetic engineering of herbicide resistance in higher plants. *CRC Crit. Rev. Plant Sci.* 9:1–15.

Scopel, A. L., C. L. Ballaré, and C. M. Ghersa. 1988. Role of seed reproduction in the population ecology of *Sorghum halepense* in maize crops. *J. Appl. Ecol.* 25: 951–962.

Scopel, A. L., C. L. Ballaré, and S. R. Radosevich. 1994. Photostimulation of seed germination during soil tillage. *New Phytol.* 126:145–152.

Scopel, A. L., C. L. Ballaré, and R. A. Sánchez. 1991. Induction of extreme light sensitivity in buried weed seeds and its role in the perception of soil cultivations. *Plant Cell Environ.* 14:501–508.

Scott, C. S., D. E. Bunker, and W. P. Carson. 2006. A null model of exotic plant diversity tested with exotic and native species-area relationships. *Ecol. Lett.* 9:136–141.

Seabloom, E. W., E. T. Borer, V. L. Boucher, R. S. Burton, K. L. Cottingham, L. Goldwasser, W. K. Gram, B. E. Kendall, and F. Micheli. 2003a. Competition, seed limitation, disturbance, and reestablishment of California native annual forbs. *Ecol. Appl.* 13:575–592.

Seabloom, E. W., W. S. Harpole, O. J. Reichman, and D. Tilman. 2003b. Invasion, competitive dominance, and resource use by exotic and native California grassland species. *Proc. Natl. Acad. Sci. USA* 100:13384–13389.

Searle, S. R. 1966. *Matrix Algebra for the Biological Sciences.* Wiley, New York.

Selman, M. 1970. The population dynamics of *Avena fatua* (wild oats) in continuous spring barley—desirable frequency of spraying with triallate. Pp. 1176–1188 in *Proceedings 10th British Weed Control Conference.* British Crop Protection Council, Nottingham, United Kingdom.

Seshu, D. V. and M. E. Sorrells. 1986. Genetic studies on seed dormancy in rice. Pp. 369–382 in *Rice Genetics.* IRRI, Los Banos, Laguna, Philippines.

Shainsky, L. J. and S. R. Radosevich. 1991. Analysis of yield-density relationships in experimental stands of Douglas-fir and red older seedlings. *For. Sci.* 37:574–592.

Shainsky, L. J. and S. R. Radosevich. 1992. Mechanisms of competition between Douglas-fir and red alder seedlings. *Ecology* 73:30–45.

Shaner, D. L. 2000. The impact of glyphosate-tolerant crops on the use of other herbicides and on resistance management. *Pest Manag. Sci.* 56:320–326.

Shaw, I. C. and J. Chadwick. 1998. *Principles of Environmental Toxicology*. Taylor and Francis, New York.

Shaw, R. H. and R. Milne. 2000. The use of molecular techniques in the classical biological control programme against an invasive *Ligustrum* species in La Reunion. Pp. 303 in N. R. Spenser (Ed.), *X International Symposium on Biological Control of Weeds*, Bozeman, MT. Advanced Litho Printing, Great Falls, MT.

Sheldon, J. C. and F. M. Burrows. 1973. The dispersal effectiveness of the achene-pappus units of selected Compositae in steady winds with convection. *New Phytol.* 72: 665–675.

Sheley, R. and J. Petroff. 1999. *Biology and Management of Noxious Rangeland Weeds*. Oregon State University Press, Corvallis, OR.

Sheley R., J. Petroff, and M. Borman. 1999. *Introduction to biology and management of noxious rangeland weed*. Pp. 1–3 in R. Sheley and J. Petroff (Eds.). Biology and Management of Noxious Rangeland Weeds. Oregon State University Press.

Sheley, R. L., J. S. Jacobs, and T. J. Svejcar. 2004. Integrating disturbance and colonization during rehabilitation of invasive weed-dominated grasslands. *Weed Sci.* 53:307–314.

Sheley, R. L. and J. M. Krueger-Mangold. 2003. Principles for restoring invasive plant-infested rangeland. *Weed Sci.* 51:260–265.

Sheley, R. L., J. M. Mangold, and J. L. Anderson. 2006. Potential for successional theory to guide restoration of invasive plant-dominated rangeland. Ecol. Monogr. 76:365–379.

Sheley, R. L., and M. J. Rinella. 2001. Incorporating biological control into ecologically based weed management. Pp. 211–228 in E. Wajnberg, J. K. Scott, and P. C. Quimby (Eds.), *Evaluating Indirect Ecological Effects of Biological Control*. CAB International, Wallingford, United Kingdom.

Sheley, R. L., T. J. Svejcar, and B. D. Maxwell. 1996. A theoretical framework for developing successional weed management strategies on rangeland. *Weed Technol.* 10:766–733.

Shinozaki, K. and T. Kira. 1956. Intraspecific competition among higher plants. VII. Logistic theory of the C-D effect. *J. Inst. Polytech.*, Osaka City Univ. Ser. D 7, pp. 35–72.

Shipley, B., P. A. Keddy, and L. F. Lefkovitch. 1991. Mechanisms producing plant zonation along a water depth gradient: A comparison with the exposure gradient. *Can. J. Bot.* 69:1420–1424.

Shogren, J. F. and T. Crocker. 1991. Risk, self-protection, and ex ante economic value. *J. Environ. Econ. Manag.* 20:1–15.

Short, P. and T. Colborn. 1999. Pesticide use in the U.S. and policy implications: A focus on herbicides. *Toxicol. Ind. Health* 15:240–275.

Silander, J. A. and S. W. Pacala. 1990. The application of plant population models to understanding plant competition. Pp. 67–92 in J. B. Grace and D. Tilman (Eds.), *Perspectives on Plant Competition*. Academic, San Diego, CA.

Silvertown, J. W. 1987. *Introduction to Plant Population Ecology.* 2nd ed. Blackwell Science Publications, Longman, Harlow.

Silvertown, J. W. and J. Lovett Doust. 1993. *Introduction to Plant Population Biology.* Blackwell Science Publications, Oxford, GB.

Silvertown, J. W. and D. Charlesworth. 2001. *Introduction to Plant Population Biology,* 4th ed. Blackwell Science, Oxford.

Simard, S. W., S. R. Radosevich, D. L. Sachs, and S. M. Hagerman. 2006. Evidence for competition and facilitation tradeoffs: Effect of Sitka alder density on pine regeneration and soil productivity. *Can. J. For. Res.* 36:1286–1298.

Simmonds, F. J., J. M. Franz, and R. I. Sailer. 1976. History of biological control. Pp. 17–39 in C. B. Huffaker and P. S. Messenger (Eds.), *Theory and Practice of Biological Control.* Academic, New York.

Simpson, G. M. 1990. *Seed Dormancy in Grasses.* Cambridge University Press, New York. Pp. 195–231.

Skellam, J. G. 1951. Random dispersal in theoretical populations. *Biometrika* 38:196–218.

Slabaugh, W. H. and T. D. Parsons. 1966. *General Chemistry.* Wiley, New York.

Smith, D. E., B. C. Larson, M. J. Kelty, and P. M. S. Ashton. 1997. *The Practice of Silviculture: Applied Forest Ecology,* 9th ed. Wiley, New York.

Smith, D. 1986. *The Practice of Silviculture,* 8th ed. Wiley, New York.

Smith, H. 1982. Light quality, photoperception, and plant strategy. *Annu. Rev. Plant Physiol.* 33:481–518.

Smith, M. D. and A. K. Knapp. 1999. Exotic plant species in a C_4-dominated grassland: Invasibility, disturbance and community structure. *Oecologia* 120:605–612.

Smith, R. S., R. S. Sheil, D. Millward, P. Corkhill, and R. A. Sanderson. 2002. Soil seed banks and the effects of meadow management on vegetation change in a 10-year meadow field trial. *J. Appl. Ecol.* 32:279–293.

Smith, R. F. and R. Van Den Bosch, 1967. Pp. 295–293 in W. W. Kilgore and R. L. doutt (Eds.), *Integrated Control in Pest Control: Biological, Physical and Selected Chemical Methods.* Academic Press, New York, NY.

Solbrig, O. T. (Ed.). 1980. *Demography and Evolution in Plant Populations.* Botanical Monographs 15. University of California Press, Berkeley, CA.

Solbrig, O. T. and R. C. Rollins. 1977. The evolution of autogamy in species of the mustard genus *Leavenworthia. Evolution* 31:265–281.

Soons, M. B. and G. W. Heil. 2002. Reduced colonization capacity in fragmented populations of wind-dispersed grassland forbs. *J. Ecol.* 90:1033–1043.

Soriano, A., R. J. C. León, O. E. Sala, R. S. Lavado, V. A. Deregibus, M. A. Cahuépé, O. A. Scaglia, C. A. Velázquez, and J. H. Lemcoff. 1992. Río de la Plata grasslands. Pp. 367–407 in R. T. Coupland (Ed.), *Ecosystems of the World 8A. Natural Grasslands: Introduction and Western Hemisphere.* Elsevier, New York.

Spencer, L. J. and A. A. Snow. 2001. Fecundity of transgenic wild-crop hybrids of *Cucurbita pepo* (Cucurbitaceae): Implications for crop-to-wild gene flow. *Heredity* 86:694–702.

Spencer, N. R. (Ed.). 2000. *X International Symposium on Biological Control of Weeds,* Bozeman, MT. Advanced Litho Printing, Great Falls, MT.

Spitters, C. J. T. 1983a. An alternative approach to the analysis of mixed cropping experiments. 1. Estimation of competition effects. *Neth. J. Agric. Sci.* 31:1–11.

Spitters, C. J. T. 1983b. An alternative approach to the analysis of mixed cropping experiments. 2. Marketable yield. *Neth. J. Agric. Sci.* 31:143–155.

Spitters, C. J. T. 1989. Weeds: Population dynamics, germination and competition. Pp. 182–216 in R. Rabingge, S. A. Ward, and H. H. van Laar (Eds.), *Simulation and Systems Management in Crop Protection*, Vol. 32. Simulation Monographs. Pudoc, Wageningen, Netherlands.

Spitters, C. J. T. and J. P. Van den Bergh. 1982. Competition between crops and weeds: A system approach. Pp. 137–146 in W. Holzner and M. Numata (Eds.), *Biology and Ecology of Weeds*. W. Junk, The Hague.

Stacey, G., R. H. Burris, and H. J. Evans (Eds.). 1992. *Biological Nitrogen Fixation.* Chapman and Hall, New York.

Stadler, J., A. Trefflich, S. Klotz, and R. Brandl. 2000. Exotic plant species invade diversity hot spots: The alien flora of northwestern Kenya. *Ecography* 23:169–176.

Stallings, G. P., D. C. Thill, C. A. Mallory-Smith, and L. Lass. 1995. Plant movement and seed dispersal of Russian thistle *(Salsola iberica). Weed Sci.* 43:63–69.

Stebbins, G. L. 1959. The role of hybridization in evolution. *Proc. Am. Philos. Soc.* 103:231–251.

Stebbins, G. L. 1970. Adaptive radiation of reproductive characteristics in angiosperms, I: Pollination mechanisms. *Annu. Rev. Ecol. Syst.* 1:307–326.

Steiner, R. 1983. *The Boundaries of Natural Science.* Translated from the German text (1969). Anthroposophic Press, Spring Valley, NY.

Stern, V. M., R. F. Smith, R. Van Den Bosch, and K. S. Hagen. 1959. The integrated control concept. *Hilgardia* 29:81–99.

Stevens, O. A. 1954. Weed seed facts. N.D. Agric. Coll. Ext. Ctr. P. A-128. North Dakota State University, Fargo, ND.

Stevens, O. A. 1957. Weights of seeds and numbers per plant. *Weeds* 5:46-55.

Stewart, R. E., L. L. Gross, and B. H. Honkala. 1984. Effects of competing vegetation on forest trees: A bibliography with abstracts. *Gen. Tech. Rep.* WO-43. USDA Forest Service, Washington, DC.

Stigliani, L. and C. Resina. 1993. SELOMA: Expert system for weed management in herbicide-intensive crops. *Weed Technol.* 7:550–559.

Stöcklin, J. and M. Fischer. 1999. Plants with longer-lived seeds have lower local extinction rates in grassland remnants 1950–1985. *Oecologia* 120:539–543.

Stohlgren, T. J., D. T. Barnett, and J. T. Kartesz. 2003. The rich get richer: Patterns of plant invasions in the United States. *Front. Ecol. Environ.* 1:11–14.

Stohlgren, T. J., D. Binkley, G. W. Chong, M. A. Kalkhan, L. D. Schell, K. A. Bull, Y. Otsuki, G. Newman, M. Bashkin, and Y. Son. 1999. Exotic plant species invade hot spots of native plant diversity. *Ecol. Monogr.* 69:25–46.

Stohlgren, T. J, C. Crosier, G. W. Chong, D. Guenther, and P. Evangelista. 2005. Life-history habitat matching in invading non-native plant species. *Plant Soil* 277:7–18.

Stohlgren, T. J., M. Lee, K. A. Bull, Y. Otsuki, and C. A. Villa. 1998. Riparian zones as havens for exotic plant species in the central grasslands. *Plant Ecol.* 138:113–125.

Stoll, P. and D. Prati. 2001. Intraspecific aggregation alters competitive interactions in experimental plant communities. *Ecology* 82:319–327.

Stowe, L. C. 1979. Allelopathy and its influence on the distribution of plants in an Illinois old-field. *J. Ecol.* 67:1065–1085.

Stowe, L. G. and M. J. Wade. 1979. The detection of small-scale patterns in vegetation. *J. Ecol.* 67:1047–1064.

Strauss, S., K. Raffa, and P. List. 2000. Ethical guidelines for assessing genetically engineered plantations. *J. For.* 98:47–48.

Strebig, J. C. 1988. Herbicide bioassay. *Weed Res.* 28:479.

Streibig, J. C., M. Rudemo, and J. E. Jensen. 1993. Dose-response curves and statistical models. P. 29 in J. C. Streibig and P. Kudsk (Eds.), *Herbicide Bioassays*. CRC Press, Boca Raton, FL.

Stubbendieck, J., C. H. Butterfield, and T. R. Flessner. 1992. *An Assessment of Exotic Plants at Pipestone National Monument and Wilson's Creek National Battlefield*. Final Report. U.S. National Park Service. Omaha, NB.

Stucky, J. M. 1984. Forager attraction by sympatric *Ipomoea hederacea* and *Ipomoea purpurea* (Convolvulaceae) and corresponding forager behaviour and energetics. *Am. J. Bot.* 71:1273–1244.

Stuefer, J. F., H. J. During, and H. DeKroon. 1994. High benefits of clonal integration in two stoloniferous species in response to heterogeneous light environments. *J. Ecol.* 82:511–518.

Suarez, S., E. de la Fuente, C. M. Ghersaand, and R. J. C. Leon. 2001. Weed community as an indicator of summer crop yield and site quality. *Agron. J.* 93:524–530.

Suding, K. N., K. L. Gross, and G. R. Houseman. 2004. Alternative states and positive feedbacks in restoration ecology. *Trends Ecol. Evol.* 19:46–53.

Sun, M. and H. Corke. 1992. Population genetics of colonizing success of weedy rye in northern California. *Theor. Appl. Genet.* 83:321–329.

Sutherland, S. 2004. What makes a weed a weed: Life history traits of native and exotic plants in the USA. *Oecologia* 141:24–39.

Sutherst, R. W. and G. F. Maywald. 1985. A computerized system for matching climates in ecology. *Agric. Ecosyst. Environ.* 13:281–299.

Sutton, R. F. 1985. Vegetation management in Canadian forestry. Information Rep. 0-X-369. Great Lakes Forest Research Centre, Canadian Forestry Service, Sault Ste. Marie, Canada.

Svejcar, T. 2003. Applying ecological principles to wildland weed management. *Weed Sci.* 51:266–270.

Swanton, C. J., D. R. Clements, and D. A. Derksen. 1993. Weed succession under conservation tillage: A hierarchical framework for research and management. *Weed Technol.* 7:286–297.

Swanton, C. J., A. Shrestha, D. R. Clements, B. Booth, and K. Chandler. 2002. Evaluation of alternative weed management systems in a modified no-tillage corn-soybean-winter wheat rotation: Weed densities, crop yield, and economics. *Weed Sci.* 50:504–511.

Swift, M. J. and J. M. Anderson. 1993. Biodiversity and ecosystem function in agricultural systems. Pp. 15–38 in E. D. Schulze and H. A. Mooney (Eds.), *Biodiversity and Ecosystem Function*. Springer-Verlag, Berlin, Germany.

Symonides, E. 1988. On the ecology and evolution of annual plants in disturbed environments. *Vegetatio* 77:21–31.

Symstad, A. J. and D. Tilman. 2001. Diversity loss, recruitment limitation, and ecosystem functioning: Lessons learned from a removal experiment. *Oikos* 92:424–435.

Taylorson, R. B. and S. B. Hendricks. 1977. Dormancy in seeds. *Annu. Rev. Plant Physiol.* 28:331–354.

Teasdale, J. R. 1998. Influence of corn (*Zea mays*) population and row spacing on corn and velvetleaf (*Abutilon theophrasti*) yield. *Weed Sci.* 46:447–453.

Tecchi, R. A. 1983. Contenido de silicofitolitos en suelos del sector sudoriental de la Pampa. *Ondulada Ciencia del Suelo* 1:75–82.

Tewksbury, J. J. and J. D. Lloyd. 2001. Positive interactions under nurse-plants: Spatial scale, stress gradients and benefactor size. *Oecologia* 127:425–434.

Thomas, S. H., J. Schroeder, and L. W. Murray. 2005. The role of weeds in nematode management. *Weed Sci.* 53:923–928.

Thomas, W. L., Jr. (Ed.). 1956. *Man's Role in Changing the Face of the Earth.* An international symposium under the co-chairmanship of J. D. Sauer, M. Bates, and L. Mumford. Sponsored by the Wenner-Gren Foundation for Anthropological Research. University of Chicago Press, Chicago, IL.

Thompson, D. 2005. Measuring the effects of invasive species on the demography of a rare endemic plant. *Biol. Invas.* 7:615–624.

Thompson, K., J. P. Bakker, and R. M. Bekker. 1998. Ecological correlates of seed persistence in the soil in the NW European flora. *J. Ecol.* 86:163–170.

Thompson, K. and J. P. Grime. 1979. Seasonal variation in the seed banks of herbaceous species in ten contrasting habitats. *J. Ecol.* 67:893–921.

Thomsen, M. A., C. M. D'Antonio, K. B. Suttle, and W. P. Susa. 2006. Ecological resistance, seed density, and their interactions determine patterns of invasion in a California coastal grassland. *Ecol. Lett.* 9:160–170.

Thrall, P. H., S. W. Pacala, and J. A. Silander. 1989. Oscillatory dynamics in populations of an annual weed species, *Abutilon theophrasti.* *J. Ecol.* 77:1135–1149.

Thuiller, W., D. M. Richardson, P. Pyšek, G. F. Midgley, G. O. Hughes, and M. Rouget. 2005. Niche-based modeling as a tool for predicting the risk of alien plant invasions at a global scale. *Glob. Change Biol.* 11:2234–2250.

Thurston, J. M. 1961. The effect of depth of burying and frequency of cultivation on survival and germination of seeds of wild oats (*Avena fatua* L. and *Avena ludoviciana*). *Weed Res.* 1:19–31.

Tilman, G. D. 1982. *Resource Competition and Community Structure.* Princeton University Press, Princeton, NJ.

Tilman, G. D. 1985. The resource-ratio hypothesis of plant succession. *Am. Nat.* 125:827–852.

Tilman, G. D. 1988. *Plant Strategies and the Dynamics and Structure of Plant Communities.* Princeton Monographs, Princeton, NJ.

Tilman, G. D. 1990. Mechanisms of plant competition for nutrients: The elements of a predictive theory of competition. Pp. 117–141 in J. B. Grace and D. Tilman (Eds.), *Perspectives in Plant Competition.* Academic, San Diego, CA.

Tilman, G. D. 1997. Community invasibility, recruitment limitation and grassland biodiversity. *Ecology* 78:81–92.

Tivy, J. 1990. A collective contribution to the understanding of plant domestication and agricultural evolution. *J. Biogeogr.* 17:711–712.

Tollenaar, M., S. P. Nissanka, A. Aguilera, S. F. Weise, and C. J. Swanton. 1994. Effect of weed interference and soil nitrogen on four maize hybrids. *Agr. J.* 86:596–601.

Tomback, D. F. and B. Linhart. 1990. The evolution of bird-dispersed pines. *Evol. Ecol.* 4:185–219.

Tremmel, D. C. and F. A. Bazzaz. 1993. How neighbor canopy architecture affects target plant performance. *Ecology* 74:2114–2124.

Trenbath, B. R. 1974. Biomass productivity of mixtures. *Adv. Agric.* 26:177–210.

Trenbath, B. R. 1976. Plant interactions in mixed crop communities. Pp. 129–170 in R. I. Papendick, P. A. Sanchez, and G. B. Triplett (Eds.), *Multiple Cropping*. ASA Special Publication No. 27. American Society of Agronomy, Madison, WI.

Tripp, R. 1996. Biological diversity and modern crop varieties: Sharpening the debate. *Agric. Human Values* 13:48–62.

Tuesca, D., E. C. Puricelli, and J. E. Papa. 1995. Changes in the weed community in contrasting tillage systems. *Actas del XII Congreso Latinoamericano de Malezas*, ALAM. INIA, Montevideo Uruguay.

U.S. Department of Agriculture. 1999. *Agricultural Chemical Usage: 1998 Field Crops Summary*. Publication Ag Ch 1-99. U.S. Department of Agriculture National Agricultural Statistics Service, Washington, DC.

U.S. Department of Agriculture IPM Center. 2006. *Crop Profiles*. Available: http://www.ipmcenters.org/Crop Pro-files/.

U.S. Economic Research Service. 2002. *Agricultural Biotechnology: Adoption of Biotechnology and Its Production Impacts*. U.S. Department of Agriculture, Economic Research Service, Washington, DC. Available: http://www.ers.usda.gov/whatsnew/issues/biotech.

U.S. Environmental Protection Agency. 1993. *Agricultural Atrazine Use and Water Quality: A CEEPES Analysis of Policy Options*. U.S. Environmental Protection Agency, Office of Program and Policy Evaluation, Water and Agricultural Policy Division, Agricultural Policy Branch, Washington, DC.

U.S. Forest Service. 2001. *2000 RPA Assessment of Forest and Range Lands*. FS-687. U.S. Department of Agriculture, Forest Service, Washington, DC.

Valladares, F. and F. I. Pugnaire. 1999. Tradeoffs between irradiance capture and avoidance in semi-arid environments assessed with a crown architecture model. *Ann. Bot.* 83:459–469.

Van Baarlen, P., P. R. J. van Dijk, R. F. Hoekstra, and J. H. de Jong. 2000. Meiotic recombination in sexual diploid and apomictic triploid dandelions (*Taraxacum officinale* L.). *Genome* 43:827–835.

Van Clef, M. and E. W. Stiles. 2001. Seed longevity in three pairs of native and non-native congeners: Assessing invasive potential. *North East Nat.* 8:301–310.

Van Den Bosch, R. 1992. Alternatives to pesticides: IPM. Pp. 261–262 in S. A. Briggs (Ed.), *Basic Guide to Pesticides: Their Characteristics and Hazards*. Hemisphere Washington, DC.

Vandermeer, J. 1989. *Ecology of Intercropping*. Cambridge University Press, New York.

Van der Pijl, L. 1982. *Principles of Dispersal in Higher Plants*. Springer-Verlag, New York.

Van Esso, M. L. and C. M. Ghersa. 1989. Dynamics of *Sorghum halepense* (L.) Pers. seeds in the soil of an uncultivated field. *Can. J. Bot.* 67:940–944.

Van Esso, M. L., C. M. Ghersa, and A. Soriano. 1986. Cultivation effects on the dynamics of a Johnsongrass seed population in the soil. *Soil Tillage Res.* 6:325–335.

Van Gressel, M. J. 2001. Glyphosate-resistant horseweed from Delaware. *Weed Sci.* 49:703–705.

Vasek, F. 1980. Creosote bush: Long-lived clones in the Mojave Desert. *Am. J. Bot.* 67:246–255.

Vasey, D. E. 1992. *An Ecological History of Agriculture: 10,000 BC–AD 10,000.* Iowa State University Press, Ames.

Vencill, W. K. (Ed.). 2002. *Herbicide Handbook*, 8th ed. Weed Science Society of America, Lawrence, KS.

Vere, D. T., J. A. Sinden, and M. H. Campbell. 1980. Social benefits of serrated tussock control in New South Wales. *Rev. Mark. Agric. Econ.* 48:123–138.

Verhoog, H. 1996. Genetic modification in animals: Should science and ethics be integrated? *The Monist* 79:247–264.

Vesk, P. A. and M. Westoby. 2001. Predicting plant species' responses to grazing. *J. Appl. Ecol.* 38:897–909.

Vibrans, H. 1999. Epianthropochory in Mexican weed communities. *Am. J. Bot.* 84:476–481.

Vila, M., M. Williamson, and M. Lonsdale. 2004. Competition experiments on alien weeds with crops: Lessons for measuring plant invasion impact? *Biol. Invas.* 6:59–69.

Vila Aiub, M. M. and C. M. Ghersa. 2001. The role of fungal endophyte infection in the evolution of *Lolium multiflorum* resistance to diclofop-methyl herbicide. *Weed Res.* 41:265–274.

Vila Aiub, M. M. and C. M. Ghersa. 2005. Building up resistance by recurrently exposing target plants to sub-lethal doses of herbicide. *Eur. J. Agric.* 22:195–207.

Vila Aiub, M. M., M. A. Martinez-Ghersa, and C. M. Ghersa. 2003. Evolution of herbicide resistance in weeds: Vertically transmitted fungal endophytes as genetic entities. *Evol. Ecol.* 17:441–456.

Vitousek, P. M., C. M. D'Antonio, L. L. Loope, and R. Westbrooks. 1996. Biological invasions as global environmental change. *Am. Sci.* 84:648–478.

Volterra, V. 1926. Fluctuations in the abundance of a species considered mathematically. *Nature* 118:558–560.

Von Holle, B. and D. Simberloff. 2005. Ecological resistance to biological invasion overwhelmed by propagule pressure. *Ecology* 86:3212–3218.

Wade, M. 1997. Predicting plant invasions: Making a start. Pp. 1–18 in J. H. Brock, M. Wade, P. Pyšek, and D. Green (Eds.), *Plant Invasions: Studies from North America and Europe.* Backhuys, Leiden, Netherlands.

Wagenius, S. 2006. Scale dependence of reproductive failure in fragmented *Echinacea* populations. *Ecology* 87:931–941.

Wagner, R. G., K. M. Little, B. Richardson, and K. McNabb. 2006. The role of vegetation management for enhancing productivity of the world's forests. *Forestry* 79:57–80.

Wagner, R. G., T. D. Petersen, D. W. Ross, and S. R. Radosevich. 1989. Competition thresholds for the survival and growth of ponderosa pine seedlings associated with woody and herbaceous vegetation. *New For.* 3:151–170.

Wagner, R. G. and S. R. Radosevich. 1998. Neighborhood dynamics for quantifying inter-specific competition in Coastal Oregon Forests. *Ecol. Appl.* 8:779–794.

Wagner, R. G. and S. R. Radosevich. 1991. Neighborhood predictors of interspecific competition in young Douglas-fir plantations. *Can. J. For. Res.* 21:821–828.

Walker, L. R. 1999. Patterns and processes in primary succession. Pp. 585–610 in L. R. Walker (Ed.), *Ecosystems of Disturbed Ground.* Elsevier, Amsterdam, Netherlands.

Walker, L. R. and F. S. Chapin III. 1987. Interactions among processes controlling successional change. *Oikos* 50:131–135.

Walker, L. R., D. B. Thompson, and F. H. Landau. 2001. Experimental manipulations of fertile islands and nurse plant effects in the Mojave Desert, USA. *West. N. Am. Nat.* 61:25–35.

Waller, G. R., D. Kumari, J. Friedman, N. Friedman, and C. Chou. 1986. Caffeine autotoxicity in *Coffea arabica* L. Pp. 243–269 in A. R. Putnam and C. S. Tang (Eds.), *The Science of Allelopathy.* Wiley, New York.

Walstad, J. D. and F. N. Dost. 1986. All the king's horses and all the king's men: The lessons of 2,4,5-T. *J. For.* 84:28–33.

Walstad, J. D. and P. J. Kuch. 1987. *Forest Vegetation Management for Conifer Production.* Wiley, New York.

Walstad, J. D., M. Newton, and R. J. Boyd. 1987. Forest vegetation problems in the Northwest. Pp. 157–202 in J. D. Walstad and P. J. Kuch (Eds.), *Forest Vegetation Management for Conifer Production.* Wiley, New York.

Wardle, D. A. 2001. Experimental demonstration that plant diversity reduces invasibility: Evidence of a biological mechanism or a consequence of sampling effect? *Oikos* 95:161–170.

Wardle, D. A., M. C. Nilsson, C. Gallet, and O. Zackrisson. 1998. An ecosystem-level perspective of allelopathy. *Biol. Rev.* 73:305–319.

Warwick, S. I. and L. D. Black. 1983. The biology of Canadian weeds. *Sorghum halepense* (L.) Pers. *Can. J. Plant Sci.* 63:997–1014.

Waser, N. M. 1993. The natural history of inbreeding and outbreeding. Pp. 173–199 in N. W. Thornhill (Ed.), *Theoretical and Empirical Perspectives*, University of Chicago Press, Chicago, IL.

Watkins, A. J. and J. B. Wilson. 1994. Plant community structure and its relation to the vertical complexity of communities: Dominance, diversity and spatial rank consistency. *Oikos* 70:91–98.

Watkinson, A. R. 1985. On the abundance of plants along an environmental gradient. *J. Ecol.* 73:569–579.

Watson, A. K. 1985. Integrated management of leafy spurge. Pp. 93–104 in A. K. Watson (Ed.), *Leafy Spurge.* Weed Science Society of America, Champaign, IL.

Watson, S. 1871. *Botany.* Pp. 1–525 in C. King, Report U.S. Geological Exploration of the Fortieth Parallel, Vol. 5. Government Printing Office. Washington, DC.

Way, M. F. 1977. Pest and disease status in mixed stands *vs.* monocultures: The relevance of ecosystem stability. Pp. 127–138 in J. M. Cherrett and G. R. Sagar (Eds.), *Origins of Pest, Parasite, Disease and Weed Problems.* Blackwell Scientific, Oxford.

Weber, E. and B. Schmid. 1998. Latitudinal population differentiation in two species of *Solidago* (Asteraceae) introduced into Europe. *Am. J. Bot.* 85:1110–1121.

Weeda, E. J. 1987. Invasions of vascular plants and mosses into the Netherlands. Pp. 19–29 in W. Joenje, K. Bakker, and L. Vlijm (Eds.), *Proceedings of the Koninklijke Nederlandse, Series C: Biological and Medical Sciences*. Akademie van Wetenschapen, Amsterdam, The Netherland.

Weed Science Society of America. 1956. Terminology Committee Report-WSSA. *Weeds* 4:278–287.

Weinig, C. 2000. Differing selection in alternative competitive environments: Shade-avoidance responses and germination timing. *Evolution* 54:124–136.

Weins, D. 1978. Mimicry in plants. In M. K. Hecht, W. C. Steere and B. Wallace (Eds.), *Evolutionary Biology*, Vol. 11. Plenum, New York.

Weldon, C. W. and W.L Slausen. 1986. The intensity of competition versus its importance: An overlooked distinction and some implications. *Q. Rev. Biol.* 61:23–44.

Werner, D. and W. E. Newton (Eds.). 2005. *Nitrogen Fixation in Agriculture, Forestry, Ecology, and the Environment*. Springer-Verlag, Dordrecht, Netherlands.

Werner, P. A. 1975. The biology of Canadian weeds. 12. *Dipsacus sylvestris* Huds. *Can. J. Plant Sci.* 55:783–794.

Werner, P. A. and J. D. Soule. 1976. The biology of Canadian weeds. 18. *Potentilla recta* L., *P. norvegica* L., and *P. argena* L. *Can. J. Plant Sci.* 56:591–603.

Wesson, G. and P. F. Wareing. 1969a. The induction of light sensitivity in weed seeds by burial. *Exp. Bot.* 20:413–425.

Wesson, G. and P. F. Wareing. 1969b. The role of light in the germination of naturally occurring populations of buried weed seeds. *J. Exp. Bot.* 20:402–413.

Westerman, P. R., M. Liebman, A. H. Heggenstaller, and F. Forcella. 2006. Integrating measurements of seed availability and removal to estimate weed seed losses due to predation. *Weed Sci.* 54:566–574.

Westerman, P. R., J. S. Wes, M. J. Kropff, and W. Van Der Werf. 2003. Annual losses of weed seeds due to predation in organic cereal fields. *J. Appl. Ecol.* 40:824–836.

Westoby, M., B. H. Walker, and I. Noy-Meir. 1989. Opportunistic management for rangelands not at equilibrium. *J. Range Manag.* 42:266–274.

Weston, L. A. 2005. History and current trends in the use of allelopathy for weed management. *Hort. Technol.* 15:529–534.

Whitford, F. (Ed.). 2002. *The Complete Book of Pesticide Management: Science, Regulation, Stewardship, and Communication*. Wiley, New York.

Whittaker, R. H. 1975. *Communities and Ecosystems*, 2nd ed. Macmillan, New York.

Wiens, D. 1978. Mimicry in plants. *Evol. Biol.* 11:365–403.

Wijesinghe, D. K. and S. N. Handel. 1994. Advantages of clonal growth in heterogeneous habitats: An experiment with *Potentilla simplex. J. Ecol.* 82:495–502.

Wiles, L. and E. Schweizer. 2002. Spatial dependence of weed seed banks and strategies for sampling. *Weed Sci.* 50:595–606.

Wiles, L. J. 2004. Economics of weed management: Principles and practices. *Weed Technol.* 18:1403–1407.

Wilkerson, G. G., S. A. Modena, and H. D. Coble. 1991. HERB: Decision model for post-emergence weed control in soybean. *Agron. J.* 83:413–417.

Wilkerson, G. G., L. J. Wiles, and A. C. Bennett. 2002. Weed management decision models: Pitfalls, perceptions, and possibilities of the economic threshold approach. *Weed Sci.* 50:411–424.

Williams, J. T. and J. L. Harper. 1965. Seed polymorphism and germination. I. The influence of nitrates and low temperatures on the germination of *Chenopodium album*. *Weed Res.* 5:141–150.

Williams, P. A., E. Nicol, and M. Newfield. 2001. Assessing the risk to indigenous biota from exotic plant taxa not yet in New Zealand. Pp. 100–116 in R. H. Groves, J. G. Panetta, and J. G. Virtue (Eds.), *Weed Risk Assessment.* Commonwealth Scientific and Industrial Research Organization (CSIRO) Plant Industry Canberra, Australia.

Williams, R. F. 1975. *The Shoot Apex and Leaf Growth: A Study in Quantitative Biology.* Cambridge University Press, Cambridge, UK.

Williamson, M. H. and A. Fitter. 1996. The characters of successful invaders. *Biol. Conserv.* 78:163–170.

Williamson, B. G. 1990. Allelopathy, Koch's postulates, and the neck riddle. Pp. 143–164 in J. B. Grace and D. Tilman (Eds.), *Perspectives on Plant Competition.* Academic, San Diego, CA.

Williamson, M. 1996. *Biological Invasions.* Chapman and Hall, London.

Williamson, M. H. 1991. Are communities ever stable? Pp. 353–372 in A. J. Gray, M. J. Crawley, and P. J. Edwards (Eds.), *Colonization, Succession and Stability.* Blackwell Scientific, London.

Wilson, B. J. 1972. Studies on the fate of *Avena fatua* seeds on cereal stubble as influenced by autumn treatment. Pp. 242–247 in *Proceedings 11th British Weed Control Conference.* British Crop Protection Council, Brighton, United Kingdom.

Wilson, J. B., R. B. Allen, and W. G. Lee. 1995. An assembly rule in the ground and herbaceous strata of a New Zealand rain forest. *Funct. Ecol.* 9:61–64.

Wilson, K. and G. E. B. Morren, Jr. 1990. *Systems Approaches for Improvement in Agriculture and Resource Management.* Macmillan, New York.

Wilson, P. W. 1940. *The Biochemistry of Symbiotic Nitrogen Fixation.* University of Wisconsin Press, Madison, WI.

Wilson, R. G. 1988. Biology of weed seeds in the soil. Pp. 25–39 in M. A. Altieri and M. Liebman (Eds.), *Weed Management in Agroecosystems: Ecological Approaches.* CRC Press, Boca Raton, FL.

Wilson, R. G. and P. Westra. 1991. Wild proso millet (*Panicum miliaceum*) interference in corn (*Zea mays*). *Weed Sci.* 39:217–220.

Wiser, S. K., R. B. Allen, P. W. Clinton, and K. H. Platt. 1998. Community structure and forest invasion by an exotic herb over 23 years. *Ecology* 79:2071–2081.

Woodmansee, R. G. 1984. Comparative nutrient cycles of natural and agricultural ecosystems: A step toward principles. Pp. 145–156 in R. Lowrance, B. R. Stinner, and G. J. House (Eds.), *Agricultural Ecosystems: Unifying Concepts.* Wiley, New York.

Woodward, F. I. 1987. *Climate and Plant Distribution.* Cambridge University Press, New York.

Worsham, A. D., G. G. Nagabhushana, and J. P. Yenish. 1995. The contribution of no-tillage crop production to sustainable agriculture. *Proc. I(A) 15th Asian-Pac. Weed Sci. Soc. Conf.* 15:104–111.

Wyatt, R. 1984. The evolution of self-pollination in granite outcrop species of *Arenaria* (Caryophyllaceae). III. Reproductive effort and pollen-ovule ratios. *Syst. Bot.* 9: 432–440.

Yenish, J. P. D. Worsham, and A. C. York. 1996. Cover crops for herbicide replacement in no-tillage corn (*Zea mays*). *Weed Technol.* 10:815–821.

Yoda, K., T. Kira, H. Ogawa, and K. Hozumi. 1963. Self-thinning in overcrowded pure stands under cultivated and natural conditions. *J. Biol. Osaka City Univ.* 14:107–129.

Young, J. A. and R. A. Evans. 1976. Response of weed populations to human manipulations of the natural environment. *Weed Sci.* 24:186–190.

Zamora, D. L. and D. C. Thill. 1999. Early detection and eradication of new weed infestations. Pp. 73–84 in R. L. Sheley and J. K. Petroff (Eds.), *Biology and Management of Noxious Rangeland Weeds.* Oregon State University Press, Corvallis, OR.

Zavaleta, E. S., R. J. Hobbs, and H. A. Mooney. 2001. Viewing invasive species removal in a whole-system context. *Trends Ecol. Evol.* 16:454–459.

Zimdahl, R. L. 1988. The concept and application of the critical weed-free period. Pp. 145–155 in M. A. Altieri and M. Liebman (Eds.), *Weed Management in Agroecosystems: Ecological Approaches.* CRC Press, Boca Raton, FL.

Zimdahl, R. L. 1993. *Fundamentals of Weed Science.* Academic, San Diego, CA.

Zimdahl, R. L. 1999. *Fundamentals of Weed Science*, 2nd ed. Academic, San Diego, CA.

Zimdahl, R. L. 2004. *Weed Crop Competition: A Review*, 2nd ed. Blackwell Publishing, Ames, IA.

Zimmerman, C. A. 1976. Growth characteristics of weediness in *Portulaca oleracea* L. *Ecology* 57:964–974.

Zossimovich, V. P. 1939. Evolution of cultivated beet, *Beta vulgaris. C. R. (Doklady) Acad. Sci. URSS* 24(1):73–76.

Zouhar, K. 2003. *Potentilla recta. In Fire Effects Information System.* USDA Forest Service, Rocky Mountain Research Station, Fire Sciences Laboratory. Available: http://www.fs.fed.us/database/feis/.

Zouhar, K. L., J. Kapler-Smith, S. Sutherland, and M. L. Brooks (Eds.). 2007. *Wildland Fire in Ecosystems: Fire and Nonnative Invasive Plants.* Gen. Tech. Rep. RMRS-GRT-42-vol. 6. USDA Forest Service, Rocky Mountain Research Station, Ogden, UT.

INDEX

Absorption, 53, 243, 252, 331, 340, 347.
  See also Herbicide(s), absorption
Abutilon theophrasti, 113, 120, 159, 219, 239
A. circinatum, 4
Acer macrophyllum, 121
Acroptilon repens, 300
Adaptation(s), 5–7, 59, 64–65, 88, 94–95, 119,
  145–148, 244–245, 256, 266, 363–366
  and human activities, 122–126
  and selection, 113–114
  also see safesites
Addition series, 207–210, 221, 224, 255
Additive experiment(s), 201–202. *See also*
  Competition, methods to study
Adsorption. *See* Herbicide(s), adsorption
Aeschynomene virginica, 243, 301
Afterripening, 157, 167, 172
Agamospermy, 103–104, 109, 114–115,
  117–118, 363
Aggressivity, 219
Agrestals, 4–6, 7–8, 19, 21, 90
Agroecosystem(s):
  deterioration in, 379–381
  early, 20–21
  modern, 8–9, 36, 43–44, 69–70, 87,
    124–125, 127, 143, 158, 183, 295, 301

weed management in. *See* Weed management,
  and agroecosystems
Agropyron, 28
  A. repens, 19, 23, 120, 236, 239, 241
  A. spicatum, 28
Agrostemma githago, 149, 239, 252
Agrostis canina ssp. *canina*, 19
  A. stolonifera, 19, 236
  A. tenuis, 19, 153, 160, 236
Ailanthus, 121
Aircraft application, 334, 337
Albinism, 327–328, 346
Alchemilla arvensis, 23
Alien. *See* Exotic
Aliphatic acid, 327–328
Allelopathy, 185, 187, 188, 237–243, 256,
  301–302, 306, 356, 369, 373
  allelochemicals, 5, 37, 189, 238, 241–242,
    256, 301–302
  effects on seed, 157, 242
  exudates, 241–242, 252
  leachates, 241–242
  methods of study, 242–243
Allium vineale, 239
Allogamy, 114, 115
Alnus rubra, 24, 121, 152, 221, 223, 254, 289

Ecology of Weeds and Invasive Plants. By Steven R. Radosevich, Jodie S. Holt, and
Claudio M. Ghersa
Copyright © 2007 John Wiley & Sons, Inc.

Amaranthus, 8, 108
  A. albus, 147
  A. dubius, 239
  A. hybridus, 8
  A. retroflexus, 163, 220, 239, 283
  A. spinosus, 8, 239
Ambrosia, 24
  A. artemisiifolia, 120, 239
  A. cumanensis, 239
  A. psilostachya, 239, 252
  A. trifida, 239
Amensalism, 185, 187–188, 204, 237–238,
    256, 301
Amsinckia, 23
Anabena, 254
Animal dispersal. *See* Dispersal, animal
Annonaceae, 171
Annual bluegrass, 156, 236
Annuals, 13–14, 18–19, 28, 53, 65, 103, 106,
    116–117, 120, 137, 166, 168, 251, 287
Anthriscus caucalis, 59
Apiaceae, 171–172, 242
Apomixis, 5, 115, 116
Arceuthobium, 244–245
Arctostaphylos patula, 24, 121, 127
Arenaria uniflora, 115
Artemisia, 97–98, 363
  A. absinthium, 239
  A. tridentata, 26, 27, 28, 29, 98, 189, 363,
      372, 380
  A. vulgaris, 239
Asclepiadaceae, 172
Asteraceae, 12, 106, 108, 115, 146–147, 168, 242
Atrazine, 320, 325, 376–377, 383
Autogamy, 5, 103–104, 114–115, 117
Autopolyploidy, 116
Avena barbata, 112
  A. fatua, 8, 23, 31, 58, 110, 112, 122–123,
      143, 154, 177, 180–181, 219, 233, 239,
      247, 266
Azolla–Anabena, 254

Baker's rule, 103, 117
Beneficial insects, 267–268
Benefin, 320, 332
Beta sp., 122
Betula, 121
Biennials, 14, 19, 32, 65
Biodiversity, 48–49, 75, 86, 90–93, 183, 197,
    268, 278, 375, 378, 382

impacts of invasive plants, 3, 9, 30, 80, 268,
    278
impacts of weeds, 80, 224, 250, 278
Biological control. *See* Weed control, biological
Birch, 24, 152
Brassica spp., 12, 63, 109, 147, 252, 291
  B. campestris, 22, 63, 233
  B. napus, 109, 294
  B. nigra, 239
Brassicaceae, 63, 168
Breeding systems, 104–105, 114–117, 126,
    363–364
Bromoxynil, 328
Bromus erectus, 19
  B. japonicus, 239
  B. mollis, 153
  B. sterilis, 19
  B. tectorum, 26–31, 55–56, 84–85, 98, 107,
      196–198, 239, 363–364
  and fire, 28–29
Buds, 60, 66, 290, 327
Bulb(s), 14, 117, 131
Bunchgrass, 27–29, 72, 98, 159, 249, 372
Bur chervil, 59

Cakile maritima, 89
Camelina alyssum, 239
Camelina sativa, 123, 239
  C. sativa var. *linicola*, 123
  C. sativa var. *sativa*, 123
Canellaceae, 171
Capsella bursa-pastoris, 112, 156, 196, 198
Carbamate, 320, 327–328, 338, 340
Carbohydrate starvation, 177, 284, 287
Cardaria draba, 289
Carriers. *See* Herbicide(s), carriers
Carrying capacity, 54–55, 62, 64–65, 87, 140
Ceanothus, 24, 121
Celastrus spp., 159
Cell division, 133, 326–327
Centaurea spp., 98, 238
  C. diffusa, 239, 297
  C. maculosa, 239, 297, 302
  C. nigra, 19
  C. repens, 239, 289, 300
  C. solstitialis, 117, 143, 146, 159
Cephalosporium diospyri, 243
Cereals, 6, 122, 149, 291, 292, 330
Chaining, 279, 289, 306
Chaparral, 280, 289

Chenopodiaceae, 168
Chenopodium album, 8, 16, 19, 155–156, 220,
  163, 164, 169, 236, 239
  C. rubrum, 160
Chloroacetamide, 327
Chlorophyll, 244, 246, 327–328
Chlorosis, 327–328, 346
Chondrostereum purpureum, 243
Chrysothamnus viscidiflorus, 28
Cirsium arvense, 22, 26, 120, 143, 146, 239
  C. discolor, 239
  C. vulgare, 19
Classification of weeds. *See* Weed(s),
  classification
Clearcut logging, 26, 280, 370, 376
Climate, 14, 24, 29, 36–37, 48, 59, 64, 71,
  111, 112, 113, 222, 229, 231, 270, 273,
  289, 373
Climax, 49–50, 53, 214
Clone(s), 68, 117, 133. *See also* Reproduction,
  vegetative
Cohort, 67–68, 165, 182
Coleoptile, 172
  herbicide effects on, 327
Colletotrichum gloeosporoides, 243
Colonization, 17, 59, 93–94. *See also* Invasion,
  colonization phase
Colonizer(s), 4–6, 7–9, 14, 16, 52, 122, 144
Commensalism, 185, 188, 237, 251–252, 256
Community, 39–54, 120, 144, 213–217,
  229, 248, 251, 273–274, 278, 305,
  356–364, 383
  assembly, 52
  attributes, 47–50
  definition, 6
  invasability, 84–87, 89–101
  microbial, 254
  stability, 48–50
  structure, 47–49
  succession, 9, 26
Competition, 68, 185–256, 292–294
  apparent, 187, 188, 249–250, 256
  asymmetrical, 222
  exploitation, 187, 188
  importance of, 68, 199, 216–217, 219,
    221–223, 255
  intensity of, 217, 218, 221–222, 255
  intraspecific *vs.* interspecific, 221–222
  for light, 231
  mechanisms of, 222, 230–237

role of plant traits, 231–237
  theories, 230–231
methods to study, 201–216
  additive, 201–202, 207, 218
  addition series and additive series,
    207–209, 221, 224, 255
  case studies, 214–216
  descriptive studies, 213–214
  diallel, 202, 206–207, 255
  gradient studies, 216
  intensity indices, 218–222
  in natural and managed ecosystems,
    212–216
  neighborhood, 201, 209–212, 222, 255
  Nelder, 205–206
  replacement series, 203–205
  retrospective studies, 214
  substitutive, 202–207, 217
  systematic, 201, 207–209, 210, 218, 255
  vs. other types of interference, 237
  thresholds, 224–230
  for weed control, 292–294
Competitive exclusion principle, 56, 189, 361
Competitive production principle, 189
Competitive ruderals, 20, 65, 119–120
Competitors, 17–20, 49, 65, 119, 121
  stress-tolerant, 49, 231, 361
Complexity, 38, 39, 41, 44, 50, 95, 97,
  100, 137, 213, 216, 224, 229, 238,
  256, 357
Compositae. *See* Asteraceae
Conifers, 11, 24–25, 146, 289, 381
Conservation tillage. *See* Tillage, conservation
Containment, 60, 71, 73, 76, 261, 278, 291.
  See also Weed control
Continuum theory, 40
Controlled burn, 280–281
Convolvulaceae, 244
Convolvulus, 13
  C. arvensis, 8, 236
Conyza canadensis, 196–198
Corm, 14, 117
Cost-benefit analysis, 73–76, 269. *See also* Risk
  assessment
Cover crops. *See* Crop(s), cover
Critical period, 227, 228, 354. *See also*
  Thresholds
Crop(s):
  competition, 180, 200, 216, 225, 290, 292,
    303, 356

Crop(s): (*Continued*)
  cover, 223–224, 241, 248, 287, 290,
    293–294, 301, 306, 355, 373
  herbicide resistant. *See* Herbicide(s),
    resistance, in crops
  in intercropping. *See* Intercropping
  mimics, 8, 103, 113, 122, 123
  response to weed control, 261–262
  rotation, 80, 290, 291, 292, 303, 306, 342,
    355, 360, 369
  smother, 290, 292–293, 301, 306
  trap, 246
Cultivation. *See* Tillage
Cultivators, 282
Cuscuta spp., 22, 149, 244–247, 291
  C. campestris, 246, 247
Cybernetics, 379
Cynara cardunculus, 144, 147
Cynodon dactylon, 8, 147, 239
Cyperus esculentus, 8, 239
  C. rotundus, 8, 80, 239
Cytisus scoparius, 26

2,4-D, 267, 303, 326
Datura ferox, 178
Death, 13–14, 24, 32, 150, 164, 173, 176–179,
  195, 246, 248, 284, 312, 328
Decomposers, 42, 52, 358
Decomposition, 241, 253, 302, 342
  chemical, 343, 344
  herbicide, 335–338, 342, 344, 346
  microbial, 241, 344
  photochemical, 343
Degeneriaceae, 171
Delphinium, 23
Demography, 67–68, 81, 85, 129–182
  models of. *See* Model(s), demographic
  vital rates, 130
Density, 41, 58, 105, 114, 141, 144, 157, 176,
  253, 255, 271, 293, 306, 344, 356, 369
  density-dependent mortality, 193, 194, 195
  effects on growth, 55, 192
  effects on mortality, 162, 171, 176–181,
    195–198
  effects on population growth, 60–64, 87–88,
    130–134
  effects on reproduction, 176–181, 196–198
  effects on size, 193–196, 199–200
  effects on yield, 200–224
  thresholds, 224–225, 249, 262–266

Developing countries, 279, 356
Diallel experiments. *See* Competition, methods
  to study
Dicamba, 325, 327, 343
Diclofop-methyl, 111
Dicots, 11, 12, 172
Difference equations. *See* Model(s), difference
  equations, 134, 138
Diffusability, 87–88, 101
Digitaria sanguinalis, 8, 122, 147, 239
Dinitroaniline, 320, 327
Diodia virginiana, 243
Dipole, 319
Dipsacus sylvestris, 145–146
Diquat, 325, 328
Directed sprays, 332
Disk, 215, 216, 265
Dispersal, 5, 17, 85, 94–95, 142–149, 166,
  178–179, 244, 274, 285, 358, 362, 373
  agents of, 94, 146–149
  animal, 94–95, 147
  distance estimates, 144–146
  by humans, 94, 148–149
  through space, 142
  by tillage, 285
  through time, 366
  in water, 147
  wind, 146–147, 346
Dispersing agent. *See* Herbicide(s), dispersing
  agent, 342
Disturbance, 9, 24–27, 58, 68, 71–72, 81,
  95–99, 126, 214, 248–249, 368–369, 379,
  380, 381
  and cultivation, 6, 20, 53, 98, 143–144,
    176–177, 261, 265, 281–284
  and ecosystem disfunction, 379–380
  and evolution, 17, 20, 65, 108, 119, 140
  and habitats, 15, 30, 90–93, 101, 214, 371
  and land use, 46, 81, 96
  and propagule pressure, 95
  and reproduction, 115, 119–122
  and risk, 273–275, 277–278
  and safesites, 88–89, 143
  and seed banks, 152, 156–157, 159–164,
    168, 172
  and stress, 49, 96, 98–99, 143–144,
    176–177, 261, 265, 281–284
  and succession, 49–50, 52–53, 97–98, 360,
    365, 369, 373
  and thresholds, 228–275, 381

Diversity. *See* Species, diversity
Domesticates, 122
Dominants, 48, 92, 195, 214
Dormancy, 63, 94–95, 104, 110–111, 142, 159,
    162, 164, 176, 178, 365
    conditional, 167–168
    descriptions of, 166–171
    and management, 171
    physical, 166, 168–170, 182
    physiological, 167–168, 170, 283, 293
    seed banks, 157, 159–60, 164
    underdeveloped embryo, 166, 171, 182
Dredging, 293
Drift, 73, 104, 118, 126, 334, 337, 338, 346, 372
    genetic. *See* Genetics, drift
    methods to reduce, 337
    use of buffer zones, 338
Drosophila, 144
DSMA, 328

Early detection, 69, 76, 229, 271
    and rapid response, 229–230
Echinochloa crus-galli, 8, 112–113, 123–126,
    149, 163, 220, 239, 289, 291, 293
    E. crus-galli var. *crus-galli*, 123, 125–126
    E. crus-galli var. *oryzicola*, 123, 125
    E. crus-galli var. *phyllopogon*, 149
Echium plantagineum, 31, 107, 117, 125
Ecological agriculture, 45, 353, 354, 383
Ecological law of thermodynamics, 41–42
Ecosystem(s), 8, 23, 26, 29, 76, 96, 178, 181,
    250, 254, 349, 355, 360, 375
    agricultural. *See* Agroecosystem(s)
    community structure, 46–48, 276
    and evolution, 101–103, 107
    and human systems, 77, 80, 355–360
    management in. *See* Weed management,
        agroecosystems and natural systems
    natural, 30, 43, 48, 70, 83–101, 107, 143,
        149–150, 159, 178, 180, 187–188, 213,
        216–218, 224, 226, 228–230, 238, 243,
        250, 256, 272, 349, 374
    novel, 53, 99–101, 366–374, 381–383
    reproduction, 107, 115, 363
    riparian, 30, 89, 91, 97, 230, 300, 338, 341,
        372, 381
    and scale, 39–48
    and succession, 49–53
Ecotype, 112, 122, 123, 126, 177, 266, 287
EDRR. *See* Early detection, and rapid response

Education, 41, 43, 79, 80, 249, 268, 352,
    369, 381
Eichhornia crassipes, 8
    E. paniculata, 95
Emergence, 135, 162–164, 171–174, 178–180,
    200, 222, 227, 228, 233, 283–285,
    292–293, 304–305, 323, 326, 330
Emulsifiable concentrate, 322
Emulsifier, 321, 322, 332, 346
Endothall, 325, 328
Environment, 35–36, 59, 62–65, 71, 73–75,
    77–78, 81, 84, 86–88, 91, 98–101,
    103–106, 110–115, 117, 119–121, 123,
    127, 129–131, 139, 141, 142, 150, 157,
    163–170, 176, 177, 178, 180–189, 192,
    193, 196, 204, 206, 209, 213, 214, 216,
    217, 222, 224, 227, 230–231, 233, 238,
    241, 243, 250–253, 255, 259–265,
    267–268, 270–274, 279, 282, 284–285,
    291, 298, 307–309, 311, 314–318, 323,
    328–330, 332, 335–339, 341–343,
    345–347, 351–353, 356, 358–359,
    362–365, 368–369, 371, 373–376,
    379–383
    conditions of, 5, 21, 29, 37, 87, 88, 93–95,
        103, 110, 111, 113, 114, 126, 157, 161,
        163, 164, 166, 167, 170, 172, 182, 204,
        228, 234, 262, 292, 359, 363
    gradients, 46, 47, 112–113, 206, 209, 216
    macroenvironment, 36
    microenvironment, 36, 41, 47, 51, 91, 92,
        119, 121, 142, 241, 292, 329, 371
    resources of, 20, 37, 52, 54, 92, 188, 230, 355
Environmental Protection Agency (EPA), 22,
    298, 308, 376
Enzyme inactivation, 308
Ephemerals, 137, 152, 163
Epidemics of weeds and invasive plants,
    178–182
Epilobium angustifolium, 19, 110, 152
Epinasty, 327
Epiphytes, 251
Eradication, 69, 70, 259, 269, 369
Eriogonum longifolium, 23, 236, 293
Euphorbia esula, 31, 60–61, 131, 139, 239, 297
Evolution, 62–65, 103–127, 156–157, 159,
    255, 351–354, 357, 363, 365, 366, 379,
    383
    convergent, 48
    influence of humans on, 43–44, 119–126

Evolution (*Continued*)
    strategy, 17–20, 65–67, 119–121
    of weeds and invasive plants, 8, 33, 43, 44,
        63–65, 103–128, 130, 291
Evolutionary genetics, 104–111. *See also*
    Genetics
Exotic, 5, 7–9, 10–11, 14–17, 23–26, 28–30,
    70–73, 76, 83–84, 89–101, 107–109, 118,
    140–141, 159–160, 171–172, 229, 230,
    237, 238, 248–250, 254–255, 270, 272,
    274–277, 290, 299, 351
Externalities, 74
Extirpation, 140

Fabaceae, 108, 154, 172
Facilitation, 51, 187–190, 204, 224, 250–255,
    373
    model. *See* Model(s), facilitation
Facilitative production principle, 189, 250
Factorial design, 207–208
FDCA. *See* Federal Laws, Food, Drug, and
    Cosmetic Act
Fecundity, 113, 132, 137, 139, 140, 145, 248
Federal Laws, 308
    Federal Insecticide, Fungicide and
        Rodenticide Act, 308
    Federal Noxious Weed Act, 291
    Federal Seed Act, 291
    Food, Drug, and Cosmetic Act, 308
Feedback, 27, 39, 41, 43, 44, 45, 345, 351, 353,
    379, 380, 381
Fertilizer, 282, 322, 379–380
Festuca, 19, 28, 153–154, 236
FIFRA. *See* Federal Laws, Federal Insecticide,
    Fungicide and Rodenticide Act
Fire, 28–29, 37, 121, 159, 280–281, 363–364,
    367, 380–381. *See also* Weed control
Fisher–Skellam equation, 87
Fitness, 10, 59, 63, 105–107, 110, 115, 117,
    222, 363
Flail, 287
Flaming, 281
Flax, 123, 239
Flooding, 36, 279, 306, 360
Food chain, 42, 308, 311, 314, 315, 358–359
Food, Drug, and Cosmetic Act. *See* Federal
    Laws, Food, Drug, and Cosmetic Act
Forest(s):
    managed, 24–25, 200, 355, 368, 369–371,
        383

plantations, 16, 21, 53, 217, 218, 226, 265,
        279, 305, 307, 338, 355, 369
    regeneration, 25–26, 121, 266
Forestry, 25, 264, 273, 288, 335, 350–351, 353,
        360, 367, 376, 379, 382, 383
Formulation. *See* Herbicide(s), formulation
Founder effects, 107, 111, 117, 118. *See also*
    Breeding systems and Genetics
Fragmentation, 50, 86, 90, 93, 119, 133,
        140, 360
Fruit fly, 144

Galium, 19, 214
    G. saxatile, 214
    G. sylvestre, 214
Gause's competitive exclusion principle. *See*
    Competitive exclusion principle
Gene flow. *See* Genetics, gene flow
General purpose genotype, 10, 59, 126
General Systems Theory, 38
Genet(s), 14, 184
Genetics, 62–63, 67–68, 104–118, 122–123,
        126–127, 130, 133–134, 137, 144, 166,
        207, 292, 302, 329, 346, 360, 363–365,
        373, 377–378, 382. *See also* Evolutionary
        genetics
    drift, 104, 118, 126
    epigenetics, 110–111
    gene flow, 94, 104, 105, 109, 114–116, 118,
        122, 127, 140, 142
    variation, 62, 104, 105, 107, 109, 111,
        113–115, 117, 118, 127, 363, 364
        epigenetic, 111, 114
        epistatic, 109–110, 114
        heritable, 62, 63, 103, 104–111, 363
    of weeds and invasive plants, 62–67,
        104–111
Genotype, 5, 59, 62–63, 94, 104–107,
        109–119, 177, 306, 363–364
Geotropism, 346
Germination, 13, 37, 68, 88, 94–95, 110–113,
        129, 131, 150, 159–160, 162, 165–180,
        184, 189, 228, 242, 244–246, 250,
        281–286, 290, 293, 294
    light requirement for, 110, 168, 172–176,
        282–284, 293
Gleditsia triacanthos, 146
Glomus spp., 254
Glufosinate, 325, 329, 377
Glyphosate, 325, 328, 329, 377

Gramineae. *See* Poaceae
Granules, 322, 346
Grass(es), 12–13, 26–28, 97–98, 120,
    151–153, 159–162, 238, 252, 256, 330,
    363
Grassland, 28–30, 72–73, 84–86, 91, 94,
    96, 97–98, 111, 120, 144, 157, 162, 216,
    230, 256, 280, 290, 360, 363–364,
    372, 380
Grazing, 20, 23, 28–29, 65, 96–98, 100,
    119–120, 133, 188, 248–249, 294, 301,
    363, 371–373, 380
Great Basin, 26–29, 84, 249
Growth:
    analysis, 234–237
    population, 59–60
    of weeds and invasive plants, 234–237, 304
    modular, 133–134
Guild, 39, 56, 379

Habitat, 14–15, 20, 30–33, 36–37, 46, 52,
    59, 71, 83, 84, 86–98, 103–104, 112–113,
    120–121, 122, 140, 150, 159–160,
    169, 172, 176–177, 216, 232–234,
    248, 250, 259, 267–268, 269, 272,
    273, 275, 359, 360, 362, 364, 365, 372,
    379, 381
    invasibility. *See* Invasibility
    management, 248. *See also* Weed
        management
Half-life, 163, 165, 315, 316, 342, 343
Hand pulling, 279. *See also* Weed control
Hard seed. *See* Hardseededness
Hardseededness, 159, 162, 170, 176
Harvesting, 21–22, 24–25, 148, 178, 182, 266,
    287, 294–295, 360, 370, 371
Haustorium, 246–247
Hazard, 74, 165, 178, 300, 308–309, 311, 314,
    316–317, 322
Herbicide(s), 303–307, 307–347
    absorption, 331, 340, 346
    active ingredient, 315, 316, 318, 321–323,
        334, 346
    additives, 322, 338, 346
    adsorption, 315, 331, 335–342, 344, 346
    application, 322
        aerial, 334, 335, 338
        method of, 333–335
    aquatic, 325, 326
    biological magnification, 311, 314–316, 346

classification, 304–305, 322–327
contact, 324, 325, 328, 330
decomposition, 302, 335, 336, 337, 338,
    342–346
degradation, 308, 327, 332, 335, 340, 342,
    343, 344, 346, 368
desorption, 339, 341
development, 308
displacement of, 336, 338, 341, 346
dose. *See* Herbicide(s), rate
effects on plants, 322, 332
fate of, 335–346
foliage-applied, 324, 330
formulations, 319, 321, 322, 323, 332, 338
inactivation, 331, 332
injury due to, 326, 328, 329, 330, 332, 335,
    337, 338, 340, 345
labels, 309, 338, 340
laws governing, 308, 309, 345
leaching, 320, 333, 334, 340, 341, 342, 346
list of, 323, 325, 326
mechanism of action, 324, 327
metabolism, 314, 326, 331
mode of action, 302, 325, 327
persistence, 315, 340
problems with, 303
properties of, 311, 318–322, 345, 346
rate, 322
residues, 337, 342
resistance, 109, 111–114, 267, 306, 328, 365
    in crops, 329–330, 377–378
    restrictions on, 309–311
    tolerance, 308, 323, 327, 328, 332, 342,
        372
safety, 311
selectivity, 305, 308, 322–324, 327,
    328–335, 337, 341
    in soil, 323, 338, 340, 342
soil-applied, 323–325, 333, 338
structure, 318–319, 342, 346
symptoms, 327–328, 331
timing and use of, 304
toxicity, 309–314
    acute, 311–312
    chronic, 312–314
transformations of, 314, 332
transport of, 303, 321, 332
uptake of, 320, 340, 341
volatility, 318, 319, 320, 321, 334, 338, 346
voluntary selection criteria, 315–317

Herbicide(s) (*Continued*)
  water solubility of, 318–319, 321, 322,
    341–342
Herbivory, 19, 40, 44, 111, 133, 185, 188, 222,
    237–238, 243–244, 248–249, 278, 369,
    373
Heterosis, 106, 107, 109, 363
Hierarchy theory, 38
Hilaria jamesii, 254
Hoeing, 227, 279, 306
HRC. *See* Herbicide(s), resistance, in crops
Human institutions, 80, 375
Humus, 340
Hybridization, 59, 63, 105–109, 115, 122, 364
Hypericum perforatum, 23, 113, 147, 299

Impatiens, 160, 246, 247
Imperata cylindrica, 8, 240
Inbreeding, 94, 109, 115, 116, 117, 363
Income, 54, 377
Infestations, 31, 178, 179, 182, 274, 275, 370
Inheritance, 105, 110, 111, 115, 329, 346
  epigenetic, 105, 110, 111, 115
Inhibition model. *See* Model(s), inhibition
Insects, 8, 20, 22, 44, 246, 249–250, 267, 268,
    295, 352, 353, 359–360
Integrated Pest Management, 44, 45, 224, 249,
    268, 245, 351–366. *See also* Weed
    management
  approaches for, 353–366
  levels of, 354–355
  modern, 351–352
Integrated Weed Management, 354–366. *See
    also* Integrated Pest Management
  and community dynamics, 358–360
  future directions, 357–366
  and population dynamics, 357–358
  principles, 355–357
  and succession, 360–363
Interactions, 10, 35–37, 39–40, 42, 44, 47, 50,
    80–81, 90, 99, 109, 113, 122, 130, 133,
    140, 157, 177, 180, 183–256, 262, 295,
    338, 357–358, 379. *See also* Interference
  negative, 188, 214, 237, 244, 256, 295
  positive, 185, 187, 189, 201, 237, 253, 256
Intercropping, 186, 190, 218, 223, 251, 253,
    256, 294, 355
  and weed suppression, 223–224
Interference, 6, 184–256, 301–302, 356, 372,
    373. *See also* Interactions

critical period, 45–46
  effect and response, 185, 186, 187
  methods to study. *See* Competition, methods
    to study
  negative, 185, 187, 237–250, 301
  positive, 250–257
Introduction, 3–4, 9, 10, 16–17, 30, 53, 56–60,
    77, 83–84, 88, 90, 93–94, 97–98,
    140–141, 224, 269–270, 351. *See also*
    Invasion, introduction phase
  and genetics, 105–117
  and thresholds, 224, 229, 364, 367
Introgression, 59, 106, 107, 109, 114
Invasion:
  colonization phase, 17, 59, 93–94
  detection, 70, 76, 271
  documenting, 271–275
  impacts, 49, 70, 352
  introduction phase, 17, 56–57, 58, 93
  lag phase, 59
  local, 87–89
  management. *See* Weed management in
    natural ecosystems
  naturalization phase, 17, 60, 62, 95, 357
  predicting, 71
  process, 17, 36, 44, 56–62, 76, 83–84, 85,
    88, 93, 94, 95, 357
  in production systems, 20–29, 53, 96
  resistance, 91–92
  risk assessment for, 69–70, 272–277, 305
  role of disturbance, 58
  screening for, 272–273
Invasive plants:
  definitions, 3–4, 9–10, 16–17
  evolution of. *See* Evolution of weeds and
    invasive plants
  expansion rates, 139, 145–146
  impacts, 9, 229, 277, 351
  in production systems, 20–29
  traits, 9–10, 36
  in wildlands, 30–33
Invasibility, 37, 83–102, 129, 270, 273, 278,
    380, 381
  community, 84, 87, 101, 278
  and community structure, 90–93
  and disturbance, 95–99
  and diversity, 91–93
  and evolutionary history, 89–90
  factors in, 89–99
  habitat, 86

and invasiveness, 99–101
and propagule pressure, 93–95
risk assessment for, 273–275
role of plant size, 92–93
Invasiveness, 10, 71, 84, 99–101, 106–107, 129, 236, 237, 238, 270
IPM. *See* Integrated Pest Management
Ipomoea hederacea, 115
Irrigation, 147, 290, 292, 332, 340, 341, 342, 379
    to incorporate herbicides, 340, 341, 342
IWM. *See* Integrated Weed Management

Juglans nigra, 241–242
Juglone, 241–242

Keystone species, 40, 367
K selection. *See* Selection, r and K

Labor, 21, 75, 80, 149, 160, 199, 308, 369, 379
Lactoridaceae, 171
LAI. *See* Leaf Area Index
Lamium amplexicaule, 122
Land equivalent ratio, 221
Landscape, 8, 24, 46–47, 69, 83, 91, 99, 248, 271, 277, 289, 357, 359, 361, 362, 382
Land use, 21, 26–28, 30, 36–37, 70, 89, 90, 93, 96–99, 101, 140, 165, 166, 260, 335, 350–353, 362, 373
    cycles of for production, 350–353
Lantana camara, 31–32
LAR. *See* Leaf Area Ratio
Latin binomial, 12
Law of constant final yield, 192, 255
Lay-by, 305
Leaching. *See* Herbicide(s), leaching
Leaf Area Index, 41, 120, 233
Leaf Area Ratio, 10, 234, 235
Leptochloa, 240, 289
LER. *See* Land Equivalent Ratio
Leslie model. *See* Model(s), Leslie
Life cycles, 8, 123, 149, 227, 352, 356, 362
Life history, 13–14, 35, 48, 49, 53, 60, 63, 65–67, 69, 92, 105, 113, 115, 119, 163, 236, 373
Life tables, 131, 139, 180, 182
Light, 15–16, 47, 90, 98, 110, 113–114, 157, 160, 168, 172–176, 187–188, 196, 231, 233, 236, 237, 282–283, 284, 285, 289, 293, 343

competition for, 231
quality, 113, 172
requirement for germination, 168, 172, 176, 282, 284
Linaria spp., 98
Lithocarpus densiflora, 121
Litter, 19, 241, 242, 362, 373
Living mulch. *See* Crop(s), cover
Logging, 25–26, 46, 53, 76, 96, 121, 171, 280, 371, 380–381
Logistic equation, 54, 55, 64
Lolium, 19, 153, 160, 267
    L. multiflorum, 104, 111, 120, 154, 209, 240, 293
    L. rigidum, 267
Loranthaceae, 244
Loranthus, 244
Lotka–Volterra equation, 542
Lotus corniculatus, 19
Lythrum salicaria, 31, 297

Macroenvironment. *See* Environment, macroenvironment
Magnoliaceae, 171, 242
Matricaria matricarioides, 19, 155, 236
Matrix model. *See* Model(s), matrix
Mechanism of action. *See* Herbicide(s), mechanism of action
Meristems, 133, 281, 327, 330
Mesocotyl, 172
Metabolites, 232, 254
Metapopulation(s), 57, 129, 139–140
    dynamics, 141–142
    risk of extinction, 140–141
Microenvironment. *See* Environment, microenvironment
Millet, 12, 221, 222, 293
Mimics. *See* Crop(s), mimics
Mineral nutrition, 67, 244
Minimum tillage. *See* Tillage, minimum
Mistletoe, 244, 245, 346
Mitosis, 111, 327
Model(s):
    age specific, 137
    APSIM, 70
    demographic, 179–181, 346
    difference equations, 134, 138
    facilitation, 51
    HERB, 70
    inhibition, 52

Model(s): (*Continued*)
INTERCOM, 70
invasion, 71, 75–76
Leslie, 138, 139
matrix, 138, 139
population dynamics, 60, 134–139,
    179–181
SELOMA, 70
of succession, 49–53
tolerance, 51
threshold, 70–71
weed management, 180–181
Mode of action. *See* Herbicide(s), mode of
    action
Monocots, 11, 12, 13, 172
Morrenia odorata, 243, 301
Mortality, 17, 64, 85, 88, 130, 132–140, 142,
    164, 174, 180, 191, 193, 195–198, 238,
    241, 246, 248, 282, 290, 351, 369–370
risk of death, 176–178
Mower, 287, 288
Mowing, 29, 65, 127, 279, 286–288, 355
MSMA, 325, 326, 328
Mulches, 248, 279, 289–290, 294, 301,
    306, 371
Mutualism, 10, 185, 237, 251, 252, 253,
    255, 256
Mycoherbicide(s), 301, 306
Mycorrhizae, 90, 252, 253, 254

Naptalam, 325, 328
NAR. *See* Net Assimilation Rate
Nassella trichotoma, 73, 74
Natality, 130, 181
Natural enemies, 10, 114, 187, 188, 268,
    274, 295, 298, 300, 306, 351, 358,
    372, 373
Naturalization, 17, 60, 62, 95, 357. *See also*
    Invasion, naturalization phase
Naturalized species, 4–5, 6, 7, 14–16
Necrosis, 327, 328, 346
Neighbors, 184–185, 190, 192, 194, 200, 206,
    209, 211–213, 217, 230, 233, 237, 238
Neighborhood experiments. *See* Competition,
    methods to study
Nelder designs. *See* Competition, methods to
    study
Nematodes, 22, 268, 269, 294, 358, 359, 360
Neotyphodium spp., 111
Net Assimilation Rate, 234, 235

Niche, 36, 49, 54–56, 62, 71, 86, 89, 91–93,
    187, 214, 270, 355, 363, 365
Night tillage. *See* Tillage, night
Nitrile, 327
Nitrogen fixation, 10, 48, 254, 257
Nitrophiles, 176
Nostoc, 254
No-tillage. *See* Tillage, no-tillage
Novel ecosystems. *See* Ecosystem(s), novel
Noxious weeds, 16, 73, 83, 291
Nurse plants, 252
Nutrient(s), 36, 37, 42, 53, 65, 98, 119,
    186–188, 196, 216, 233, 238, 251–253,
    255, 344, 365, 379–380

Old-fields, 96, 98, 144, 380
Opuntia stricta, 36, 297
Orobanchaceae, 244
Orobanche spp., 244–246
    O. ramosa, 301
Outbreeding depression, 106, 107
Outcrossing, 103, 106, 109, 114, 115, 117, 363.
    See also Breeding systems

Panicum miliaceum, 221–222
Papaver, 156, 157
Pappus, 146
Paraquat, 324, 325, 328
Parasitic weeds. *See* Weed(s), parasitic
Parasitism, 127, 185, 188, 237, 243, 244,
    246, 256
Parthenocissus spp., 159
Pathogens, 22, 44, 94, 114, 171, 250, 281, 294,
    295, 299, 301, 345, 351, 359
Pellets, 322, 346
Perennials, 14, 19, 53, 65, 104, 106, 109,
    116, 117, 120, 159, 163, 166, 168, 282,
    289, 362
Peromyscus, 163
Persistence. *See* Herbicide(s), persistence
Persistent seed, 10, 36, 67, 95, 159, 160,
    162, 165
Pest management. *See* Integrated Pest
    Management
pH, 18, 37, 238, 241, 315, 329, 340, 341, 344
Phenology, 41, 48, 66, 105, 126, 149,
    363, 365
Phenotypic variation, 63, 105, 114, 363
Phenoxys, 267, 326, 328
Phloem, 246, 324

Phoradendron, 244
Photosensitization, 23
Photosynthesis, 15–16, 67, 112, 188, 236, 282, 289, 324, 326, 327
  C_4 pathway, 16, 236
  Calvin–Benson cycle, 15
Phototropism, 328
Physiognomy, 40, 41, 48, 86
Phytochrome, 174–176, 282
  and daylength, 16
Phytophthora palmivora, 243
Phytotoxins, 24, 188, 243
  from microorganisms, 243
Picloram, 60, 61, 325, 327
Pinaceae, 115
Pinus spp., 237
  P. ponderosa, 40, 215, 216, 226
Pioneer, 24, 49, 53, 150, 152, 156, 157, 165, 172, 352
Plant development, 132, 328
Plantago, 236, 240
  P. lanceolata, 14, 19, 155, 196–198, 236, 256
  P. major, 196–198
Plasticity, 103, 105, 113–114, 120, 150, 165, 193, 196, 197, 287
Plow, 22, 95, 97, 98, 241, 265
Poa, 19, 28, 156, 194, 236, 240
  P. annua, 19, 156, 194, 236
Poaceae, 12, 28, 106, 108, 147, 168, 172, 194
Poisonous weeds, 16, 24
Polycropping, 253, 256
Polygonum spp., 159
  P. pensylvanicum, 120, 240
Polymorphism, 107, 111, 113, 157, 168, 169, 172, 176
Polyploidy, 63, 105, 109, 116
Population(s)
  exponential growth, 44, 59–60, 130, 178, 345
  growth, 54, 55, 56, 57, 59, 60, 62, 64, 87, 130, 132, 139, 140, 141, 145, 222, 270, 300, 345
  satellite, 60–62, 85, 86, 93, 101, 140, 271
  source, 60–62, 71, 127, 140
Population dynamics
  birth, 57, 68, 130–134, 137, 139, 141
  death, 57, 68, 95, 129–134, 138–141
  emigration, 57, 68, 130, 131, 133–134, 139, 142, 181

immigration, 57, 68, 104, 130, 131, 133–134, 139, 140–142
  models of. *See* Model(s), population dynamics
Populus, 24, 121
Portulaca oleracea, 8, 219, 240
Postemergence, 180, 304, 305, 323
Potentilla recta, 14, 98, 144, 275, 277–278
3/2 Power law, 195, 196. *See also* Self-thinning
Precautionary principle, 79–80, 330, 378
Predation, 54, 94, 114, 150, 157, 178, 185–188, 237, 243–248, 284–285, 359
  of seed, 162–163, 165, 182, 246–248
Preemergence, 227, 304, 323, 330
Pre-harvest, 305
Preplant, 227, 304, 323
Prescribed fire, 278, 280, 281
Press perturbation, 53
Prevention, 69, 70, 75–76, 184, 230, 259, 260, 269, 272, 281, 291, 352, 355
Primary consumers, 42
Primary succession, 49, 51
Primary tillage. *See* Tillage, primary
Primulaceae, 115
Producers, 8, 42, 267, 317, 352, 358, 377
Propagule pressure, 93–95. *See also* Invasibility, and propagule pressure
Proportion of species or population, 131, 132, 139, 189, 199–210, 218, 221, 223–224, 231, 265
Protocooperation, 185, 237, 251, 252, 253, 256
Proximity factors, 202, 207
Pruning, 184
Prunus serotina, 243
Pseudotsuga douglasii, 25, 162–163, 211, 221, 223, 380, 381
Pteridium aquilinum, 23
"The public", 79, 376
Purshia tridentata, 254

Quarantines, 260, 290, 291, 306, 381

Race, 122, 123, 125, 273
Ramet(s), 14, 68, 117, 134, 135, 137, 171, 177, 178, 184
Rangeland(s), 26–29, 35, 53, 56, 65, 120, 200, 238, 248, 287, 294, 300, 372
Ranunculaceae, 171
Ranunculus bulbosus, 156
  R. repens, 19, 136, 137, 155, 156, 236

Raphanus sp., 122
Reciprocal yield law, 192, 196, 199, 207, 262
Recruitment, 60, 68, 85, 88, 94, 137, 162,
    165–166, 171–180, 293, 365. *See also*
    Germination
Red:far red ratio, 114
Reduced tillage. *See* Tillage, reduced
Reforestation, 25, 121, 243, 264
Relative Growth Rate, 10, 67, 231, 234, 235,
    236, 256
of weeds, 234
Relative yield, 204, 218, 219, 220
Relative yield total, 219
Replacement series, 202, 203, 204, 205, 207,
    209, 219, 220, 237, 255. *See also*
    Competition, methods to study
Reproduction, 13, 17, 44, 53, 62–65, 68, 94,
    103–104, 120, 123, 134–138, 178–179,
    196–198, 290, 298, 314, 361, 363
asexual. *See* Reproduction, vegetative
sexual, 114, 115, 117, 118, 134, 135
vegetative, 5, 53, 68, 104, 117–118, 134,
    137, 363
    occurrence of, 68, 117, 134, 137
    risk of death, 177–178
Resistance, 21, 49–50, 91–92, 105, 109, 111,
    113, 114, 126, 267, 306, 328, 365, 375
herbicide. *See* Herbicide(s), resistance
Resources, 37–38, 51–52, 54, 56, 58, 62–67,
    75–76, 86, 91–92, 119–121, 140–141,
    176, 184–190, 213, 219, 228, 230, 231,
    233, 253, 262, 355–356, 361–362, 379
allocation of, 17, 63–65, 114, 119
limitation of, 130, 234
Resource-based theory, 231
Resource ratio hypothesis, 373
Respiration, 41, 188
Restoration, 70, 72, 75, 140, 178, 184, 197,
    229–230, 252, 264, 266, 278, 290, 293,
    307, 351, 368, 371, 372, 374
R:FR ratio. *See* Red:far red ratio
RGR. *See* Relative Growth Rate
Rhizobium, 254
Rhizomes, 14, 23, 117, 131, 178, 241, 282, 285
Rhododendron catawbiense, 106
    R. ponticum, 106
    R. radicans, 24
Rice, 12, 110, 113, 123, 124, 125, 126, 148,
    149, 243, 254, 255, 289, 291
Ridge tillage. *See* Tillage, ridge

Ring cleavage, 34
Riparian ecosystems. *See* Ecosystem(s), riparian
Risk, 24, 62, 77, 78, 81, 97, 112, 118, 140, 162,
    181, 191, 229, 230, 255, 265, 269, 275,
    276, 277, 309, 353, 357, 358, 372, 374,
    375, 376, 378, 381
assessment, 69–76, 271, 272, 273, 274,
    278, 372
of death, 176–179. *See also* Mortality
economic, 74, 75
endogenous, 75–76
environmental, 95, 311, 376
from weeds and invasive plants, 69–70
Risk-benefit analysis, 78, 81, 375
Rolling cultivator, 265, 266, 282
Rolling Pampas, 97, 98, 980
Root(s), 14–15, 117, 133, 177, 231, 233, 234,
    241–242, 246, 252–256, 285, 323,
    327, 333
hairs, 253
moisture extraction by, 172
r selection. *See* Selection, r and K
Rubiaceae, 108, 115
Rubus, 15, 121
Ruderals, 17–20, 65, 119, 120, 121, 163, 361,
    362
Rule of tens, 58
Rumex acetosella, 19
    R. crispus, 143, 147, 155, 240
RY. *See* Relative yield
RYT. *See* Relative yield total

Saccharum spontaneum, 240
Safe sites, 88–89, 93, 101, 141, 176–177, 182
Sagittaria, 147
Salsola iberica, 146, 147, 240
Scale, 9, 30, 38–45, 91–92, 140, 222, 231,
    247–248, 357, 381, 382
in ecological systems, 39–42
in human production systems, 43–45
Scrophulariaceae, 108, 244
Secale cereale, 122
Secondary consumers, 42
Secondary succession, 24, 46, 49, 52, 53, 95,
    97, 228, 284
Secondary tillage. *See* Tillage, secondary
Sedges, 12, 13
Seed(s), 10, 19, 66–67, 152, 158, 165,
    167–168, 170, 179, 197
cleaning of, 21, 149

decay of, 161, 242
dispersal, 5, 17, 85, 94–95, 142–149. *See also* Dispersal
dormancy, 88, 95, 110–111, 159, 166–174, 365. *See* also Dormancy
fate in soil, 162–164
germination, 37, 88, 94, 113, 129, 171–172, 176, 242, 282, 294, 302, 330
laws, 149, 290, 291
light requirement. *See* Germination, light requirement for
longevity, 152–157
predation of, 162–163, 165, 246–248
senescence, 163–164, 182
viability, 162
Seed bank, 10, 36, 67, 88, 95, 104, 134–137, 145, 149–168, 171, 177, 179–181, 197, 222, 242, 246–247, 260, 281–282, 355, 359, 381
in agricultural soils, 157, 159–162
density and composition of, 157
entry into, 150
fate of seed, 162–164
longevity, 152
in natural ecosystem soils, 159
predation. *See* Predation
and propagule pressure, 95
recruitment from, 162
and weed occurrence, 165–166
Seed bed preparation, 281
Seedling recruitment, 85, 88, 165, 171, 179, 293
Selection, 6, 12, 52, 62–65, 90, 103–105, 107, 112–113, 115, 117–119, 122–123, 126–127, 172, 259, 267, 269, 285, 287, 292, 302–303, 352, 355–356, 363, 365–367, 372
C, R, and S, 17–20, 49, 65–67
r and K, 49, 64–65
Selective herbicide(s). *See* Herbicide(s), selective
Selectivity. *See* Herbicide(s), selectivity
Self-pollination, 114, 115, 364. *See also* Autogamy and Breeding systems
versus outcrossing, 115–116
Self-thinning, 195, 197
Senecio jacobaea, 19, 26, 159, 249, 297, 299
S. sylvaticus, 196–197
S. viscosus, 197
S. vulgaris, 19, 156, 194, 236

Seral stage, 49, 50, 53, 121, 214, 371
Sesquiterpene lactones, 242
Setaria spp., 106, 291, 365–366
S. faberi, 159, 240
S. viridis, 106, 219, 240
Sexual reproduction. *See* Reproduction, sexual
Shredding, 279, 286
Shrub(s), 14, 18, 20, 24–26, 28–29, 32, 53, 59, 65, 97, 104, 121, 152, 159, 177, 189, 211, 215–216, 236, 251, 254, 280, 288, 289, 372, 380
Sickle, 287, 288
Silene alba, 115
S. angelica, 197
S. gallica, 197
Sisymbrium altissimum, 147
Size classes, 138, 194, 196
Smother crops. *See* Crop(s), smother
Social principles, 77–81, 350
conflict and resolution, 79
precautionary principle, 79–80
Soil:
biological phase, 339
characteristics of, 339
colloids, 336, 339, 340, 341, 344, 346
enrichment, 59
fertility, 243, 253, 292, 353, 361, 379
microorganisms, 36, 238, 255, 308, 344
effect of herbicides on, 344
type, 36, 214, 339
water, 326, 341
Soil-block washing technique, 231
Solanaceae, 267, 329
Solanum spp., 12, 13, 16, 22, 23, 220, 240, 267, 283, 289, 329
S. elaeagnifolium, 289
S. nigrum, 283
S. nodiflorum, 220
Solarization, 279, 289–290, 355
Solidago, 112, 152, 240
S. altissima, 112
S. gigantea, 112
Somatic polymorphism, 168, 169
Sorghum halepense, 8, 23, 62, 94, 120, 165, 178, 240, 289, 360
Space, 37–38, 40, 41, 44, 46, 51, 52, 60, 91, 114, 142, 182, 188, 189–191, 205–206, 294, 303, 360
capture, 191, 194, 355

Spacing, 47, 184, 195, 206, 211, 292, 293, 354, 355
Spartina anglica, 109
Spatial arrangement, 41, 189, 200–203, 209, 210, 253, 255
Species composition, 41, 48, 49, 50, 126, 156–157, 188, 214, 217, 222, 228, 238, 371, 372
  shifts in, 126, 183, 204, 356, 361
  effects of weed control on, 21, 265, 266, 267, 306
Species diversity, 41, 48–49, 56, 91, 199, 229, 268, 365, 361, 378. *See also* Biodiversity
Spergula arvensis, 156, 194
Stellaria media, 19, 156, 194, 236, 240, 283
Stipa spp., 28, 108, 254
  S. hymenoides, 254
Stolons, 14, 117, 131, 137, 282
Stress, 17, 20, 49–50, 65–68, 84, 90, 96, 112, 191–193, 195, 230–231, 292, 314, 356, 361, 369, 370, 372–73, 381, 383
  and disturbance, 98–99
  Stress-tolerant competitors, 17–20, 49, 65, 119, 121
Striga asiatica, 242, 244–246, 291
Substitutive experiments. *See* Competition, methods to study
Succession, 49–56
  effects of fragmentation, 360–361
  mechanisms of, 50–52
  models of. *See* Models, succession
  primary, 49, 51
  in production systems, 52–54
  secondary, 24, 46, 49, 52, 53, 95, 97, 228, 384
  stages of, 19, 20, 52, 65, 214, 229, 233
Survival, 57, 62–65, 85, 88, 93, 95, 109–110, 131–133, 142, 146, 157, 163, 164, 166, 169, 172, 176–179, 193, 195, 197, 201, 210, 215–216, 226–227, 232, 245, 268, 323, 356, 370
Symbiosis, 10, 204, 219, 253
Systematic experiments. *See* Competition, methods to study

2,4,5-T, 79, 375–376, 383
Taeniatherum asperum, 147
  T. caput-medusae, 98
Taraxacum officinale, 14, 109, 122, 143, 146, 236

Temperature, 15, 20, 36–37, 65, 86, 98, 104, 110, 112, 126, 157, 167, 172, 174, 216, 251, 282, 289–290, 315, 332, 334, 337–338, 342, 344
Terbacil, 320, 325
Terminology, 10–11, 17, 49, 93, 166, 187, 231
Themeda triandra, 159
Thiocarbamate, 327, 328, 338, 340
Thresholds, 45, 217, 224–230, 256, 261, 296, 345
  action, 224, 225, 229, 256, 345
  of competition, 224–230. *See also* Competition, thresholds
  critical period, 227–228, 354
  damage, 224–226
  density/biomass, 225
  ecological, 228–230
  economic, 45, 217, 225, 229–230, 261, 296
  models using. *See* Model(s), thresholds
  in natural ecosystems, 228–230
  period, 224, 225, 227
Tillage, 21, 281–286, 293, 303, 382
  alternatives to, 285
  conservation, 285
  effects on seed, 66–67
  equipment used for, 284
  night, 282
  no-tillage, 21, 284, 285, 293, 303, 382
  perennial weed suppression from, 281, 284
  primary, 284, 285
  problems with, 284–285
  reduced, 21, 285
  ridge, 285–286
  secondary, 284–285
  seedling suppression from, 282
Tolerance model. *See* Model(s), tolerance
Toxicity. *See* Herbicide(s), toxicity
Tragopogon pratensis, 143, 146
Transient seed, 159, 161
Transition matrices, 138
Translocation, 324, 331
Transport, 94–95. *See also* Dispersal
Transportation, 23–24, 94, 290, 353
  of herbicides. *See* Herbicides, transport of
Trap crops, 246
Triazines, 320, 376
Triclopyr, 325, 327
Trifluralin, 326, 332
Trifolium repens, 19, 154, 170, 236, 294
Trophic levels, 8, 42, 358–360

Tubers, 14, 117
Tussilago farfara, 19, 120

ULR. *See* Unit Leaf Rate
Unit Leaf Rate, 234, 235
Uracil, 328
Urea, 328

Value systems, 77, 78, 81, 374, 375, 377, 379, 381, 383
Vascular plants, 36, 58, 83, 97, 131, 133, 134, 251
Vegetative reproduction. *See* Reproduction, vegetative
Ventenata dubia, 98
Veratrum californicum, 23
Verbascum, 152, 156
    V. densiflorum, 156
    V. thapsus, 156
Veronica, 19, 283
Very-low fluence, 175
Viscum, 244
VLF. *See* Very-low fluence
Volatility. *See* Herbicide(s), volatility

Water:
    as herbicide carrier, 322
    quality of, 73, 75, 78, 308, 376, 377, 383
    as a resource, 216
    seed dispersal in. *See* Dispersal, in water
    in soil, 326, 341, 342
Water dispersible liquids, 322
Water soluble liquid, 322
Water soluble powder, 322
Waxes, 327, 328
Weed(s):
    abundance of, 6, 70, 242, 265, 356
    anthropomorphic perspective, 1–5, 10–11
    aquatic, 14, 15, 23, 289, 301
    and biodiversity, 8–9
    biology of, 260, 352
    breeding systems, 114, 115, 117
    classification of, 11–20
    definitions of, 6–8
    distribution, 178–181, 361–362
    as ecological good, 8–9
    epidemic nature of, 178–179
    evolution of. *See* Evolution of weeds and invasive plants
    in forests, 24–26

    ideal characteristics of, 5
    interactions with other organisms, 183, 225, 250
    invasive. *See* Invasive plants
    laws about, 31, 260, 290, 291, 306
    noxious, 16, 73, 83, 291
    parasitic, 243–246, 256
    poisonous, 16
    as product of humans, 53, 358
    in production systems, 20–29, 368
    in rangelands, 26–29
    in regional and global context, 30–32
    relative growth rate of, 10, 67, 231, 234, 235, 236, 256
    weed shifts. *See* Species composition, shifts in
    as specialists, 5, 6, 10, 19
    as strategists, 119
    in wildlands, 30
Weed control:
    biological, 44, 76, 90, 244, 295–302, 306, 352, 354, 358–360, 372
        methods for implementing, 299–300
        procedures for developing, 296–298
    chemical, 26, 113, 292, 302–305, 329, 354. *See also* Herbicide(s)
    crop response to, 261–262
    cultural, 44, 80, 260, 266, 279, 290–295, 306, 342
    districts, 291
    economics of, 260–265
    effects on density, 265
    effects on other organisms, 267–269
    effects on species composition, 265–267
    physical methods of, 279–290, 306
    profitability, 35, 70, 261, 262, 264
    reasons for, 21–24
    tools for, 21, 279–306
    value of, 73, 225, 262–265
    weed response to, 261
Weed management, 69–77, 259–305. *See also* Integrated Pest Management
    in agriculture, 369
    in agroecosystems, 260–269, 276–277, 305
    ecological, 353–357, 369–372, 382
    in forests, 369–372
    future challenges, 381–382
    and evolutionary patterns, 363–366
    in natural ecosystems, 269–279
        approaches, 76–77, 269–270
        framework for, 71, 277–279, 357, 372–374

Weed management (*Continued*)
    novel, 368–374
    prioritizing, 71–73, 269
    risk assessment and, 69–71, 272,
        275–277
    risk of action and inaction, 275–277
    integrated (IWM). *See* Integrated Weed
        Management
    market-driven considerations, 73–77
    methods and tools for, 70, 353–366
    models. *See* Model(s), weed management
    in modern society, 80–81
    in rangeland, 372–374
    as selective force, 364–366
    principles, 69–77, 355–357
        risk assessment and, 69–70, 275–277
        risk of action and inaction, 275–277
    socioeconomic influences, 381
    systems approaches, 349–383
    value systems, 374–382

Weed science, 79, 376
    evolution of, 352–353
Weed Science Society of America, 6, 38, 57,
        206, 208, 210, 322, 343, 364, 373
Weeding, 23, 43, 113, 184, 225, 227, 303
Wettable powder, 322
Wetting agent, 322
Wild plants, 3, 88, 122, 337
Wilderness. *See* Ecosystem(s), natural
Wildlands. *See* Ecosystem(s), natural
Wind dispersal. *See* Dispersal, wind
Winteraceae, 171
Woody plants, 14, 98, 362
WSSA. *See* Weed Science Society of America
WUE. *See* Water Use Efficiency

Xanthium, 16, 117, 143, 169, 240, 297
    X. pensylvanicum, 240
    X. strumarium, 117, 297
Xylem, 246